Long-Wavelength Semiconductor Devices, Materials, and Processes

Long-Wavelength Semiconductor Devices, Materials, and Processes

Symposium held November 26-29, 1990, Boston, Massachusetts, U.S.A.

EDITORS:

A. Katz
AT&T Bell Laboratories, Murray Hill, New Jersey, U.S.A.

R. M. Biefeld
Sandia National Laboratories, Albuquerque, New Mexico, U.S.A.

R. L. Gunshor
Purdue University, West Lafayette, Indiana, U.S.A.

R. J. Malik
AT&T Bell Laboratories, Murray Hill, New Jersey, U.S.A.

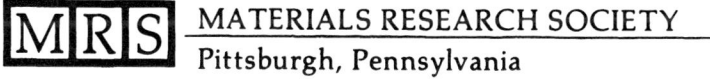

MATERIALS RESEARCH SOCIETY
Pittsburgh, Pennsylvania

CAMBRIDGE UNIVERSITY PRESS
Cambridge, New York, Melbourne, Madrid, Cape Town,
Singapore, São Paulo, Delhi, Mexico City

Cambridge University Press
32 Avenue of the Americas, New York NY 10013-2473, USA

Published in the United States of America by Cambridge University Press, New York

www.cambridge.org
Information on this title: www.cambridge.org/9781107409934

Materials Research Society
506 Keystone Drive, Warrendale, PA 15086
http://www.mrs.org

© Materials Research Society 1991

This publication has been registered with Copyright Clearance Center, Inc.
For further information please contact the Copyright Clearance Center,
Salem, Massachusetts.

First published 1991
First paperback edition 2012

Single article reprints from this publication are available through
University Microfilms Inc., 300 North Zeeb Road, Ann Arbor, MI 48106

CODEN: MRSPDH

ISBN 978-1-107-40993-4 Paperback

Contents

*Invited Paper

*Invited Paper

*Invited Paper

*Invited Paper

*Invited Paper

Preface

The Symposium on Long-Wavelength Semiconductor Devices, Materials, and Processes was held on November 26-29, 1990, in Boston, Massachusetts, U.S.A. as Symposium T of the 1990 Fall Meeting of the Materials Research Society. The symposium was directed as an overview of devices, materials and material processes having application for long-wavelength detectors and sources for fiber optics communication systems and thermal imaging. The aim of the symposium was to provide an appropriate forum for discussions on advanced III-V, II-VI and IV-VI material processes for applications at wavelength above 1 μm. More than 120 papers from 15 countries were presented on fundamental and applied aspects of long-wavelength semiconductor devices, materials, and processes.

The symposium included sessions on overview and concerns of long-wavelength semiconductors; effects of quantum size in semiconductor structures; mercury cadmium telluride devices and processing; growth of mercury cadmium telluride; narrow-gap compounds; narrow gap (In,Ga)Sb III-V compounds; III-V-based long-wavelength materials, processes and devices; and a large poster session. More than 80 papers, which are included in this proceedings, illustrate both the present state of knowledge in this field and the thrust of the current research and development.

The editors wish to thank the sponsors, invited and contributing authors, session chairmen, reviewers, and MRS staff and program officials for their help in organizing this successful symposium and publishing this volume.

A. Katz
R.M. Biefeld
R.L. Gunshor
R.J. Malik

February 1991

Acknowledgments

The editors of this proceedings are grateful to all the following organizations for sponsoring the symposium and raising their kind support:

Aixtron
AT&T Bell Laboratories
Blake Industries, Inc.
Chorus Corporation
Hughes Aircraft Company
MR Semicon Inc.
Perkin-Elmer Corporation
Sumitomo Electric

MATERIALS RESEARCH SOCIETY SYMPOSIUM PROCEEDINGS

Earlier Materials Research Society Symposium Proceedings listed in the back.

PART I

Mercury Cadmium Telluride— Growth and Characterization

$Hg_{1-x}Cd_xTe$: DEFECT STRUCTURE OVERVIEW

M.A. BERDING*, A. SHER*, AND A.-B. CHEN**
*SRI International, Menlo Park, California 94025
**Auburn University, Auburn, Alabama 36849 – 3501

ABSTRACT

Native point defects play an important role in HgCdTe. Here we discuss some of the relevant mass action equations, and use recently calculated defect formation energies to discuss relative defect concentrations. In agreement with experiment, the Hg vacancy is found to be the dominant native defect to accommodate excess tellurium. Preliminary estimates find the Hg antisite and the Hg interstitial to be of comparable densities. Our calculated defect formation energies are also consistent with measured diffusion activation energies, assuming the interstitial and vacancy migration energies are small.

INTRODUCTION

Sophisticated infrared devices are currently made from HgCdTe. While the material has entered a manufacturing phase, there remain many unanswered questions about its nature. Native point defects appear to play a far more important role in the properties of HgCdTe than they do in other technologically important semiconductor materials. The formation energies of these defects and their associated localized states in the band gap are still not completely characterized. Similarly, the characterization of impurities in HgCdTe is far from complete. Beyond point defects it is well known that a number of extended defects are important. The best studied are tellurium inclusions and dislocations which form at heterojunctions and at HgCdTe-substrate interfaces. As grown material is tellurium-rich, the excess Te is accommodated in the form of Hg vacancies and tellurium inclusions. As such, post growth anneals in a Hg vapor overpressure are typically performed to reduce the Hg vacancies prior to other device processing; this converts high purity undoped material from p-type to n-type. The donor causing the n-type doping has not been identified, although it is believed to be a residual impurity, rather than a native defect.

One of the more perplexing properties of HgCdTe stems from the fact that it is an alloy. Until a few years ago it was thought that all tetrahedrally coordinated zincblende semiconductor alloys of the form $A_{1-x}B_xC$ were random, with the C atoms occupying one fcc sublattice and the A and B atoms sharing the other sublattice at random. This has now been demonstrated to be incorrect in many semiconductor alloys. The nature of the correlations can be characterized by counting the populations of those clusters consisting of a centered C atom and its four surrounding A or B atoms. There are five types of clusters of the form $A_{4-n}B_nC$ with n=0, 1, 2, 3, or 4. If the alloy is random the cluster occupation probabilities, p_n, form the binomial distribution, i.e. $p_{n,random} = \binom{4}{n}(1-x)^{4-n}x^n$. In correlated material, deviations from this distribution are found. For example, if at x=0.25 the material tends towards a regular compound, then p_1 is increased while the other cluster occupation probabilities are reduced relative to $p_{n,random}$. If the material tends towards spinodal decomposition, then p_0 and p_4 are enhanced in separate regions of space, and the other cluster occupation probablilities are reduced. In most semiconductor alloys, the major forces for correlations are strain energies resulting from bond length differences between the AC and BC constituents. Thus theory predicts that HgCdTe will be nearly random because HgTe and CdTe are nearly lattice matched. Calculations that include chemical energies and charge shifts, in addition to the strain energies, have also predicted HgCdTe to be a nearly random alloy. Yet several recent experiments on alloys with x=0.20-0.25 have suggested that HgCdTe is correlated, with $p_n > p_{n,random}$ for n=0 and 1, and $p_n < p_{n,random}$ for n=2, 3, and 4. If this is the case it may have important consequences on diffusion and transport properties. The consequences of such correlations on the defect populations are discussed further below.

NATIVE DEFECTS

The doubly ionized cation vacancy is believed to be the dominant native defect in HgCdTe, that dictates the electrical behavior of the undoped material [1, 2]. We have recently reported the results of calculations of native and impurity defect total energies in HgTe, CdTe and ZnTe [3]. The energies for the formation of various native point defects in HgTe are summarized in Table I; the HgTe solid and the Hg in the vapor are used as references [4]. These calculations used the linearized muffin tin orbital (LMTO) method within the local density approximation (LDA) to the exchange correlation energy. Large supercells containing one defect per cell were repeated periodically, and from the difference in total energies per cell, with and without the defect, the defect formation energy was calculated. To expedite the calculations, the energies were calculated within atomic spheres approximation (ASA) with a small (spd) basis set. In the ASA, an approximation to the exact density functional is evaluated; as a result, an error is introduced which is larger than in other LDA methods [5], and relaxation energies cannot be accurately calculated. Thus only those differences in energies here that are ≥ 0.5 eV here should be viewed as significant for these calculations.

An appropriate set of mass action constants for the neutral defect reactions is also given in Table I. The notation in that table is as follows. A_B corresponds to an A species occupying a B site, where I corresponds to an interstitial, and V to a vacancy. No subscript on a species indicates it is occupying the correct lattice site, e.g., $Hg = Hg_{Hg}$. Square brackets, [ ], refer to concentrations. A subscript "g" indicate the species in the gaseous, or vapor, phase and P_{Hg} is the Hg vapor pressure. Most of the reactions in Table I involve the creation or destruction of one or more unit cells. Because the resulting change in volume is accommodated at the surface, the change in the number of unit cells will enter into the determination of the defect equilibrium through the surface entropy. Additionally, surface preparation and orientation will affect the surface free energy. We have assumed for the present that such surface effects are neglible, i.e., that the volume expansions and contractions can occur with negligible changes in the surface properties.

To correctly complete the defect equilibrium determination one must include the equilibration of the electronic charges of the system. To do so, one must have knowledge of the dominant charge states of the defects, and their activation energies with respect to the neutral defect. Such calculations are complicated by the fact that most *ab initio* calculations of the electronic band structure of semiconductors predict an incorrect band gap, E_G, a shortcoming of the local density approximation (LDA.) As such, we shall focus

Table I. Defect reactions and formation energies

Defect reaction	Defect concentration	Energy (eV)
$E_{V_{Hg}} + HgTe \leftrightarrow V_{Hg}Te + Hg_g$	$[V_{Hg}] = P_{Hg}^{-1} K_{V_{Hg}}^0 \exp(\frac{-E_{V_{Hg}}}{kT})$	2.24*
$E_{Te_{Hg}} + 2HgTe \leftrightarrow Te_{Hg}Te + 2Hg_g$	$[Te_{Hg}] = P_{Hg}^{-2} K_{Te_{Hg}}^0 \exp(\frac{-E_{Te_{Hg}}}{kT})$	4.53
$E_{Te_i} + HgTe \leftrightarrow Te_i + Hg_g$	$[Te_i] = P_{Hg}^{-1} K_{Te_i}^0 \exp(\frac{-E_{Te_i}}{kT})$	4.96
$E_{V_{Te}} + Hg_g \leftrightarrow HgV_{Te}$	$[V_{Te}] = P_{Hg} K_{V_{Te}}^0 \exp(\frac{-E_{V_{Te}}}{kT})$	3.12**
$E_{Hg_{Te}} + 2Hg_g \leftrightarrow HgHg_{Te}$	$[Hg_{Te}] = P_{Hg}^2 K_{Hg_{Te}}^0 \exp(\frac{-E_{Hg_{Te}}}{kT})$	-0.42
$E_{Hg_I} + Hg_g \leftrightarrow Hg_I$	$[Hg_I] = P_{Hg} K_{Hg_I}^0 \exp(\frac{-E_{Hg_I}}{kT})$	0.84, 0.98

* Experimental number from Vydyanath [1, 2].
** Calculated using tight binding Hamiltonian[6].

on the neutral native defects here, and the established or expected charge states of these defects. In wide band gap materials the defects equilibration can be substantially affected by the Fermi level; for example the formation energy of a donor will decrease when the Fermi energy is near the valence band edge, since the donor electron can drop into a vacant state near the valence band, thereby lowering the energy by $\sim E_G$. Because we are discussing HgCdTe with a narrow band gap, we expect the Fermi effects to be small, but not insignificant at high temperatures. Because of the small conduction band effective mass, though, in n-type material the filling of the conduction band states by electrons can shift the Fermi energy significantly; combined with the increase in the band gap for the high temperatures at which most defect studies are done, the effective band gap can be significantly larger than the usual 77 K bandgap associated with a given concentration of HgCdTe.

First we consider the defects which accommodate excess tellurium, the first three defects in Table I. The mass action constants are given by a product of the form

$$K_i = K_i^0 e^{\frac{-\Delta E_i}{kT}} \tag{1}$$

where ΔE_i is the change in the enthalpy for the ith reaction. For the first three equations, the K_i^0 are given by

$$K_{V_{Hg}}^0 = C_0(kT)^{\frac{5}{2}}(2\pi m_{Hg})^{\frac{3}{2}}h^{-3}\exp(\frac{\Delta S_{V_{Hg}}}{k}) \quad , \tag{2}$$

$$K_{Te_{Hg}}^0 = C_0(kT)^5(2\pi m_{Hg})^3 h^{-6}\exp(\frac{\Delta S_{Te_{Hg}}}{k}) \quad , \tag{3}$$

and

$$K_{Te_I}^0 = C_0(kT)^{\frac{5}{2}}(2\pi m_{Hg})^{\frac{3}{2}}h^{-3}\exp(\frac{\Delta S_{Te_I}}{k}) \quad . \tag{4}$$

Here T is the temperature in Kelvin, k is Boltzmann's constant, m_{Hg} is the mass of the mercury vapor atoms, h is Planck's constant, ΔS_i is the change in vibrational entropy upon formation of the defect, and C_0 converts from site fraction to volume concentrations. Because two unit cells of HgTe are destroyed when a tellurium antisite is created, compared to one unit cell when a mercury vacancy is created, we do not expect that $\exp((\Delta S_{Te_{Hg}} - \Delta S_{V_{Hg}})/kT) \sim 1$. While we have not completed the evaluation of these entropy terms, our preliminary estimates indicate this ratio is $\sim 10^{-4}$. For the tellurium interstitial and the mercury vacancy we expect that $\exp((\Delta S_{Te_i} - \Delta S_{V_{Hg}})/kT) \sim 1$ will be correct within a factor of 10. Evaluating the numerical constants, we find

$$\frac{[Te_{Hg}]}{[V_{Hg}]} \simeq 10^{-10} \qquad (5)$$

and

$$\frac{[Te_i]}{[V_{Hg}]} \simeq 10^{-18} \qquad (6)$$

for T=500 C and $P_{Hg} = 1$ atm. The conclusion from Equations 5 and 6 that the mercury vacancy is the dominant native defect is consistent with experimental observation. This conclusion is unchanged if we include the possibility that the species may be ionized at the equilibration temperature where the material is expected to be intrinsic. Although the tellurium antisite density decreases more rapidly with decreasing Hg pressure than does the mercury vacancy density, the point at which the concentrations are comparable is at less than $P_{Hg} \simeq 10^{-10}$ atm, and certainly the HgTe phase boundary is reached before such low Hg pressures can be achieved. This is also consistent with the fact that no p-to-n conversion is observed for low mercury pressures, as would be expected if tellurium antisites became the majority native defect. Because the pressure dependences of the tellurium interstitial and the mercury vacancy concentrations are the same, the above conclusion will hold independent of the mercury pressure.

We have checked the sensitivity of the calculated concentration ratios to the magnitude of the reaction enthalpy. Because the enthalpies enter in the exponents, small changes in the enthalpies will result in large changes in the predicted defect concentrations. For example, let us assume that our calculated antisite formation enthalpy is in error by 0.5 eV; in this case the ratio of antisite to vacancy concentrations (at 500 C) will be reduced to $\sim 1 \cdot 10^{-11}$. For an antisite formation enthalpy in error by 1.0 eV, this ratio is reduced to $\sim 1 \cdot 10^{-7}$. We do not expect the ASA errors to exceed 0.5 eV [5].

If the HgCdTe is not completely annealed, and tellurium precipitates are still present, the defect equilibrium will not be that predicted by the mass action equations given in Table I. For example, near the inclusions we can assume that the defects will be nearly in equilibrium with the tellurium solid; thus

$$E'_{V_{Hg}} + Te_s \leftrightarrow V_{Hg}Te \qquad (7)$$

and

$$E'_{Te_{Hg}} + 2Te_s \leftrightarrow Te_{Hg}Te \qquad (8)$$

will be the appropriate reactions. The formation energies for a tellurium antisite and a Hg vacancy from the tellurium solid are 1.63 eV and 1.15 eV, respectively. Although the difference in the formation energies isless than when both defects are referenced to the mercury vapor (~ 0.5 eV compared to ~ 2 eV), the gas phase entropy factor does not enter into the ratio of the defect concentrations. Using the same estimate of the entropy ratio, the defect concentration ratio using tellurium solid as the reference state is

$$\frac{[\text{Te}_{\text{Hg}}]}{[V_{\text{Hg}}]} \simeq 10^{-8} \quad . \tag{9}$$

Thus, near the inclusions we expect higher relative concentration of tellurium antisites, as compared to the rest of the material equilibrated with the Hg vapor. Additionally, the absolute defect concentrations may differ substantially in the two regions of the material. A better calculation of the vibrational entropy is needed before we can predict these absolute defect concentrations. Differences in the defect concentrations arising from different equilibration condition are a possible source of spatial variability of the HgCdTe material. If the material is not fully annealed to equilibrium, for example because of an abundance of tellurium precipitates, this history may affect subsequent processing.

In the above we have discussed the defect energies for HgTe and applied them directly to the small x, $\text{Hg}_{1-x}\text{Cd}_x\text{Te}$ system. Because we are dealing with the native defects of an alloy we expect a number of complexities to affect the above analysis. First the variation of the defect formation energies for vacancies is sensitive to the alloy environment, in particular for the vacancies of the nonsubstituted species, such as tellurium in HgCdTe [6]. Even for vacancies of the substituted species, we have found the formation energies may vary by several tenths of an electron volt. Because of this variation in the formation energy, the fraction of defective sites will vary by as much as a factor of 100 from one class of sites to the next. Consider various classes of Hg sites in ideal HgCdTe, which can be distinguished to first order by specifying the number of Hg and Cd atoms in the second neighbor shell (the four first neighbors are always tellurium), $\text{Hg}_{12-j}\text{Cd}_j$, with a concentration given by [j]. The total vacancy concentration is given by

$$[V_{\text{Hg}}] = \sum_{j=1}^{12} [j] P_{\text{Hg}}^{-1} K_j^0 \exp\left(\frac{-E_j}{kT}\right) \tag{10}$$

where E_j is the vacancy formation energy for the jth cluster. The populations of vacancies in each class of cluster, j, can be expected to differ because of differences in the cluster populations, [j], and the formation energies, E_j. Additionally, the defect energy levels may differ in the various classes of sites, possibly leading to different ionization states for vacancies in different classes of sites. If the cations in the alloy are randomly arranged, such differences may be difficult to infer experimentally. If, on the other hand, the cations are correlated, exhibiting short range order, more complex behavior may be present. Such short range order has been demonstrated using Raman scattering [7, 8], infrared reflectivity [9], and nuclear magnetic resonance [10]. In these cases, the contribution to the vacancy densities from the dominant class of clusters will be increased. Because the studies finding short range order focus on the tellurium centered 5-atom clusters of the form $\text{Hg}_{4-n}\text{Cd}_n$ rather than on cation centered clusters of the form $\text{Hg}_{12-n}\text{Cd}_n$, higher level 5-atom cluster-cluster correlations must be known to predict the effects on the vacancy populations.

Next we examine the defects which accommodate excess Hg in the solid. The existence region for HgCdTe is always tellurium rich and thus the native defects which accommodate excess tellurium are expected to dominate. For these equations in Table I, K_i^0 is given by

$$K^0_{V_{Te}} = C_0^{-1}(kT)^{-\frac{5}{2}}(2\pi m_{Hg})^{-\frac{3}{2}}h^3 \exp(\frac{\Delta S_{V_{Te}}}{k}) \quad , \tag{11}$$

$$K^0_{Hg_{Te}} = C_0^{-1}(kT)^{-5}(2\pi m_{Hg})^{-3}h^6 \exp(\frac{\Delta S_{Hg_{Te}}}{k}) \quad , \tag{12}$$

and

$$K^0_{Hg_I} = C_0(kT)^{-\frac{5}{2}}(2\pi m_{Hg})^{-\frac{3}{2}}h^3 \exp(\frac{\Delta S_{Hg_I}}{k}) \quad . \tag{13}$$

If we assume the change in entropy is comparable for all three defects, we find

$$\frac{[Hg_{Te}]}{[V_{Te}]} \simeq 10^{+14} \tag{14}$$

and

$$\frac{[Hg_i]}{[V_{Te}]} \simeq 10^{+14} \quad , \tag{15}$$

for T=500 C and $P_{Hg} = 1$ atm. From Equations (14) and (15) we see that the tellurium vacancy, $[V_{Te}]$, is a minority defect species. For the pressure and temperature considered, the density of Hg antisites is predicted to be comparable to the density of Hg interstitials. Because the ratio of $[Hg_{Te}]$ to $[Hg_i]$ is nearly unity, any errors in the calculation of the activation energy could push the balance toward one side or the other. Thus we must depend on the next generation of calculations, with the ASA removed and full relaxation included plus a quantitative comparison of the entropy differences between the mercury antisite and the tellurium interstitial, to determine the dominant defect in this class.

As mentioned above, we have shown that the tellurium vacancy formation energy varies significantly with the alloy environment. Because the tellurium vacancy is not expected to be a dominant defect in HgCdTe, and the tellurium diffuses by an interstitial mechanism, we do not expect any measurable manifestation of this variation. On the other hand, the Hg antisite may be the dominant Hg-excess defect, and its formation energy may vary significantly with the alloy environment. We are currently calculating the magnitude of this variation.

DIFFUSION

HgCdTe exhibits a complex tracer diffusion profile, with both a fast and a slow branch. The fast branch is attributed to a vacancy and interstitial diffusion in parallel mechanism where the dominant diffuser is determined by the pressure and temperature, while the slow component fits a vacancy and interstitial in series mechanism [11]. The activation energies for the fast branch are 2.10 eV and 0.61 eV for the vacancy and interstitial mechanisms, respectively. Our calculated formation energy for the mercury interstitials are 0.89 and 0.98 eV for the anion- and cation-surrounded tetrahedral interstitial sites, respectively, and the experimental formation energy for the mercury vacancy is 2.24 eV. Comparing these energies to the experimental activation energies we find close agreement, indicating that the migration energy contribution to the diffusion activation energies are small for both interstitials and vacancies.

In a recent experiment on mercury diffusion in ion-implantation damaged HgCdTe, an activation energy of several tenths of an electron volt was measured [12]. The disparate result can be interpreted as a measure of only the defect migration contribution to the diffusion activation energy, since defects in excess of the equilibrium concentration were likely formed during implantation. It is not evident that the measured activation energy corresponds to the vacancy or the interstitial mechanism. The conclusion that the diffusion activation energies are largely defect formation energies, with the migration energies being much smaller, is in agreement with the above interpretation of the Richter and Kalish [12] experiment.

CONCLUSIONS

We have incorporated our calculated defect energies into the mass action equations for the neutral defects in HgCdTe. In agreement with experiment, we find the mercury vacancy to be the dominant native defect in tellurium rich material. We also find the mercury antisite and interstitial defect densitities to be comparable, although a better calculation of the vibrational entropy is needed to confirm this result. Comparing the defect formation energies to the diffusion measurements by Tang and Stevenson [11], we find agreement with their diffusion activation energies for both the vacancy and the interstitial mechanism, if we assume that the migration energy is small. Further work is in progress to incorporate the defect charge states into the calculation, and also to calculate the fully relaxed defect energies with the full potential LMTO.

The work was supported by NASA contract NAS1-18226, by ONR contract N00014-88-C0096, and by AFOSR contract F49620-88-K-0009.

REFERENCES

[1] H. R. Vydyanath, J. Electrochem. Soc. 128, 2609 (1981).

[2] H. R. Vydyanath, J.C. Donovan, and D.A. Nelson, J. Electrochem. Soc. 128, 2625 (1981).

[3] M.A. Berding, M. van Schilfgaarde, A.T. Paxton, and A. Sher, J. Vac. Sci. Technol. A 8, 1103 (1990).

[4] A different reference is used here than was used in Table III in Reference 3. An error appears in that table owing to the incorrect use of an energy of $2E_b$ per unit cell rather than $4E_b$. The defect energies in Table I in Ref. 3 from which the energies in Table III were derived, are correct.

[5] A. Sher, M. van Schilfgaarde, and M.A. Berding, presented at the 1990 HgCdTe Workshop in San Francisco, CA, and accepted for publication in J. Vac. Sci. Technol.

[6] M.A. Berding, A. Sher, and A.-B. Chen, J. Appl. Phys. 68, 5064 (1990).

[7] P.M. Amirtharaj and F.H. Pollak, Appl. Phys. Lett. 45, 789 (1984).

[8] A. Compaan, R.C. Bowman, and D.E. Cooper, Appl. Phys. Lett. 56, 1055 (1990).

[9] L.K. Vodopyanov, S.P. Kozyrev, Y. A. Aleshchenko, R. Triboulet, and Y. Marfaing, Appl. Phys. Lett. 56, 1057 (1990).

[10] D. Zamir, K. Beshah, P. Becla, P.A. Wolff, R.G. Griffin, D. Zax, S. Vega, and N. Yellin, J. Vac. Sci. Technol. 6, 2612 (1988).

[11] M.-F.S. Tang and D.A. Stevenson, J. Vac. Sci. Technol. 7, 544 (1990), and references therein.

[12] V. Richter and R. Kalish, J. Appl. Phys. 67, 6578 (1990).

INTEGRATED MSM-FET PHOTORECEIVER FABRICATED ON

MOCVD GROWN $Hg_{1-x}Cd_xTe$

Patrick W. Leech[*], Peter J. Gwynn[*], Geoffrey N. Pain[*], Novica R. Petkovic[*], James Thompson[*] and David N. Jamieson[**],
[*]Telecom Australia Research Laboratories, 770 Blackburn Road, Clayton 3168, Victoria, Australia
[**]School of Physics, University of Melbourne, Parkville, 3052, Victoria, Australia.

ABSTRACT

We report on progress in the monolithic integration of a metal-semiconductor-metal (MSM) detector and transimpedence amplifier and of a photoconductive detector (PCD) with a metal-semiconductor field effect transistor (MESFET) in $Hg_{1-x}Cd_xTe$. The layers of $CdTe/n$-type $Hg_{1-x}Cd_xTe$ were grown by MOCVD on semi-insulating GaAs substrates (2° misoriented 100). Fabrication of the devices was by an FET planar process; with a standard lift-off used to form Schottky metallization on both the interdigitated electrodes of the MSM detector (2μm width, 2μm spacing) and the gate of the MESFETs (5μm length, 100μm width). The MSM photodetectors exhibited breakdown voltages in the range 60 to 80V, a dark current of 10na at 5V bias, and responsivities of >1.0 A/W measured at 40V using CW 1.3um illumination. The integrated devices have been characterised by electrical and micro RBS techniques; the results were found to be strongly dependent on the stoichiometric x ratio of the $Hg_{1-x}Cd_xTe$. This initial work demonstrates the suitability of $Hg_{1-x}Cd_xTe/GaAs$ structures in the fabrication of integrated optoelectronic circuits.

INTRODUCTION

There is considerable current research into the development of optoelectronic integrated circuits (OIEC's) for application in optical communications and data processing. Particular emphasis has been given to the monolithic integration of a photodetector and preamplifier as an optoelectronic receiver circuit. Several examples of integrated optical receivers comprising a metal-semiconductor field effect transistor (MESFET) and a photoconductive detector (PCD) have been reported in GaInAs [1], GaAs [2] and CdTe [3]. Also, recent progress in metal-semiconductor-metal (MSM) detectors has resulted in the fabrication of high performance integrated recievers comprising an MSM detector and MESFETs in GaAs [4] and GaInAs [5]. MSM structures have become a promising detector alternative for use in optoelectronic integrated circuits; with faster response than PCDs because of photocurrent flow through a high impedence depletion region formed by the Schottky-Schottky contacts.

The semiconductor $Hg_{1-x}Cd_xTe$ is an attractive material for use in integrated receiver circuits because of its low intrinsic carrier concentration, high absorption coefficient and flexibility of bandgap tuning allowing 1.3, 1.5 or 2.5µm detection [6,7]. Circuit elements necessary for the fabrication of optical integrated receivers in $Hg_{1-x}Cd_xTe$, such as the MESFET [8] and MSM detector [9], have recently been realised. In addition, advances in the growth of $Hg_{1-x}Cd_xTe$ on substrates including GaAs [10] have increased the potential for integration of the detector element and other device functions.

In this paper, we report on the initial fabrication of receiver circuits comprising a transimpedence amplifier-MSM detector and an FET-PCD detector in $Hg_{1-x}Cd_xTe$. Alloy stoichiometries were in the range x=0.5 to 0.7, corresponding to the near infrared spectral region.

DEVICE FABRICATION

The $CdTe/Hg_{1-x}Cd_xTe$ layers were grown using an MR Semicon Quantax 226 reactor on substrates of semi-insulating GaAs (2° misoriented 100). Precursors used in the growth process were dimethyl cadmium, diethyl tellurium and an elemental Hg source. The epitaxial $Hg_{1-x}Cd_xTe$ was formed by an Interdiffused Multilayer Process (IMP) [10] in which alternate layers of HgTe/CdTe of period thickness 0.1um were grown at 310-330°C and then annealed in-situ in a flow of H_2 at 380°C. A 30nm capping layer of CdTe was subsequently deposited on the surface of the $Hg_{1-x}Cd_xTe$ to produce an enhancemennt of Schottky barrier characteristics. Carrier concentrations for the $Hg_{1-x}Cd_xTe$ layers were in the range N_d = 5 x 10^{15} to 1 x 10^{16} cm^{-3} as determined from Hall measurements at room temperature. The stoichiometric x-ratio was found to vary from x=0.3 to 0.7 in the direction of gas flow across the wafer.

Arrays of the devices were fabricated over the surface of quarter wafers of 50mm diameter. Initially, mesa isolated regions were formed by selective etching of $CdTe/Hg_{1-x}Cd_xTe$ down to the GaAs [11] with planarising of the structure using a polyimide film. Ohmic contacts (30nm In/400nm Au) for the source/drain of the MESFET and for the electrodes of the photoconductor were then formed by a lift-off procedure. Chemical etching through the CdTe layer was necessary prior to the formation of ohmic contacts. A further lift-off process was used to form Schottky metallisation of either Au, Pd or Pt on the interdigitated fingers of the MSM detector (2µm width, 2µm spacing) and on the gate of the MESFETs (2µm length, 100µm width). A final sequence of electron beam deposition (Ti 30nm/Au 400nm) and lift-off was used to form the interconnections and bonding pads which were overlaid on the polyimide.

Figure 1 shows the elements of the two photoreceiver circuits which comprised a) an interdigitated MSM detector and trans-impedance amplifier containing four cascade MESFETs and level shifting diodes and b) a PCD with a single FET depletion mode n-channel MESFET.

The level shifting in the larger circuit was intended to produce an output voltage close to zero. These circuits were part of a larger chip with dimensions of 3mm x 3mm including the bonding pads.

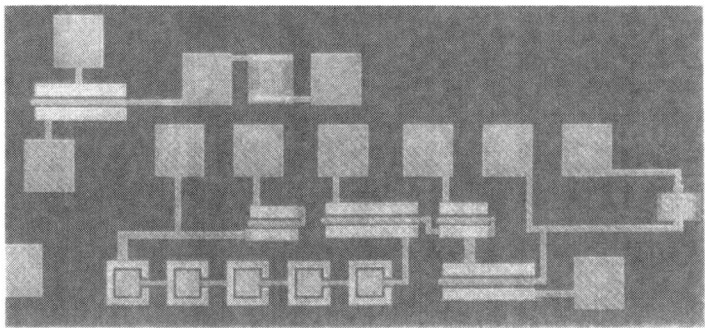

Figure 1   Scanning electron micrograph showing the MSM-transimpedence
amplifier and PCD-FET integrated structures on $Hg_{1-x}Cd_xTe/GaAs$.

RESULTS AND DISCUSSION:

The current-voltage characteristics of the MSM photodetectors were symmetrical with respect to the zero bias and exhibited breakdown voltages in the range from 60 to 80Volts. These values of breakdown voltage were significantly greater than the 20V measured for $Hg_{1-x}Cd_xTe$ without the CdTe overlayer, an effect which has been attributed to a reduction in the magnitude of surface leakage current. The breakdown voltages for the $CdTe/Hg_{1-x}Cd_xTe$ detectors were also considerably greater in magnitude than for the equivalent devices in GaAs [4] and InAlAs/GaInAs [12]. The dark current of the MSM detectors at 5V bias was typically 0.2µA for Pt with values as low as 1.0nA at 1V as shown in Figure 2. Pt and Au metallizations were found to give the lowest values of dark current, corresponding to a similar ordering as the reverse current of Schottky barrier diodes fabricated with these metals. The photoresponsivity of the detectors was measured by laser diode at 1.3um CW illumination as a function of level of optical power (1.85µW and 4.63µW) and increasing bias voltage. For the MSM detectors, responsivities of 5.0 A/W have been measured with 20V bias voltage at room temperature (Figure 3). In comparison, the PCDs fabricated on the same chip exhibited responsivities of 40 A/W at 20V bias voltage. Also, the frequency response of the detectors was determined by illumination with a 50um core optical fibre irradiated by a 1.3um laser with a measured bandwidth of 1.3GHz. The response of the MSM-laser was observed using a sampling oscilloscope, with recorded times of 200ns. The depletion mode MESFETs which were incorporated in the OIEC receiver structure exhibited a peak transconductance of $1mS.mm^{-1}$ with saturation current typically in the range $V_{DSS} > 8V$.

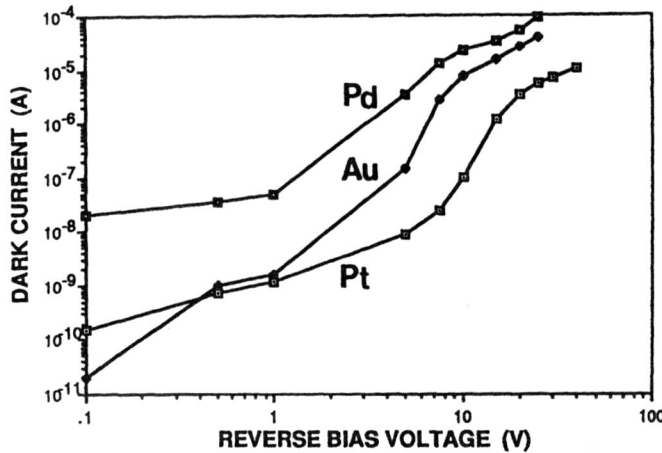

Figure 2   Dark current versus bias voltage for MSM detectors with either Pd, Au or Pt metallization.

This small magnitude of transconductance may be attributed to the low carrier concentration and high sheet resistivity of the $Hg_{1-x}Cd_xTe$ layers.

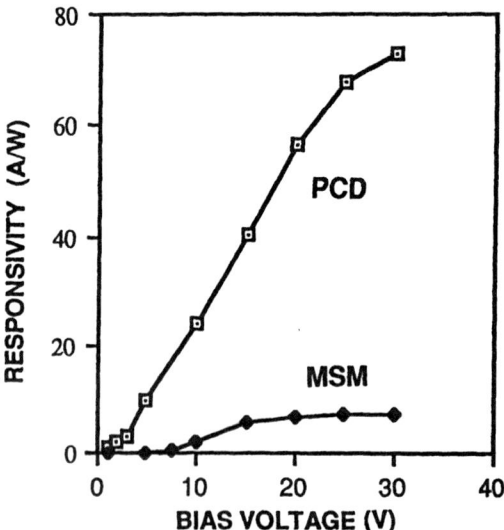

Figure 3   Responsivities of MSM detector and PCD in $Hg_{1-x}Cd_xTe$ under 1.3μm, 1.85μW CW illumination.

Probe testing of the wafers was carried out to determine the receiver characteristics. For the PCD-FET structure, 1.3μm illumination resulted in an increase in I_{ds} from 510μA to 760μA at 10V. But functioning of the FET-MSM was evident only over limited regions of the wafers. In order to investigate these effects, selected devices at locations across the wafers were mapped using the Melbourne Nuclear Microprobe [13]. With a scanned, focussed 2MeV He$^+$ ion beam of 3um width, the microprobe allows the use of traditional ion beam analysis techniques such as Rutherford Backscattering (RBS) [14], and Particle Induced X-Ray Emission (PIXE) [15] to map buried structures (RBS) or elemental distributions (RBS, PIXE). Figure 4 shows a comparison of RBS spectra from a) a region of polyimide/GaAs and b) an adjacent MSM device.

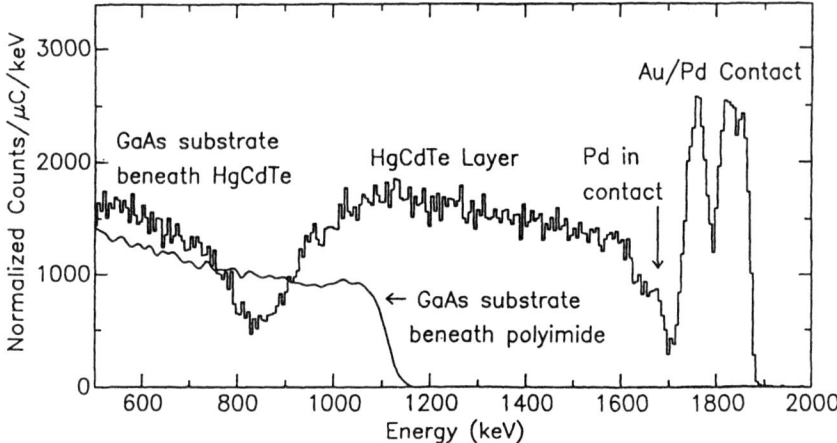

Figure 4   RBS Spectra from a) a region of polyimide/GaAs substrate (smooth curve) and b) an MSM detector with layer structure 50nm Au/15nm Pd/35nm Au/1.5um Hg$_{1-x}$Cd$_x$Te/ GaAs (Histogram). These data were obtained with 0.2μC of 2MeV He$^+$ with a detector angle of 155^0.

Spectral features in Figure 4 associated with the substrate, the epitaxial Hg$_{1-x}$Cd$_x$Te and the contact metals were clearly delineated. In the metallisation, the Au layer was thinner than expected and contained the Pd layer within it rather than at the interface with the Hg$_{1-x}$Cd$_x$Te.

This processing error was not critical in the determination of device characteristics, since Au has been shown to form Schottky contacts with Hg$_{1-x}$Cd$_x$Te of having a large barrier height φ_b and suitable for MSM detectors. As a further monitor of the elemental characteristics of the same contact regions, PIXE spectra have revealed a negligible contamination of the devices by unexpected metals or from other sources during processing.

Moreover, channeling experiments have indicated the single crystal character of the $Hg_{1-x}Cd_xTe$ with chi-min <8% (ratio of surface backscattered particle yield from a single crystal aligned with the particle beam /yield of randomly aligned crystal). However, the observed absence of Schottky barrier formation across regions of the wafers was found to be associated with a stoichiometric x value of >0.5. Only regions in which x >0.5 were the devices functioning. This effect is consistent with the predictions of major theoretical models for Schottky barrier formation [16] that rectifying contacts will occur in $Hg_{1-x}Cd_xTe$ only at x values greater than 0.4-0.5.

The permission of the General Manager, Research, Telecom Australia, to publish this paper is acknowledged.

REFERENCES

1. J.Barnard, H.Ohno, C.E.Wood and L.F.Eastman, IEEE Electron Device Lett., EDL-2, (1981),7.
2. H.Matsuo, H.Ohno and H.Hasegawa, Jpn.J.Appl.Phys., 23, (1984), 648.
3. J.L.Filippozzi, F.Therez, D.Esteve, M.Fallahi, D.Kendil, M.Da Silva, M.Barbe and G.Cohen-Solal, J.Crystal Growth, 101, (1990), 1013.
4. H. Pohjonen and M.Anderson, Sensors and Actuators, A21-23, (1990), 1124.
5. L.Yang, A.S.Sudbo, W.T.Tsang, P.A.Garbinski and R.M.Carmada, IEEE Photonics Technology Letters, 2, (1), (1990), 59.
6. B.Orsal, R.Alabedra, M.Valenza, G.Lecoy, J.Meslage and C.Y.Boisrobert, IEEE Trans. Electron Devices, ED-35, (1988), 101.
7. J.Thompson, P.Mackett, G.T.Jenkin, T.Nguyen Duy and P.Gori, J.Crystal Growth, 86, (1988), 917.
8. P.W.Leech, P.J.Gwynn, G.N.Pain, N.Petkovic and J.Thompson, Electronics Letters, 26, (4), (1990), 221.
9. P.W.Leech, N.Petkovic, P.J.Gwynn, G.N.Pain and J.Thompson,Electronics Letters, 26, (22), (1990), 1848.
10. G.N.Pain, N.Bharatula, T.J.Elms, P.Gwynn, M.Kibel, M.S.Kwietniak, P.W.Leech, N.Petkovic, C.Sandford, J.Thompson, T.Warminski, D.Gao, S.R.Glanvill, C.J.Rossouw, A.K.Stevenson, S.W.Wilkins and L.Wielunski, J.Vac.Sci.Technol., A8(2), (1990), 1067.
11. P.W.Leech, P.J.Gwynn and M.Kibel, App.Surface Science, 37, (1989), 291.
12. J.B.D.Soole, H.Schumacher, H.P.Leblanc, R.Bhat and M.A.Koza, IEEE Photonics Technology Letters, 1, (8), (1989), 250.
13. G.J.F.Legge, C.D.McKenzie, A.P.Mazzolini, R.M.Sealock, D.N.Jamieson, P.M.O'Brien, J.C.McCallum, G.L.Allan, R.A.Brown, R.A.Coleman, B.J.Kirby, M.A.Lucas, J.Zhu and J.Cerini, N.I.M., B15, (1986),669.
14. W.K.Chu, J.W.Mayer and M-A.Nicolet, "Backscattering Spectrometry", Academic Press, (1978).
15. S.A.E.Johansson and T.B.Johansson, N.I.M., 137, (1976), 473.
16. W.E.Spicer, D.J.Friedman and G.P.Carey, J.Vac.Sci.Technol., A6, (1988), 2746.

HgTe-CdTe SUPERLATTICE AND HgCdTe EPILAYER DEVICE STRUCTURES GROWN BY PHOTON-ASSISTED MOLECULAR BEAM EPITAXY

T.H. MYERS, R.W. YANKA, L.M. MOHNKERN, K.A. HARRIS, D.W. DIETZ,
G.K. DUDOFF, K.M. GIROUARD, AND S.C.H. WANG
Electronics Laboratory, GE Company, Syracuse, New York 13221

ABSTRACT

HgCdTe grown by photon-assisted molecular beam epitaxy is now suitable for use in high performance detector fabrication. These are the preliminary results for infrared detectors which have been fabricated in HgCdTe grown using this technique at GE. The detectors were fabricated using a modified Hg-diffused diode process. In addition, the first high quantum efficiency infrared detectors based on the HgTe-CdTe superlattice material system is presented as an example of the sophisticated structures obtainable with photon-assisted molecular beam epitaxy. The superlattice detectors exhibited quantum efficiencies as large as 66% (at 140K) at the peak wavelength of 4.9μm and an average quantum efficiency over the 3-5μm waveband of 55%.

INTRODUCTION

HgCdTe is the most useful material for fabricating infrared (IR) detectors. Molecular beam epitaxy (MBE) is the most promising technique for the growth of HgCdTe for advanced detector applications and is rapidly approaching maturity based on recently published results[1,2,3,4]. The geometric nature of the MBE growth process allows simple solutions for obtaining lateral uniformity. The relatively low growth rates provide strict monolayer control over composition and doping profiles while the low growth temperatures assure the stability of these profiles throughout the growth of the epilayer. At GE's Electronics Laboratory, a variant of photon-assisted MBE (PAMBE) growth technique—compositionally modulated structures (CMS)[1,5,6]—has been shown to produce the material quality necessary for detector fabrication in general as well as the control necessary for advanced device structure development.

This paper treats the compositional uniformity obtainable by the PAMBE/CMS technique and then presents results from two applications of PAMBE growth which represent the broad versatility of device fabrication provided by this technique. Simple Hg-diffused planar diodes, which have been fabricated at GE using a modified version of the SAT[7] process, represent the potential for use of PAMBE HgCdTe in a current, production-ready device fabrication technology. Results of the first high quantum efficiency (over 65%) HgTe-CdTe superlattice detector demonstrate the highly sophisticated structures that can only be fabricated using the monolayer control allowed by PAMBE.

EXPERIMENTAL CONDITIONS

The growth of HgCdTe epilayers was carried out in two custom MBE machines modified for PAMBE growth. An argon ion laser equipped with broadband optics was the illumination source. Epilayers were grown without buffer layers directly onto CdTe substrates obtained from II-VI, Inc. GE's proprietary substrate preparation, which includes a combination of standard wet-etching techniques and thermal processing prior to growth, has been shown to produce high-quality growth interfaces without the added complexity of a buffer layer[1,5]. Epilayers were grown at temperatures ranging from 170-180°C, and a Varian beam flux monitor provided precise determination of the flux from the thermal sources, thus enabling reproducible growth conditions.

IR transmission, IR photoluminescence, and fabricated detector response determined the composition of HgCdTe epilayers grown for this study. Defect etching was accomplished through the use of standard etches. X-ray diffraction rocking curves were measured at GE's Corporate Research and Development Center. Transmission electron microscopy (TEM) characterizations were performed at Purdue University using iodine thinning techniques. Secondary ion mass spectrometry (SIMS) analysis was conducted by Charles Evans and Associates.

Detectors were fabricated at the GE-Electronics Laboratory's HgCdTe processing facility. A mesa diode structure was used for the HgTe-CdTe superlattice detector while the Hg-diffused diodes were formed using planar techniques. After processing, all wafers were tested in a 78K cold-probe station to evaluate I-V characteristics. Selected arrays were then packaged and mounted in a variable-temperature test station for more extensive evaluation, including spectral response, D*, quantum efficiency, and response uniformity.

COMPOSITIONAL UNIFORMITY

Due to the geometric nature of MBE growth, lateral composition uniformity should be achievable without the added complexity of substrate rotation. At GE, both a custom MBE system and custom multiple-zone MBE sources for Te and Hg have been designed and built specifically to achieve this goal. The Hg source was designed to have a flat profile over a 2-in substrate area while maintaining a constant flux (<1% variation) for growth periods exceeding 30 hrs. The Te source uses a true effusion cell that leads to constant flux for constant temperature, allowing both uniform depth composition and run-to-run reproducibility. Combining a longer source-to-substrate distance (25cm) with the use of multiple sources to minimize flux asymmetry results in a highly uniform molecular flux distribution at the substrate.

Figure 1 illustrates the uniformity obtainable with this approach. Data for the compositional map were measured using the microbeam capability of our Nicolet 60x Fourier-transform IR (FTIR) spectrophotometer for x-value determination of a 10μm-thick mercury-cadmium-telluride (MCT) epilayer grown on a 1x1in-square (211)B GaAs substrate. The measurements indicated that compositional uniformity values of Δx=±0.001 are obtainable without substrate

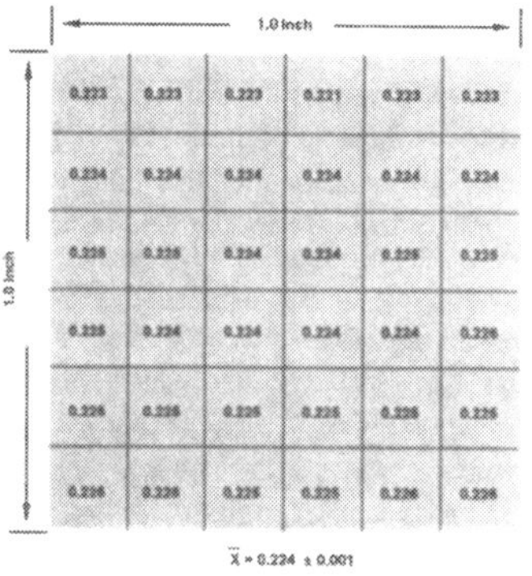

Figure 1. Compositional Map of a 10μm-thick HgCdTe Epilayer Grown on a 1x1in CdTe/GaAs Substrate. The map illustrates the high degree of lateral uniformity in composition (x-value) obtainable over large areas without substrate rotation.

rotation over a 1.5in-diameter epilayer. The systems in use at GE can grow on up to a 2in-diameter wafer. Preliminary measurements suggest that $\Delta x = \pm 0.002$ can be achieved over the entire 2-in diameter.

Hg-DIFFUSED DIODES IN PAMBE/CMS HgCdTe

Since licensing the SAT process in 1985, GE has concentrated on Hg-diffused diodes for its near-term IR detector needs. Initial difficulties were encountered in transferring the process from bulk HgCdTe to epitaxial MBE layers resulting in overdiffusion of the diodes. However, recent process improvements allow devices to be processed on PAMBE/CMS layers, resulting in high-quality diodes. To date extensive testing has only taken place on PAMBE/CMS HgCdTe with cutoff wavelengths near 8μm, to determine how material properties influence the device process. While of shorter wavelengths than typical in GE's IR detector applications, these devices have provided a technology baseline for Hg-diffused device fabrication in PAMBE/CMS HgCdTe.

Diodes fabricated on 8.3μm cutoff (80K) material averaged >25mΩ resistance at 200mV reverse bias and had breakdown voltages >3V. The 80K R_oA product averaged 200Ω-cm^2 for a random sampling of 50 diodes with a maximum of 380Ω-cm^2. A randomly selected 64-element array from another wafer was connected to a multiplexer for detailed testing. Histograms of cutoff wavelength and quantum efficiency are shown in Figure 2. The array was highly uniform, with an average λ_c of 7.44μm and a standard deviation of only 0.01μm, which corresponds to our measurement uncertainty. The wavelength translates to an x-value of 0.2514 with a standard deviation of 0.0002 across the array, which is approximately 0.4cm in length. This result agrees with x-value variations of 0.0005/cm (0.001/in) measured on 1x1in MCT layers grown on GaAs substrates. The average quantum efficiency for this array was 66% with a standard deviation of 2%. D* values could not be directly measured because the diode noise was completely masked by the (low) kTC noise from the output stage of the multiplexer. Analysis of the noise, however, indicates that the performance of these diodes would be background-limited for virtually any application. The diodes were reverse biased out to 150mV without exhibiting measurable 1/f noise through the multiplexor. Similar test results were obtained from a second test array from the same wafer.

Preliminary results have also been obtained for long-wave infrared (LWIR) diodes. An example of the spectral response of a LWIR diode is shown in Figure 3. Preliminary testing of devices from this wafer indicate R_oA products comparable to the best ever reported for n-on-p

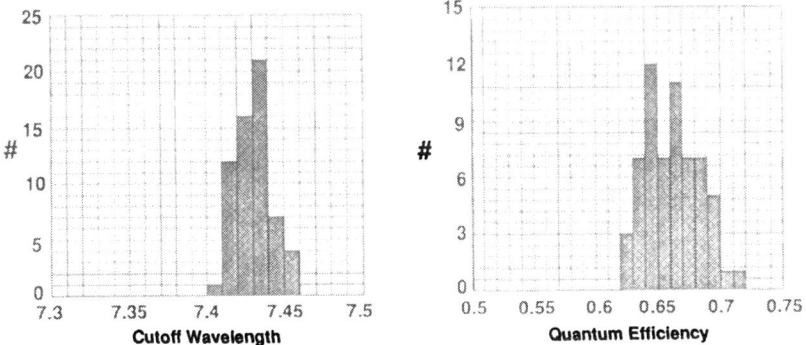

Figure 2. Histograms of Cutoff Wavelength and Quantum Efficiency at 80K for a 64-Element Test Array Fabricated in PAMBE/CMS HgCdTe. The average wavelength is 7.44±0.01μm. The average quantum efficiency is 66%.

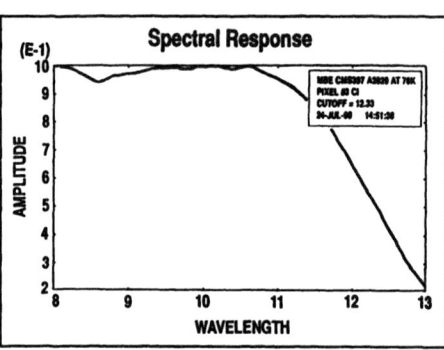

Figure 3. 0° Field-of-View I-V Curve for a LWIR Diode Fabricated in 30μm-thick PAMBE/CMS HgCdTe. The 78K cutoff wavelength is 12.3μm with an R_oA product of 1.5Ω-cm².

homojunctions, with indications of high device performance. Importantly, detailed spatial photo-response measurements indicate control of the Hg-diffusion process. The results to date are consistent with GE's plan to insert MBE material into its standard process line during 1991.

MWIR HOMOJUNCTION SUPERLATTICE DETECTOR

The successful fabrication of midwave IR (MWIR) superlattice detectors which exhibit high quantum efficiencies serves to demonstrate the flexibility and high degree of control given by PAMBE. Figure 4 is a representative TEM micrograph of the MWIR superlattice used to fabricate the detector. The layer thicknesses for this superlattice are 30Å (HgTe) and 50Å (CdTe). MBE is the only growth technique that permits the growth of Hg-based structures consisting of such thin layers. All observed areas of the superlattice exhibited a highly regular layer spacing with a high degree of interfacial sharpness and layer uniformity. Line dislocation counts were consistent with <10⁶cm⁻³, the resolution of the TEM technique.

The device structure was grown on a (211)B CdTe substrate and consisted of a 3μm n-type absorbing (base) superlattice followed by the growth of a 1μm-thick p-type capping superlattice layer. Dopant incorporation was verified by SIMS analysis on a small piece of the grown layer. Indium was used as the n-type dopant and arsenic as the p-type dopant. Arsenic was used only in the CdTe layers to minimize problems associated with Hg-deficient growth conditions[8].

The measured spectral response for one of these devices with a cutoff wavelength of about 4.5μm at 140K is shown in Figure 5. Direct measurements yielded peak quantum efficiencies as high as 66% (at 140K) at the peak wavelength and an average over the 3-5μm waveband of 55%. The variation in cutoff wavelength varied by 0.28meV/K. At 78K, the measured peak quantum efficiency was found to decrease to 45-50% with a cutoff wavelength of 4.9μm. These devices represent the first HgTe/CdTe superlattice-based IR detectors to exhibit significant quantum efficiencies.

The optically sensitive area was determined by IR spot scan. The device response was uniform over the entire detector area for this front-side-illuminated device except in the region of the contact. The optically active area was in good agreement with that of the physical area of the mesa device.

Figure 6 shows a representative I-V measurement obtained at 78K of the MWIR superlattice detectors. The p-on-n nature of the homojunction is clearly shown as the I-V curve appears to be reversed from that obtained by the more conventional n-on-p diode. The measured variation of R_oA versus inverse temperature indicated that tunneling processes begin to dominate the diffusion dark-current limited behavior at temperatures less than 90K. Investiga-

Figure 4. TEM Cross-Section Micrograph Obtained from a Small Section of the p-on-n Homojunction Superlattice. The layer uniformity and interface sharpness demonstrates the deposition control provided by MBE.

Figure 5. Electrical Behavior of the Superlattice Devices is Illustrated by the I-V Curve of a Representative Device. The I-V curve is reverse of that obtained from the more conventional n-on-p diode. Measured R_oA values are $5 \times 10^5 \Omega$-cm² at 78K.

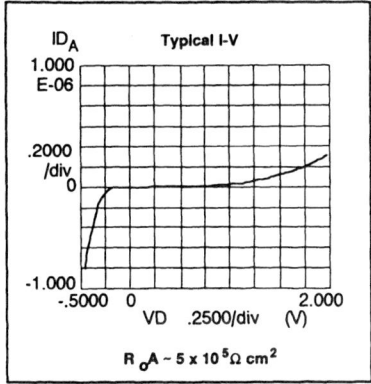

Figure 6. Representative Spectral Response of a Superlattice Detector with λ_c=4.53µm @ 140K Illustrating a Sharp Profile. The peak response corresponds to 66% peak quantum efficiency.

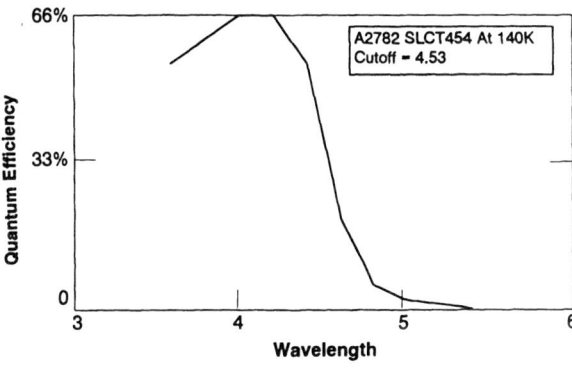

tions of the bias dependence of the guard ring on gated diodes indicated that there was a surface contribution to the tunneling currents, indicative of a surface limited by the passivation. In spite of this, the R_oA values for these superlattice detectors were typically $5 \times 10^5 \Omega$-cm^2 which is comparable to that achieved in the corresponding alloy.

SUMMARY

The results presented support the contention that HgCdTe grown by PAMBE/CMS is suitable for use in fabricating IR detectors. The HgTe-CdTe superlattice detector results illustrate the high degree of sophistication available to engineering device structures when using PAMBE. Both sets of results confirm that the PAMBE growth of HgCdTe and related materials can be used for near-term IR detector applications.

ACKNOWLEDGEMENTS

The authors would like to thank Professor J.F. Schetzina and his staff at North Carolina State University for performing the near-term IR transmission analysis of the interdiffused superlattice. The work summarized in this paper was supported by GE Aerospace Internal Research and Development funding.

REFERENCES

1. K.A. Harris, T.H. Myers, R.W. Yanka, L.M. Mohnkern and N. Otsuka. Proceedings of the 1990 Workshop on the Physics and Chemistry of HgCdTe and Related Materials (to be published in J. Vac. Sci. Technol. A.)

2. J.M. Arias, S.H. Shin, M. Zandian, J.G. Pasko and R.E. DeWames, Proceedings of the 1990 Workshop on the Physics and Chemistry of HgCdTe and Related Materials (to be published in J. Vac. Sci. Technol. A.)

3. R.J. Koestner, M.W. Goodwin and H.F. Schaake, Proceedings of the 1990 Workshop on the Physics and Chemistry of HgCdTe and Related Materials (to be published in J. Vac. Sci. Technol. A.)

4. R.J. Koestner and H.F. Schaake. J. Vac. Sci. Technol. A, 6, 2834 (1988).

5. T.H. Myers, R.W. Yanka, K.A. Harris, A.R. Reisinger, J. Han, S. Hwang, Z. Yang, N.C. Giles, J.W. Cook, Jr., J.F. Schetzina, R.W. Green, and S. McDevitt. J. Vac. Sci. Technol. A, 7, 300 (1989).

6. R.W. Yanka, K.A. Harris, L.M. Mohnkern and T.H. Myers, Proceedings of the Sixth International Conference on Molecular Beam Epitaxy (to be published in J. Crys. Growth.)

7. Societe Anonyme De Telecommunications [SAT], Paris, France.

8. J.W. Han, S. Hwang, Y. Lansari, R.L. Harper, Z. Yang, N.C. Giles, J.W. Cook, Jr., and J.F. Schetzina. J. Vac. Sci. Technol. A, 305 (1989).

RBS AND PIXE ANALYSIS OF $Hg_{1-x}Cd_xTe$ grown by MOCVD.

S.P. Russo, R.G. Elliman and P.N. Johnston
Microelectronics and Materials Technology Centre,
Royal Melbourne Institute of Technology, 3001, Australia.
G.N. Pain
Telecom Australia Research Laboratories, 3168, Australia.

ABSTRACT

The techniques, Particle Induced X-ray Emission (PIXE) and Rutherford Backscattering Spectrometry (RBS) have been used to investigate compositional and thickness uniformity of $Hg_{1-x}Cd_xTe$ (MCT) grown on GaAs substrates by Metal Organic Chemical Vapour Deposition (MOCVD). Composition and thickness variations are reported for orientations perpendicular and parallel to gas flow in the MOCVD reactor. Crystalline quality of the MCT layer was also determined by RBS channelling analysis.

INTRODUCTION

The ternary semiconductor $Hg_{1-x}Cd_xTe$ (MCT) is presently the most widely used semiconductor for infrared detectors. Recent progress in the development of heteroepitaxial growth of MCT heterostructures by Molecular Beam Epitaxy (MBE) and Metal Organic Chemical Vapour Deposition (MOCVD) has expanded the potential opto-electronic applications of this material. MCT based solar cells, optical waveguides, optical switches, light emitting diodes and infrared lasers have all been demonstrated.

For production of low cost devices MOCVD offers the capability of high quality epitaxial growth on various substrates with high wafer throughput. However the device performance of MOCVD grown MCT heterostructures depends critically on the composition (including dopant concentration) and layer thickness. Therefore, materials characterization is essential for proper optimization of the processing sequence and to understand the observed device performance.

In this paper the composition, thickness and crystalline quality of MOCVD grown MCT epitaxial thin films was examined using Rutherford Backscattering (RBS) and Particle Induced X-ray Emission (PIXE). By simultaneously counting backscattered particles and x-rays the composition (PIXE), thickness (RBS) and crystal quality (channelling RBS) can be determined at any point on the wafer. This enables the calculation of the thin film growth rate as a function of position on the wafer.

EXPERIMENTAL

MCT was grown on 5cm (100) oriented GaAs wafers using a M.R. Semicon Quantax 226 MOCVD reactor at Telecom Australia Research Laboratories. Dimethyl cadmium (Me_2Cd), diethyl tellurium (Et_2Te) and mercury (99.999 99%) in high purity hydrogen carrier gas were used. The substrates were placed in a quartz reactor cell which houses a rectangular SiC coated graphite susceptor suspended above the wafers. The temperature was monitored by a thermocouple placed in the centre of the cracking susceptor, further details of reactor cell design, typical growth rates and wafer preparation are given by Pain et al[1].

For this study an MCT layer was grown by the interdiffused multilayer process[2]. Ten successive layers of CdTe and HgTe were deposited at a temperature between 300-330°C. The layers were then annealed at approximately 400°C for 45 minutes to enable the last layers to interdiffuse. A HgTe capping layer was then deposited at the original growth temperature.

RBS and PIXE data were obtained using 1.8 MeV H$^+$ or 2.0 MeV He^{2+} ions. Measurements on the wafer were taken at 5mm intervals both parallel and perpendicular to the direction of gas flow in the MOCVD reactor.

Backscattered particles were measured at a scattering angle of 170° by a silicon surface barrier detector with an energy resolution of 14keV (FWHM). Ion beam currents of 5nA were used to reduce pulse pileup, beam damage and heating. Beam heating is of particular concern since this can cause loss of Hg from the surface and hence change sample composition in the region under investigation.

Emitted x-rays were collected using a Princeton Gamma Tech SiLi detector with an energy resolution of 154eV for the MnK$_\alpha$ line (5.90 keV), using a 25.4 μm Be window. The detector was located at an angle of 135° with respect to the incident beam direction. The average composition of the epilayer at a particular point was determined by peak fitting to the HgL$_\gamma$ (13.83 keV), CdK$_\alpha$ (23.17 keV) and TeK$_\alpha$ (27.47 keV) x-rays. Peak fitting and x-ray yield calculations were performed with a PIXE analysis package, PIXAN, developed by Clayton[3] and modified at RMIT. The results were normalized using a certified 6.3μm thick CdTe layer on sapphire as a standard. A 50 μm aluminium filter was placed in front of the SiLi detector to suppress secondary electron bremsstrahlung and lower energy HgM$_\alpha$, CdL$_\alpha$, TeL$_\alpha$ (and to a lesser extent GaK$_\alpha$ and AsK$_\alpha$) x-rays.

Measurement errors were calculated by propagation of uncertainties in quadrature[4]

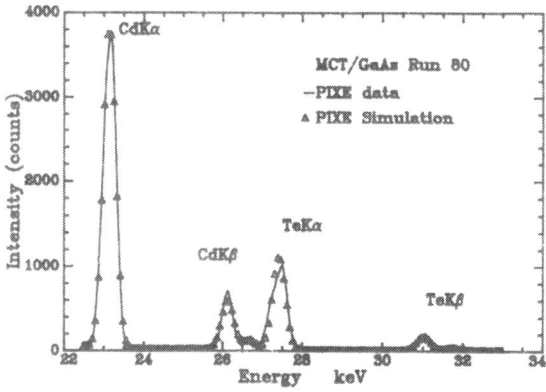

FIGURE 1(a)

PIXE spectrum of MCT showing CdK$_\alpha$ and TeK$_\alpha$ peaks.

RESULTS AND DISCUSSION

Figures 1(a) and (b) show a typical a PIXE spectrum along with the PIXAN simulation fits to the HgL$_\gamma$, Cdk$_\alpha$ and TeK$_\alpha$ peaks. The HgL$_\gamma$ line is chosen because the L$_\alpha$(9.99 keV) and L$_{\beta 1}$(11.82keV) lines are mixed with the GaK$_{\alpha/\beta}$ (9.24keV, 10.26keV) and AsK$_{\alpha/\beta}$ (10.53, 11.72keV) lines which can lead to inaccuracies in the peak fitting algorithms. The CdL$_\alpha$ (3.13 keV) and TeL$_\alpha$ (3.76 keV) x-rays emitted from deep within the epilayer undergo significant absorption by the Hg closer to the surface. To avoid this problem K x-rays were used to calculate the Cd and Te composition. The error in the Hg, Cd and Te peak areas between data and fit in these figures ranged from 1-2% which was representative of the error for all points sampled. Composition was calculated using the peak area and x-ray yield results determined by PIXAN[4]. The use of a CdTe standard allowed the relative concentrations of Cd, Te and Hg in the epilayer to be measured to an accuracy of 5-7%, the main sources of error being the calculation of x-ray yields.

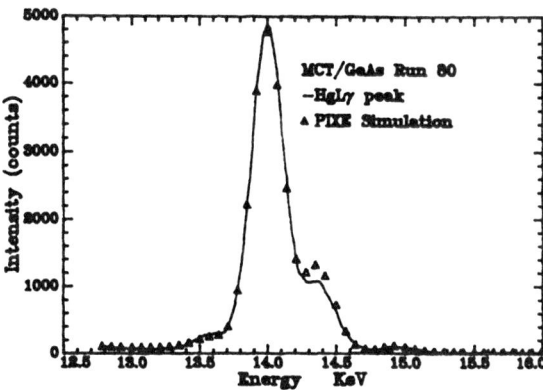

FIGURE 1(b)

PIXE spectrum of MCT showing HgL_γ peaks.

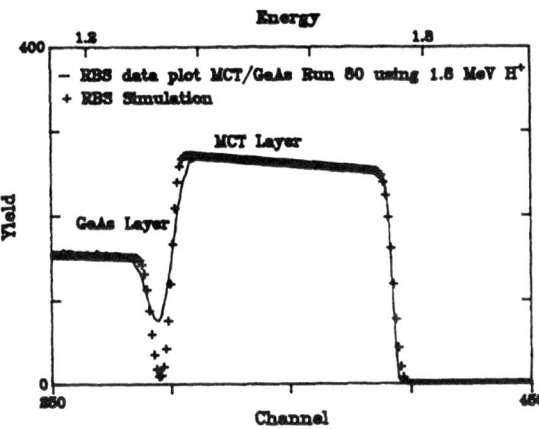

Figure 2

RBS simulation fit to MCT using 1.8 Mev H^+

Having determined composition at a particular point, the thickness at that point was calculated from the RBS spectrum[5]. The thickness was determined by simulation of the depth profile using the average composition measured by PIXE. The simulation was performed by RUMP[6] an RBS analysis package. Figure 2 shows a typical simulation fit to the MCT layer at a particular point. The error in the thickness measurements for all points sampled ranged from 7-10%

Figures 3(a) and 4 show the composition and thickness variation at 5mm intervals in the direction of gas flow. The composition of Te remains constant (50% within experimental error) while the Cd and Hg levels decrease and increase respectively, along the wafer. There is also a steady increase in the thickness in the deposited layer along the wafer.

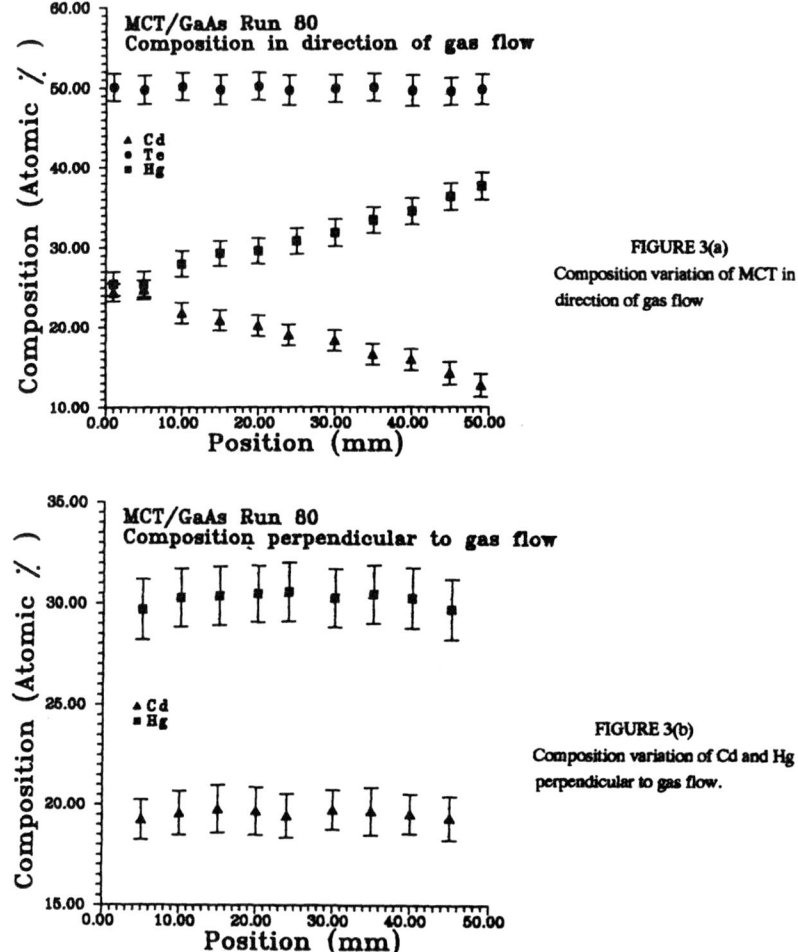

FIGURE 3(a)

Composition variation of MCT in direction of gas flow

FIGURE 3(b)

Composition variation of Cd and Hg perpendicular to gas flow.

The average growth rate (in μm/hr) of CdTe and HgTe were determined at each point by multiplying the composition and thickness data and dividing by the total growth time for each layer. Figure 5 shows the respective growth rates as a function of position in the direction of gas flow. The variation of CdTe and HgTe growth rates in the direction of flow is due to different energy requirements for Et_2Te cracking in the presence of Me_2Cd or Hg . Et_2Te decomposes at a temperature as low as 330^oC in the presence of Me_2Cd. Thus CdTe has a high deposition rate at the leading edge of the wafer. As the carrier gas moves along the reactor the Cd deposition rate decreases due to concentration depletion in the gas phase. The temperature for deposition of HgTe on the substrate is higher than that for CdTe because Et_2Te is not as easily decomposed in the absence of Me$_2$Cd. Therefore, the growth efficiency of HgTe is kinetically limited, increasing as carrier gas temperature rises further into the deposition zone.

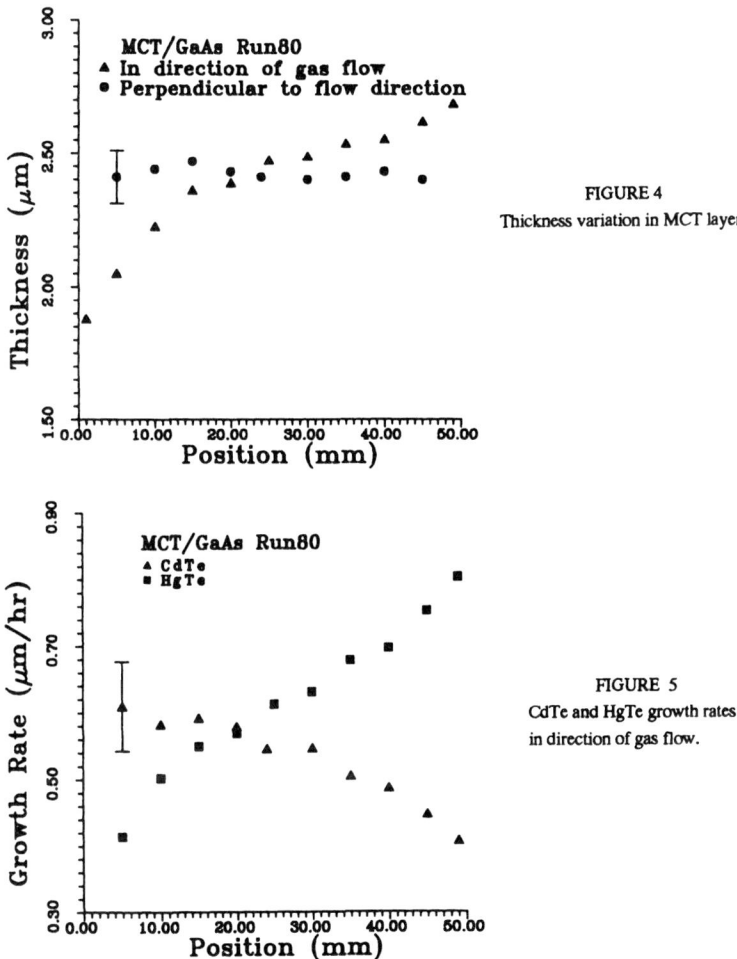

FIGURE 4

Thickness variation in MCT layer

FIGURE 5

CdTe and HgTe growth rates
in direction of gas flow.

The concentration of Et$_2$Te, Me$_2$Cd and Hg in the gas phase is approximately uniform, along the wafer, in the direction perpendicular to gas flow. Therefore there is little deviation in the MCT layer thickness or Cd and Hg composition as shown in figures 3(b)and 4.

In addition to composition and thickness, a measurement of crystalline quality and epitaxial relationship of the MCT layer can be determined by channelling the ion beam into the (100) axis of the GaAs substrate and measuring the channelled to random backscattered yield in the near surface region (the minimum yield). Figure 6 shows a typical channelled and random spectra for the MCT layer with a minimum yield of approximately 18%. Characterisation of defects and calculation of defect depth distributions are also possible by the channelling technique[7] however these meaurements are not discussed here.

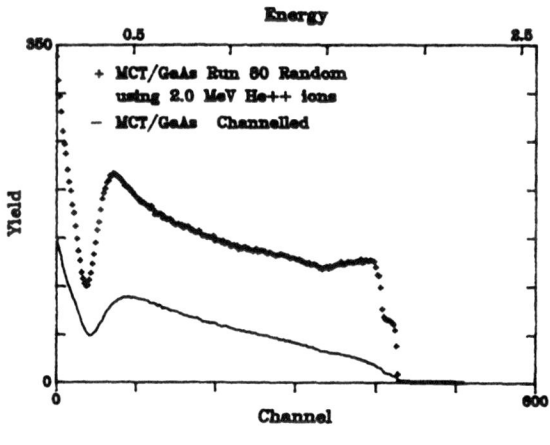

FIGURE 6

RBS random and channelled sprctra of MCT using 2.0 MeV He^{2+} ions.

CONCLUSION

RBS and PIXE are powerful techniques for the analysis of composition, thickness and quality of epitaxial thin films. For MCT deposited on GaAs by MOCVD, it was shown that composition and thickness of the MCT layer is uniform perpendicular to gas flow in the MOCVD reactor. In the direction of gas flow the Cd and Hg concentrations decrease and increase respectively, while the Te concentration remains at approximately 50% This is due to the relative ease of cracking Et$_2$Te in the presence of Me$_2$Cd and Hg. The minimum yield of the MCT layer averaged approximately 18%·indicating high epitaxial quality..

ACKNOWLEDGEMENTS

The permission of the Executive General Manager, Research, of Telecom Australia to publish this paper is acknowledged.

REFERENCES

[1] G.N. Pain,D. Gao,S. Glanvill,P. Gwynn,M. Kibel, M. Kwietniak,P. Leech,C. Rossouw,A. Stevenson,J. Thompson,T. Warminski,L. Wielunski and S. Wilkins, J.Vac.Sci.Technol.A 8(2), 1990.
[2] J.B. Mullin,J. Giess,S.J.C. Irvine,J.S. Gough and A. Royle in Materials for Infrared Detectors and Sources, edited by R.F.C. Farrow, J.F. Schetzina and J.T. Cheung (Mater.Res.Soc.Proc. 90, Pittsburgh, PA 1987) pp. 367-378.
[3] E. Clayton,P. Duerden and D.D. Cohen, Nucl.Instr. and Meth. 180, 541 (1981); B22 (1-3), 64-67 (1987).
[4] K. Debertin and R.G. Helmer in Gamma and X-ray Spectrometry with Semiconductor Detectors (Elsevier Science Publishers, New York, 1988), p. 48.
[5] W. Chu,J.W Mayer and M.Nicolet in Backscattering Spectrometry (Academic Press, New York, 1978).
[6] L.R.Doolittle, Nucl.Instr.and Meth., B9, 422-351 (1985); B15, 227 (1986).
[7] L.C. Feldman,J.W. Mayer and S.T. Picraux in Materials Analysis by Ion Channeling (Acedemic Press, New York, 1982).

ORGANOMETALLIC EPITAXY OF EXTRINSIC n-TYPE HgCdTe USING TRIMETHYLINDIUM

N.R. TASKAR, K.K. PARAT, I.B. BHAT and S.K. GHANDHI
Electrical, Computer and Systems Engineering Department, Rensselaer Polytechnic
Institute, Troy, New York 12180.

ABSTRACT

Indium doping of mercury cadmium telluride, grown by organometallic epitaxy, has been accomplished using trimethylindium (TMIn) as the dopant source. Layers, grown by the Direct Alloy Growth (DAG) process, exhibited a linear doping vs TMIn partial pressure characteristic over the 5×10^{16} to 3×10^{18} cm^{-3} range. A maximum doping concentration of 5×10^{18} cm^{-3} was obtained in these layers.

The optical band edge in these layers was observed to move to higher energy with increased doping. It is shown that this is caused by the Burstein-Moss shift of the bandedge with doping, and that the Cd fraction is independent of the doping concentration.

INTRODUCTION

Mercury cadmium telluride layers, grown by organometallic vapor phase epitaxy (OMVPE), have shown considerable promise for use in far infrared detector applications in recent years [1-3]. Layers with uniformity in composition and thickness, that are required for present day device structures, can be grown by this method [4, 5]. In this paper we report on the doping characteristics of indium in $Hg_{1-x}Cd_xTe$. These layers were grown by the Direct Alloy Growth (DAG) process, which does not involve the interdiffusion of sequential layers of HgTe and CdTe.

Both Group III elements, incorporated on the metal sublattice, and Group VII elements on the Te sublattice [6], behave as n-type dopants. Group VII dopants are undesirable in OMVPE growth because of possible reaction with alkyls. Of the Group III dopants, In is preferred over others since it is a slower diffusing species, by a factor of more than 10. In our work [7], we have used trimethylindium (TMIn) as the dopant source.

Indium doping of HgCdTe layers, grown by the interdiffused multilayer process (IMP) has been reported previously [8]. However, the doping uniformity through these layers depends on the relative incorporation efficiencies of In in CdTe and HgTe, and has not been established. The doping characteristic in these layers showed a very abrupt increase from 3×10^{16} to 3×10^{18} cm^{-3}, with changes in the partial pressure of the In species. This makes it difficult to control doping in the intermediate range, using this approach. In our work, we have shown that controllable indium doping can be achieved using a trimethylindium (TMIn) source, when HgCdTe is grown by the DAG process.

$Hg_{1-x}Cd_xTe$ layers were grown in an atmospheric pressure, horizontal reactor by the simultaneous introduction of elemental mercury, dimethylcadmium (DMCd), and diisopropyltelluride (DIPTe). Substrates were (100) 2° → (110) oriented semi-insulating GaAs. After cleaning in hot organic solvents, followed by a 10 minute etch in $H_2SO_4:H_2O_2:H_2O$ (10:1:1), a 2.5 μm thick CdTe buffer layer was grown at 350°C. The subsequent HgCdTe layers were grown at 370°C and were typically 6 μm thick. This was followed by the growth of a 1 μm thick CdTe cap. For the doped layers, the TMIn bubbler was maintained at -10°C and was operated in the conventional bubbler mode, with the hydrogen flow through it providing a partial pressure of TMIn over the range 3×10^{-7} to 1.5×10^{-5} atm. As seen in Fig. 1, controlled n-type doping, with a linear variation of electron concentration with flow through the TMIn bubbler, was achieved using this approach.

Ohmic contacts were made by etching the CdTe cap near the contact area and evaporating a 400 Å thick layer of gold. Hall measurements were made with magnetic field strength values from 0.5 to 6 kG during the course of this study. In all cases, the Hall coefficient showed classical extrinsic n-type behavior over the entire 300 to 10K temperature regime, and was independent of the magnetic field strength. In two experiments, the carrier concentration was estimated to be 5×10^{16} cm^{-3} for the layer with $x = 0.28$ and 3×10^{16} cm^{-3} for the layer with $x = 0.23$. The low temperature mobility values of 3.3×10^4 and 7.3×10^4 cm^2/Vs for cadmium compositions of 28% and 23% respectively, are consistent with these carrier concentration values.

The linear dependence of the donor concentration on the dopant flux, combined with the high values for mobility, implies an electrical activation of almost 100% for the indium incorporated in the layers, until a doping level of 4×10^{18} cm^{-3} is reached. At higher values of the dopant flux, however, it is possible that indium gets increasingly incorporated in electrically inactive form as In_2Te_3, as has been reported for In doping of bulk HgCdTe in the melt.

The mercury pressure dependence of the donor concentration was investigated in the linear doping regime. Mercury partial pressure values of 0.03 and 0.008 atm. were used with a TMIn partial pressure of 3.7×10^{-7} atm. The net donor concentration increased from 2×10^{17} to 7.5×10^{17} cm^{-3} as the Hg partial pressure was reduced from 0.03 to 0.008 atm.

Indium, occupying sites on the metal sublattice, is donor-like in character. The donor concentration is therefore expected to be proportional to the number of vacant mercury sites, thus, the observed inverse proportionality between the mercury partial pressure and the donor concentration is as predicted.

Figure 2 shows the variation of electron mobility at 20K as a function of donor concentration for a number of samples. The donor concentration in these samples varies over the entire range of the doping characteristic from 3×10^{16} to 5×10^{18} cm^{-3}. Samples corresponding to two different values of the cadmium fraction x, namely 0.22 to 0.25 and 0.27 to 0.30 are shown. Also shown are MBE grown samples corresponding to the lower values of x [9]. The theoretical curves represent the mobility values for $x = 0.25$ and $x = 0.30$, computed using the disorder alloy and ionized impurity scattering.

Optical transmission characteristics of the indium doped layers were determined by Fourier Transform Infrared Spectroscopy at 300K. The optical bandgap was determined as the energy corresponding to $\alpha = 500$ cm^{-1}. Figure 3 shows the transmission characteristics of two indium doped layers with carrier concentration values of 1×10^{17} and 3×10^{18} cm^{-3}. The two layers were grown under identical conditions, except that the heavier doped layer had a higher TMIn partial pressure. The optical bandedge for the 3×10^{18} cm^{-3} doped layer occurs at a higher energy value than for the 1×10^{17} cm^{-3} doped layer. The optical bandgap values correspond to $x = 0.23$ and $x = 0.30$ respectively. The transmission characteristic of the 1×10^{17} cm^{-3} layer was similar to that of an undoped layer. A further shift in the bandedge was observed in layers which were even more heavily doped. The possibility of a Burstein-Moss shift was

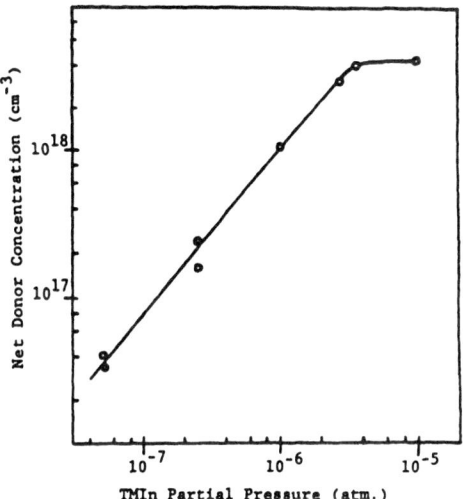

Fig. 1 Electron concentration vs. partial pressure of TMIn.

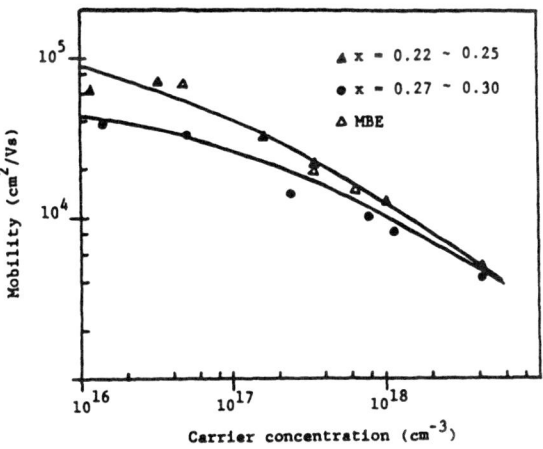

Fig. 2 Electron mobility (20K) vs. electron concentration.

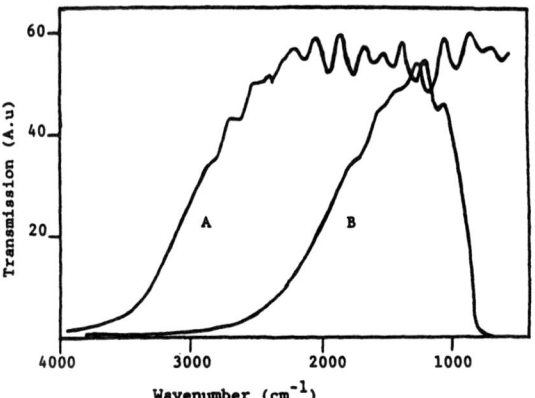

Fig. 3 Transmission characteristics for two HgCdTe layers: (a) $3 \times 10^{18}/\text{cm}^3$ and (b) $1 \times 10^{17}/\text{cm}^3$.

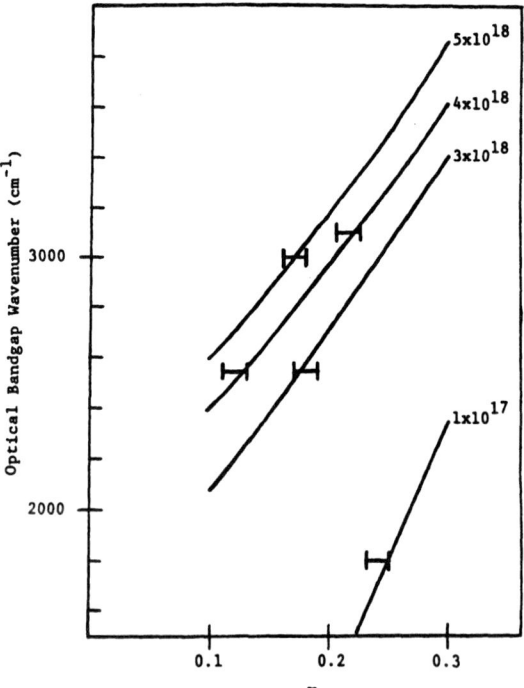

Fig. 4 Predicted optical energy gap as a function of x and electron concentration.

investigated in order to explain this effect.

Figure 4 shows the predicted variation of the optical energy gap due to the Burstein-Moss shift, as a function of the cadmium fraction x, with carrier concentration values as a parameter. Also shown are the data points for the layers with measured values of carrier concentration and optical energy gap. The value of x in these layers was determined by quantitative Energy Dispersive X-ray (EDAX) measurements. The carrier concentration dependence of the experimentally determined optical energy gap for layers with $x = 0.17$ and $x = 0.23$ is listed in Table 1, for different values of doping concentration. The fair agreement between the theoretical values and the data indicates that the shift in the absorption characteristic is primarily due to the Burstein-Moss effect, which has been observed in bulk indium doped HgCdTe [10].

In conclusion, we have shown that indium doped n-type HgCdTe layers can be grown, by the DAG process, with carrier concentrations as high as 5×10^{18} cm^{-3} and with a linear variation of the donor concentration with TMIn partial pressure, up to 3×10^{18} cm^{-3}. A saturation in the incorporation of indium has been observed at higher TMIn concentrations. The optical bandgap of these layers was found to remain unchanged until an electron concentration of 10^{17} cm^{-3} was attained, corresponding to a TMIn partial pressure 1.5×10^{-6} atm. At higher values of TMIn partial pressure, this bandgap was found to increase. This increase was shown to be caused predominantly by a Burstein-Moss shift of the bandedge with doping, and not due to an increase in the Cd fraction.

ACKNOWLEDGEMENT

The authors would like to thank J. Barthel for technical assistance on this program and P. Magilligan for manuscript preparation. We are indebted to Dr. H.F. Schaake of Texas Instruments for technical discussions, and for making precision composition measurements on HgCdTe layers provided by us. Partial funding for this program, from the Raytheon Corporation, is hereby acknowledged. This work was sponsored by the Defense Advanced Research Projects Agency (Contract No. N-00014-85-K-0151), administered through the Office of Naval Research, Arlington, VA. This support is greatly appreciated.

TABLE 1

Carrier Concentration Dependence of Experimentally Determined

Optical Energy Gap

| | | | | $E_{optical}$ (meV) | |
x	E_g (meV)	n (cm^{-3})	$E_F - E_c$ (meV)	Theoretical	Observed
0.17	113	3×10^{18}	262	311	316
0.17	113	5×10^{18}	319	372	372
0.23	200	1×10^{17}	18	200	220
0.23	200	4×10^{18}	259	390	384

REFERENCES

1. S.J. Irvine and J.B. Mullin, J. Crys. Growth, 55, 107 (1981).
2. S.K. Ghandhi and I.B. Bhat, Appl. Phys. Lett., 44, 779 (1984).
3. W.E. Hoke, P.J. Lemonias and R. Taczewski, Appl. Phys. Lett., 45, 1092 (1984).
4. S.K. Ghandhi, I.B. Bhat and H. Fardi, Appl. Phys. Lett., 52, 392 (1988).
5. J. Thompson, P. Mackett, L.M. Smith, D.J. Cole-Hamilton and D.V. Shenai-Khatkhate, J. Crys. Growth, 86, 233 (1988).
6. H.R. Vydyanath and F.A. Kroger, J. Electron. Mater., 11, 111 (1982).
7. S.K. Ghandhi, N.R. Taskar, K.K. Parat and I.B. Bhat, Appl. Phys. Lett., 57(3), 252 (1990).
8. J.S. Whiteley, P. Koppel, V.L. Conger and R.E. Owens, J. Vac. Sci. Tech., A6(4), 2804 (1988).
9. M. Boukerche, J. Reno, I.K. Sou, C. Hsu and J.P. Faurie, Appl. Phys. Lett., 48, 1733, 1986.
10. D. Qian, W. Tang, J. Shen, J. Chu and C. Zheng, Sol. St. Comm., 56, 813, 1985.

THE EPITAXIAL GROWTH BY MOVPE OF (Hg.Mn)Te
ON (001) GaAs SUBSTRATES

H. M. Al-Allak, A. W. Brinkman, P. A. Clifton and P. D. Brown
University of Durham, School of Engineering and Applied Science,
South Road, Durham, DH1 3LE, U.K.

Abstract

Epitaxial thin films of the dilute magnetic semiconductor $Hg_{1-x}Mn_xTe$ have
been grown by MOVPE on (100) GaAs substrate with or without buffer
layers. Deposition took place in a horizontal, atmospheric pressure
reactor at temperatures in the range 350–400°C, using tricarbonyl
(methyl cyclopentadienyl) manganese, di–isopropyl tellurium and
elemental mercury. Buffer layers consisted of thin layers of ZnTe and
an upper thick layer (~1μm) of CdTe. The good crystallinity was
confirmed by RHEED and double crystal x-ray diffraction with rocking
curve widths of 500–600 arc. sec. for non–buffered layers, and 315 arc.
sec. for buffered layers. TEM investigations show that layers grown on
buffered substrates had improved microstructure defect content.
Electrical transport measurements revealed that as–grown layers were p-
type with Hall mobilities in excess of $0.1 \ m^2V^{-1}s^{-1}$.

Introduction

Dilute magnetic semiconductors (DMS) are obtained by substituting
a magnetic transition metal ion such as Mn or Fe for some fraction of
the cations in certain semiconducting compounds [1]. The resulting
mixed crystal systems then display both semiconducting and magnetic
(i.e. as a disordered magnetic alloy) behaviour. They also possess
some unique properties that arise from the spin–spin exchange
interactions between the localized magnetic moments on the magnetic
ions and the conduction band electrons [2]. This affects the energy
band parameters giving rise to properties such as magnetically tuneable
energy gaps (important in narrow gap long wavelength detector systems)
and large Faraday rotations (potentially useful for magneto–optic
modulation).

Most of the research into the preparation and properties of DMS
materials has been carried out on Mn containing II–VI semiconducting
compounds. These were first prepared nearly thirty years ago in bulk
crystal form [1] and much of the physics of these materials was
discovered from such bulk crystal samples. Epitaxial material has been
grown by both MBE [3] and MOVPE [4] but this has been predominantly in
the wider gap Cd and Zn based semiconductors. There has been rather
less reported work on the epitaxial growth of narrow gap Hg containing
DMS, inspite of their potential in long wavelength applications.

This paper reports on recent work on the epitaxial growth of
(Hg,Mn)Te by MOVPE. Although (Hg,Mn)Te has been grown epitaxially by
MBE [5] both as single layers and as part of a (Hg,Mn)Te–HgTe
superlattice [6], its growth (in this laboratory) [7] by MOVPE was
first reported only in the past few months. Since then the growth
procedures have been refined and there has been rather more detailed

structural and electrical characterisation of the grown layers. This on-going work constitutes the substance of this paper.

Epitaxial Growth

The epitaxial layers of (Hg,Mn)Te were grown in a specially adapted horizontal atmospheric reactor that had been originally designed for the growth of (Hg,Cd)Te [8]. In this reactor, the substrates were placed on a stainless steel, resistively heated substrate heater immediately downstream of a small silica glass boat which contained elemental Hg. The boat and the reactor were heated independently of the substrate to obtain the desired vapour pressure of Hg in the reactor and to prevent condensation of the Hg vapour within the reactor. Di-isopropyl tellurium (DIPTe) (supplied by Epichem Limited) was used as the Te precursor and tricarbonyl cyclopentadienyl methyl manganese (TCMn) (supplied by Morton CVD Limited) was used as the Mn precursor. The low volatility of TCMn at room temperature meant that the bubbler had to be heated to a temperature of $\approx 75°C$ in order to obtain a useable vapour pressure (≈ 1.9 Torr). The Mn lines also had to be heated to $\approx 100°C$ to avoid condensation of TCMn along them. The inlet switching manifold could not be heated and so the quartz reactor had to be modified to allow injection of the hot TCMn vapour. A small bore tube was connected to the narrow part of the reactor inlet, immediately after the inlet manifold and at right angles to the main gas stream. The injection tube was located off-centre to impart rotational motion to the gas stream to improve gas mixing.

Growth was carried out on (100) GaAs substrates with and without buffer layers. The substrates were prepared in the conventional manner using a (4:1:1) $H_2SO_4:H_2O_2:H_2O$ etch at 40°C after degreasing in organic solvents. They were then refluxed in isopropyl alcohol, dried and loaded directly into the reactor.

For layers grown on substrates without buffer layers, no additional in-situ treatment was employed and growth commenced about 10 mins. after the substrate had reached growth temperature. Buffer layers consisted of a thin layer of ZnTe over which was grown a thick (≈ 1 μm) layer of CdTe. The ZnTe layer is used to force the CdTe into the (100) orientation, while CdTe is almost latticed matched to Hg rich (Hg,Mn)Te. Both CdTe and ZnTe were grown at 350°C using respectively the reaction between dimethyl zinc (DMZn) and dimethyl cadmium (DMCd) with DIPTe. The DMZn bubbler was cooled to −12°C, the DMCd to 0°C while the DIPTe was held at room temperature (16°C). Once the buffer layers were grown, the alkyls were switched off and the substrate temperature was raised to 400°C for $\approx$ 10 mins. to anneal the structure immediately prior to growth.

The (Hg,Mn)Te was grown over a range of temperature between 350°C and 400°C, under otherwise nominally constant conditions that delivered equal quantities of Te and Mn precursor to the reactor (total flow 1400 cm^3 min^{-1}). The Hg bath was maintained initially at 205°C although more recently this has been raised to 215°C. Growth lasted usually for 1 hour after which the sample was cooled rapidly in H_2 to room temperature. Typically, this produced a $\approx$ 5 μm thick layer of (Hg,Mn)Te.

Composition

The composition of as-grown layers was determined by calibrated EDX at RSRE. Layers grown at or near 400°C were found to contain only Mn and Te with little or no Hg. Conversely layers grown at 350°C contained only a few percent of Mn. The fraction of Mn incorporated, increased monotonically but not linearly with temperature over the range 350°C – 400°C. Thus layers of $Hg_{1-x}Mn_xTe$ grown had $x \approx 0.04$ at 350°C, $x \approx 0.13$ at 375°C and x had increased to $x \approx 1.0$ at 400°C.

While the pyrolysis temperature of TCMn is not known precisely, previous studies where TCMn has been used as a dopant source, suggest that full pyrolysis takes place at temperatures well in excess of 400°C [9]. The progressive reduction in Hg incorporation as the temperature is raised is consistent with experience in the growth of other Hg containing systems such as (Hg,Cd)Te or HgTe-ZnTe superlattices where the presence of either DMCd or DMZn is known to suppress the incorporation of Hg in the layers [10,11]. In these other systems it is necessary to discriminate against the DMCd or DMZn in some way in order for the Hg-Te reaction to proceed. This would appear to be the case here, where in effect, reduction of the temperature to well below that required for full pyrolysis, discriminates against the TCMn. It is interesting to note that the results of this study suggest that TCMn is decomposed at lower temperatures in the presence of DiPTe.

Structure and Morphology

The morphology of as-grown layers was quite flat, but characterised by a low density of elongated rectangular features as illustrated in the Nomarski photograph shown in fig. 1.

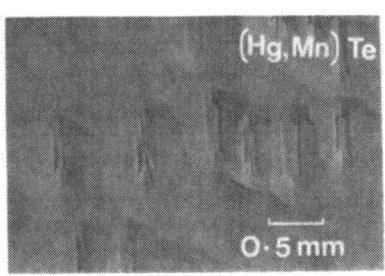

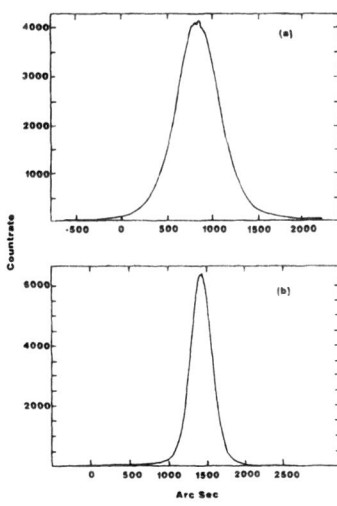

Fig. 1. Normarski micrograph of surface of epitaxial $Hg_{0.87}Mn_{0.13}Te$.

Fig. 2. Rocking curve for (a) unbuffered (Hg,Mn)Te layer, (width = 533 arc. sec) and (b) buffered layer (width=315 arc. sec.)

They were always observed on the surfaces of low Mn content layers irrespective of whether the layers had been grown on buffer layers or not. The very high degree of asymmetry of these features, typically 60 μm long but only 15 μm wide, differs from the pyramidal structures commonly observed on {100} (Hg,Cd)Te, which are square in plan. In addition, the way in which the rectangular features in the (Hg,Mn)Te all lie along the same direction, suggests that there is some pronounced anisotropy in the growth and/or defect structure of the epilayers.

Layers grown at temperatures near 400°C, and which contained little or no detectable levels of Hg, were invariably polycrystalline. However, layers grown below 390°C and which were Hg rich, were epitaxial.

The good crystallinity was confirmed by RHEED and double crystal x-ray diffraction. Rocking curve widths from layers grown directly onto the GaAs, without buffer layers, were typically 500–600 arc sec. for a ≈5 μm thick layer (fig. 2a). Depending on the precise composition, the lattice parameter mismatch between (Hg,Mn)Te and GaAs is about 13%, and thus there will be a substantial dislocation network, which would lead to significant broadening of the rocking curve. Fig. 3 shows a transmission electron micrograph of a cross–section through a (Hg,Mn)Te layer grown directly onto GaAs.

Fig. 3.   TEM micrograph of   Fig. 4.   TEM micrograph of
           (Hg,Mn)Te grown directly   (Hg,Mn)Te grown on
           on {100} GaAs substrate   hybrid substrate
           (g - 220, bar - 0.2 μm).   (g - 220, bar - 1 μm).

This shows the expected dense threading dislocation tangle near the interface. However, in common with other similar compounds, for example (Hg,Cd)Te, the threading dislocation density is progressively and rapidly reduced with distance away from the interface. A marked improvement in the microstructural defect content of the (Hg,Mn)Te layers was demonstrated for growth onto buffered substrates. Fig. 4 shows a TEM graph of a ≈2μm layer of (Hg,Mn)Te on a 1μm buffer layer of CdTe. The high threading dislocation content with the CdTe buffer layer is dramatically reduced at the (Hg,Mn)Te/CdTe interface with the observation of just isolated dislocations within the (Hg,Mn)Te epilayers. The x-ray rocking curve widths of (Hg,Mn)Te layers grown on CdTe/ZnTe buffered GaAs substrates were roughly half those recorded on un–buffered layers. Fig. 2b shows a typical x-ray rocking curve obtained from a buffered layer, for which the full width at half maximum is 315 arc sec. The slight asymmetry in the curve is probably due to the CdTe buffer layer. The lattice mismatch between low x

(Hg,Mn)Te and CdTe is very small (in the case of the layer in fig. 2b, it is about 0.05%).

The x-ray analysis also revealed that all the layers were symmetrically relaxed. Rotation of the sample through 90°C yielded identical rocking curves.

Electrical and Optical Properties

Infra-red absorption measurements of semiconducting layers generally revealed sharp absorption edges and long wavelength fringes. Fig. 5 shows the infra-red absorption spectrum from a 5.5 μm layer of $Hg_{0.87}Mn_{0.13}Te$ grown on a GaAs substrate without a buffer layer. The transmission is broadly similar to that reported previously in the literature for MBE grown material of similar composition.

Hall and resistivity data for a semiconducting layer are shown in fig. 6. The as-grown layers are p-type as expected. The conditions of growth are such that the Hg vapour pressure over the substrate is well below the saturation vapour pressure at the growth temperature. This gives rise to a high Hg vacancy concentration in the as-grown layers which result in p-type material. This is the same situation as occurs in (Hg,Cd)Te growth.

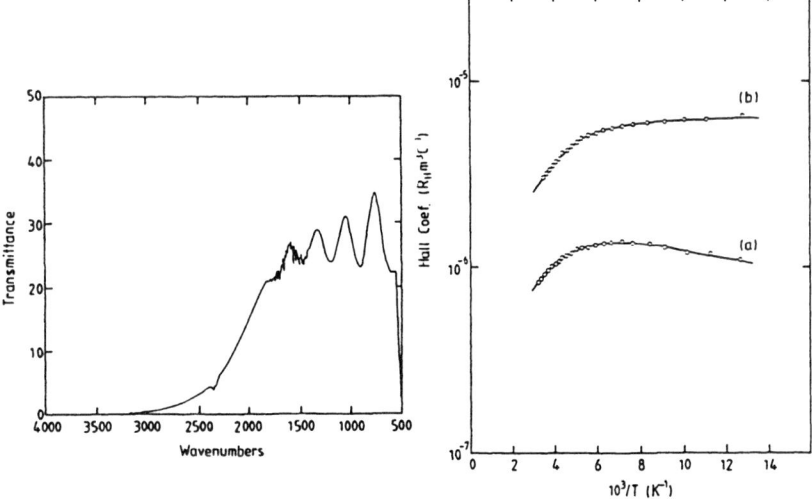

Fig. 5   Infra-Red transmission   Fig. 6.   Hall coefficient vs
          spectra of $Hg_{0.87}Mn_{0.13}Te$ $10^3/T$ for (a)
                                                 $Hg_{0.96}Mn_{0.04}Te$ and
                                                 (b) $Hg_{0.87}Mn_{0.13}Te$

Analysis of the Hall coefficient characteristic (fig. 6) shows that at liquid nitrogen temperatures conduction is by holes alone and the semiconductor behaves extrinsically with a constant Hall coefficient for this particular sample of $6.4 \times 10^{-6} m^3 C^{-1}$ and resistivity of $2.6 \times 10^{-5} \Omega m$. Assuming single carrier behaviour in this temperature regime gives values for the hole density (and by implication Hg vacancy concentration) of about $1 \times 10^{24} m^{-3}$ and for the hole mobility of

$0.24\ m^2V^{-1}s^{-1}$. At higher temperatures the behaviour begins to take on a two carrier character with a consequent reduction in Hall coefficient as the influence of the increasing and much more mobile intrinsic electron population becomes more significant. However, at room temperature the (Hg,Mn)Te is still p-type and the transition to intrinsic behaviour (with its associated change of conductivity type) was not observed. This made the analysis difficult since there was no available data with which to estimate the intrinsic carrier population at room temperature. However, attempts at curve fitting suggest that the electron populations and mobilities were in the range of 10^{18}-$10^{19}m^{-3}$ and $5 - 20\ m^2V^{-1}s^{-1}$. These values of carrier concentration and mobility are comparable with those published for both epitaxial [5,6] and bulk material [1,2]. Taken together with the rocking curve results they are an indication of the high quality material that may be grown by MOVPE.

Summary

The MOVPE growth of $Hg_{1-x}Mn_xTe$ has been shown to give low x semiconducting epitaxial layers of high quality. When grown on suitably buffered substrates rocking curve widths of 315 arc sec. have been recorded. As grown the layers are p-type with a high concentration of Hg vacancies and good hole mobilities.

References

1.   R. T. Delves and B. Lewis, J. Phys. Chem. Solids, 24, 549 (1964).

2.   J. K. Fardyna, J. Vac. Sci. Technol., A4, 2002 (1986).

3.   L. A. Kolodziejski, T. Sakamoto, R. L. Gunshor and S. Datta, Appl. Phys. Lett., 44, 799 (1984).

4.   A. Nouhi and J. Stirn, Appl. Phys. Lett., 51, 2251 (1987).

5.   J. Reno, I. K. Sou, P. S. Wijewarnasuriya and J. P. Faurie, Appl. Phys. Lett., 47, 1168 (1985).

6.   K. A. Harris, S. Hwang, Y. Lansari, J. W. Cook Jnr., J. F. Schetzina and M. Chu, J. Vac. Sci. Technol., A5, 3085 (1987).

7.   P. A. Clifton, A. W. Brinkman and H. M. Al-Allak, Semicond. Sci. Technol. 5, 1067 (1990).

8.   J. E. Hails, G. J. Russell, A. W. Brinkman and J. Woods, J. Cryst. Growth, 79, 940 (1986).

9.   P. J. Wright, B. Cockayne, A. F. Cattell, P. J. Dean, A. D. Pitt and G. W. Blakemore, J. Cryst. Growth, 59, 155 (1982)

10.   J. Tunnicliffe, S. J. Irvine, O. D. Dosser and J. B. Mullin, J. Cryst. Growth, 68, 245 (1984).

11.   P. A. Clifton, J. T. Mullins, P. D. Brown, N. Lovergine, A. W. Brinkman and J. Woods, J. Cryst. Growth, 99, 468 (1990).

EFFECT OF SUBSTRATE ORIENTATION ON CdTe FILM GROWTH BY OMVPE

D.W. SNYDER*, E.I. KO**, S. MAHAJAN***, and P.J. SIDES**
*Alcoa Electronic Packaging, Inc., Alcoa Center, PA 15069, **Department of Chemical Engineering, ***Department of Metallurgical Engineering and Materials Science, Carnegie Mellon University, Pittsburgh, PA 15213.

ABSTRACT

Most CdTe epilayers grown by OMVPE have been deposited on two low index orientations: $\{111\}_{Te}$ and $\{100\}$. In this study we have examined the OMVPE of CdTe as a function of deposition temperature on CdTe substrates having the four low index orientations ($\{111\}_{Te}$, $\{111\}_{Cd}$, $\{110\}$, and $\{100\}$) and on the $\{211\}_{Te}$ and $\{211\}_{Cd}$ orientations. Results are presented for deposition rate, surface morphology, and structural quality of the epilayers.

INTRODUCTION

An understanding of the organometallic vapor phase epitaxy (OMVPE) of CdTe and HgCdTe is needed for the fabrication of infrared detectors and imaging systems based on these materials. Most studies of the OMVPE of CdTe and HgCdTe have utilized two substrate orientations: $\{111\}_{Te}$ and $\{100\}$. Recent work has shown that the growth orientation plays an important role in determining properties such as deposition rate [1], surface morphology [2], and crystal structure [2,3]. A systematic study of the low index orientations is important because the properties of higher index orientations generally fall between the primary low index planes [4].

CdTe is a compound semiconductor having the zincblende structure with the polar $\{111\}_{Te}$ and $\{111\}_{Cd}$ faces ideally terminated with single dangling bond sites and the polar $\{100\}$ face ideally terminated with double dangling bond sites [4]. The non-polar $\{110\}$ face is ideally terminated with equal numbers of Cd and Te atoms each having a single dangling bond [4].

Growth on the $\{111\}_{Te}$ face results in a smooth surface morphology [3] but a lamellar twinned structure [3,5]. Films deposited on the $\{111\}_{Cd}$ face contain large numbers of triangular growth facets (referred to as hillocks) with a double-positioning twinned structure [3]. CdTe $\{100\}$ films are free of twins [3] but contain a high density of square hillocks [6].

The $\{211\}$ surfaces represent 19.5° misorientation from the $\{111\}$ towards the nearest $\{100\}$ and consists of combination of $\{111\}$ and $\{100\}$ planes. The $\{211\}$ orientation is of interest as recent investigations of the molecular beam epitaxy of CdTe on the $\{211\}_{Te}$ surface yields epilayers free of hillocks and microtwins [7,8].

EXPERIMENTAL

The OMVPE system and impinging jet reactor used for this study have been previously described [6,9]. Dimethylcadmium (DMCd) and diethyltelluride (DETe) were used as the organometallic precursors. The experiments were conducted at a reactor pressure of 100 torr in an impinging jet reactor which had a nozzle diameter of 2.54 cm and a nozzle height above the substrates of 2.54 cm. Four 1 cm x 1 cm CdTe substrates, positioned in a symmetric pattern on a susceptor directly beneath the nozzle, were used for each growth run. The deposition rate was found to vary by less than 2% over the region occupied by the substrates [9].

CdTe substrates were provided by II-VI, Inc., Saxonburg, PA. The substrates were lapped using 0.1 μm alumina and chemi-mechanically polished using a 4% (v/v) bromine in a 5:3 methanol:ethylene glycol solution. A brief dip etch was conducted in 1% (v/v) bromine:methanol solution immediately prior to loading the substrates into the reactor. Deposition rate was determined by mass difference using a Mettler model AE163 balance accurate to 0.01 mg. The surface morphology of the CdTe films was examined using Nomarski microscopy. A Blake Industries double-crystal x-ray diffractometer was used to characterize the crystalline quality of the epilayers. The full width at half maximum height (FWHM) of the rocking curves for the $\{111\}$,

{110}, and {100} films were measured over an area of 4 mm² against InSb reference crystals for the (333), (440), (400) Bragg diffraction angles using CuKα radiation. The CuKα (400) reflection from a Si reference crystal was used to measure the FWHM for the {211} films.

RESULTS AND DISCUSSION

The deposition rate, surface morphology, and crystalline quality of the CdTe OMVPE films were found to be strongly dependent on the substrate orientation and deposition temperature. Figure 1 is an Arrhenius plot for CdTe films deposited on the four low index orientations. At growth temperatures below ≈ 480°C the deposition process is limited by the intrinsic chemical reaction kinetics as indicated by the strong, linear dependence on reciprocal temperature. In this temperature regime the deposition rates on the (111)$_{Te}$ surface were approximately twice as high as those for the other three orientations whose growth rates were similar in magnitude. At temperatures above 480°C the slope of the Arrhenius plot decreases with the deposition rate becoming weakly dependent on temperature, an indication of a transition to mass transfer limited deposition. Also, the deposition rates for all of the low index orientations approached the same magnitude at temperatures greater than 530°C, further evidence of mass transfer limitations.

Figure 2 shows the CdTe deposition rates at 400°C versus the angular separation of the crystallographic orientations of the substrate. The highest deposition rates were obtained on the {211}$_{Te}$ planes which were slightly higher than deposition rates on the {111}$_{Te}$ planes. Similarly, growth rates on the {211}$_{Cd}$ were slightly higher than on the {111}$_{Cd}$.

The surface morphologies of CdTe films are shown in Figure 3. The surface of the film grown on the {111}$_{Te}$ at 400°C was smooth and specular while {111}$_{Te}$ films grown at temperatures ≥ 430°C were coarse and hazy. The lack of crystallographic hillocks and transition in surface morphology are believed to be related to the high growth rate on the {111}$_{Te}$. CdTe films deposited on the {111}$_{Cd}$ orientation had a high density of triangular hillocks whose size slightly increased with increasing growth temperature for films of comparable thicknesses. The surface morphologies of CdTe films grown on the {110} and {100} substrates were dominated by high densities of square and rectangular hillocks. The shape of the hillocks on the {111}$_{Cd}$, {110}, and {100} replicated the respective three-fold, two-fold, and four-fold symmetries of these orientations. The {211}$_{Cd}$ and {211}$_{Te}$ surface morphologies for films grown at 400°C, shown in Figure 4, were rough but lacked any distinct crystallographic growth features. The data from Figures 2 and 3 suggest that hillock formation occurs for orientations which correspond to local minima in the deposition rate plot. This observation agrees with the work of Shaw who observed similar behavior for vapor phase growth of GaAs [10].

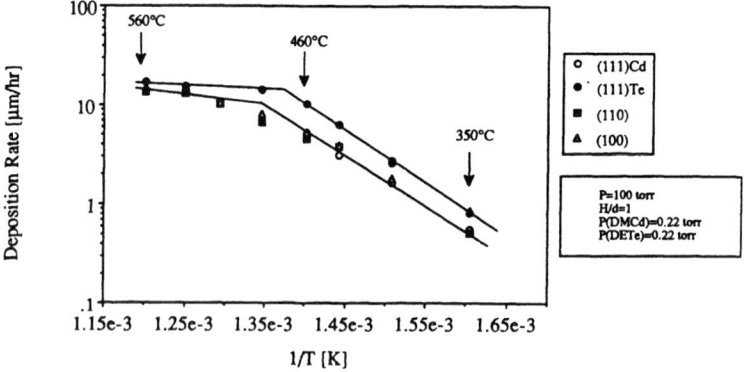

Figure 1. Arrhenius plots for CdTe OMVPE on the {111}Te, {111}Cd, {110} and {100} substrate orientations.

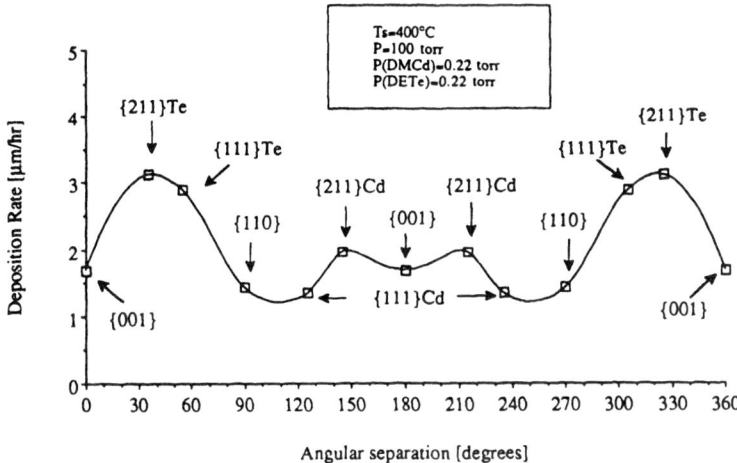

Figure 2. CdTe deposition rates at 400°C versus the angular separation of the crystallographic orientations of the substrates.

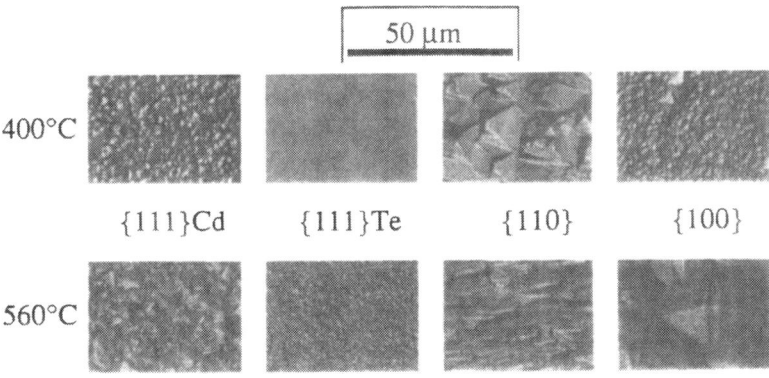

Figure 3. Nomarski micrographs showing the surface morphologies for CdTe films deposited on the low index substrates at 400°C and 560°C.

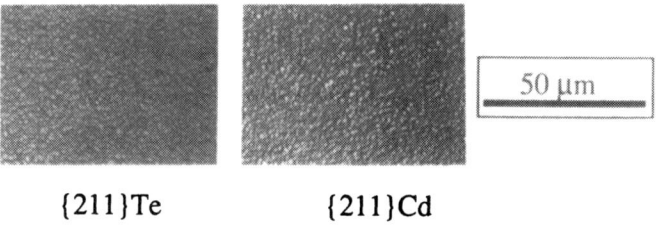

{211}Te   {211}Cd

Figure 4. Nomarski micrographs illustrating the surface morphologies of CdTe films deposited on the {211}$_{Te}$ and {211}$_{Cd}$ orientations at 400°C.

The size of the hillocks increased with increasing film thickness and was relatively independent of deposition temperature for films of comparable thicknesses for growth temperatures up to 480°C (i.e. when the deposition was kinetically limited). Above 480°C when the deposition process became mass transfer limited, hillock size increased with increasing deposition temperature for comparable film thicknesses as is clearly illustrated in Figure 3 for growth on the {100}.

The crystallographic alignment of the hillocks on the {100} film was determined from the location of the {110} cleavage plane which is perpendicular to the {100} surface as shown in Figure 5 as well as from the alignment of etch pits [11,12]. From the alignment of the hillocks, we believe that the planes which comprise the outer surface of the hillocks are {110} or possibly a higher index multiple. Also, the symmetries of the {111}$_{Cd}$ and {110} hillocks are consistent with the outer surfaces of these hillocks also being {110} or a higher index multiple. The similar deposition rates on the {111}$_{Cd}$, {110}, and {100} may be related to the nature of the planes exposed on the surface of the hillocks. Work is underway to clarify this issue.

{110} Cleavage Plane

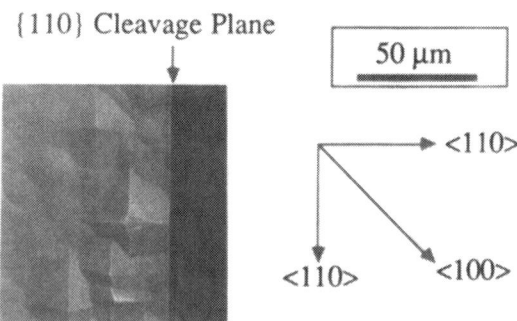

Figure 5. Nomarski micrograph showing the alignment of hillocks on a CdTe{100} film with respect to a {110} cleavage plane.

Table I shows the double crystal x-ray rocking curve (DCRC) full widths at half maximum height (FWHM) of the three points on each sample for several growth temperatures. The {100} film consistently had FWHM near 35 arc seconds for growth temperatures from 400°C to 560°C. This was comparable to the average FWHM of 32.4 arc seconds measured for a set of 35 substrates and indicated that the crystal structure of the substrate was replicated in the epitaxial films. FWHM for the {111}$_{Cd}$ and {110} films were near 65 arc seconds and showed no significant temperature dependence. CdTe {111}$_{Te}$ films grown at 400°C had broad FWHM (>200 arc seconds) and were found to have a slight tilt of ≈ 90 arc seconds with respect to the substrate and to have a 0.11% distortion in the lattice parameter [9]. Increasing the deposition temperature to 440°C decreased the FWHM and suppressed the tilt and lattice parameter distortion. Further increases in deposition temperature decreased the FWHM to near 40 arc seconds at 560°C. The {211}$_{Cd}$ and {211}$_{Te}$ films, grown at 400°C, had FWHM near 50 arc seconds and 41 arc seconds respectively. No evidence of tilt or lattice parameter distortion was observed for {211}$_{Te}$ films. The presence of microtwinning and the effect of processing parameters on CdTe {211}$_{Te}$ films is currently under investigation. The high deposition rates, narrow DCRC FWHM, and lack of hillock formation cause the {211}$_{Te}$ to be a promising substrate orientation.

Table I. Full widths at half maximum height of double crystal x-ray rocking curves.						
Temperature (°C)	{111}Te	{111}Cd	{110}	{100}	{211}Te	{211}Cd
400	>200	67.2	52.1	48.6	36.1	55.7
	>200	59.2	70.1	27.5	47.4	51.1
	>200	63.6	65.0	36.5	39.3	42.4
440	79.6	44.6	72.7	32.8	---	---
	96.8	45.8	68.4	28.6		
	80.9	53.7	60.5	56.4		
527	49.5	56.4	75.6	43.5	---	---
	50.0	74.9	63.4	37.0		
	66.7	70.8	89.3	28.9		
560	38.7	51.4	33.1	28.6	---	---
	48.7	52.3	50.4	41.7		
	41.4	50.6	60.5	32.1		

In summary the effect of substrate orientation on CdTe OMVPE has been systematically studied. Of the four low index orientations, growth rates on the {111}$_{Te}$ surface were approximately twice as large as the other low index orientations which had similar growth rates. The {111}$_{Te}$ film had a smooth, specular surface morphology when grown below ≈ 430°C and was coarse and hazy at higher temperatures. The {111}$_{Cd}$, {110}, and {100} surfaces were dominated by hillocks. The best crystal structure was obtained for the {100} films which had DCRC FWHM near 35 arc seconds. Limited studies at 400°C on {211}$_{Te}$ and {211}$_{Cd}$ substrates showed that the growth rates were higher than on the corresponding {111} orientations, the surface morphologies did not have distinct hillocks, and DCRC FWHM were narrow.

ACKNOWLEDGEMENTS

The authors gratefully acknowledge Two-Six, Inc. for the substrates and use of their x-ray diffraction equipment and the financial support of the National Science Foundation, Rockwell International, the Western Pennsylvania Advanced Technology Center, and the AT&T Foundation. The authors also acknowledge the helpful discussions with Dr. Malcolm Bevan of Westinghouse Electric and the technical assistance of S. Riedinger and S. Jogun.

REFERENCES

1. G. Cinader, A. Raizman, and A. Sher, Workshop on the Physics and Chemistry of Mercury Cadmium Telluride, San Francisco, CA, 1990.

2. J.E. Hails, G.J. Russel, P.D. Brown, A.W. Brinkman, and J. Woods, J. Cryst. Gr. 86, 516 (1988).

3. M. Oron, A. Raizman, H. Shtrikman, and G. Cinader, Appl. Phys. Lett. 52, 1059 (1988).

4. J.M. Aries, S.H. Shin, and E.R. Gertner, J. Cryst. Gr. 86, 362 (1988).

5. P.D. Brown, J.E. Hails, G.J. Russel, and J. Woods, Appl. Phys. Lett. 50, 1144 (1987).

6. D.W. Snyder, E.I. Ko, S. Mahajan, and P. Sides, Appl. Phys. Lett. 56, 1166 (1990).

7. R.J. Koestner, H.Y. Liu, and H.F. Schaake, J. Vac. Sci. Technol. A7, 517 (1989).

8. R.J. Koestner and H.F. Schaake, J. Vac. Sci. Technol. A6, 2834 (1988).

9. D.W. Snyder, Metalorganic Chemical Vapor Deposition of CdTe, Ph.D. Thesis, Carnegie Mellon University, Pittsburgh, PA , 1990.

10. D.W. Shaw, Proceedings of the International Symposium on GaAs, 6, 50 (1968).

11. M. Inoue, I. Teramoto, and S. Takayanagi, J. Appl. Phys., 33, 2578 (1962).

12. Y.-C. Lu, R.K. Route, D. Elwell, and R.S. Feigelson, J. Vac. Sci. Technol., A3, 264 (1985).

A STUDY OF THE INTERFACE BETWEEN THM GROWN MCT CRYSTALS AND CdTe SEEDS.

I. SHILO*, E. KEDAR* AND D. SZAFRANEK**
*Semi-Conductor Devices, Jerusalem 91082.
**Institute of Earth Sciences, The Hebrew University of Jerusalem, 91904 Israel.

ABSTRACT

HgCdTe crystals have been grown from CdTe seeds by the travelling heater method (THM). Three different kinds of interfaces between the growing crystal and the seed were found: a sharp planar interface; a non-planar interface which is caused by meltback near the ampoule walls; a diffuse zone, where an interface cannot be discerned. A correlation was found between the initial growth parameters and the interface structure. Microprobe analysis revealed a boundary layer which, in the case of the planar interface, had an unusual shape. This layer had a thickness varying from 400 to 800 microns. It consisted mostly of HgTe, and contained holes as defects.

INTRODUCTION

Mercury cadmium telluride is well established as the most important semiconductor material for the fabrication of infrared detectors. Recently, there has been increasing interest in large-area single crystals for two-dimensional focal plane arrays and other advanced photovoltaic devices.

Thin film technology which holds a significant share of the detector market still suffers from high substrate cost and other difficulties. Among the many published bulk methods, THM has been found to be the most advantageous and powerful technique for the growth of large-diameter, homogeneous HgCdTe crystals [1,2].

In this paper, we study the interface between the seed and the grown crystal, a subject which so far has not been reported by others. Many papers describe the means of controlling the shape and position of the interface between the crystal and the melt, (e.g. Bridgman) [3-6], or between the crystal and the solution (THM) [7,8]. There is no doubt that this is a subject which affects the quality of the crystal and hence, from a practical point of view, is very important. However, the interface which is studied in this paper can lead to a better understanding of (a) the initial step of crystal growth from a seed and (b) the processes of convection and diffusion which occur in the solution near the seed.

EXPERIMENTAL CONDITIONS

HgCdTe crystals were grown in a single-zone furnace shown schematically in Fig. 1. Two types of furnaces which differed in their temperature gradients were used to grow the crystals which are dealt with in this paper. The solvent composition was determined from the Te-rich corner of the ternary phase diagram and was experimentally optimized to yield the desired composition of the grown crystal.

The CdTe seed was a non-oriented crystal which had one or two grains and few twins. For the purpose of this study, an oriented seed was not needed. The crystals were grown at a rate of about 2 mm/day at a temperature of about 550°C. The first few millimeters of each ingot were cut in half longitudinally to study the cross-section of the interface.

Electron microprobe analysis (EMA) was carried out to determine the composition across the interface. Linear scans were made using a Jeol JAX-8600 Superprobe, equipped with Tracor Northern automation system. Calibration of WDS spectrometers was done using CdTe and HgTe standards. Measurements of cadmium, tellurium and mercury concentrations were done in 20-micron steps close to the interface and in 300-micron steps further away from the interface, using a focused beam.

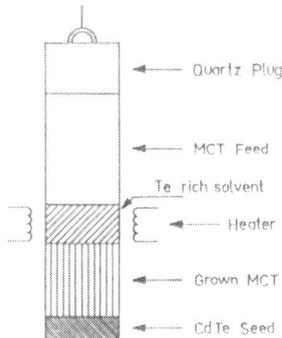

Fig. 1   Schematic diagram of the crystal growth furnace and the ampoule containing the constituent materials.

METALLOGRAPHY

Two different types of interfaces are shown in Fig. 2 and Fig. 3. In both cases, the propagation of the seed structure into the crystal is clearly seen. A severe dissolution of the CdTe seed near the ampoule walls can be seen in the non-planar interface (see Fig.3). The asymmetry of this meltback phenomenon is probably a result of the eccentricity of the ampoule in the Furnace.

A distribution of the defects near the planar interface is shown in Fig. 4. A layer of holes inside the MCT can be seen, located about 100 microns from the optical interface.

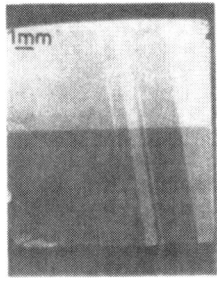

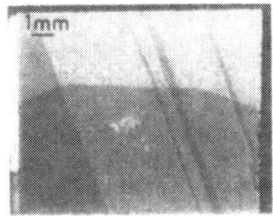

Fig. 2 Planar interface between CdTe seed (bottom) and MCT crystal.

Fig. 3 Non-planar interface

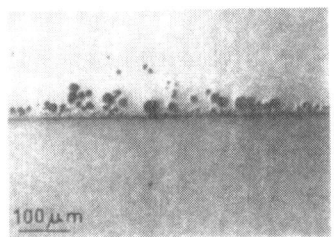

Fig. 4 Defects (holes) in the MCT near the interface

INTERFACE STUDIES

Fig. 5. shows the composition variation of mercury, cadmium and tellurium measured with EMA across the interface of crystal C-23 which had a planar interface (e.g. Fig. 2). Since the distance between individual measurements was about 600 microns, the results shown in Fig. 5 are somewhat misleading because of inadequate resolution. Hence, a line scan with 20-micron steps was performed in the vicinity of the interface, which is shown in Fig. 6. In this figure where only the Cd concentration is shown, a boundary layer with an unusual profile is revealed. In this layer, the Cd concentration has a "flat minimum" of about 3 at%; whereas at steady state, which is reached after 500 microns, the Cd concentration is 22 at% . A general scan of the planar interface of crystal C-25 is shown in Fig. 7. In this case the step between measurements was reduced to 300 microns. A boundary layer whose thickness is 800 microns can be seen.

A different situation was found with the non-planar interface (e.g., Fig. 3). A typical composition profile across the non-planar interface is shown in Fig. 8. The decrease in Cd concentration is gradual and even at the distance of 3 mm. from the interface the composition of the crystal is not yet close to steady state. The scans of five crystals, performed in 20-micron steps near the interface, are presented together in Fig. 9. There are two distinct groups of data: (1) the planar interface where a boundary layer with increasing thickness is seen, and (2) the non-planar interface where a different behavior is observed.

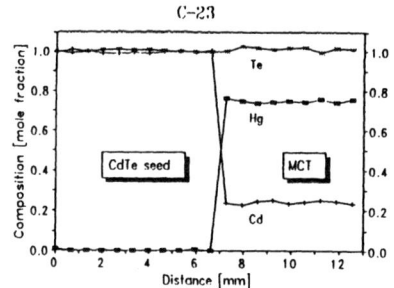

Fig. 5 Planar interface: composition profile along the axial direction in C-23

Fig. 6 Cd concentration profile in the vicinity of the planar interface

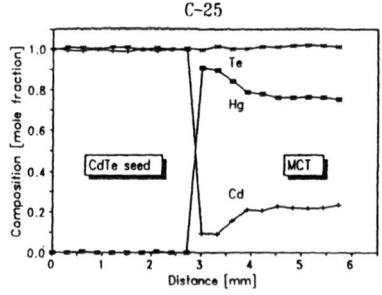

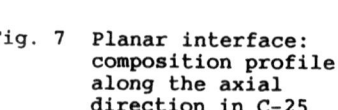

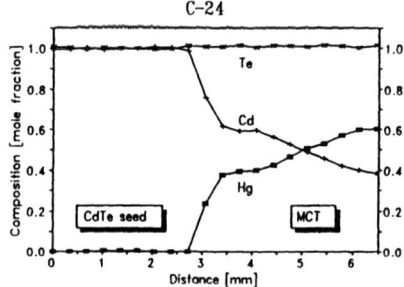

Fig. 7 Planar interface: composition profile along the axial direction in C-25

Fig. 8 Non-planar interface: composition profile along the axial direction in C-24

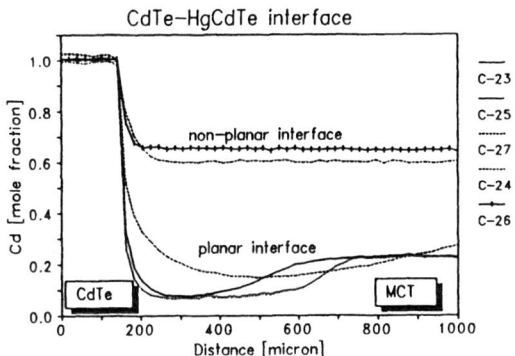

Fig. 9   1000 micron line scan performed in
          20 micron steps across the interface
          of the planar (C-23, C-25, C-27) and
          the non-planar (C-24, C-26) crystals

DISCUSSION AND CONCLUSIONS

The shape of the interface between the seed and the grown
crystal depends on the location of the ampoule in the furnace.
Fig. 10 shows various location arrangements in which a planar
interface was obtained . Crystals C-23, C-25 and C-27 were
located in the furnace as shown in Fig. 10 a,b,c respectively.
The "minimum flat" layer which was observed in C-23 and C-25 is
composed mainly of HgTe which is the heavier component and is
known to be rejected due to density-driven convection. The
increase in the thickness of this boundary layer (δ), from 350
microns in C-23 to 500 microns in C-25 is due probably to
differences in (a) the rate of growth and (b) the location of
the ampoule. Crystal C-23 was grown at 1.8 mm/day while C-25
was grown at 2.5 mm/day. When the growth rate is increased,
there is a shift in the liquid zone in the direction of travel
which maximizes the difference (Td-Tg), so that the Cd content
at the dissolving interface will be higher. However at the
initial stage of crystal C-25 the relative temperatures of the
dissolving and growing interfaces were inverted, i.e. Tg>Td, as
shown in Fig. 10b, which led to Cd depletion near the growing
interface. Through that mechanism, HgTe became the main species
in front of the interface.
     The layer of holes, a "piping" phenomenon, which was
observed only near the planar interface (e.g. Fig. 4) could be
the result of a sudden freeze of the material with lower
melting point that grows near the interface.

52

Crystal C-27 was grown in a furnace with smaller temperature gradient at a rate of 2.0 mm/day. In order to prevent meltback of the seed, the ampoule was located in a non-mixing configuration as shown in Fig. 10C, and as a result the unusual layer was not observed. Capper et.al. [9] found that in Bridgman growth the thickness of the layer decreased when the growth rate increased. The opposite has been observed here, which implies that the parameters that govern the shape of the interface between the seed and the crystal are not the same as for the interface between the melt and the crystal.

The fact that the layer thickness in the "mixing" temperature profile is greater than in the "non-mixing" profile, along with the unusual shape of these layers, suggests that these boundary layers are not diffusional layers.

The non-planar interface was observed when the ampoule was located in the furnace so that the temperature at the interface between the seed and the solvent was close to Tm. Since dissolution of the seed occurs, the solvent was enriched with Cd and a normal boundary layer was observed.

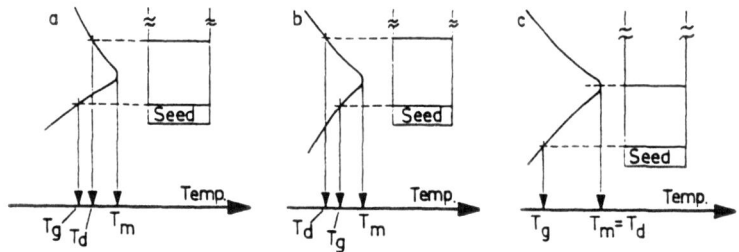

Fig. 10 Location of the ampoule with regard to the
temperature gradient before growth starts
a. Td >Tg (C-23) b. Tg > Td (C-25) c. Td = Tm (C-27)

REFERENCES

1. R. Triboulet, T. Nguyen Duy, and A. Durand, J. Vac. Sci. Technol. A3(1), (1985).
2. L. Colombo, R.R. Chang, C.J. Chang, and B.A. Baird, J. Vac. Sci Technol. A6(4), 2795 (1988).
3. Do Hyun Kim and R.A. Brown, J. Cryst. Growth 96, 609 (1989).
4. L. R. Holland, J. Cryst. Growth 96, 577 (1989).
5. Y. Huang, W.J.Debnam and A.L. Fripp, J. Cryst. Growth 104, 315 (1990).
6. F.R. Szofran and S.L. Lehoczky, J. Cryst. Growth 70, 349 (1984).
7. V.F.S. Yip, C.E. Chang and W.R. Wilcox, J. Cryst. Growth 29, 69 (1975).
8. J. Chang, B. Baird, Pok-Kai Liao, R. Chang and L. Colombo, J. cryst. Growth 98, 595 (1989).
9. p. Capper, J.J.G. Gosney, C.L. Jones and M.J.T. Quelch, J. Cryst. Growth 63, 154 (1983).

LARGE AREA (100)CdTe AND CdZnTe LAYERS ON (100)GaAs GROWN BY HOT WALL BEAM EPITAXY FOR HgCdTe PHOTOVOLTAIC MWIR–DETECTORARRAY APPLICATION

J.HUMENBERGER*, K.H.GRESSLEHNER*, W.SCHIRZ*, K.LISCHKA**, H.SITTER**

*TOPLAB, a Division of HAINZL Industriesysteme, Industriezeile 56, A–4020 Linz, Austria
**Research Institute for Optoelectronics, Linz University, A–4040 Linz, Austria

ABSTRACT

The newly developed Hot Wall Beam Epitaxy was used for the growth of (100)CdTe and (100)$Cd_{1-y}Zn_yTe$ layers ($y \approx 0.04$) on 1 inch (100)GaAs wafers. The structural properties of the epilayers were investigated by means of high resolution x-ray rocking curve measurements. The results demonstrate the high crystalline perfection of the layers. The CdTe and $Cd_{1-y}Zn_yTe$/GaAs wafers were used as substrates for the growth of $Hg_{1-x}Cd_xTe$ by close space vapor phase epitaxy. The layers show excellent homogeneity. Across a 1 inch wafer the composition x varies by ± 0.0026, and the width of the rocking curve (FWHM) by ± 10 arcsec. Values of the FWHM as low as 50 arcsec for $Hg_{1-x}Cd_xTe$ ($x = 0.32$) on $Cd_{1-y}Zn_yTe$/GaAs were obtained. Linear arrays of planar photovoltaic detectors were fabricated. Arrays of 32 elements show high uniformity of the cutoff wavelength and the responsivity. The variation of the cutoff wavelength is $\pm 1.5\%$ ($\pm 0.04\mu$m) and the variation of the responsivity is $\pm 3\%$.

INTRODUCTION

The epitaxy of $Hg_{1-x}Cd_xTe$ on large area high quality substrates stimulated the interest in the growth on alternative substrates like GaAs. In a typical structure for the deposition of $Hg_{1-x}Cd_xTe$ a buffer layer of CdTe is grown, serving as a substrate for $Hg_{1-x}Cd_xTe$ epitaxy. It is evident, that highest quality CdTe is essential for this purpose.

In our laboratory growth of CdTe has been performed by Hot Wall Epitaxy (HWE) [1] and by the newly developed Hot Wall Beam Epitaxy (HWBE) [2] on GaAs. An improvement of the substrate quality for $Hg_{1-x}Cd_xTe$ epitaxy was achieved by growing $Cd_{1-y}Zn_yTe$ layers with low Zn–content on GaAs using HWBE [2]. In this paper we present data on the growth of $Cd_{1-y}Zn_yTe$, and we demonstrate the application of this material as substrate for epitaxy of $Hg_{1-x}Cd_xTe$ by close space vapor phase epitaxy (CSVPE) for the fabrication of linear photovoltaic detector arrays, which are sensitive in the mid infrared spectral region.

HOT WALL BEAM EPITAXY OF CdTe AND CdZnTe ON GaAs

Fig. 1 shows the schematic arrangement of the apparatus used in HWBE. It essentially consists of an evaporator and a substrate heater, which are placed in an UHV–chamber. The substrate position is at a certain distance above the opening of the evaporator. The substrate diameter and the diameter of the evaporator are equal. In this way the near–field angular distribution of the effusing flux provides an uniform impingement rate on the substrate [3]. HWBE combines some advantages of Molecular Beam Epitaxy (control of fluxes and flux interruption, low substrate temperature, application of in situ diagnostic facilities) and HWE (higher growth rate, minimal waste of material) with the additional advantage of providing a uniform impingement rate on the substrate.

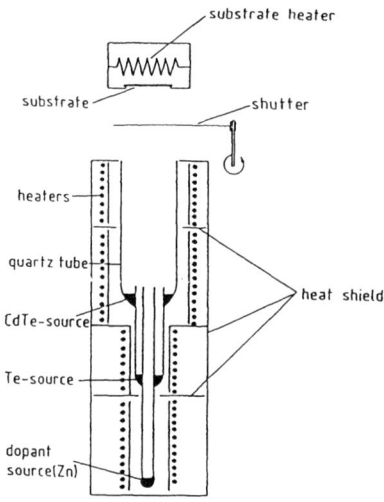

substrate heater

substrate

shutter

heaters

quartz tube

CdTe-source

heat shields

Te-source

dopant source(Zn)

Fig. 1. Schematic diagram of the Hot Wall Beam Epitaxy setup

The hot wall beam evaporator containing CdTe and Zn as source was designed for high material capacity. It allows continuous growth of more than $10^3 \mu m$ of $Cd_{1-y}Zn_yTe$ epilayers of 1 inch diameter. All CdTe and $Cd_{1-y}Zn_yTe$ epilayers were grown on commercially available (100)GaAs which was misoriented 2° towards the next (110) direction. Before growth the substrates were cleaned in organic solvents, etched in $H_2SO_4 : H_2O_2 : H_2O$ (5:1:1), rinsed and blow dried with nitrogen prior to loading into the preparation chamber. The substrates were mechanically fixed in the sample holder without Indium solder. Preheating of the substrates was performed in a seperate preheating station at 540 °C for 30 min. The films were grown with a rate of $2 - 3\mu m/h$ at 350 °C substrate temperature.

HgCdTe EPITAXY AND PHOTOVOLTAIC DEVICE FABRICATION

Epilayers of $Hg_{1-x}Cd_xTe$ were grown on CdTe/GaAs or $Cd_{1-y}Zn_yTe/GaAs$ using mercury pressure controlled CSVPE [4]. The substrate composition used most frequently was $y = 0.03 - 0.04$, $Hg_{1-x}Cd_xTe$ layers of 1 inch size with an adjustable x of 0.28 − 0.32 were grown. The composition of the layers was determined from IR transmission measurements.

The $Hg_{1-x}Cd_xTe$ layers were used for the preparation of linear arrays of planar photovoltaic detectors with 32 elements. The cutoff wavelength was chosen between 2.8 and $5.5\mu m$. The $n^+ - p$ junctions were formed by ion implantation with $100\,KeV\,B^+$ ions and subsequent annealing of the radiation damage. Devices were passivated with a layer of ZnS. Metallic contact pads to the p and n contacts were made by evaporation of Gold and Indium respectively.

RESULTS AND DISCUSSION

CdTe and $Cd_{1-y}Zn_yTe$ Layers on GaAs

All our CdTe and $Cd_{1-y}Zn_yTe$ layers ($y \approx 0.04$) were (100) oriented. The surface was mirrorlike and under an interference microscope no facetting was visible. Only shallow terraces occured, which are most likely induced by the 2° misorientation of the substrate. For example, films of $50\mu m$ thickness exhibited a mean terrace height of about $0.05\mu m$ as determined from stylus type measurements. By visual inspection of our films we found no trend for an increase of surface roughness with increasing film thickness. The crystalline perfection of the epitaxial films was investigated by measurement of x–ray rocking curves using a high resolution x–ray diffractometer with a four–crystal monochromator [5] and a spot size of $0.5 * 1.6mm^2$. The resolution of the spectrometer is less than ±1 arcsec.

The rocking curves of our epilayers are single peaked, indicating that no low angle grain boundaries are present.

Fig. 2 shows the full width at half maximum (FWHM) of the (400) Bragg–reflex rocking curves of HWBE–grown CdTe and $Cd_{1-y}Zn_yTe$ epilayers versus layer thickness. It is evident from the figure, that $Cd_{1-y}Zn_yTe$ films show a substantial improvement of the crystalline quality when compared with HWBE–grown CdTe layers.

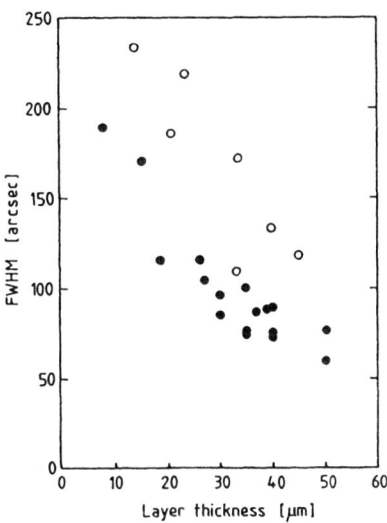

The FWHM decreases with increasing thickness of the epilayers. Since the density of dislocation is proportional to $(FWHM)^2$ [6], the data indicate a decreasing density of dislocations in epilayers with a thickness exceeding $10\mu m$. Previous investigations of the strain in CdTe/GaAs epilayers showed, however, that the misfit strain is accomodated in a dense dislocation network in a region of about $3\mu m$ thickness close to the GaAs epilayer interface [7].

The observation of a substantial dislocation density in thick epilayers is considered to indicate that misfit dislocations generated close to the interface are bent towards the epilayer surface during growth. The decrease of the dislocation density is approximately proportional to the inverse epilayer thickness: this is deemed to be a strong motivation to use growth methods, which allow relatively large growth rates and high sample through–put, like the HWBE, for the growth of CdTe and $Cd_{1-y}Zn_yTe$ on GaAs.

Fig. 2. The width of the (400) rocking curve from HWBE grown CdTe (o) and $Cd_{1-y}Zn_yTe$ (•) epilayers as a function of layer thickness ($y \approx 0.04$)

$Hg_{1-x}Cd_xTe$ Layers on $Cd_{1-y}Zn_yTe$/GaAs

The $Hg_{1-x}Cd_xTe$ layers were grown with a thickness of 20 – $50\mu m$ by CSVPE. The thickness of the $Hg_{1-x}Cd_xTe$ layer is equal to the thickness of that region of the layer which contains Hg. The CSVPE–process is based on interdiffusion of Hg and Cd or Zn. Although we use $Cd_{1-y}Zn_yTe$ substrates it was established by microprobe measurements that the Zn–concentration on the surface was below the detection limit of the instrument. This result is consistent with the fact that (a) only $\approx 4\%$ Zn is present in the substrate and (b) the interdiffusion coefficient was reported to be much lower for Hg–Zn than for Hg–Cd [8]. Thus the layer can be considered as ternary compound $Hg_{1-x}Cd_xTe$ — at least within the surface region of several microns. The distribution of mercury along the growth direction was determined by a step etching technique combined with IR–transmission measurements. The compositional gradient in a region of about 10 μm below the epilayer surface was less than $0.001\mu m^{-1}$.

Fig. 3 shows the results of composition mapping of a 1 inch $Cd_{1-x}Hg_xTe$/ /$Cd_{1-y}Zn_yTe$/GaAs wafer. The spot size for the determination of x using an FTIR–

spectrometer was 2mm in diameter and the accuracy of the x-measurement is ± 0.0005. The mean value of x of the layer is 0.292, the variation of x is ± 0.0026 across the wafer.

At the surface of the $Hg_{1-x}Cd_xTe$ layers we found defects ("hillocks") of irregular shape and $10 - 20\mu m$ in diameter. The mean height of these hillocks is $2 - 3\mu m$. The mean density of the defects was determined in a statistical manner from several different areas on the layer. Fig. 4 is a histogram of the hillock–density. A random sample of 56 $Hg_{1-x}Cd_xTe$ layers has been used for the statistics. The figure shows that the most frequent hillock density in HWBE–grown layers is $20 - 60cm^{-2}$.

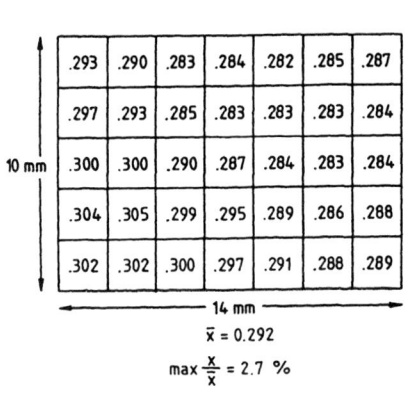

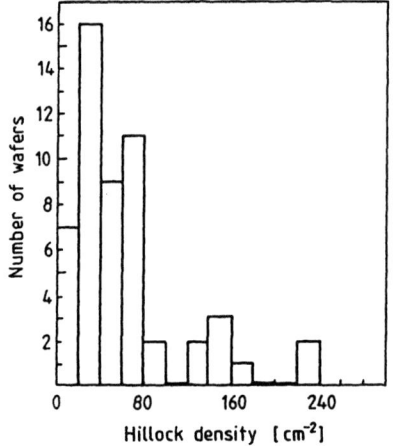

Fig. 3. Composition mapping of a CSVPE grown $Hg_{1-x}Cd_xTe$ epilayer on $Cd_{1-y}Zn_yTe/GaAs$

Fig. 4. Histogram of the hillock density obtained from a random sample of 56 CSVPE–grown HgCdTe layers (3 layers had a hillock density exceeding $300cm^{-2}$)

The $Hg_{1-x}Cd_xTe$ layers show good crystalline quality as demonstrated by the FWHM of the (400) reflex x–ray rocking curve. The FWHM of rocking curves was obtained from different positions along the diameter of an epilayer. For a typical layer we found an average value of 60 arcsec and a variation of only ± 10 arcsec across a 1 inch layer.

Photovoltaic Device Fabrication

Our $Hg_{1-x}Cd_xTe/Cd_{1-y}Zn_yTe/GaAs$ layers with a p-type concentration of $2 - 4 * 10^{16}cm^{-3}$ (for $x = 0.28 - 0.32$) and a mobility of $250 - 350cm^2/Vs$ at 77 K were used for the fabrication of linear arrays of planar, photovoltaic detectors.

For the electrical characterization of devices with different cutoff wavelengths the differential resistance at zero bias (R_oA) was measured at 192 K. The dots in Fig. 5 show the R_oA-product of our diodes as a function of the cutoff wavelength. The corresponding x-value of the $Hg_{1-x}Cd_xTe$ layer can be read from the upper scale of Fig. 5. For comparison R_oA-values from diodes on bulk and epitaxially grown $Hg_{1-x}Cd_xTe$ layers [9,10,11] are included in the figure. The solid curve represents the results from a model calculation of the R_oA-product in the spectral range from $3\mu m$ to $6\mu m$ ($x = 0.38$ to 0.25) [12]. In this calculation a constant hole concentration of $2 * 10^{16}cm^{-3}$ and a hole mobility of

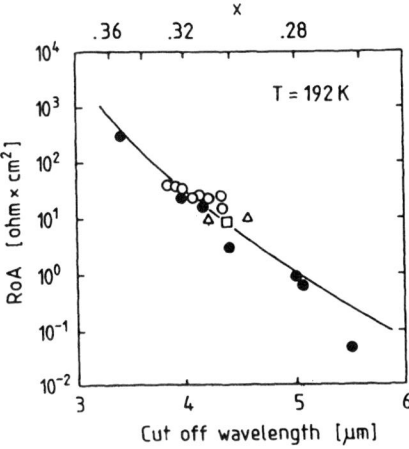

Fig. 5. Experimental values of the R_oA-product at 192 K from p–n junctions in $Hg_{1-x}Cd_xTe$ vs. their cutoff wavelength. Dots are data from our diodes in $Hg_{1-x}Cd_xTe$, (o) data from [9], ($\triangle$) data from [10], ($\square$) data from [11].

$250 cm^2/Vs$ was assumed. The lifetime of minority carriers in the p–region at 200 K was assumed 100 ns in agreement with values from photoconductive decay measurements [13]. The carrier concentration in the n^+ region was assumed $2 * 10^{17} cm^{-3}$. For the calculation the following reverse current mechanisms were taken into account: (a) diffusion current in the p–region, (b) generation recombination current in the depletion region and (c) band to band tunneling current.

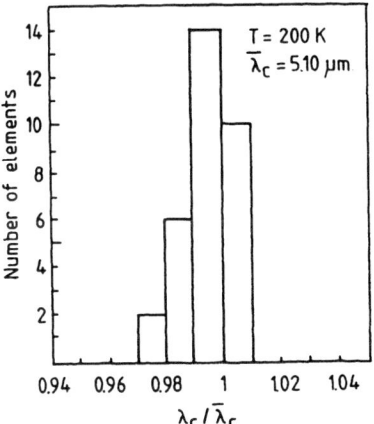

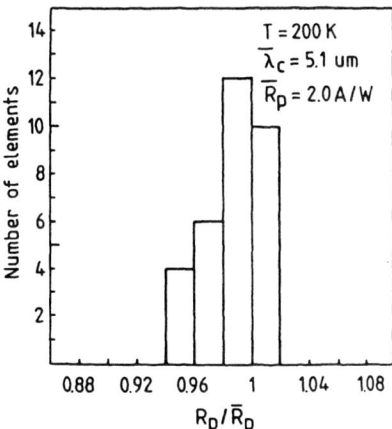

Fig. 6a. Histogram of the variation of the cutoff wavelength in a 32 element linear array (element size is $50 * 50\mu m$, pitch $100\mu m$, λ_c is the cutoff frequency, $\bar{\lambda}_c$ is the mean value of λ_c)

Fig. 6b. Histogram of the variation for the peakresponse for a 32 element linear array (element size is $50 * 50\mu m$, pitch $100\mu m$, R_p is the response at the peak wavelength, $\bar{R}_p$ is the mean value of R_p)

A typical variation of the cutoff wavelength as well as the variation of the reponsivity of a 32 element linear detector array is shown in Fig. 6a and Fig. 6b, respectively. The element size is $50 * 50\mu m^2$, the space between the elements is $50\mu m$. The mean value of the cutoff wavelength is $5.10\mu m$ with a maximal deviation of $\pm 0.04\mu m$. As can be seen in Fig. 6b, the homogeneity of the responsivity of the detector elements is quite good, it varies by about ±3%. The quantum efficieny of the elements is about 60 %.

CONCLUSIONS

It was demonstrated that the HWBE growth method is able to supply high quality epilayers of $Cd_{1-y}Zn_yTe$ on GaAs to be used as a substrate for subsequent epitaxy of $Hg_{1-x}Cd_xTe$. The HWBE method offers the inherent advantages of high growth rate together with an optimum use of valuable source material. By growing thick layers it is possible to benefit from the fact that crystalline quality increases with increasing layer thickness. The (100) oriented $Cd_{1-y}Zn_yTe$ epilayers with a Zn content of $y \approx 0.04$ have been used for $Hg_{1-x}Cd_xTe$ growth by close space vapor phase epitaxy. The $Hg_{1-x}Cd_xTe$ layers show high crystalline quality and homogeneity. Linear photovoltaic detector arrays of 32 elements with state of the art performance of the diodes at temperatures achievable by thermoelectric coolers (200 K) show remarkable uniformity of responsivity and cutoff wavelength.

ACKNOWLEDGEMENTS

This work is partly supported by the Austrian "Fonds zur Förderung der wissenschaftlichen Forschung" and by "Forschungsförderungsfonds für die gewerblichen Wirtschaft".

References

[1] D.Schikora, H.Sitter, J.Humenberger and K.Lischka, Appl.Phys.Lett. **18**, 1276 (1986)

[2] J.Humenberger, H.Sitter, A.Pesek, K.Lischka and H.Pascher, Proc. 20th Int.Conf. on the Physics of Semiconductors, Thessaloniki (1990) (to be published)

[3] J.Humenberger and H.Sitter, Thin Solid Films **163**, 241 (1989)

[4] P.Becla, P.A.Wolff, R.L.Aggarwal and S.Y.Yuen, J. Vac. Sci. Technol. **A3**, 119 (1985)

[5] W.J.Bartels and W.Nij, J.Chryst. Growth **44**, 518 (1978)

[6] P.Gay, P.B.Hirsch and A.Kelly, Acta Metall **1**, 315 (1953)

[7] J.Petruzzello, D.Olego, S.K.Ghandhi, N.R.Taskar and I.Bhat, Appl.Phys.Lett. **50**, 1423 (1987)

[8] R.Triboulet, J.Cryst. Growth **86**, 79 (1988)

[9] M.B.Reine, A.K.Sood and T.J.Tredwell, in Semiconductors and Semimetals, Vol 18, (Academic Press, New York, 1981) p 262

[10] W.Tennant, SPIE **217** (1980)

[11] J.Aries, S.Shin, J.Pasko, R.DeWames and E.Gertaer, J. Appl.Phys. **65**, 4 (1989)

[12] A.Rogalski, Infrared Phys. **28**, 139 (188)

[13] K.H.Greßlehner, W.Schirz, J.Humenberger, H.Sitter, J.Andorfer and K.Lischka, SPIE **1361** (1990) (to be published)

INTRINSIC CARRIER CONCENTRATIONS IN LONG WAVELENGTH HgCdTe BASED ON THE NEW, NONLINEAR TEMPERATURE DEPENDENCE OF $E_g(x,T)$ [†]

D. G. Seiler*, J. R. Lowney*, C. L. Littler**, and I. T. Yoon**

*Semiconductor Electronics Division, National Institute of Standards and Technology, Gaithersburg, MD 20899

**Department of Physics, University of North Texas, Denton, TX 76203

Abstract

Intrinsic carrier concentrations of narrow-gap $Hg_{1-x}Cd_xTe$ alloys ($0.17 \leq x \leq 0.30$) have been calculated as a function of temperature between 0 and 300 K by using the new nonlinear temperature dependence of the energy gap obtained previously by two-photon magneto-absorption measurements for samples with $0.24 \leq x \leq 0.26$. We report here experimental values for $E_g(x,T)$ for samples with $x = 0.20$ and 0.23 obtained by one-photon magneto-absorption measurements. These data confirm the validity of the new $E_g(x,T)$ relationship for these x values. In this range of composition and temperature, the energy gap of mercury cadmium telluride is small, and very accurate values are needed for the gap to obtain reliable values for the intrinsic carrier density. Large percentage differences exist between our new calculations and previous values for n_i at low temperatures. Even at 77 K, differences approaching 10 percent exist, confirming the importance of using the new n_i results for materials and device characterization and a proper understanding of device operation in long-wavelength materials.

Introduction

The intrinsic carrier concentration, n_i, of mercury cadmium telluride is an important quantity that must be known accurately for understanding and characterizing a wide variety of material and device properties. A great deal of effort has gone into calculating n_i from measured band parameters [1-3]. The calculations of Madarasz et al. [2] and Hansen and Schmit [1] are in reasonably good agreement, but are only as good as the values for the parameters that define them. The most important parameter that must be known is the band gap, which varies greatly with temperature T and composition x. Recently, two-photon magneto-absorption measurements [4] on samples of HgCdTe with $0.24 \leq x \leq 0.26$ were used to accurately determine the nonlinear temperature dependence of $E_g(T)$ for T <77 K, yielding a new $E_g(x,T)$ relation:

$$E_g = -0.302 + 1.93x - 0.810x^2 + 0.832x^3 + 5.35 \times 10^{-4}(1-2x)(-1822+T^3)/(255.2+T^2), \quad (1)$$

where E_g is in eV, T in K, and $0.2 < x < 0.3$. For samples with 10-μm cutoff wavelengths (i.e., x $\approx$ 0.2), the predicted maximum deviation of the new relation from that of Hansen, Schmit, and Casselman (HSC) [5] is approximately 3 to 4 meV at 10 to 12 K. Here we also report on further, experimental values of $E_g(x,T)$ for samples with x = 0.201 and x = 0.229 that confirm the use of the new relationship for lower x-value samples.

We have calculated the intrinsic carrier density, n_i, as a function of T between 0 and 300 K and as a function of x between 0.17 and 0.30 for use in materials-characterization studies and for the prediction of device operation. The model we have used to determine the nonparabolicity of the conduction band is Kane's three-band $k \cdot p$ model [6]. This

---

[†] Contribution of the National Institute of Standards and Technology; not subject to copyright.

model has also been used in the most recent calculation of n_i, which was performed by Madarasz et al. [2]. Their calculation was an improvement on earlier work [1,3,7] by not making any approximations other than those inherent in the $k \cdot p$ theory itself. We have used full Fermi-Dirac statistics and a momentum matrix element that did not vary with composition or temperature over our range of calculations. The results show values of n_i that differ from those of Refs. [1] and [2] by more than an order of magnitude at 5 K for all x-values studied. Differences approaching 10 percent occur even at 77 K for samples throughout the compositional range studied. Thus, it is important to include the nonlinear temperature dependence of the energy gap of $Hg_{1-x}Cd_xTe$ when modeling the low temperature operation of long-wavelength detectors.

Theory

Our calculations follow the general outline of Madarasz et al. [2] except that the band-gap dependence given by the HSC relation is used in Ref [2]. As in Madarasz et al. [2], the theory of Kane [6] was used to calculate the $E(k)$ relation for the conduction band of mercury cadmium telluride. This method uses $k \cdot p$ perturbation theory to calculate the conduction band in the vicinity of the Γ point. Interactions with the light-hole and split-off valence bands are included. To first order the conduction band does not interact with the heavy-hole band, and therefore, this band is not included. The value of Δ, the energy of the split-off band, is taken to be 1 eV [8], and the value of P, the momentum matrix element, to be 8.49×10^{-8} eV cm [8]. Madarasz et al. [2] vary the value of P with composition by a few percent to obtain better agreement with measured effective masses. However, we have kept P constant in order to deal directly with the effect of the nonlinearity of the temperature dependence of the energy gap and because any effective variation in P is very small over our range of x.

The intrinsic carrier density is determined by solving the following equation, from Eq. (1) of Ref. [2], where energy is in Ry and length is in atomic units:

$$\frac{2\beta^{3/2}}{\pi^{1/2}} \left(\frac{m_{hh}^*}{m_0}\right)^{-3/2} e^{\eta} \times \int_{\beta E_g}^{\infty} \frac{\gamma^{1/2}(x/\beta)\frac{d\gamma(x/\beta)}{dx}dx}{1+e^{(x-\eta)}} = 1, \qquad (2)$$

where $\beta = 1/k_B T$, m_{hh}^* is the heavy hole mass, $\gamma = k^2$, $\eta = \beta E_F$, E_F is the Fermi energy, and k is the electron wavenumber. Note that the lower limit of the integral in Eq. (2) is the reduced band gap because the zero of energy is the top of the valence band. The heavy-hole mass is taken to be 0.55 as in Ref. [2]. The value of 0.55 was used by Madarasz et al. because it gives better agreement with experiment. This value is also within the range of uncertainty for the spatially averaged effective mass determined from Weiler's cyclotron resonance data [8].

The computation time for the integral in Eq. (2) can be significantly reduced by performing an integration by parts to remove the derivative from the integrand. The resulting equation is:

$$\frac{4\beta^{3/2}}{3\pi^{1/2}} \left(\frac{m_{hh}^*}{m_0}\right)^{-3/2} e^{\eta} \times \int_{\beta E_g}^{\infty} \gamma^{3/2}(x/\beta)f(x)(1-f(x))dx = 1, \qquad (3)$$

where $f(x) = 1/(1+e^{(x-\eta)})$, and $f(x)$ is the Fermi-Dirac distribution function.

The function $\gamma(x/\beta)$ is found by inverting Kane's secular equation which is cubic in energy and solving for γ. The resulting cubic equation for γ is solved directly. A Newton-iteration technique finds the value of η. Note that full Fermi-Dirac statistics have been used for the conduction band while nondegenerate statistics have been used for the valence band because the valence-band edge is more than the required $4k_BT$ below the Fermi energy.

Once the Fermi energy is found, the intrinsic carrier density is computed by calculating either the hole or electron density. As in Eq. (2) of Ref. [2]:

$$n_i = p = \frac{(\pi\beta)^{-3/2}}{4} \left(\frac{m_{hh}^*}{m_0}\right)^{3/2} e^{-\eta}. \tag{4}$$

Results and Discussion

One-photon magneto-absorption data were taken at various wavelengths in order to extract accurate values for the band gap for a particular temperature. A modified Pidgeon-Brown energy-band model and a free exciton binding energy of 2 meV were used to determine E_g [9]. First we show comparisons between the energy gaps obtained from the data and the predictions of Eq. (1) for the energy gap as a function of composition x and temperature T. Figures 1(a) and (b) show the excellent agreement between predictions and experiment for the temperature dependence of the energy gap for samples with x=0.229 and 0.201. The slope of the curves of E_g-vs-T initially rises rapidly and then decreases slightly beyond about 20 K to reach the asymptotic value of the HSC relation at temperatures above about 100 K. The agreement between prediction and experiment is excellent. The agreement that has been obtained between Eq. (1) and data over the compositional range of 0.20 to 0.26 from this and our earlier work now gives us the confidence to use Eq. (1) to predict the intrinsic carrier density over the narrow-gap compositional range of 0.17 to 0.30.

The result of our n_i calculation between 4 and 100 K for x=0.17 is shown in Fig. 2a. We plot the logarithm of n_i for both Eq. (1) (solid) and the HSC relation (dotted). One can see that at 5 K the difference is more than an order of magnitude while the differences become small above 77 K. We show the percentage difference between n_i computed with our new relation and with the HSC relation, relative to the values computed with our new relation in Fig. 2b for the same composition. Even at 77 K the percentage difference is about seven percent. We show the same set of graphs in Fig. 3 for an x-value of 0.22. It is not possible to see the difference between our results and those based on the HSC relation directly in Fig. 3a because of the rapid variation of n_i with T. However, from the percentage differences given in Fig. 3b, one observes that the overall differences are similar to those of Fig. 2b, which shows that our relation is important throughout the long-wavelength region of mercury cadmium telluride. Similar results are also obtained for x=0.30.

The differences between the n_i results of Hansen and Schmit [1] and Madarasz et al. [2] are small in the range of x-values between 0.17 and 0.30. Part of the difference is due to the larger effective mass used in Ref. [2], which leads to an approximately fifteen percent increase in n_i. However, there are compensating effects due to the use of full Fermi-Dirac statistics in Ref. [2] that reduce the differences to generally less than ten percent. Thus, whether one compares the results from our calculations with either those of Refs. [1] or [2], the overall behavior is similar. At temperatures below about 50 K, our new relation gives significantly larger values of n_i than either previous one, while at temperatures above 77 K, the differences are small among all three sets of calculations. Analytic functions are being fit to the n_i curves for use in modeling.

These results have important implications for HgCdTe material characterization and device operation. In device operation it is usually necesary to compute diffusion currents for minority carriers, which depend on n_i^2. Recombination is also sensitive to n_i with radiative and Auger recombination varying as n_i^2. A quantity of great importance to the operation of infrared detectors, the resistance-area product $R_0 A$, also depends on n_i^2. Tunneling currents and impact ionization depend exponentially on the energy gap so that there is a definite need to use Eq. (1) in expressions for these quantities, especially at low temperatures. The reason that such quantities depend so critically on the value of E_g at low temperatures is that the thermal energy, $k_B T$, is so small.

Conclusions

We have computed the intrinsic carrier density of $Hg_{1-x}Cd_x Te$ for x between 0.17 and 0.30 as a function of temperature between 4 and 300 K. A new and highly accurate relation for the energy gap has been used, which was determined from one- and two-photon magneto- absorption spectroscopy. These data are presented in this and earlier work. Kane's theory has been used to treat the nonparabolicity of the conduction band. The results of the calculations show the need to include the nonlinearity of the temperature dependence of the energy gap, even though the nonlinearities are only several millivolts. Therefore, our new relation for the energy gap has important consequences in HgCdTe material characterization and device operation at temperatures below 77 K.

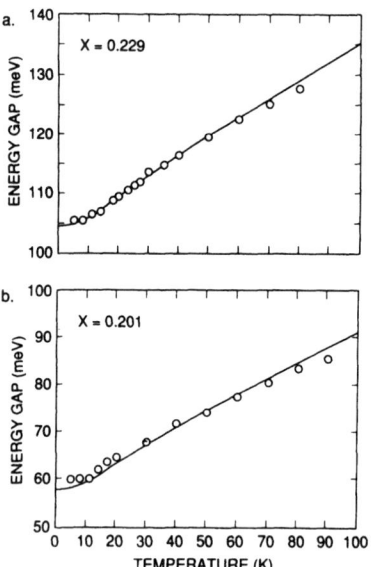

Figure 1. a.) Energy gap of sample with $x = 0.229$ as a function of temperature. The fit by our new relation with $x = 0.2285$ is the solid line. b.) Energy gap of sample with $x = 0.201$ as a function of temperature. The fit by our new relation is the solid line.

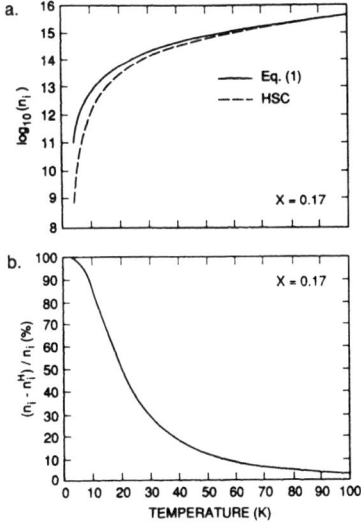

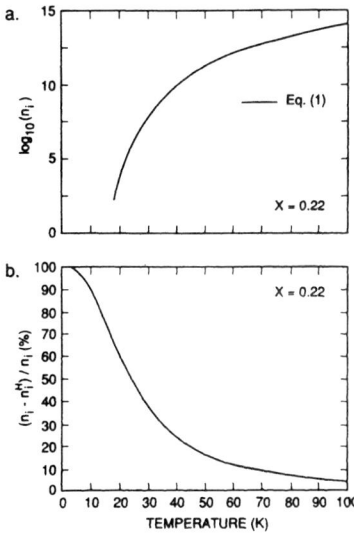

Figure 2. a.) Logarithm of the intrinsic carrier density, n_i, (cm^{-3}) as a function of temperature for $x = 0.17$. The solid line is with our new relation and the dotted line is with the HSC relation. b.) The percentage difference between n_i from our new relation and n_i^H from the HSC relation, relative to our new relation.

Figure 3. a.) Logarithm of the intrinsic carrier density, n_i, (cm^{-3}) as a function of temperature for $x = 0.22$. The solid line is with our new relation. b.) The percentage difference between n_i from our new relation and n_i^H from the HSC relation, relative to our new relation.

References

1. G. L. Hansen and J. L. Schmit, *J. Appl. Phys.* **54**, 1639 (1982).
2. F. L. Madarasz and F. Szmulowicz, *J. Appl. Phys.* **58**, 2770 (1985).
3. Y. Nemirovsky and E. Finkman, *J. Appl. Phys.* **50**, 8107 (1980).
4. D. G. Seiler, J. R. Lowney, C. L. Littler, and M. R. Loloee, *J. Vac. Sci. Technol.* A **8**, 1237 (1990).
5. G. L. Hansen, J. L. Schmit, and T. N. Casselman, *J. Appl. Phys.* **53**, 7099 (1982).
6. E. O. Kane, *J. Phys. Chem. Solids* **1**, 249 (1957).
7. J. L. Schmit, *J. Appl. Phys.* **41**, 2876 (1970).
8. M. H. Weiler, *Semiconductors and Semimetals*, edited by Willardson and Beer (Academic, New York, 1981), Vol. 16, p. 119.
9. D. G. Seiler, C. L. Littler, M. R. Loloee, and S. A. Milazzo, *J. Vac. Sci. Technol.* A **7**, 370 (1989).

TEM ASSESSMENT OF DEFECTS IN (CdHg)Te HETEROSTRUCTURES

S.G. Lawson-Jack*, I.P. Jones*, D.J. Williams** and M.G. Astles**
*School of Metallurgy and Materials,University of Birmingham,
Birmingham, United Kingdom.
**Royal Signals and Radar Establishments,
Great Malvern, Worcs, United Kingdom.

ABSTRACT

Transmission electron microscopy has been used to assess the defect contents of the various layers and interfaces in (CdHg)Te heterostructures. Examination of cross sectional specimens of these materials suggests that the density of misfit dislocations at the interfaces is related to the layer thicknesses, and that the high density of dislocations which are generated at the GaAs/CdTe interface are effectively prevented from penetrating into the CdHgTe epilayer by a 3um thick buffer layer. The majority of the dislocations in the layers were found to have a Burgers vector $\underline{b} = a/2<110>$ and either lie approximately parallel or inclined at an angle of $\sim 60^{o}$ to the interfacial plane.

INTRODUCTION

Progress in the development of infra-red (8 - 14 um waveband) imaging devices in which $Cd_x Hg_{1-x} Te$ is used as the detector depends upon the availability of this material in a sufficiently large area and with a high degree of compositional uniformity and structural quality. The initial choice of substrate for the epitaxial growth of (CdHg)Te is CdTe primarily because of the low mismatch (~0.21% for x~0.3) between the two lattices. However, examination of the CdTe/CdHgTe structure by transmission electron microscopy and other techniques has shown that the CdHgTe epilayer is fairly defective, containing an unacceptably high density of misfit and threading dislocations, precipitates of Te, and pyramid-like surface features [2] all of which adversely affect the performance of the devices. The origins of these defects have been traced to the poor structural quality of the CdTe substrate which may contain as many as 10^5 dislocations per cm^2 of material [1] , and to the damage done to the substrate surface during preparation for epitaxial growth.

As a result of these problems, thin epitaxial layers of HgTe and CdTe are being grown onto the CdTe substrate before the final growth of the CdHgTe. Such buffer layers are required to prevent the grown-in defects in the substrate from penetrating into the epilayer, as well as providing a clean and uniform surface for the growth of the CdHgTe.

Also, there has been an increasing interest in growing MOVPE CdHgTe on other substrates such as CdZnTe, CdSeTe and GaAs. The attraction of the CdZnTe and CdSeTe is that the Zn and Se contents can be varied to obtain a near-perfect lattice match with the CdHgTe, while the GaAs has the advantage of being less expensive and is available in large areas with good structural quality. However, the large lattice mismatch between GaAs and CdHgTe (~ 15%) necessitates the growth of a CdTe buffer layer onto the substrate to contain the misfit dislocations which would be generated at the interface, as well as preventing Ga diffusion into the epilayer. As the epitaxial growth techniques are improved upon and various combinations of complex CdHgTe structures are grown, it has become increasingly important to conduct a detailed structural assessment of the new materials as a component of the development process. Transmission electron microscopy is one of the major techniques for conducting such a study.

In this work, we have examined in the TEM MOVPE CdHgTe grown onto CdTe and GaAs substrates. The principal objective of the study is to carry out a comparative assessment of the structural quality of the CdHgTe epilayer. We have assessed the densities and distributions of the dislocations in the different layers and interfaces in these structures, as well as the effectiveness of the CdTe buffer layers (of varying thicknesses) in the containment of the misfit dislocations. Cross sectional specimens have been examined primarily since these offer the advantage of directly assessing the interfaces.

MATERIALS

The details of the materials which have been examined are given below:

J/304 $CdTe/Cd_xHg_{1-x}Te/CdTe/HgTe/CdTe$ <100>

Layer thicknesses t(um)   0.1   10   0.5   0.1   substrate

                              $x \sim 0.31$, In doped CdHgTe, grown at 360°C

EPI/0/369 $CdTe/Cd_xHg_{1-x}Te/CdTe/GaAs$ <100>

           t(um)   0.1   3   3   substrate

                              $x \sim 0.23$, grown at 360°C

SVG 185 $Cd_xHg_{1-x}Te/CdTe/GaAs$ <100>

           t(um)   10   2   substrate

                              $x \sim 0.21$, grown at 350°C

The epitaxial materials were grown at RSRE, Malvern, and Philips Components,Southampton in the UK. The MOVPE CdHgTe was grown by the interdiffusion multilayer process (IMP) in which alternate layers of CdTe and HgTe are deposited, followed by a short period of annealing to bring about a complete interdiffusion of the binary layers to form a uniform alloy. The substrates are all misoriented by 2° towards a <110> direction.

EXPERIMENTAL TECHNIQUES

The cross sectional specimens were prepared by a standard technique the details of which are given in [5]. It has been observed that using Ar^+ beam for the milling until the final 30 minutes when I^+ beam is used at a low beam current (~ 1uA) improved the quality of the specimens obtained. Also, sometimes it has been necessary to pre-thin the specimen using a chemical technique by which a jet of the etchant (2% Br in ethanediol) is directed onto the centre of a 3mm disc specimen until a dimple is observed. The specimens were examined in a JEOL 4000FX TEM at an operating voltage of 400kV.

RESULTS and DISCUSSION

Bright field TEM micrographs which show the layers and interfaces in the materials we have examined are presented in Figures 1-8.

J/304

It is shown in Figure 1(a) that the interfaces in this material are well-defined and that there is no evidence of incomplete diffusion of the original IMP layers. Figure 1(b) shows dislocation tangles in the CMT epilayer at about 1 μm from the CdTe(CT)/CdHgTe (CMT) interface, where the dislocation density is ~ 10^8 cm^{-2}. This is about 2-3 orders of magnitude higher than in the CT substrate as shown in Figure 1(c). Figure 2 shows a montage of a cross section of this material. It is evident that the different layers are dislocated to varying extents. Of particular interest is the observation that the CT/MT(HgTe) interface is much more dislocated than the MT/CT interface. This result is unexpected since geometrically the two interfaces should be identical. Also, since the stiffnesses of the two materials are quite similar (At 300K, C_{11} for CT is 53.6 GNm^{-2} whereas C_{11} for MT is 54.1 GNm^{-2}. C_{12} and C_{44} are similarly close) it is unlikely that the possibility of one layer providing an easier source of dislocations would explain the difference in the dislocation densities.

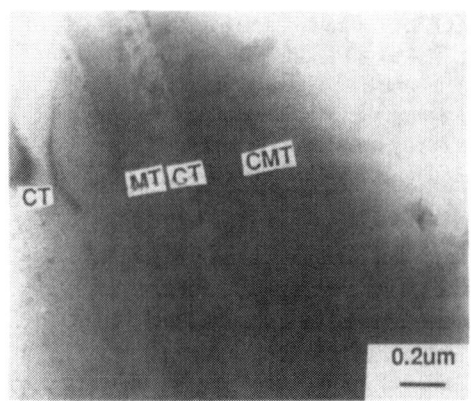

Figure 1.

(a) XTEM IMAGE SHOWING THE
INTERFACES IN THE MATERIAL
J/304.

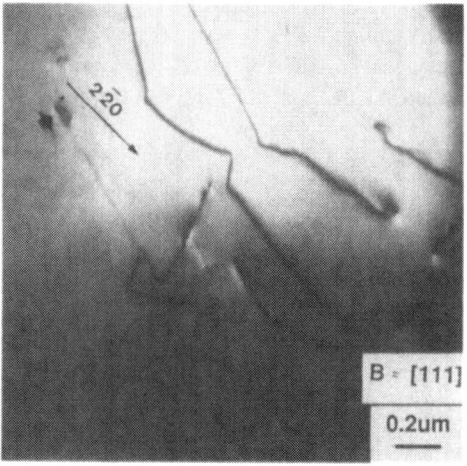

(b) DISLOCATION TANGLES IN THE
CMT LAYER IN J/304.

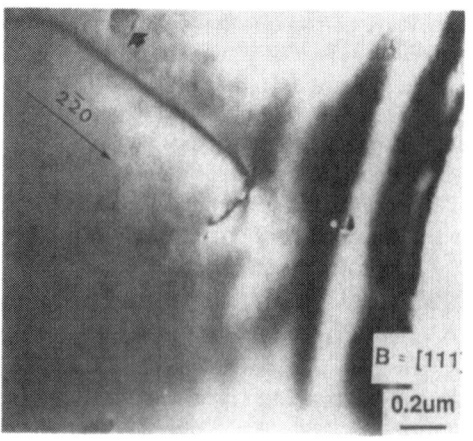

(c) DISLOCATIONS IN THE CT
SUBSTRATE.

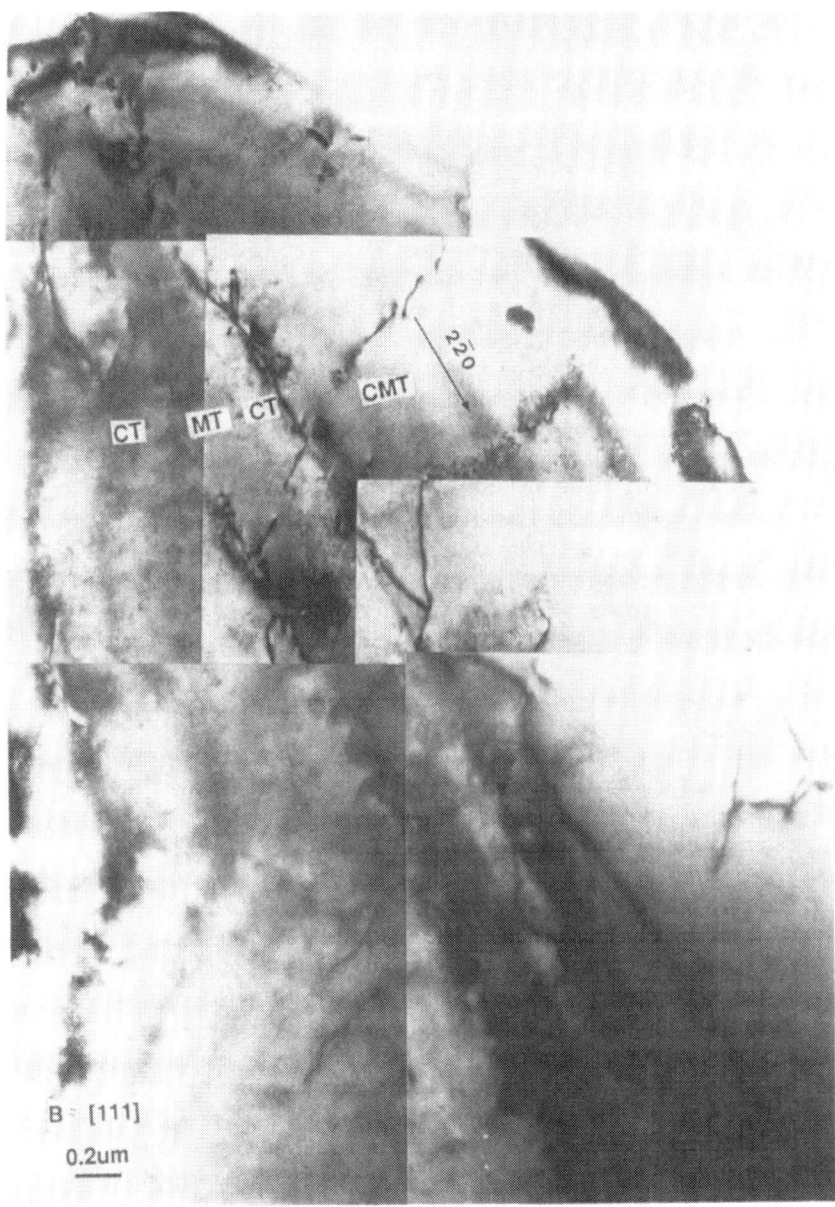

Figure 2. XTEM IMAGE SHOWING THE DIFFERENCES IN DISLOCATION
DENSITIES AT THE VARIOUS INTERFACES IN J/304.

However, in a previous work by Nouruzi-Khorasani et al. [8], an examination of rather thick (2 μm MT and 5 μm CT) layers of IMP CT/MT/CT material showed the CT/MT interface to be much less dislocated than the MT/CT interface.   It would seem therefore that the difference in dislocation densities at the two interfaces is due to the effect of the layer thicknesses in the generation of interfacial dislocations.   It is observed also in Figure 2 that some of the dislocations have propagated from both the substrate and the buffer layers into the CMT epilayer which suggests that the total thickness ($\sim$ 0.6 μm) of buffer layers is insufficient to prevent this from occurring.   In Figure 3, which shows a large area of the CMT layer, it is observed that the dislocation density decreases rapidly with distance from the CT/CMT interface so that at a distance of $\sim$ 1. 5 μm the dislocation density falls below what could be easily measured in the TEM ($\sim$ 10^6 cm^{-2}).   However, the 0.1 μm thick CT cap in this material has produced, in some areas, substantial dislocation densities in the top surface of the CMT layer as shown in Figure 4. Clearly, this observation will have a consequence for the capping of of CMT layers.

Apart from dislocations, no other defects have been observed in this material.   Some isolated particles were found in the CMT layer (arrowed in Figure 1(b)) and also in the CT substrate (arrowed in Figure 1(c)) but their compositions have not been determined to verify whether they are precipitates or debris deposited on the specimen during ion beam thinning.

EPI/0/369

Figure 5 shows a montage of a cross section of this material.   The networks of misfit dislocations which were generated at the GaAs/CT interface are seen to propagate into the CT buffer layer, their density falling off rapidly with distance from the interface.   Thus, we observe that the 3 μm thick buffer layer has effectively prevented the GaAs/CT misfit dislocations from pentrating into the CMT layer.   The large particle which is shown in this figure is believed to be debris deposited on the surface of the specimen during ion beam thinning.   Figure 6 shows the dislocation networks $\sim$ 1 μm into the CT buffer layer where the density is estimated to be $\sim$ 10^7 lines cm^{-2}.   The majority of the dislocations have a Burgers vector $\underline{b}$ = a/2 <110> and lie either parallel or at an angle of $\sim$ 60° to the interface.   The misfit dislocations which were generated at the CdTe/CdHgTe interface in this material were observed to be largely confined to a region $\sim$ 1 μm into the CMT layer.   Above this region the dislocation density was sufficiently low to be difficult to estimate by TEM.   The particles of which one is shown (arrowed) in Figure 6 were very rarely found in the CT buffer layer.   It has a similar morphology and

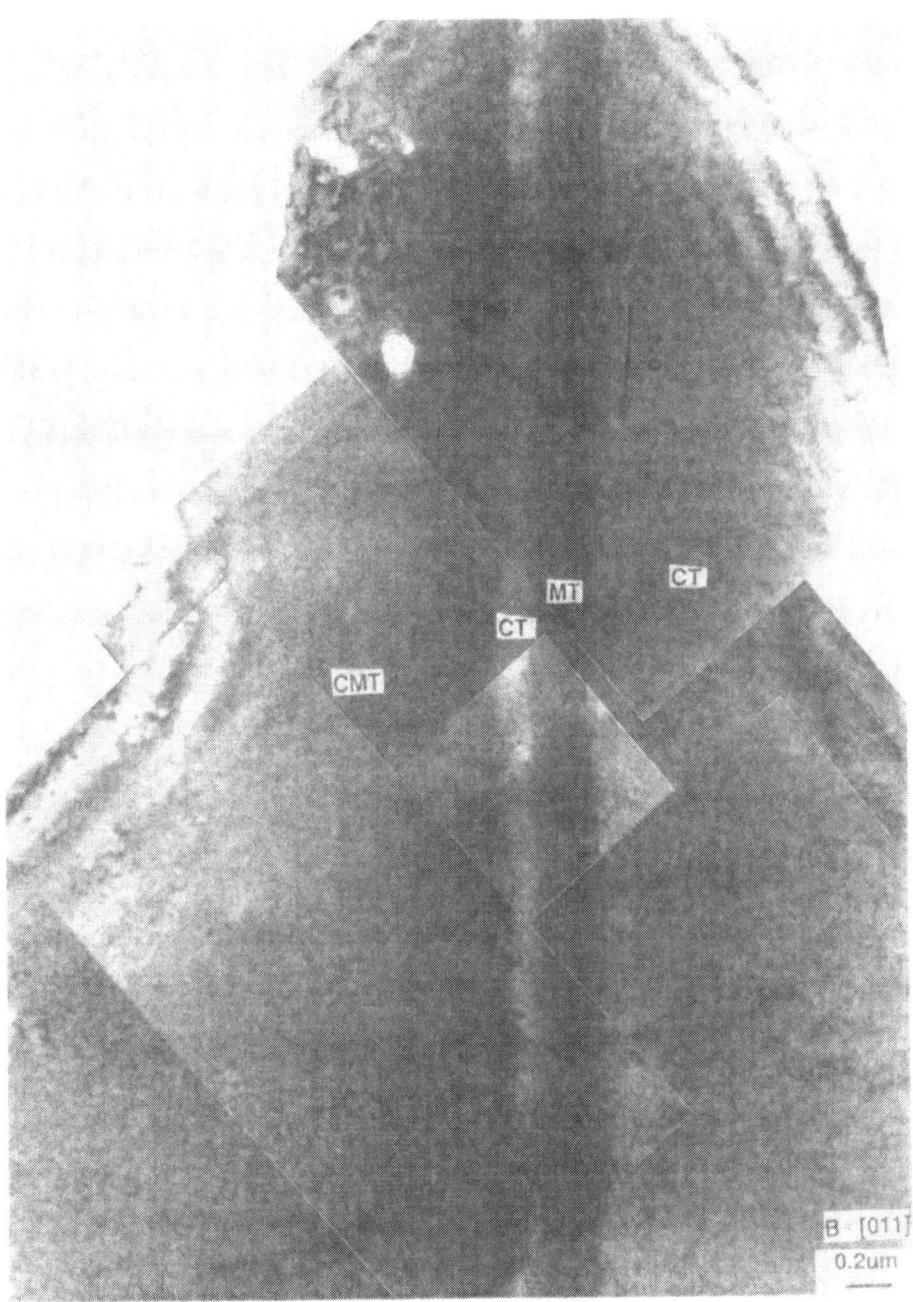

Figure 3. XTEM IMAGE SHOWING THE DIFFERENCES IN DISLOCATION
DENSITIES AT THE VARIOUS INTERFACES IN J/304.

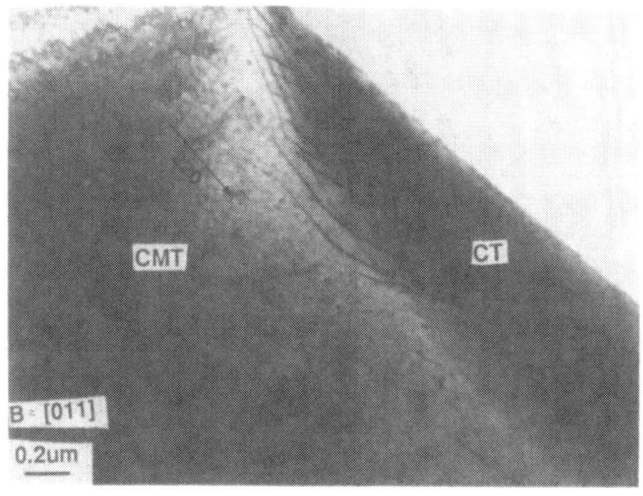

Fig. 4. XTEM IMAGE SHOWING THE MISFIT DISLOCATIONS AT
CMT/CT(CAP) INTERFACE IN THE MATERIAL J/304.

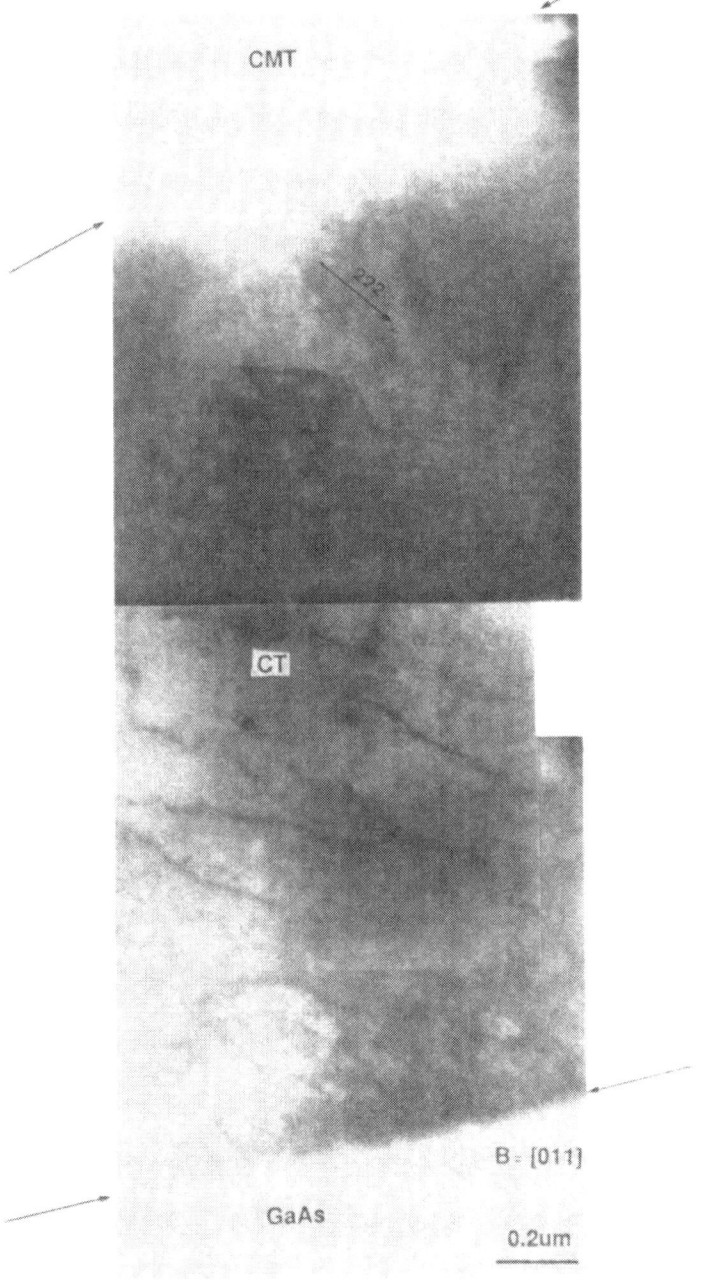

Figure 5. XTEM IMAGE SHOWING THE VARIOUS LAYERS IN THE
MATERIAL EPI/0/369.

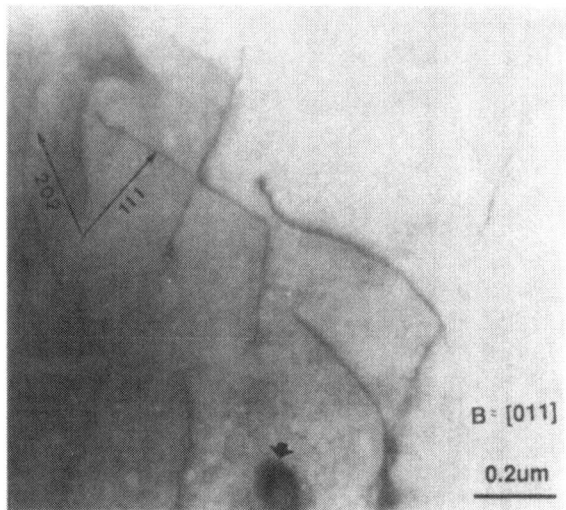

Figure 6. DISLOCATION NETWORKS IN THE CT BUFFER LAYER
IN THE MATERIAL EPI/0/369.

contrast to the Te-rich precipitates which have be reported by other
workers [6,12,13] to be present in CT substrates and LPE CMT layers
grown under a Te-rich atmosphere. However, such particles have not been
reported in MOVPE grown layers before. Further investigation involving
compositional analysis and stereomicroscopy (to verify whether the
particle is inside the layer or on the surface) is being carried out in order
to determine the identity of the particles.

<u>SVG185</u>

Figure 7 is a montage of a cross section of this material showing the
layers and interfaces. No defects were observed in the GaAs substrate
which indicates that it is generally of a high structural quality. However,
as expected (because of the large (~ 15%) lattice mismatch) the GaAs/CT
interface is shown to be grossly dislocated. Although the dislocation
density in the CT buffer layer does decrease with distance away from the
interface, a substantial number of such dislocations have propagated

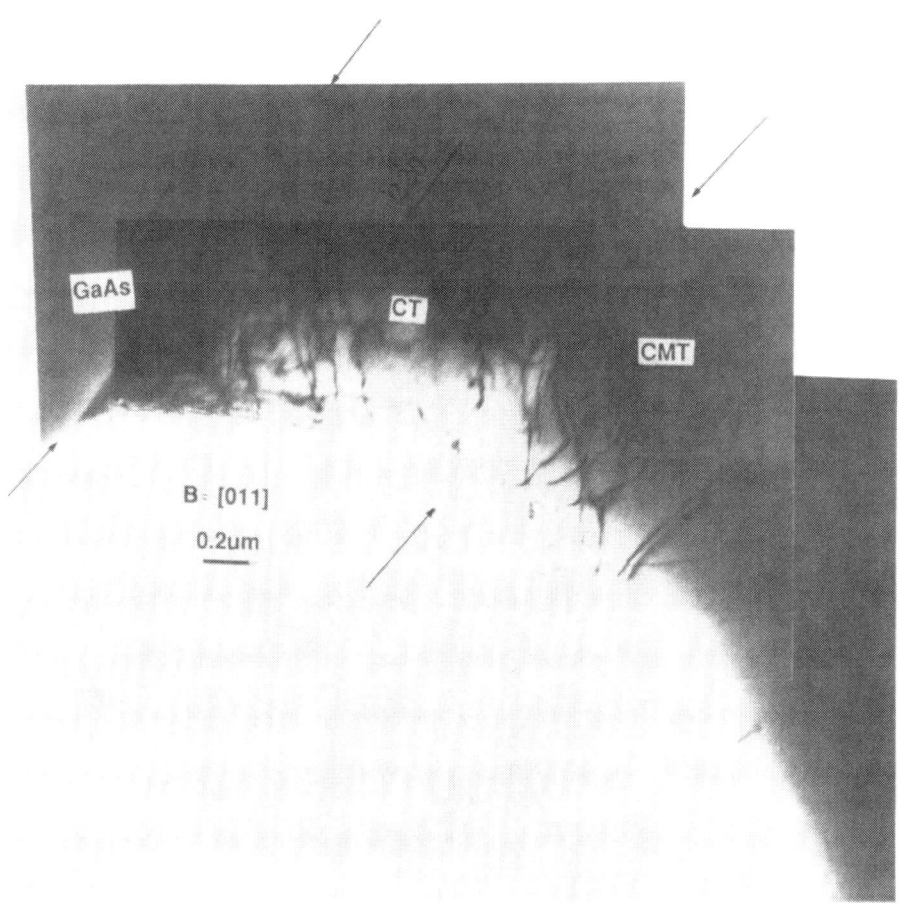

Figure 7. **XTEM IMAGE SHOWING THE VARIOUS LAYERS IN THE
MATERIAL SVG185**

straight through into the CMT layer.   This clearly shows that the buffer
layer of 2μm thickness is insufficient to contain all the misfit
dislocations generated at the GaAs/CT interface.   The dislocation density
about halfway (1μm) into the buffer layer is estimated to be in excess of
10^8 lines cm^{-2}.   The CT/CMT interface in this material is observed to be
substantially dislocated too.   However, the majority of the misfit
dislocations are shown to be confined to a region of <1 μm into the CMT

layer beyond which the dislocation density falls off quite rapidly. Figure 8 shows the top layer of the CMT (~ 2µm below the surface) in this material. It is evident that the layer is virtually free of dislocations.
However, it is shown to contain a well-defined subgrain boundary the origin of which is believed to be in the GaAs substrate. Subgrain boundaries which grow to accommodate tilt misorientations have been reported to occur in LEC GaAs [7,9]. Such subgrain boundaries penetrate into the CMT layer where they become enlarged.

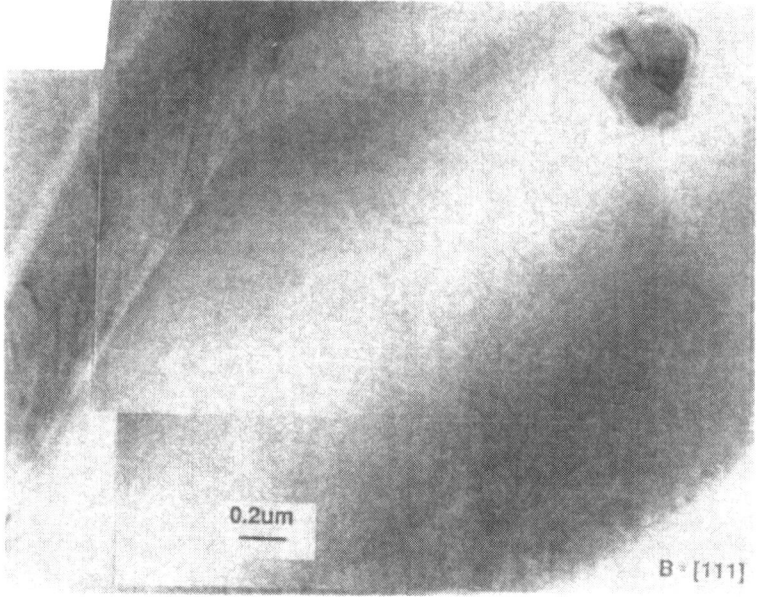

Figure 8. XTEM IMAGE OF THE TOP LAYER OF THE CMT IN THE MATERIAL SVG185 SHOWING A SUBGRAIN BOUNDARY.

CONCLUSIONS

The following conclusions may be drawn from the results which are presented in this report:
The predominant defect in the MOVPE (CdHg)Te is interfacial dislocations.
In the material J/304 the total thickness (~ 0.6um) of buffer layers is insufficient to prevent both substrate and misfit dislocations from penetrating into the CdHgTe epilayer whereas in the GaAs/CdTe/CdHgTe

structure the 3um thick buffer layer has effectively contained the large number of misfit dislocations generated at the GaAs/CdTe interface.

The substantial dislocation density which the 0.1um CdTe cap has produced in some areas of the CdHgTe implies that the thickness of such capping layers would require a re-evaluation.

In the GaAs/CdTe/CdHgTe structures the dislocation densities in the CdHgTe epilayer are confined to a region ~ 1um into the layer whereas in the CdTe/HgTe/CdTe/CdHgTe/CdTe this region extends to ~ 1.5um probably because of the reasons which are given in (2) and (3) above.

ACKNOWLEDGEMENTS

This work has been carried out with the support of the Procurement Executive, Ministry of Defence, United Kingdom. We are grateful to J. Gough, RSRE, Malvern and C. Ard, P. Mackett and I. Okeefe, Philips Components Limited, Southampton, for supplying the materials and taking part in the discussion of the results. We are also grateful to the Head of the School of Metallurgy and Materials, University of Birmingham, for allowing us the use of laboratory facilities.

REFERENCES

1. A. Nouruzi-Khorasani, I.P. Jones and P.S. Dobson: University of Birmingham, Project RU1-19, Annual report 1987.

2. A. Nouruzi-Khorasani, I.P. Jones and P.S. Dobson: University of Birmingham, Project RU1-19, Annual report 1988.

3. S.G. Lawson-Jack and I.P. Jones: University of Birmingham, Project RU1-23, Annual Report 1989.

4. A.G. Cullis, N.G. Chew and J.L. Hutchison: Ultramicroscopy, 17, 203, (1985).

5. N.G. Chew and A.G. Cullis: Ultramicroscopy, 23,175, (1987).

6. D.J. Williams and A.W. Vere: J. Vac. Sci. Technol., 4 (4), 2184, (1986).

7. A.M. Keir, A. Graham, S.J. Barnett, J. Geiss, M.G. Astles and S.J. Irvine: J. Crystal Growth, 101, 572, (1990).

8. A. Nouruzi-Khorasani, I.P. Jones , P.S. Dobson, Y. Etem, D.J. Williams, M.G. Astles, C. Ard and G. Coates: J. Crystal Growth,102, 819, (1990).

9. J.E. Hails, G.J. Russel, A.W. Brinkman and J. Woods: J. Appl. Phys., 60 (7), 2624, (1986).

10. S.G. Lawson-Jack: PhD. Thesis, University of London, 1989.

11. J. Gough, J. Geiss, P. Mackett and I. O` Keefe (private communication).

12. S.H. Shin, J. Bajaj, L.A. Moudy and D.T. Cheung: Appl. Phys. Lett., 43 (1), 68, (1983).

13. K. Durose, G.J. Russell and J. Woods: J. Crystal Growth, 72, 85, (1985).

PART II

Mercury Cadmium Telluride—
Processes and Devices

TRAPS AND TRAP-ASSISTED TUNNELING in HgCdTe PHOTODIODES

D. Rosenfeld and G. Bahir
Kidron Microelectronics Research Center, Department of
Electrical Engineering, Technion - I.I.T, Israel, 32000.

ABSTRACT

The performance of HgCdTe photodiodes, realized for the purpose of thermal imaging in the 8-12μm atmospheric window, is often limited by traps and by the tunneling mechanism associated with them. This mechanism dominates the leakage current and the dynamic resistance in diodes operated in reverse bias above 100mV and at temperatures below 77K. In usual working conditions, namely low reverse bias and a temperature of 77K, the presence of traps and the tunneling mechanism associated with them, degrades the performance of the diodes. Although the diffusion mechanism dominating the leakage current and the dynamic resistance in diodes operated under the above conditions is of major concern, the presence of the traps should not be ignored. The traps result in shorter electron lifetime in the bulk as well as in the generation of 1/f noise, which degrades the device's figures of merit such as NETD. Therefore, a thorough understanding of the trap-assisted tunneling mechanism is a prerequisite to studying its impact on the device performance.

In this paper we present a model which describes the connection between the leakage current associated with the traps and the trap characteristics: concentration, energy level and capture cross-section. The validity of the model is confirmed by the good fit between measured and calculated characteristics of the diodes.

INTRODUCTION

One of the critical parameters affecting the performance of HgCdTe photodiodes sensing infrared radiation in the 8-12 μm window is that of traps[1]. Traps located in the neutral regions bring about two short thermal transitions instead of one longer transition, which results in a dramatic decrease in the minority carrier lifetime (the Shockley—Read—Hall mechanism). This reduction of lifetime near the edge of the depletion region is of great importance, since it increases the diffusion leakage current and decreases the quantum efficiency of the photodiodes.

Within the depletion region, however, two additional tunneling transitions are added due to the existence of an electric field. The two added tunneling transitions are tunneling of carriers from the valence band to the traps and tunneling from the traps to the conduction band. Three more recombination paths are possible now: thermal-tunnel path, tunnel-thermal path and tunnel-tunnel path. The single thermal-thermal path associated with traps located outside the depletion region (the Shockley-Read-Hall mechanism) is therefore replaced by a 4 path mechanism, often called trap-assisted tunneling.

The temperature dependence of the band gap in HgCdTe is different than that of more common semiconductors, such as Si, Ge and GaAs. In HgCdTe, a decrease in temperature results in a decrease in the band gap. Therefore, as the temperature decreases, the tunneling processes in HgCdTe become dominant, and the difference between the "regular" Shockley-Read-Hall mechanism and the trap-assisted tunneling mechanism, becomes more significant. As a result, the performance of photodiodes operated at low temperatures is dramatically degraded. As was shown in previous papers[1,2], the negative effect of the traps is not limited to the low-temperature region. It has been clearly demonstrated that the dark current of photodiodes operated at the usual working conditions, namely low reverse bias and temperature of 77K, is dominated by the diffusion mechanism, while the 1/f noise is often dominated by the tunneling mechanisms.

In the following section we model the trap-assisted tunneling current of front illuminated bulk $Hg_{1-x}Cd_xTe$ photodiodes with $x \approx 0.22$. We proceed to confirm the validity of the model by comparing measured and calculated characteristics of the photodiodes in a wide range of temperatures and biases.

THE TRAP-ASSISTED TUNNELING MECHANISM:

In general, we model the dc behavior (leakage current as well as dynamic resistance) of our diodes with three distinct mechanisms: diffusion, band-to-band tunneling and trap assisted tunneling. The diffusion mechanism is dominant in the low bias and high temperature regions, while the two tunneling currents, trap-assisted tunneling and band-to-band tunneling currents, dominate at high reverse bias and low temperature regions[4]. The diffusion mechanism[5] and the band-to-band mechanism[6] in HgCdTe junctions are well understood. The trap assisted tunneling is therefore the unknown mechanism which requires more study.

Modeling of the trap-assisted tunneling dark current is a rather complicated task since many parameters are involved. The current is determined by the working conditions (such as applied voltage and temperature), by the bulk parameters (such as doping level and composition) and by the traps' parameters (such as the concentration of the traps, their energy level, and their capture cross-sections). Since the temperature behavior of the traps (capture cross-section as well as distance from the edges of the bands) strongly depends on the type of the traps, modeling of the temperature dependence of the current is even more complicated.

The energy bands of a reverse biased diode with traps located in the middle of the bandgap are shown in fig. (1). The different role played by the traps within the depletion region and those in the neutral regions is clearly seen. Outside the depletion region, only two types of thermal transitions are allowed: thermal transitions between the valence band and the traps, and thermal transitions between the traps and the conduction band (a typical Shockley-Read-Hall mechanism). The electric field within the depletion region, however, gives rise to two additional transitions: tunneling between the valence band and the traps and tunneling between the traps and the conduction band. Since the tunnel transitions are not always possible, the traps can be divided into 4 major groups:

(a) - Traps that can exchange carriers with the valence band by means of thermal transition only, and with the conduction band using both thermal and tunnel transitions.

(b) - Traps that can exchange carriers with both bands using thermal transitions only (Shockley-Read-Hall centers).

(c) - Traps that can exchange carriers with the valence band by means of thermal and tunnel transitions, and with the conduction band by thermal transition only.

(d) - Traps that can exchange carriers with both bands by thermal and tunnel transitions.

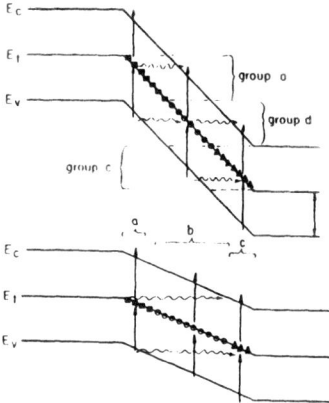

Fig. 1 - The energy bands of a diode with a mid-gap trap level in high (upper) and low (lower) reverse biases.

It should be noted that each of the four groups is associated with a different occupation probability and different probabilities for tunnel and thermal transitions. In addition, the fact that the relative number of traps depends strongly on temperature and bias, should also be considered. For example: the trap-assisted tunneling current in the diode of Fig. (1) is dominated by the contribution of group (d) in the case of high reverse bias, while in the case of low reverse bias the contribution of that group is negligible. Hence, modeling of the trap-assisted tunneling current should include detailed calculations of the relative number of traps in each group as well as their occupation probabilities and the tunnel and thermal transitions rates. We have recently published a detailed model[*] for the trap-assisted tunneling current, in both pn and n⁺p diodes. According to that model, the generation rates due to the trap-assisted tunneling process is given for the four groups by:

$$U_a = N_{Ta} \cdot \left[\frac{-\partial_n \partial_p n_i{}^2 (e^{\frac{qv}{KT}}-1) + \partial_p p_i \cdot \omega_c N_c}{\partial_n n_i + \partial_p p_i + \omega_c N_c} \right] \tag{1a}$$

$$U_b = N_{Tb} \cdot \left[\frac{-\partial_n \partial_p n_i{}^2 (e^{\frac{qv}{KT}}-1)}{\partial_n n_i + \partial_p p_i} \right] \tag{1b}$$

$$U_c = N_{Tc} \cdot \left[\frac{-\partial_n \partial_p n_i{}^2 (e^{\frac{qv}{KT}}-1) + \partial_n \cdot n_i \cdot \omega_v N_v}{\partial_n n_i + \partial_p p_i + \omega_v N_v} \right] \tag{1c}$$

$$U_d = N_{Td} \cdot \left[\frac{-\partial_n \partial_p n_i{}^2 (e^{\frac{qv}{KT}}-1) + \omega_c N_c \cdot \omega_v N_v + \partial_n n_i \cdot \omega_v N_v + \partial_p p_i \cdot \omega_c N_c}{\partial_n n_i + \partial_p p_i + \omega_c N_c + \omega_v N_v} \right] \tag{1d}$$

where v is the applied voltage, T is the temperature and N_{T1} is the density of traps in group i, and w_p and w_n are the capture coefficients for electrons and holes. The temperature dependence of the thermal emission rates is given by $p_i = N_v \cdot EXP(-E_T/kT)$ for holes and by $n_i = N_C \cdot EXP(-(Eg-E_T)/kT)$ for electrons, where N_v and N_C represent the density of states in the bands. Following Kinch[9] and Sah[7] we obtain for the carriers tunneling probabilities w_p and w_n:

$$\omega_c N_c = \frac{6 \cdot 10^5 \cdot E}{E_g - E_t} \cdot \exp\left[\frac{-1.7 \cdot 10^7 E_g^{1/2} (E_g - E_t)^{3/2}}{E} \right] \qquad (2a)$$

$$\omega_v N_v = \frac{6 \cdot 10^5 \cdot E}{E_T} \cdot \exp\left[\frac{-1.7 \cdot 10^7 \cdot E_g^{1/2} \cdot E_t^{3/2}}{E} \right] \qquad (2b)$$

Using Lax estimation[10] for the temperature dependence of the electrons and holes capture coefficients $\gamma_n = \gamma_{no} \cdot T^{-\rho n}$ and $\gamma_p = \gamma_{po} \cdot T^{-\rho p}$, the overall trap-assisted tunneling current can be calculated. By integrating the net generation rate across the depletion region (while taking into account the relative number of traps in each one of the groups), the total generation rates as well as the trap-assisted tunneling current are obtained. For high reverse bias $(V+V_o > Eg)$ such an integration yields:

$$U_{TOTAL} = \int_0^{W(\frac{E_t}{V+V_o})} U_a dx + \int_{W(\frac{E_t}{V+V_o})}^{W(1-\frac{Eg-E_t}{V+V_o})} U_d dx + \int_{W(1-\frac{Eg-E_t}{V+V_o})}^{W} U_c dx \qquad (3)$$

$$J_{TAT} = qW \cdot N_T \cdot \left[U_a(\frac{E_t}{V+V_o}) + U_d(1-\frac{Eg}{V+V_o}) + U_c(\frac{Eg-E_t}{V+V_o}) \right] \qquad (4)$$

where J_{TAT} is the trap-assisted tunneling current and Eg, ET, Vo, V and W represent the bandgap, the energy of the traps, the built-in voltage, the applied voltage and the width of the depletion region, respectively.

RESULTS AND DISCUSSION

Fig. (2) shows the dc behavior of a one sided p on n HgCdTe photodiode with x=0.219 and with $N_D = 2 \cdot 10^{15}$ cm^{-2}. The figure shows the dynamic resistance (the numerical derivative of eq. 4) associated with the trap-assisted tunneling mechanism as a function of reciprocal temperature, with the diode reverse bias as a parameter. The symbols in fig. (2) represent measured data while the solid curves represent the theoretical model presented in equations (1-4). The complicated temperature dependence of the trap-assisted tunneling process is seen very clearly. The high temperature region is dominated by the Shockley-Read-Hall thermal-thermal transitions as indicated by straight lines on a semilog scale. On the other hand, the low temperature region is dominated by the tunnel transitions as indicated by the decrease in the dynamic resistance as the temperature decreases.

A good fit between the measured data and the theoretical model was obtained for traps' concentration of $N_T = 4.3 \cdot 10^{14}$ cm^{-3}, traps energy of $E_t = 0.6 \cdot Eg$, and for electron capture cross-section of $\gamma_{no} = 2.5 \cdot 10^{-5}$ [cm^3/S/K] and $\beta_n = 1$ and hole capture cross-section of $\gamma_{po} = 3.6 \cdot 10^{-4}$ [cm^3/S/K^{-3}] and

β_p=0.3. These values are in good agreement with those published in the literature. Unfortunately, only a handful of published works (most of them by Polla and co-authors[11-13]), deal with the nature of traps in HgCdTe photodiodes. Therefore, the validity of the model should be further confirmed, by replacing the published characteristics of the traps by DLTS data measured on our samples[14].

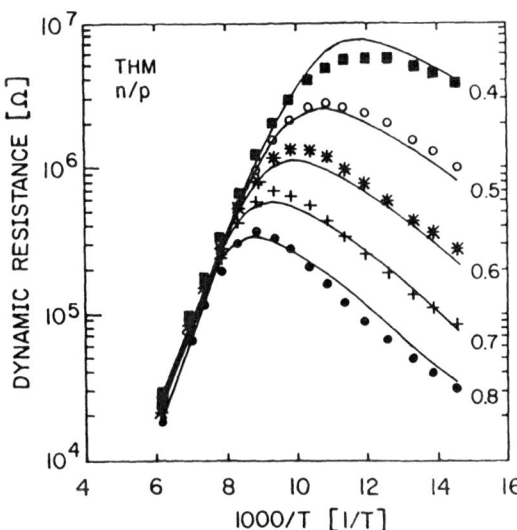

Fig. 2 - Dynamic resistance associated with the trap-assisted tunneling mechanism vs. temperature, with the diode reverse bias as a parameter. The symbols represent measured data while the solid curves represent the theory.

ACKNOWLEDGMENTS

The authors wish to thank the scientific team at Semi-Conductor Devices, Jerusalem for the fabrication of the diode of fig. (2). The technical assistance of D. Schoenmann and Y. Betser and the editorial assistance of R. Rosenfeld are also acknowledged with thanks.

REFERENCES

1. Y. Nemirovsky, R. Fastow, M. Meyassed and A. Unikovsky, accepted for publication in J. Vac. Sci. Tech., April 1991.
2. D. Rosenfeld and Y. Nemirovsky, The 16th Conference of Electrical and Electronics Engineering In Israel, March 7-9, 1989, Convention Center, Tel-Aviv Grounds, Israel.
3. Y. Nemirovsky and D. Rosenfeld, J. Vac. Sci. Technol., A(8), p. 1159, 1990.

4. Y. Nemirovsky, D. Rosenfeld, R. Adar and A. Kornfeld, J. Vac. Sci. Technol., A(7), p. 529, 1989.
5. M.B. Reine, A.K. Sood and T.J. Tredwell, in "Semiconductors and Semimetals", Edited by R.K. Willarson and A.C. Beer (Academic, New-York, 1981), vol. 18, Chap. 6.
6. W.W. Anderson, Infrared Physics, vol. 20, p. 353, 1987.
7. D. Rosenfeld and G. Bahir, submitted to IEEE Trans. Elect. Devices.
8. M.A. Kinch, in Ref. 5, Chap. 7.
9. C.T. Sah, Phys. Rev., vol. 123, p. 1594, 1961.
10. M. Lax, Phys. Rev. vol. 119, p. 1502, 1960.
11. D.L. Polla and C.E. Jones, J. Appl. Phys. vol. 52(8), p. 5118, 1981.
12. D.L. Polla, M.B. Reine and C.E. Jones, J. Appl. Phys. vol. 52(8), p. 5132, 1981.
13. C.E. Jones, V. Nair and D.L. Polla, Appl. Phys. Lett. vol. 39(3), p. 248, 1981.
14. S.J. Zackman, M.Sc. thesis, Technion, Haifa. To be published.

INFRA-RED PHOTODIODES IN $Hg_{1-x}Cd_xTe$ GROWN BY OMVPE

K.K. PARAT, N.R. TASKAR, H. EHSANI, I.B. BHAT and S.K. GHANDHI
Electrical, Computer and Systems Engineering Department, Rensselaer Polytechnic Institute, Troy, New York 12180.

ABSTRACT

$Hg_{1-x}Cd_xTe$ layers, grown by the organometallic vapor phase epitaxy (OMVPE), are p-type with carrier concentrations around $4 \times 10^{16}/cm^3$ due to the Group II vacancies in them. Following a Hg saturated anneal at 220°C, these layers become n-type with carrier concentrations around $4 \times 10^{14}/cm^3$. In order to fabricate p-n junction diodes, $Hg_{1-x}Cd_xTe$ layers were grown with a 0.5-0.8 μm thick CdTe cap. By opening windows in this CdTe cap, the underlying $Hg_{1-x}Cd_xTe$ layer was annealed in a selective manner, thus forming planar p-n junctions. The CdTe cap, which is used as the diffusion barrier for Hg during the selective anneal, also served as the junction passivant for the photodiodes. Details of device fabrication and characterization are presented in this paper.

INTRODUCTION

Mercury Cadmium Telluride ($Hg_{1-x}Cd_xTe$) is a direct bandgap semiconductor, where the bandgap can be varied from 0 eV to 1.6 eV by adjusting the alloy composition of the material [1]. This has made it the most important semiconductor for intrinsic infrared detector technology. Photoconductive as well as photovoltaic devices have been made in $Hg_{1-x}Cd_xTe$ using bulk grown materials, and in recent years using epitaxially grown materials [2-7]. OMVPE is an important technique for the growth of high quality $Hg_{1-x}Cd_xTe$ layers. Here, growth by the Direct Alloy Growth (DAG) process results in excellent crystal quality, large area uniformity, abrupt interfaces, and extrinsic p-type as well as n-type doping [8-10]. This paper reports on the fabrication and characterization of photodiodes in $Hg_{1-x}Cd_xTe$ grown by the DAG process.

As-grown undoped $Hg_{1-x}Cd_xTe$ layers are p-type with carrier concentrations around $4 \times 10^{16}/cm^3$ due to the Group II vacancies which are shallow acceptors. The vacancy concentration in these layers can be reduced to the low $10^{13}/cm^3$ range by annealing them under Hg overpressure at 220°C [11, 12]. Following such an anneal, the layers become low n-type with carrier concentrations around $4 \times 10^{14}/cm^3$, probably due to the background donor impurities. Due to the low diffusion coefficient of Hg in CdTe compared to that in $Hg_{1-x}Cd_xTe$ [12, 13], a 0.5-0.8 μm thick CdTe cap grown over the $Hg_{1-x}Cd_xTe$ layer can be used as a barrier for Hg diffusion. As a result, $Hg_{1-x}Cd_xTe$ layers annealed with a CdTe cap remain p-type with carrier concentrations in the mid $10^{16}/cm^3$ range [14]. By opening windows in this cap, the underlying $Hg_{1-x}Cd_xTe$ layer can be annealed in a selective manner. This allows the fabrication of n-regions in the as-grown material, which is p-type.

The CdTe cap, which was used for achieving selective annealing of $Hg_{1-x}Cd_xTe$ was also used as the junction and surface passivant over the p-n photodiodes. CdTe is lattice matched to $Hg_{1-x}Cd_xTe$, has a larger bandgap than $Hg_{1-x}Cd_xTe$, and can be grown with very high resistivity; thus it meets the main requirements of junction passivation. Details of the device fabrication and characterization are described in the following sections.

EXPERIMENTAL

$Hg_{1-x}Cd_xTe$ layers used in this study were grown at 370°C in a vertical reactor by the simultaneous pyrolysis of elemental mercury, dimethylcadmium, and diisopropyltelluride in hydrogen [8]. Semi-insulating CdTe was used as the substrate. Typically 10-12 μm thick layers of $Hg_{1-x}Cd_xTe$ were grown, followed by the growth of a 0.5-0.8 μm thick CdTe cap in a continuous operation. The purpose of this cap is to prevent any unintentional annealing of the underlying $Hg_{1-x}Cd_xTe$ during the post growth cool down period. The same cap is later used as the diffusion barrier for Hg during the selective annealing of $Hg_{1-x}Cd_xTe$.

In order to fabricate p-n junction photodiodes, an array of windows 600 μm in diameter and spaced 1000 μm apart, were opened in the CdTe cap using photolithography and chemical etching. An etchant consisting of HBr, H_3PO_4, and $1N\text{-}K_2Cr_2O_7$ in a volume ratio of 2:2:1 was used for this purpose. Following this, the layers were annealed in an evacuated sealed quartz ampoule along with 99.99999% pure Hg for 24 hours at 220°C. This annealing converts the entire $Hg_{1-x}Cd_xTe$ region under the open window to n-type, whereas the rest of the region under the CdTe cap remains unannealed and p-type. Thus, p-n junctions formed by this technique are cylindrical in nature, with vertical junctions.

After annealing, the samples were removed from the ampoule, and evaporated metal contacts formed on the n- and p-regions. Indium was used for contacting the n-regions, and was patterned into 400 μm diameter circular dots using a lift-off technique. Contact to the p-region were formed near the two edges of the sample using gold. Here, the CdTe cap was removed by etching in 1% bromine in methanol, prior to the deposition of gold. Completed devices were bonded in flat packs using silver epoxy, with provisions for front and backside illumination. Measurements were made in a variable temperature cryostat, over the 77-300K range.

RESULTS AND DISCUSSIONS

Figure 1 shows the I-V characteristics of a $Hg_{0.68}Cd_{0.32}Te$ photodiode at 77K under 0° field of view (FOV). Here, the thickness of the $Hg_{1-x}Cd_xTe$ layer was 10.4 μm, and the thickness of the CdTe cap was 0.8 μm. Due to the cylindrical nature of the junction, its area is approximately given by πdt, where d is the diameter of the n-region and t is the thickness of the $Hg_{1-x}Cd_xTe$ layer. The diameter of the n-region is taken to be same as the diameter of the window in the CdTe cap, which ignores the effect of any lateral diffusion of Hg under the cap. Thus the junction area for this device was 2×10^{-4} cm^2. The diode shows a forward cut-in voltage of about 0.2 volts, and a reverse breakdown voltage greater than 9 volts. The sharp and high breakdown voltage of the device can be attributed to the excellent junction passivation provided by the CdTe cap.

Figure 2 shows the ln I vs. V characteristics of this device at 77K under 0° FOV. For forward bias voltages less than 60 mV, the diode current is less than 10^{-11} A. Under forward bias, the diode exhibits an ideality factor of 1.4. At forward voltages greater than 170 mV, the diode current begins to be limited by series resistance, which is very large in this case due to the remote placing of the ohmic contact to the p-region. No attempts were made to minimize the series resistance by placing the ohmic contacts closer to the actual device, since it does not affect the measurement of the basic device parameters, such as R_oA product and spectral response. An ideality factor of 1.4 indicates that the diode dark current is limited by generation recombination in the depletion region.

The saturation current of the diode was estimated to be 1.5×10^{-14} A by extrapolating the ln I vs. V characteristic to the zero bias condition. This results in a saturation current density of 7.5×10^{-11} A/cm^2, if lateral diffusion effects are ignored. A direct measurement of the I-V around zero bias was not possible due to the very low diode leakage current. As a result, the R_oA product at 77K was estimated to be 8.8×10^7 ohm-cm^2 by dividing the thermal voltage kT/q by the diode saturation current density. It should be noted that inclusion of lateral diffusion effects would result in a larger junction area, and thereby a larger R_oA product than the one quoted here.

Figure 3 shows the R_oA of this device as a function of $1000/T$. The two straight lines in this figure correspond to the $1/n_i^2$ dependence and $1/n_i$ dependence, representing the diffusion limited and the generation recombination limited regimes respectively. At temperatures above 200K, the R_oA is diffusion limited. At temperatures lower than 120K, R_oA appears to be limited by the generation recombination in the depletion layer.

Figure 4 shows the relative spectral response of the diode at 77K under constant photon flux, for the case of frontside illumination. Due to the vertical junction, the photodiode operates in a lateral collection mode, with carriers generated within one diffusion length of the depletion layer being collected. However, due to the very high mobility of electrons in comparison to that of holes, the electron diffusion length is expected to be the dominant of the two. The electron diffusion length in this material was thus estimated to be about 90 microns. The photoresponse of the diode falls to 50% of its peak value at a wavelength of 4.5 μm, which is defined as the cut-off wavelength of the device. The spectral response remains flat in the 2.5 μm to 4.0 μm range, implying low recombination velocities and high diffusion lengths in the material.

CONCLUSION

Photodiodes were fabricated in Hg$_{1-x}$Cd$_x$Te grown by the OMVPE (DAG) process. A newly developed technique of selective annealing was used for the formation of p-n junctions, where the p-region was vacancy doped, and the n-regions were achieved by annealing under Hg overpressure. An in-situ grown layer of CdTe was used as the junction passivant for photodiodes. Diodes fabricated by this technique showed high performance with large reverse breakdown voltages and high R_oA values. The R_oA product of a photodiode with a cut-off wavelength of 4.5 μm was 8.8×10^7 ohm-cm^2 at 77K, which is comparable to the best values reported for Hg$_{1-x}$Cd$_x$Te devices with similar cut-off wavelengths [7].

ACKNOWLEDGEMENT

The authors would like to thank J. Barthel for technical assistance on this program and P. Magilligan for manuscript preparation. CdTe substrate material was kindly supplied by C.J. Johnson of II-VI, Inc., Saxonburg, PA. Partial funding for this program, from Raytheon Corporation is hereby acknowledged. This work was sponsored by Defense Advanced Research Project Agency (Contract No. N-00014-85-K-0151), administered through the Office of Naval Research, Arlington, VA. This support is greatly appreciated.

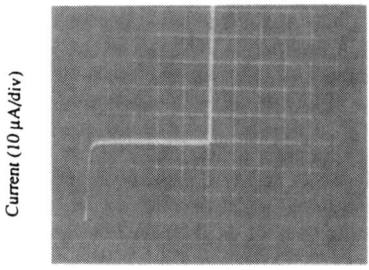

Current (10 μA/div)

Voltage (2 V/div)

Fig. 1 I-V characteristics at 77K
of an OMVPE grown
$Hg_{0.68}Cd_{0.32}Te$ p-n
junction photodiode
passivated with CdTe.

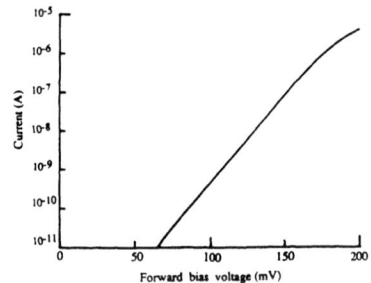

Fig. 2 Log I vs. V
characteristics at 77K of
the photodiode in Fig. 1.

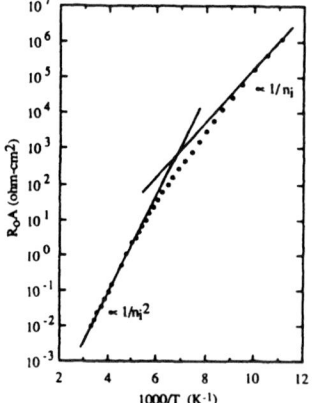

Fig. 3 Temperature dependence
of the $R_o A$ product of
the photodiode in
Fig. 1.

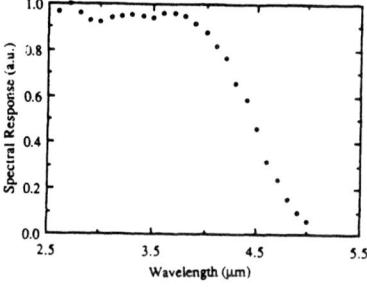

Fig. 4 Relative spectral
response per photon of
the photodiode in Fig. 1.

REFERENCES:
1. R. K. Willardson and A. C. Beer, eds. Semiconductor and Semimetals, Vol. 18, Academic Press, New York (1981).
2. M. A. Kinch, S. R. Borrello, and A. Simmons, Infrared Phys., 17, 127 (1977).
3. L. T. Specht, W. E. Hoke, S. Oguz, P. J. Lemonias, V. G. Kreismanis, and R. Korenstein, Appl. Phys. Lett., 48, 417 (1986).
4. M. Lanir and K. J. Riley, IEEE Trans. Elec. Dev., ED-29, 274 (1982).
5. E. R. Gertner, S. H. Shin, D. D. Edwall, L. O. Bubulac, D. S. Lo, and W. E. Tennant, Appl. Phys. Lett., 46, 851 (1985).
6. J. M. Arias, S. H. Shin, J. G. Pasko, R. E. DeWames, and E. R. Gertner, J. Appl. Phys., 65, 1747 (1989).
7. A. Rogalski, Infrared Phys., 28, 139 (1988).
8. S. K. Ghandhi, I. B. Bhat, and H. Fardi, Appl. Phys. Lett., 52, 392 (1988).
9. S. K. Ghandhi, N. R. Taskar, K. K. Parat, D. Terry, and I. B. Bhat, Appl. Phys. Lett., 53, 1641 (1988).
10. S. K.Ghandhi, N. R. Taskar, K. K. Parat, and I. B. Bhat, Appl. Phys. Lett. (submitted).
11. H. R. Vydyanath and C. H. Hiner,J. Appl. Phys., 65, 3080 (1989).
12. C. L. Jones, M. J. T. Quelch, P. Capper, and J. J. Gosney, J. Appl. Phys., 53, 9080 (1982).
13. K. Takita, K. Murakami, H. Otake, K. Masuda, S. Seki, and H. Kudo, Appl. Phys. Lett., 44, 996 (1984).
14. K. K. Parat, H. Ehsani, I. B. Bhat, and S. K. Ghandhi, J. Vac. Sci. Tech. (submitted).

TWO-LAYER LPE HgCdTe P-on-n 8-18μm PHOTODIODES

E. E. KRUEGER, G. N. PULTZ, P. W. NORTON, J. A. MROCZKOWSKI,
M. H. WEILER, and M. B. REINE
Loral Infrared & Imaging Systems, Lexington, Massachusetts 02173

ABSTRACT

This paper reports recent results on two-layer P-on-n LPE HgCdTe heterojunction photodiodes with cutoff wavelengths beyond 19μm. These results demonstrate the potential of photovoltaic HgCdTe detectors to satisfy the detector requirements of advanced NASA satellite instruments out to wavelengths of 17μm.

INTRODUCTION

Advanced NASA satellite missions will require LWIR (Long Wavelength InfRared) detectors with cutoff wavelengths out to 19μm and having operating temperatures of at least 60-65K [1]. In the past, photoconductive (PC) HgCdTe detectors [2] have satisfied NASA instrument requirements in this wavelength range. PC HgCdTe detectors will continue to be important for those future NASA instruments in which the number of detectors is relatively small (e.g., less than several hundred).

Many advanced NASA instruments, however, will require much larger numbers of detector elements in a two-dimensional array format. For these applications, photovoltaic (PV) HgCdTe is the preferred detector. PV HgCdTe detectors have much higher impedance than PC HgCdTe devices, which enables them to be interfaced to silicon CMOS multiplexer circuits. Additional advantages of PV HgCdTe over PC HgCdTe detectors are that PV devices do not require dc bias power and that PV devices remain linear in response to much larger signal photon flux levels.

Prior development of LWIR PV HgCdTe detectors has concentrated on the 10-12μm wavelength range [3]. We have extended the development of PV HgCdTe detectors to wavelengths out to 19μm [4]. T h e device structure is a P-on-n backside-illuminated heterostructure mesa photodiode fabricated on two-layer films grown at Loral by LPE (Liquid Phase Epitaxy), as illustrated in Figure 1. Test arrays of photodiodes processed from these films were characterized over the 60-80K temperature range, and exhibit R_0A products that are at the limit imposed by n-side diffusion current for the Auger-1 lifetime corresponding to the base layer carrier concentration. For example, an R_0A of 0.13 ohm-cm^2 was achieved at 80K for a cutoff wavelength of 18.8μm. At lower temperatures of 70K and 60K, the R_0A increased to values of 0.4 and 2.5 ohm-cm^2, following the diffusion current limit, while the cutoff wavelength shifted to 19.8 and 20.9μm.

EXPERIMENTAL

Film Growth and Characterization

Indium-doped n-type HgCdTe base layers, 15 to 20μm thick, were grown using a Te-rich LPE slider onto <111>B CdTe substrates. Base layer doping was used to achieve 77K carrier concentrations in the

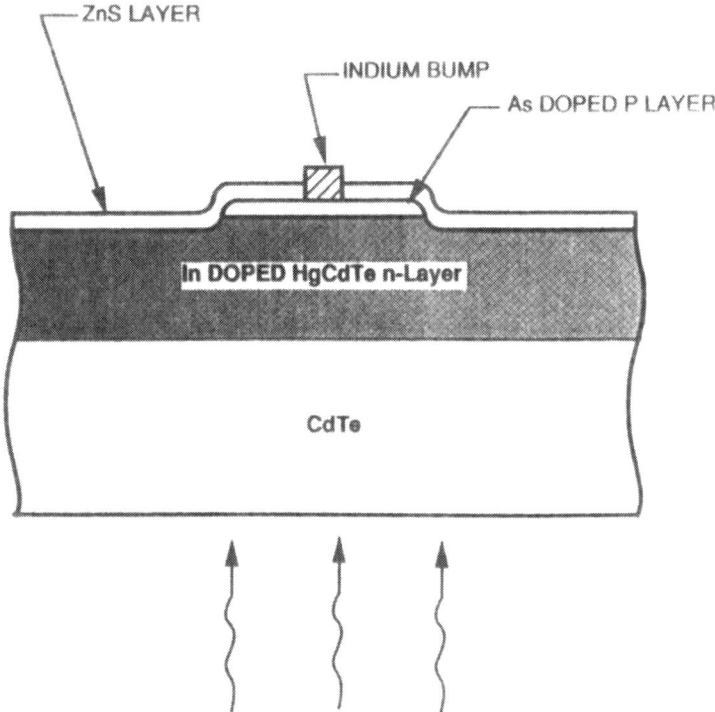

Figure 1. Cross-section of a two-layer P-on-n heterojunction film after mesa photodiode fabrication.

low-10^{14} cm^{-3} to mid-10^{15} cm^{-3} range on a controllable basis. Carrier concentrations were measured from patterned Hall samples on an annealed piece of the base layer. Capacitance-voltage (C-V) measurements were made on large-area diodes to confirm the Hall measurements. The base layer thickness was determined from interference fringes in the room temperature transmittance spectrum. The x-value of the base layer was also determined from the 300K transmittance spectrum. Minority carrier lifetime values were measured in the annealed n-type base layer piece by Loral's Optically Modulated Absorption (OMA) technique [5] or by photoconductive frequency response measurements at 85K.

After base layer growth, the wafers are prepared for cap layer growth by dicing off any damage or residual melt. The base layer is then loaded into the vertical-slide Hg-rich LPE system, and a 1 to 2µm thick p-type (arsenic-doped to 5×10^{15} cm^{-3} to 1×10^{18} cm^{-3}) cap layer is grown at a temperature around 400°C. The patterned cap layer Hall sample is on a sacrificial CdTe piece fitted in the graphite boat set during cap growth. Thickness of the cap layer is measured from an optical photograph of a cross section of the sacrificial CdTe piece. The cap layer x-value was measured by surface reflectance spectroscopy in the visible wavelength region [6].

The resulting double-layer film was then annealed under Hg-saturation conditions at a temperature of 240-260°C to reduce the native defect acceptor concentration to values well below the indium donor concentration in the base layer, thereby converting the base layer from p-type to n-type [7].

Electrical Characterization Techniques

To characterize the LPE grown junctions, test arrays of 43 circular mesa photodiodes were fabricated in the backside-illuminated configuration shown in Figure 1. The arrays were passivated with ZnS. The diodes had junction (mesa) areas in the range of 0.13-9.6×10^{-4} cm^2 and were well separated from each other. These arrays were indium bump mounted to fanout circuit boards. The relative spectral response was measured for representative photodiodes in each array at a temperature of 80K, from which the cutoff wavelength λ_{CO} (defined as that wavelength where the response has dropped to 50% of its peak value) was determined. The zero bias impedance R_0 and the quantum efficiency QE were measured by flood illumination from a chopped blackbody source. Either a 500K blackbody with an f/5 aperture or a 1000K blackbody with a 4µm spike filter was used for the measurements. Current versus voltage (I-V) measurements were made on selected diodes of each size at biases from -300mV to +180mV.

The test diodes fabricated on each array are finite in size; hence the measured values of R_0 and QE will show lateral collection effects and surface effects due to the fabrication process. To extract information about the quality of the films themselves, values for the parameters $(R_0A)_{Bulk}$ and QE_{inf} were derived from the variable area diode data. These two parameters approximate the behavior of "infinite area" one-dimensional diodes. These parameters are obtained from simple models described in detail in Reference 4. The value of $(R_0A)_{Bulk}$ is found from the intercept of the plot of $1/R_0A$ versus the diode perimeter-to-area ratio (P/A_J) [8,9]. A maximum value of the surface recombination velocity can be obtained from the slope. The value of QE_{inf} is obtained from the intercept of the plot of the square root of the measured quantum efficiency QE_{meas} (calculated using the actual junction area) versus the inverse of the junction radius [4]. From the slope of this plot, a value for the optical collection length L_{opt} can be derived, which we have found to agree well with the calculated minority carrier diffusion length in the n-type base layer.

RESULTS & DISCUSSION

Our goal was to examine LPE PV HgCdTe photodiode operation in the 80 to 60K temperature range with cutoff wavelengths extending past 19μm and to determine the mechanisms limiting performance at these temperatures. The set of films that was grown included different base layer doping levels and a range of values for the cap-base x-value difference Δx_{cb}. The following discussion focusses on the photodiode impedance and quantum efficiency as functions of temperature, cutoff wavelength and detector bias.

Base Layer Properties

The lifetimes of the n-type LWIR base layers samples were measured at a temperature of 85K by Loral's Optically Modulated Absorption technique [5] or by photoconductive frequency response. These data are plotted in Figure 3 versus the Hall carrier concentration measured at 77K. Also shown in Figure 3 for comparison are photoconductive decay lifetime data at 77K measured on high quality Loral bulk-grown [10] n-type $Hg_{0.794}Cd_{0.206}Te$ samples [11]. Both the LPE and the bulk sample data show a rapid decrease in lifetime with increasing carrier concentration, typical of the Auger-1 recombination mechanism that is known to determine carrier lifetime in well behaved n-type $Hg_{1-x}Cd_xTe$ at these temperatures [12, 13]. At carrier concentrations above $4x10^{14}$ cm^{-3}, both LPE and bulk data agree very well with the calculated Auger-1 lifetime for extrinsic n-type $Hg_{1-x}Cd_xTe$, as shown in Figure 3 by the solid line. This line is consistent with Blakemore's equation [14] for the Auger-1 lifetime with the overlap integral term $|F_1F_2|^2$ set equal to 0.41. For carrier concentrations less than $4x10^{14}$ cm^{-3}, the LPE data fall well below the Auger-1 limit, which we attribute to Shockley-Read recombination or to background-generated increase in the carrier concentration.

Photodiode Spectral Response and Quantum Efficiency

The relative spectral response data for photodiodes from two films with 80K cutoff wavelengths of 13.8μm and 18.8μm are shown in Figure 4. The upper curves are per watt; the lower curves are per photon and are relative quantum efficiency. The spectral curves are classical in shape: the quantum efficiency is independent of wavelength and the peak response is roughly 90% of the cutoff wavelength.

Figure 5 is an example of a plot of the square root of the measured quantum efficiency QE_{meas} versus the inverse of the diode junction radius $1/r$ for 80K, 70K and 60K. A linear regression of these data yields an intercept which is the square root of the infinite area quantum efficiency QE_{inf}. The slope of the line is equal to the product of the square root of QE_{inf} and the lateral collection length L_{opt}.

Photodiode Impedance Measurements

As described above, the film quality is ascertained from a

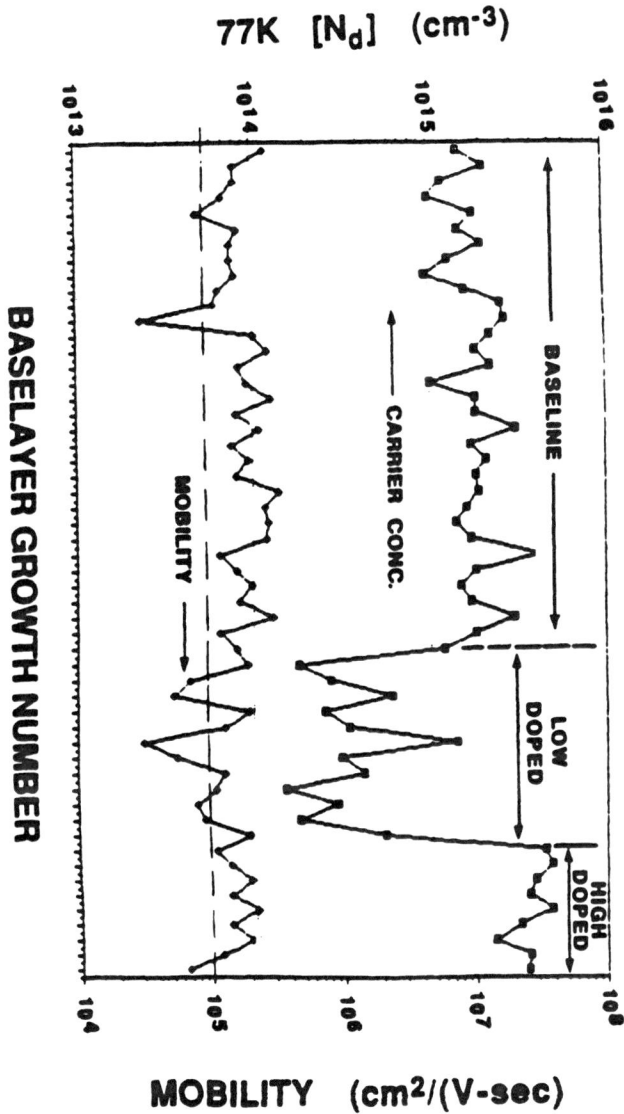

Figure 2. N-type Hg$_{0.79}$Cd$_{0.21}$Te base layer carrier concentrations and Hall mobilities at 77K for the three doping levels used in this study. The n-type dopant was indium. The magnetic field strength for the Hall measurements was 4 kGauss.

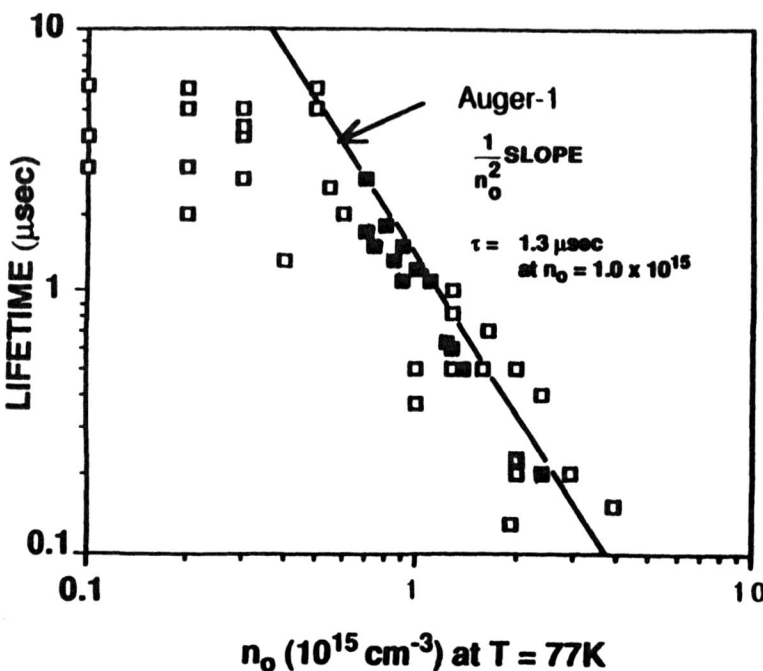

Figure 3. N-type base layer lifetime data measured by optically modulated absorption or photoconductive frequency response, plotted versus 77 K Hall carrier concentration. Solid points are from high-quality bulk crystal material (x=0.206) measured at 77K, open symbols are from LPE films (x=0.19 to 0.21) grown in Te-rich slider measured at 85K. The solid line is the calculated Auger-1 lifetime at 77K for $|F_1F_2|^2$=0.041 and x = 0.206.

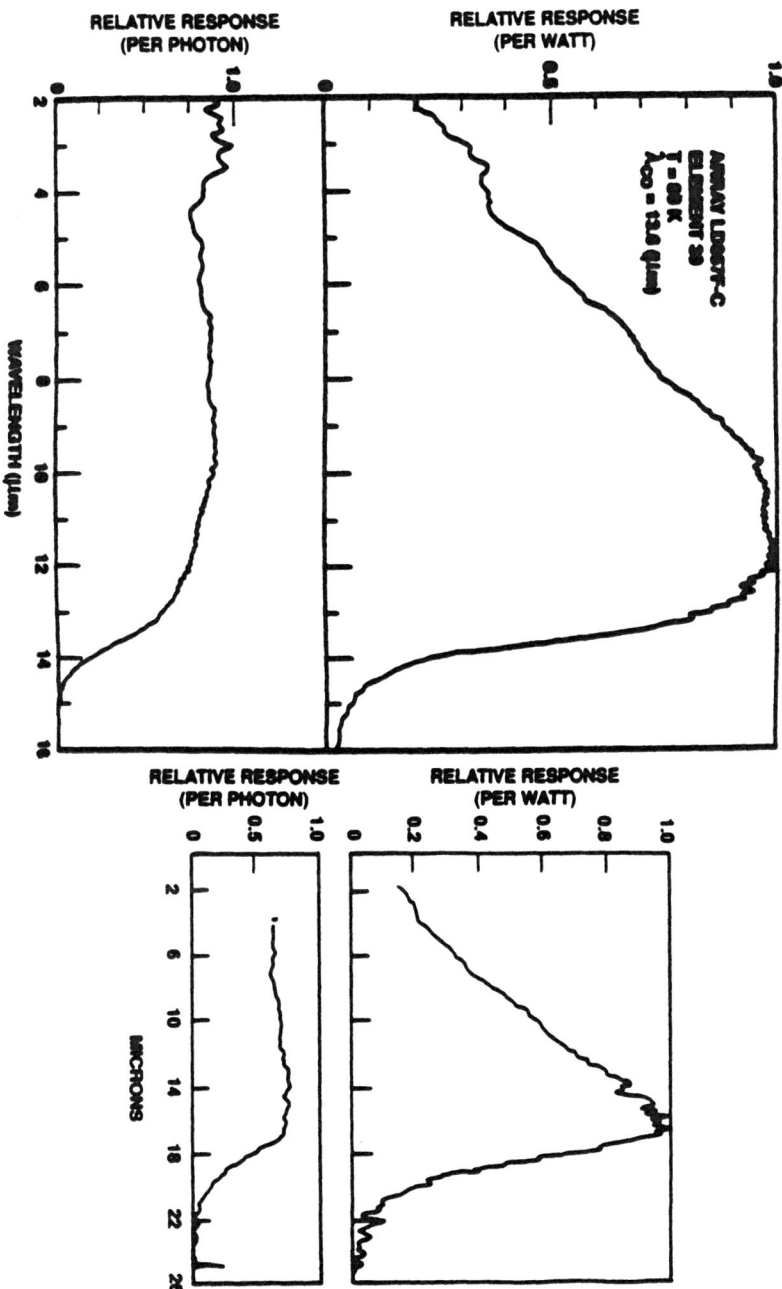

Figure 4. Relative spectral response data (per watt in upper curve and per photon in lower curve) for two films with cutoff wavelengths of 13.8μm and 18.8μm at 80 K.

plot of $1/R_0A$ values versus the perimeter-to-junction area ratio (P/A_j). Figure 6 is a set of these plots for an array with an 80K cutoff wavelength of 13.6μm (14.8μm at 60K) for the temperature of 60K. The regression analysis of the data gives an $(R_0A)_{Bulk}$ value of 84 ohm-cm^2 and a maximum surface recombination velocity of $7x10^3$ cm/sec. The solid points in the graphs were included in the regression analysis, while the open symbols were not, since the poor performance is indicative of poor signal collection, poor reverse bias impedance or morphological problems in the film surface.

These $(R_0A)_{Bulk}$ values are independent of diode geometry and surface effects from diode fabrication. This allows simple one-dimensional modeling of the temperature and cutoff wavelength dependencies of the detector impedance. Figure 7 shows a plot of RoA_{Bulk} values versus cutoff wavelength for temperatures of 60, 70 and 80K. The data are from films with cap-base x-value differences Δx_{cb} of 0.03 or less, with carrier concentrations of $0.6-2x10^{15}$ cm^{-3}, and with base layer thicknesses of 15 to 22μm. The solid lines represent calculated values of R_0A using a one- dimensional model for n-side diffusion current [3] with a carrier concentration of $2x10^{15}$ cm^{-3}, a base layer thickness of 20μm, and an Auger-1 lifetime. It is evident from the close agreement between the data and this simple calculation that these films have excellent junction quality, and that the R_0A values are determined by the Auger-1 lifetime in the base layer.

The difference Δx_{cb} between the cap x-value and the base x-value that is needed for optimum heterojunction diode performance is bounded by two limits. A cap layer with an x-value less than or equal to the base layer can be a source of diffusion current that can limit seriously the detector impedance. However, if the cap layer x-value is too high, a valence band barrier for minority carrier collection can be formed or the contact to the p-type cap can have too large a resistance.

To experimentally assess these effects for these longer cutoff wavelength devices, Base Layer LI112 was diced in half and cap layers were grown with different x-values. Arrays were then made from these films and performance was compared. Table 1 contains the results. The two double-layer films grown from this base layer are designated LD055RD and LD057FB. The x-values of the cap layers were measured with surface reflectance [6] and are listed in Table 1. The x-value (0.205) of the base layer was found from the 80K cutoff wavelength of the photodiodes fabricated on both double-layer films.

The dynamic resistance R_D versus applied bias voltage is shown in Figure 8 for Films LD057FB and LD059FB at 60 and 70K. The solid line in this figure is a plot of $\{R_0 \times exp(qV/kT)\}$, normalized to the measured R_0, and shows how the dynamic impedance would depend on voltage if diffusion current were the only current mechanism present. The area of these diodes is $4.42x10^{-5}$ cm^2 and the forward bias resistance is low (tens of ohms). Both films show a diffusion trend near zero and out to 15-20 mV reverse bias. The dynamic impedance continues to increase to a bias and temperature independent resistance. Breakdown occurs at voltages greater than 1 volt for these very long cutoff wavelength films.

CONCLUSIONS

We have demonstrated the feasibility of HgCdTe PV detectors with cutoff wavelengths out to 18-19μm as needed for advanced NASA

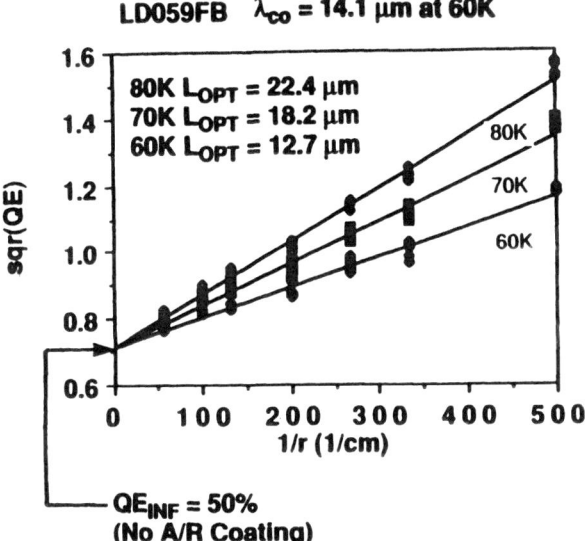

Figure 5. Square root of measured quantum efficiency QE_{meas} vs inverse diode radius $1/r$ for an array from a film with 77K carrier concentration of 5×10^{14} cm^{-3} and base layer thickness of 15.2μm. The intercept gives QE_{inf} = 50% (no A/R coating) independent of temperature. The slopes give L_{opt} values of 13, 18 and 22μm for 60, 70 and 80K. The diode cutoff wavelength at 60K was 14.1 μm.

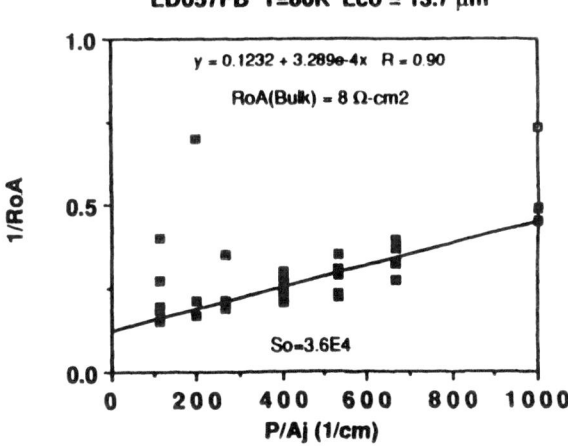

Figure 6. Plot of $1/R_0A$ versus (P/A_j). The intercept gives values for $(R_0A)_{Bulk}$ of 8 ohms-cm^2 and for surface recombination velocity of 3×10^4 cm/s for a film with an 80K cutoff wavelength of 13.8μm and a 77K carrier concentration of 9×10^{14} cm^{-3}. Impedance measurements were made at 80K.

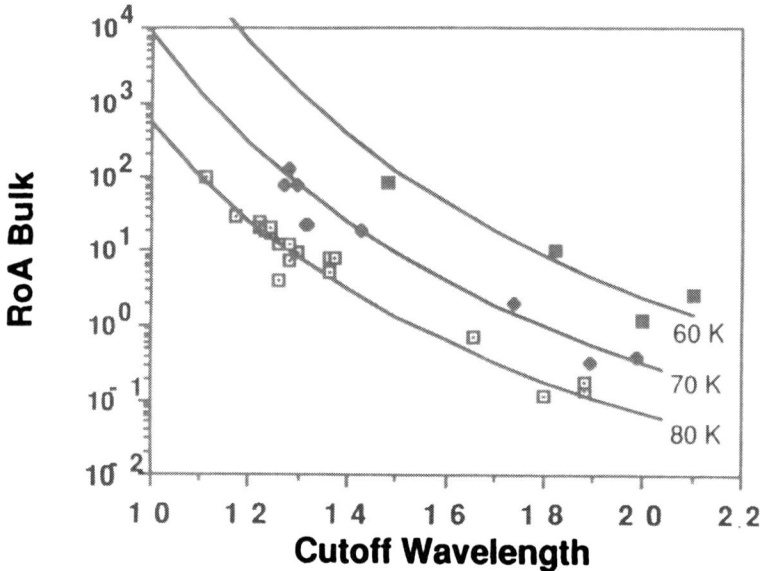

Figure 7. $(R_0A)_{Bulk}$ vs cutoff wavelength for 60K (solid squares), 70K (solid diamonds) and 80K (open squares). Solid lines are calculated from an Auger-1, n-side, 1-dimensional diffusion model for each of the three temperatures. The calculations were made using a base layer thickness of 20μm and a carrier concentration of 2×10^{15} cm^{-3}.

T (K)	80	80	70	70
Film	LD055RD	LD057FB	LD055RD	LD057FB
Δ x	0.084	0.037	0.084	0.037
R_0A_{Bulk}	5	8	17	20
QE %	45	66	33	69
QE ratio	1.4	1.0	1.6	1.0

Table 1. Comparison of photodiode electrical properties at 80K and 70K for different x-value cap layers grown on two pieces of the same base layer LI112. Δx_{cb} is the difference in the x-values of the cap layer and of the base layer (x=0.205). The QE ratio is the ratio of the quantum efficiency at -40 mV bias to the quantum efficiency at zero bias voltage.

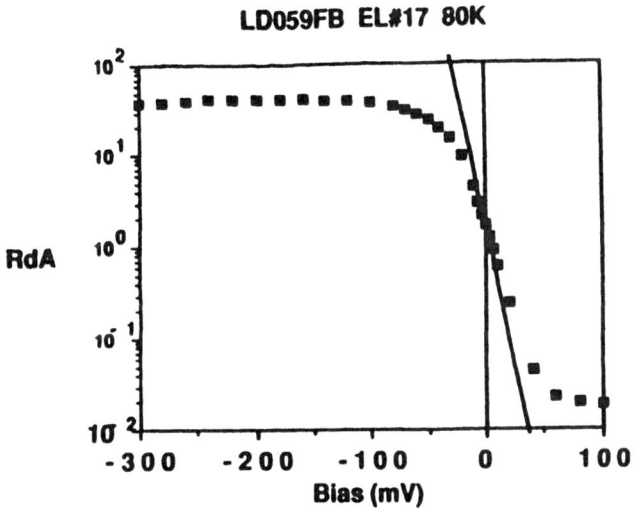

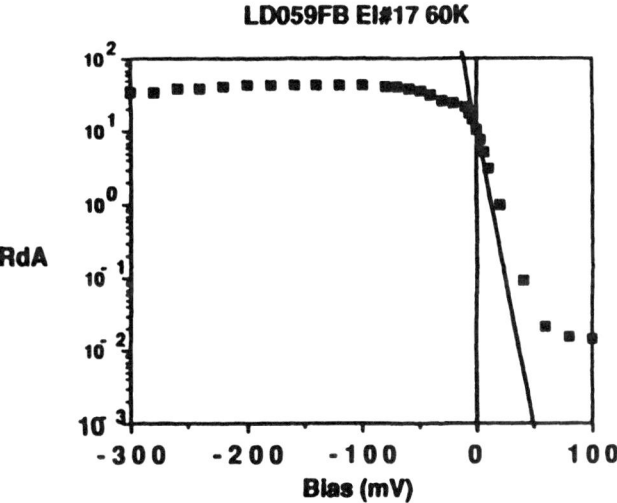

Figure 8. The 80K and 60K dynamic impedance-area product R_DA vs bias voltage for a film with an 80K cutoff of 14.1μm, an 80K R_0A product of 0.9 ohms-cm^2, and a 60K R_0A of 10 ohms-cm^2. The quantum efficiency was 50% (no A/R coating) and was independent of temperature.

satellite applications. The data show that the LPE films of areas up to 3x4 cm^2 have a uniformity in film thickness of ±10% for 15 to 20µm thick films, and a uniformity in cutoff wavelength of ±0.1µm for 11-12µm range and ±0.4µm for 18-19µm range. The base layers have Auger-1 lifetimes and the $(R_0A)_{Bulk}$ values are at the Auger-1 n-side diffusion limit. The dynamic impedance of these films shows diffusion limited behavior near zero bias and increases to a temperature and bias independent value at reverse biases of 50 to 100 mV. Good quantum efficiency is observed, 50-70% in the 10-15µm cutoff wavelength range and 20-40% in the 16-18µm range. The spectral response of photodiodes are of classical shape, with the quantum efficiency independent of wavelength and peak responsivity at about 90% of the cutoff wavelength.

ACKNOWLEDGEMENTS

This work was supported by Loral internal funds and by NASA JPL Contract 958606. The authors acknowledge the efforts of: M. Cody, L Firth and T. Dunning (LPE growth and characterization); B. White, B. Burnett, D. King (lifetime measurements and device testing); P. O'Dette (device fabrication); and N. Hartle, R. Briggs, K. Maschhoff, M. Krueger, J. Marciniec, P. Zimmermann (technical discussions).

REFERENCES

1. M. T. Chahine, "Sensor Requirements for Earth and Planetary Observations," Proc. of the Innovative Long Wavelength Infrared Detector Workshop, April 24-26, 1990 (NASA Jet Propulsion Laboratory Publication 90-22, July, 1990).

2. R. M. Broudy and V. J. Mazurczyk, "(Hg,Cd)Te Photoconductive Detectors," Chapter 5 in Semiconductors and Semimetals, Volume 18, Ed. by R. K. Willardson and A. C. Beer (Academic Press, New York, 1981).

3. M. B. Reine, A. K. Sood and T. J. Tredwell, "Photovoltaic Infrared Detectors," Chapter 6 in Semiconductors and Semimetals, Volume 18, Ed. by R. K. Willardson and A. C. Beer (Academic Press, New York, 1981).

4. G. N. Pultz, P. W. Norton, E. E. Krueger and M. B. Reine, 1990 MCT Workshop, to be published in J. Vac. Sci. Technol., 1991.

5. O. L. Doyle, J. A. Mroczkowski and J. F. Shanley, J. Vac. Sci. Technol. A3, 259 (1985).

6. M. Grimbergen and A. Szilagyi, Mat. Res. Soc. Symp. Proc. 69, 257 (1986).

7. H. R. Vydyanath and C. H. Hiner, J. Appl. Phys. 65, 3080 (1989).

8. R. J. Briggs, J. W. Marciniec, P. H. Zimmermann and A. K. Sood, IEEE International Electron Devices Meeting Technical Digest, p. 496 (1980).

9. H. K. Chung, M. A. Rosenberg and P. H. Zimmermann, J. Vac. Sci. Technol. A3, 189 (1985).

10. W. M. Higgins, G. N. Pultz, R. G. Roy, R. A. Lancaster and J. L. Schmit, J. Vac. Sci. Technol. A7, 271 (1989).

11. D. A. Nelson, Loral Infrared & Imaging Systems, unpublished work, 1981.

12. M. A. Kinch, M. J. Brau and A. Simmons, J. Appl. Phys. 44, 1649 (1973).

13. P. Capper, "Carrier Lifetimes in n-Type CdHgTe," Chapter 5.7

in _Properties of Mercury Cadmium Telluride_, edited by J. Brice and P. Capper (INSPEC, IEE, London, 1987).

14. J. S. Blakemore, _Semiconductor Statistics_ (Dover, New York, 1987), Equations 620.1 and 620.2.

BINARY OPTICS MICROLENS ARRAYS IN CdTe.

M.B. Stern*, W.F. Delaney*, M. Holz*, K.P. Kunz**, K.R. Maschhoff,** and J. Welsch**
*MIT Lincoln Laboratory, Lexington, MA 02173
**Loral Infrared and Imaging Systems, Lexington, MA 02173

ABSTRACT

Arrays of miniature focusing optics located at the focal plane can improve the performance of focal plane systems. By more completely collecting the light at the focal plane and concentrating it into a smaller spot size on the detector plane, the photodetector area can be substantially reduced. Increased gamma radiation hardening and noise reduction result from the decrease in photodetector surface area. Binary optics technology, a process for fabricating large arrays of diffractive optical elements, is especially attractive for infrared materials. In this paper, diffractive Fresnel microlens arrays containing over six thousand F/0.9 lenslets are patterned in the surface of CdTe substrates by successive photolithographic and Ar+ ion-beam-etching steps. Results on smaller arrays of *monolithically* integrated binary-optics lenslets with II-VI detectors, demonstrating enhanced photodetector responsivities, are presented for the first time.

INTRODUCTION

Binary optic microlens arrays have been used in both visible and infrared applications such as laser beam addition [1, 2], Hartmann wavefront sensors [3], and agile beam steering [4]. An especially attractive application is the integration of microlens arrays with infrared focal plane arrays (FPAs). The insertion of an optical concentrator between the objective lens and the focal plane array permits a reduction in the detector area to illuminated unit cell area ratio while maintaining the optical fill factor at 100%. Gamma induced noise, which limits sensor performance in high radiation environments, may decrease by a factor of 10 or more depending on the specific detector geometry. As thin-film HgCdTe has become the dominant material for high-performance 8-12 μm infrared photovoltaic FPAs, the ability to fabricate focusing optics in CdTe substrates acquires added importance.

Monolithically integrated CdTe binary optics structures with heterojunction HgCdTe grown epitaxially on CdTe or GaAs represent the next generation of FPAs. Besides decreasing the system size and weight by combining two elements on one substrate, the number of interfaces and the associated reflection and scattering is reduced. Furthermore, the reduction in detector area creates space for the future inclusion of on-focal-plane processing circuitry. This paper describes the design and fabrication of binary optics microlens arrays in CdTe. The first results, to our knowledge, on functional monolithically integrated optics and infrared FPAs are also reported.

BINARY OPTICS

Binary optics combines computer-generated design of diffractive optics with conventional microfabrication techniques to create phase relief structures in optical materials [5, 6]. For lenses, the surface profile is dictated by the F-number, design wavelength, and optical properties of the substrate. Binary optic elements can be readily designed to correct for aberrations such as spherical, chromatic, and astigmatic errors.

Conceptually, a diffractive microlens is constructed by subtracting integral numbers of wavelengths from the lens transmittance function which results in a phase function constrained between 0 and 2π (Figure 1). For ease of fabrication, the continuous phase profile is approximated by a multistep surface relief structure. The phase transmittance is quantized with phase step heights:

$$d = \frac{\lambda}{(n-1)N} \tag{1}$$

where λ is the incident wavelength in free space, n is the refractive index of the lens material, and N is the number of phase levels [7, 8]. In binary optics, by setting $N = 2^M$, the phase profile is fabricated from only M masks, analogous to the usual binary number coding. For an ideal holographic lens, the phase function consists of circular zones where each zone represents a phase change of $2m\pi/N$. Zone boundaries r_m are given by:

$$r_m \approx \sqrt{\left(\frac{m\lambda}{N}\right)\left(2f + \frac{m\lambda}{N}\right)} \; ; \tag{2}$$

where f is the lens focal length and m an integer. The total number of annuli is determined by setting $r_m = D/2$, where D is the aperture size (lens diameter). The distance between the last two zones, or critical dimension (CD), is $\approx 2\lambda(F/\#)/N$, where the lens speed, or F/#, is f/D.

The diffraction efficiency of a diffractive optical element gauges how much of the power of an incident plane wave is diffracted into the desired diffraction order. In the paraxial approximation, the first-order diffraction efficiency of a multilevel diffractive lens is given by $\eta = \mathrm{sinc}^2(1/N)$, where N is the number of etched phase levels [6, 7]. For 2 phase levels, η is 40.5%; for 16 phase levels, η is 99% [6, 8]. However, recent analyses by Swanson and Knowlden suggest that the scalar theory is no longer accurate when the design wavelength to period ratio is less than five, or F/4 lenses when n=4, for example[9].

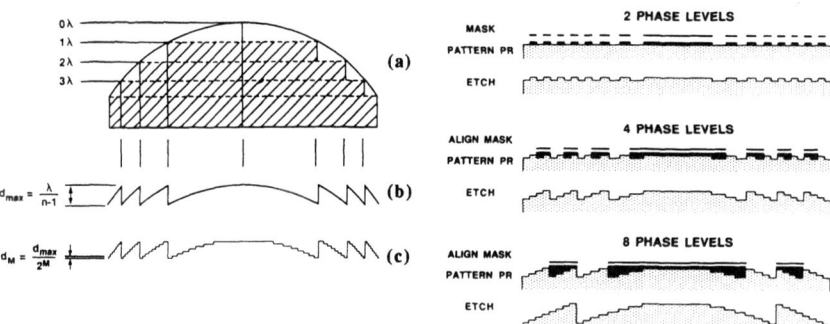

Figure 1: Comparison of refractive (a) and diffractive lenses (b, c). The diffractive surface (b) approximated by a multilevel lens in a stepwise manner (c).

Figure 2: Illustration of the fabrication of an 8-level surface relief microlens.

FABRICATION

Diffractive microlens arrays are fabricated by a series of photolithographic and etching techniques similar to those used in the manufacture of ICs. High quality aspheric lenses of arbitrary size and shape can be fabricated with this method. Large arrays of identical lenses can be produced as easily as single lenses, provided that uniformity of microlens feature sizes and etched depths is maintained across the entire array area.

The creation of binary optical elements proceeds in three stages - mask design; pattern transfer into photoresist; and high-fidelity pattern replication into the substrate (Figure 2). The desired phase profile must be specified in a form suitable for electron-beam mask generation. A single microlens pattern may be generated by a FORTRAN subroutine linked to pattern generation software; this pattern is fractured and repeated to create an array of microlenses on the Cr hardmask. As discussed above, the minimum zone width depends on λ and the F/#. Currently, annular CDs as small as 0.5 μm can be generated by photomask houses [10], which limits the highest speed lenslets realizable at short wavelengths. The CdTe microlens array shown in Figure 3, contains over six thousand F/0.9 lenslets with a 55 μm diameter. For the design wavelength of 9.0μm, each lenslet contains 14 annuli on mask layer 3; the minimum zone width is approximately 2 μm.

The Cr hardmasks are replicated by vacuum-contact photolithography into a thin layer of positive photoresist on the CdTe substrate. Zone widths on each mask layer must be accurately reproduced; a difficult task when contiguous feature sizes may vary from 0.5 μm to 20 μm across a single lenslet. The mask with the smallest features is replicated first, as illustrated in Figure 2, although the reverse order may also be used. The CdTe is then etched to the appropriate depth to produce the π/N phase shift. These processing steps are repeated until the desired number of phase levels is achieved. For an eight phase level microlens array, two separate photomask alignments are required. Alignment requirements are more stringent than those of standard VLSI processing, where tolerances of 0.25 μm are acceptable.

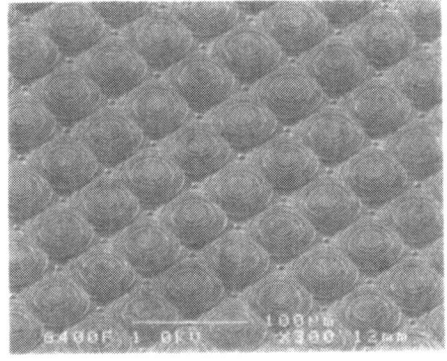

Figure 3: An SEM of a CdTe focal plane microlens array of 55 μm diameter, F/0.9 optics.

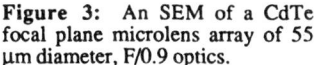

The 2 cm x 2 cm x 1 mm CdTe substrates used in these experiments were polished on both sides and had a <111> orientation. For handling purposes, the CdTe substrates were mounted on 2" carrier wafers. Substrates were etched in a Millatron ion miller with a Commonwealth Scientific Ar$^+$-ion source. Samples were mounted on a rotating planetary stage. Etch rates and sidewall anisotropy varied with the ion current, ion beam accelerating voltage, and the angle of incidence of the ion beam with respect to the substrate surface. Milling parameters were adjusted to minimize faceting and to optimize the CdTe etching rate and selectivity. Typical CdTe milling rates were between 100-200 nm/min. Selectivity between the photoresist mask and the CdTe substrate ranged from 3:1 to 10:1. As the index of refraction of CdTe at 9 μm is 2.674 [11], the necessary etch depths were 0.67 μm, 1.34 μm, and 2.67 μm, resulting in a cumulative etched depth of $\approx$ 6 μm for an 8-phase-level element. Fluctuations in Ar$^+$-ion-beam milling parameters can cause errors in phase step heights. In general, the etching depths are controlled to better than 5% of the design height. Phase step heights are measured with a stylus profilometer. For the CdTe array shown in Figure 3, the actual etched depths are 0.7 μm, 1.3 μm, and 2.6 μm.

Fabrication-related errors due to dimensional changes (zone width), misalignment (zone position), and etch depths (phase step height) all reduce the efficiency by decreasing the intensity into the desired diffraction order and blurring the focal spot.[6,12,13]. Phase step heights can be reasonably well controlled. Misalignment results in asymmetries of the concentric phase relief structures. Registration errors were evaluated by optical microscopy and SEM. The maximum misalignment between mask 3 to 2 was 0.35 μm; between mask 2 to 1, 0.55 μm, for the lens shown in Figure 3. Duty cycle errors, stemming from photolithography (exposure/development) and etching (undercutting), change the zone width and create either "ridges" or "trenches" in the etched concentric surface relief structure. Variations in resist thickness over high aspect ratios also compromise linewidth fidelity. To date, devices have been fabricated with a single layer of photoresist.

FOCAL PLANE INTEGRATION

A number of benefits accrue from placing an array of microlenses in front of a focal plane detector array, with each lenslet registered to a detector pixel element. Light that is collected over the acceptance cone of the lenslet entrance aperture is compressed into a smaller exit aperture matched to the active detector area. By reducing the detector area and increasing the optical fill factor (ratio of detector area to collection area), susceptibility to radiation-induced noise is decreased because detector noise scales with detector active area. Furthermore, the use of focusing microoptics avoids sampling artifacts related to detector edges and dead spaces.

Fast telescope objectives drive the design towards low-F/# lenslet arrays. For the CdTe lens described above, the numerical aperture was chosen to optimize the amount of light coupled into a detector element . The resulting F/0.9 lens (in air) had a focal length of 50 μm. The pitch of the lenslets in the microlens array, determined by the spacing of the photodiodes in the focal plane array, was 55 μm for the CdTe array shown in Figure 3.

Monolithic Integration

Dual-sided integration of a microlens array and an FPA on a single substrate reduces the number of interfaces and the associated scattering and reflection losses from high index materials, when compared to hybrid schemes. Optimal performance of an integrated FPA system requires accurate registration between the FPA and microlens array. Otherwise, vignetting errors will result in light loss through the system. For monolithic integration, such positioning errors can be intrinsic to the fabrication process. These positioning errors divide into three categories: translation, misfocus, and tilt. To avoid misfocusing errors, precise thinning of the substrate is required to match the focal length of the lens in the medium; a wedged substrate will result in tilt and focus errors. Careful protection of the sensitive electronic material or circuitry, depending on the fabrication sequence, is also required for monolithic integration.

Figure 4: An SEM of a monolithically integrated array of F/4 lenslets and HgCdTe photodetectors.

As an initial step in an internal program at Loral Infrared and Imaging Systems to fabricate large monolithic integrated IR focal plane arrays, 8x8 mosaics of 32x32-μm square HgCdTe mesa photodiodes on 100-μm centers have been integrated with complementary arrays of 100x100-μm square microlenses on opposite sides of a single CdTe substrate (Figure 4).

These microlenses are designed to be one component in an F/7 optical system [14] that is intended to focus radiation in the 9-12 μm band onto the co-registered mesa detectors. A two-layer LPE HgCdTe film is epitaxially grown on one side of a CdTe substrate. The HgCdTe double layer is composed of a wide-bandgap p-layer grown on a narrow-bandgap n-layer. Individual diodes are formed via wet chemical etching. The photodiodes are passivated with a layer of ZnS. A detailed description has been presented elsewhere [15,16]. A crossection of this photodiode is shown in Figure 5. The microlens arrays are fabricated on the other side of the CdTe substrate using methods similar to those described above. The microlens arrays are registered to the photodetectors using a Karl Suss IR backside aligner. Backside registration accuracy, limited by vertical travel of the focusing microscope to 1-2 μm, could be improved by refining the instrumentation. In order to compare the system efficiency, photodiodes with and without microlenses have been fabricated on the same wafer. Photodiode interconnections to a sapphire leadout board are made via indium bump interconnects.

Measurements

Integrated FPAs were evaluated from 2-14 μm. The IR source was a calibrated 500K blackbody source with an 2.5-mm aperture, chopped at 1 kHz. The HgCdTe test array with microlenses was mounted in a test dewar and cooled to 30 K to eliminate the effects of lateral collection. No antireflection coating was used for this test array. Photodiodes with microlenses showed an approximate two- to three-fold increase in detected photocurrent at 30K when compared with nearby detectors without microlenses (Figure 6). The relative spectral response for 'lensed' and 'unlensed' detectors were similar over the 8-14 μm band, indicating no significant chromatic aberrations or diffractive losses which would impact quantum efficiencies in this range. Spectral response data and blackbody responsivity data were used together to determine peak quantum efficiencies for the 32x32 μm detectors with and without a microlens. Elements without lenses showed peak responsivities and quantum efficiencies of 5.1 A/Watt and 60 %, respectively, typical of LWIR HgCdTe photodiodes. The elements with microlenses had peak responsivities and effective quantum efficiencies as high as 14.6 A/W and 170%, respectively. While the signal enhancement was less than that predicted by the scalar diffraction theory, vignetting errors, nonidealities in microlens fabrication, the absence of an AR coating, and the low operating temperature all contribute to lowered efficiency.

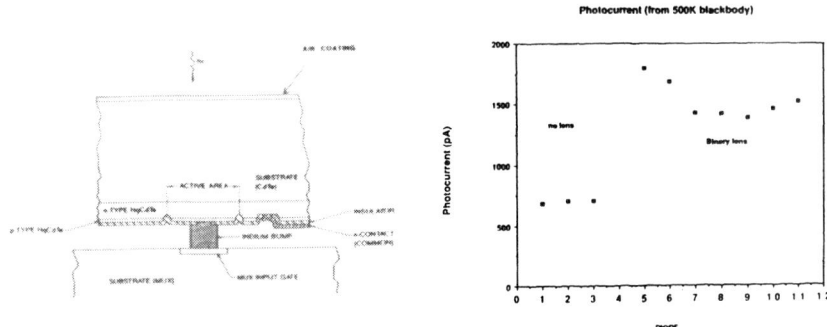

Figure 5: Crossection of a backside illuminated HgCdTe mesa photodetector.

Figure 6: Comparison of the responsivity of photodetectors with and without microlenses.

SUMMARY

Arrays of binary optic microlenses containing six thousand F/0.9, 55 μm diameter binary optic microlenses have been fabricated in CdTe. Dual sided integration of arrays of microlenses and HgCdTe detectors has been demonstrated for the first time. Microlenses increased the detector quantum efficiency by a factor of three.

ACKNOWLEDGEMENTS

This work was supported by DARPA.

REFERENCES

1. J.R. Leger, M. Holz, G.J. Swanson and W. Veldkamp, *Lincoln Lab. J.* **1**, 225-246 (1988).

2. J.R. Leger, M.L. Scott, P. Bundman, and M.P. Griswold, *SPIE Proc.* **884**, 82-89 (1988).

3. G. Swanson, R. Knowlden, and W. Veldkamp, unpublished.

4. W. Goltos and M. Holz, *SPIE Proc.* **1052**, 131-141, 1989.

5. W.B. Veldkamp and G.J. Swanson, *SPIE Proc.* **437**,54-59 (1983).

6. G.J. Swanson, *Tech. Rep.* **854**, MIT Lincoln Laboratory, Aug. 1989.

7. Damman, H., *Optik* **31.**, 95-104 (1970); *Optik* **53**, 409-417 (1979).

8. L. d'Auria, J.P. Huignard, A.M. Roy, and E. Spitz, *Opt. Comm.* **5**, 232-235 (1972).

9. G. Swanson and R. Knowlden, presented at the 1990 OSA Fall Meeting, Boston, MA, 1990 (unpublished).

10. **Micro Mask, Inc.**, Sunnyvale, Cal.

11. J.C. Brice, **Properties of Mercury Cadmium Telluride**, edited by John Brice and Peter Capper, (INSPEC, The Institution of Electrical Engineers, Publisher, New York, 1987).

12. J.A. Cox, T. Werner, J. Lee, S. Nelson, B. Fritz, and J. Bergstrom, *SPIE Proc.* **1211**, 116-124 (1990).

13. M.W. Farn and J.W. Goodman, *SPIE Proc.* **1211**, 125-136 (1990).

14. **Telic Optics, Inc.**, Marlborough, MA.

15. E.E. Krueger, G.N. Pultz, P.W. Norton, J.A. Mroczkowski, M.H. Weiler, and M.B. Reine, *Proc. MRS Symp.* **216**, (1991).

16. G.N. Pultz, P.W. Norton, E.E. Krueger, and M.B. Reine, *J. Vac. Sci. Tech.* (to be published (1991).

HgCdTe/CdTe MULTIPLE QUANTUM WELLS: GROWTH, STABILITY, AND OPTICAL PROPERTIES

R.D. FELDMAN, R.F. AUSTIN, C.L. CESAR,[*] M.N. ISLAM, C.E. SOCCOLICH, Y. KIM, AND A. OURMAZD
AT&T Bell Laboratories, Holmdel, NJ 07733
[*]Present address: University of Campinas, Campinas, Brazil

ABSTRACT

We have grown HgCdTe/CdTe multiple quantum wells by molecular beam epitaxy which show room temperature photoluminescence and sharp absorption steps at mid-infrared wavelengths. Quantitative chemical mapping, performed by transmission electron microscopy, indicates minimal interdiffusion during growth. Annealing experiments performed at higher temperatures show that the interdiffusion coefficient is a strong function of the depth of the interface below the surface. Absorption spectra have been accurately modeled with a square well/envelope function approach. The films have been used to passively mode lock color center lasers and produce pulses as short as 120 fsec near 2.7 μm.

INTRODUCTION

$Hg_{1-x}Cd_xTe$ is a nearly lattice matched material system with band gaps covering the infrared spectrum. In principle, this system should offer excellent possibilities for design of advanced multilayer structures as has been done at shorter wavelengths with the AlGaAs system. In practice, heterostructure and quantum well applications of HgCdTe have been limited by difficulties with growth and interdiffusion.

The development of low temperature growth by molecular beam epitaxy (MBE) [1] has led to a capability for producing multilayered structures with Hg-containing compounds. The most commonly studied structures have been thin barrier superlattices (SLs) with HgTe wells and nominal CdTe barriers [2]. For practical layer thicknesses, such structures are limited to wavelengths of 4 μm and beyond [3]. In this paper, we review our work on $Hg_{1-x}Cd_xTe/CdTe$ multiple quantum wells (MQWs). We have used $Hg_{1-x}Cd_xTe$ wells in order to gain access to shorter infrared wavelengths. The films are grown with thick barriers, $\approx$100 Å, in order to achieve wave functions that are truly confined in the wells. We have examined a series of MQW samples with well compositions x=0.2-0.3, and fundamental transitions ranging from 2.5-3.5 μm. The paper begins with a presentation of issues related to growth and interdiffusion. There follows a section devoted to optical studies, including linear spectroscopy and the application of HgCdTe/CdTe MQWs to subpicosecond pulse generation. Finally, there is a concluding section which includes a discussion of future prospects for these MQWs.

MATERIALS ISSUES

The MQW films were deposited in a Riber MBE system equipped for reflection high energy electron diffraction (RHEED), which was operated at 10 keV. All films were grown on GaAs substrates. A typical growth sequence begins with ≈0.3 μm of ZnTe, followed by 2 μm of CdTe, both grown at a substrate temperature (T_s) near 300 °C. T_s is then lowered to 175 °C for growth of the MQW. The $Hg_{1-x}Cd_xTe$ wells are grown using three sources, Hg, Te, and CdTe. During growth of the barrier layer, the Te cell shutter is closed. Because of the very high Hg flux, the Hg cell shutter had to remain open at all times. Under these conditions, some Hg is incorporated into the barrier layer [4]. We have measured a composition of ≈$Hg_{0.15}Cd_{0.85}Te$, but, following the convention in the literature, we will loosely refer to this layer as CdTe.

It is possible to grow either (111) or (100) CdTe buffer layers on (100) GaAs substrates. Thus a decision is required as to which orientation to use for quantum wells. For reasons to be presented below, we have chosen the (100) orientation.

Several authors have noted that (100) HgTe and HgCdTe tend to have superior Hall mobilities when grown by MBE or organometallic vapor phase epitaxy [5-8]. In our own studies [8], we showed that (100) MBE films of HgTe, grown on CdTe/ZnTe buffer layers on GaAs substrates, have room temperature Hall mobilities (μ_H) of 27,000 to 31,000 cm^2/V-s, which is similar to that of bulk HgTe samples that have been annealed for forty days in a Hg ambient [9]. HgTe films which are (111)-oriented have μ_H a factor of 2 lower. The lower μ_H of the (111) films has been attributed to twinning on the (111) growth plane [6,10], which is known to occur in (111) films when the Hg flux (F_{Hg}) is too high or T_s is too low. RHEED will show twin spots during MBE growth if there is a prolonged period in which F_{Hg} is too high. Under such conditions, the film surface roughens, and a spot pattern develops. For (111) growth, the window of conditions that gives acceptable RHEED patterns is quite limited; if T_s is too high or F_{Hg} too low, the film will be polycrystalline [1]. If F_{Hg} and T_s are well-controlled, however, the RHEED pattern will be streaked throughout the entire run, so that any information about twins or other defects will be obscured. Even under such conditions, the electrical properties of the films are relatively poor.

When (111) HgTe films are deposited under conditions that give streaked RHEED patterns, we find that twins are still present, but that they form only during nucleation [11]. A plan view transmission electron microscope (TEM) bright field image of one such film is shown in Fig. 1a. There is a cellular structure on a scale of ≈1 μm, with clusters of cells appearing either light or dark under the tilt conditions used. The contrast between the two regions occurs because they are twin-related, as has been verified by taking diffraction patterns under the proper tilt conditions. The twinning is about the (111) growth axis. The structure is suggestive of island nucleation, with the islands being (randomly) in either twin or matrix orientation. Dislocations are found where islands of the same orientation meet. We have shown elsewhere that this structure develops as the HgTe film nucleates, and it persists as the film grows [11]. That is, under conditions that yield streaky RHEED patterns, new twins are not generated after the nucleation stage, and they also do not grow out. Under conditions that yield spotty RHEED patterns, new twins are constantly being regenerated [12].

The structure shown in Fig. 1a appears to be induced by a buildup of Hg on the surface of the CdTe buffer layer prior to

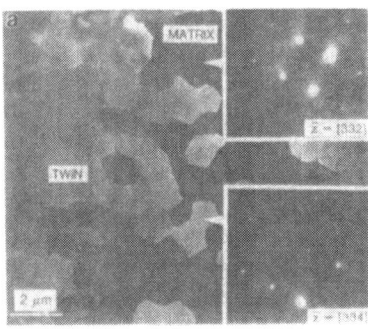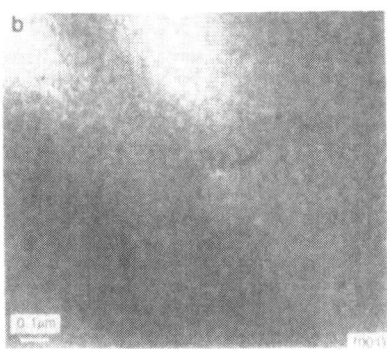

Figure 1. (a) Bright field TEM image taken from a (111) HgTe film grown under conditions that give a streaked RHEED pattern. The diffraction patterns demonstrate that the adjacent light and dark regions are twin-related. (b) Plan view, bright field image taken from a (100) HgTe sample.

the start of the HgTe deposition. The buildup has been detected both by Auger analysis and by RHEED [12]. It occurs during the time between the end of the CdTe deposition and the beginning of HgTe growth. This time interval is required because of the 100 °C difference in T_s of the two layers. In effect, the nucleation occurs under conditions of excess Hg, and causes the twins.

The defect structure seen in (111) HgTe has no analogue in (100) HgTe. Fig. 1b contains a plan view taken from the middle of a 3 μm thick (100) film. Some dislocations are present, but few of them propagate through the film.

The finding that nucleation conditions present a limitation in (111) growth suggests that there might be some set of nucleation conditions that lead to improved film structure. Some workers are now addressing this problem, with apparently good success [13]. However, the (111) orientation remains inappropriate for MQW growth. The very narrow range of conditions of F_{Hg} and T_s that will yield a good (111) $Hg_{1-x}Cd_xTe$ well will not overlap the narrow window that is optimum for the CdTe barrier. The fact that the Hg flux cannot be shuttered or modulated means that the CdTe barrier will necessarily be grown with excess F_{Hg}. With the recent development of a Hg cell with a fast flux adjustment capability [14], it may soon be possible to grow both (111) barriers and wells with optimized F_{Hg}. However, the (100) orientation is much more forgiving of the growth conditions, and we have used it exclusively in our work.

Once the choice of orientation is made, there remains a question as to how well HgCdTe/CdTe quantum wells can be grown. There has been a great deal of experience in growth of HgTe/CdTe superlattices, with typical barrier thicknesses of only 20-60 Å. True quantum wells in the mid-IR, in which the wave functions are strongly confined to the well region, require thicker barriers. The main difficulty here is that HgCdTe must be grown at T_s < 200 °C because of the low Hg sticking coefficient. CdTe is normally deposited at 280-300 °C. In order to optimize the

MQW films, we have varied T_s, growth rates, and substrate misorientation from (100).

We have found that the most critical factors are T_s and growth rate. Misorienting the (100) GaAs substrate by 2° toward <110> had little effect on the film quality, while a 2° misorientation toward <111>A yielded poor morphology. T_s in the range 170-180 °C yielded the best optical properties. Growth rates were kept low (≈ 0.6 Å/s for the barrier layers) in order to achieve a relatively smooth growth front for the CdTe layers at these low temperatures. Once the CdTe growth rate is determined, so is the HgCdTe growth rate for a given composition, since the same CdTe cell is used for both the well and barrier layers. In practice, $Hg_{1-x}Cd_xTe$ wells with compositions x= 0.2-0.3 have been grown at rates near 2 Å/s.

The slow growth rates used for these films leads to a typical growth time of about three hours for a fifty period MQW. This is a long enough time to lead to concern about interdiffusion. We used high resolution transmission electron microscopy (HRTEM) on a sample that was prepared in the <100> projection, in a mode that has been named chemical lattice imaging [15]. The sample had 50 periods with 45 Å $Hg_{0.73}Cd_{0.27}Te$ wells and 100 Å barriers. Typical HRTEM cross sections are examined in the <110> projection. It has been shown, however, that the strong contribution of chemically sensitive (200) reflections to lattice images taken in a <100> projection make this projection the superior one for observing layer intermixing and interfacial roughness [15].

The chemical lattice images show little sign of interdiffusion during growth, although the interfaces are rougher, by approximately a factor of two, than interfaces in high quality III-V quantum wells [16]. This roughness appears to have little influence on the optical properties that will be discussed below. The fluctuations measured by HRTEM are on the scale of the unit cell, while the optical studies are sensitive to sizes on the order of the Bohr radius (about 400 Å). Averaged over this larger scale, the sample appears to be quite homogeneous.

Further tests of the stability of the MQW were made by annealing portions of the same film for 30 minutes at temperatures ranging between 200 and 265 °C. The samples were sandwiched between CdTe wafers and annealed in an Ar atmosphere. The samples were then thinned and observed in the TEM.

For all anneals, interdiffusion was more rapid near the film surface than near the substrate. This effect can be observed by counting lattice fringes to measure the well width [17], as shown in Fig. 2. As grown, the well widths were constant throughout the film. After annealing, the wells were broader. For example, after a 1/2 hour anneal at 265 °C, the first-grown wells broadened by four monolayers, compared with seven monolayers for the last-grown.

These results can be made quantitative by using a digital pattern recognition technique [18] to deduce the actual composition profiles, and then using these profiles to calculate interdiffusion coefficients [19]. The composition profiles across an as-grown and an annealed interface are shown in Fig. 3. Attempts to fit the measured profiles using linear diffusion theory (i.e., composition independent diffusion coefficients) were unsatisfactory, both because the fits are poor (Fig. 3b) and because the resulting activation energy varies strongly as a function of well depth, which seems physically unreasonable.

A much better fit is achieved with nonlinear diffusion

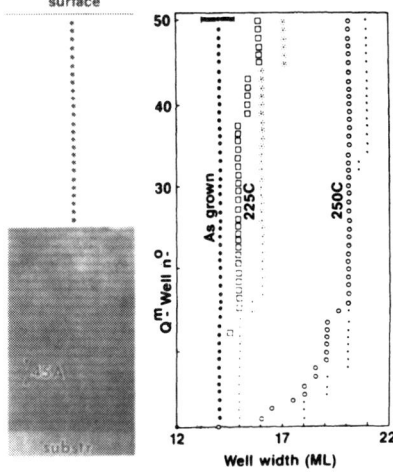

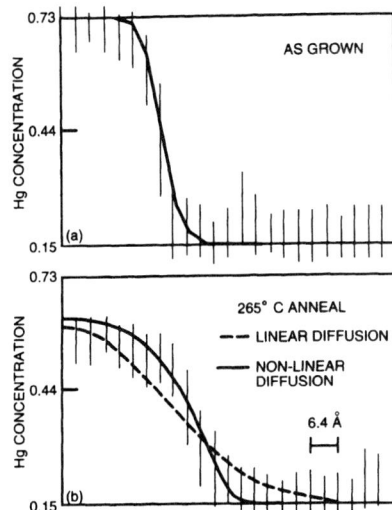

Figure 2. Low magnification image of the HgCdTe/CdTe quantum wells. The plot shows the well width as a function of the well position after 30 minute anneals at 200, 225, 250 and 265 °C.

Figure 3. Hg concentration profile across a single interface. (a) As grown. (b) After a 30 minute anneal at 265 °C. The dashed and solid lines show fits determined from linear and nonlinear diffusion theory, respectively.

theory, as shown in Fig. 3b. The resulting diffusion coefficient is written in the form $D(x_{Cd}, T) = D_z(T) \exp(\alpha/kT)$. With this expression, the activation energy is independent of depth. The exponential factor α depends linearly on concentration, and the depth dependence is contained entirely in the prefactor D_z. The variation of the diffusion coefficient with depth presumably arises because the concentration of point defects that are involved in the interdiffusion process is depth-dependent.

These interdiffusion results indicate that the location of interfaces will be an important consideration in the design of stable heterostructure devices containing Hg compounds. The experiments described here also demonstrate the necessity of examining individual interfaces rather than using averaging techniques, such as measurements of x-ray satellite peaks, to study interdiffusion at low temperatures. The strong depth-dependence of the interdiffusion rate has now also been seen in GaAs/AlGaAs MQWs [20].

OPTICAL PROPERTIES

A series of 50 period MQW films was studied by photoluminescence and absorption measurements. All of the MQWs show room temperature photoluminescence. The width of the low temperature photoluminescence peaks and the shift between the absorption and photoluminescence peak are in all cases well within the range expected from monolayer fluctuations in the well width. Absorption measurements, taken at Brewster's angle to minimize Fabry-Perot fringes, are shown as a function of temperature in Fig. 4a for a film with a well composition and thickness of x=0.24 and 50 Å, respectively. Well-defined n=1 heavy and light hole transitions are observed at all temperatures. The heavy and light hole designations were confirmed by polarization measurements [21].

We can accurately model the observed energy levels by using the envelope function approach [22] and assuming isolated square wells. The fit is shown by the solid lines in Fig. 4a. We used standard literature values for the band gaps and effective masses, as described previously [23]. The HgTe/CdTe valence band offset (ΔE_v) had been a matter of controversy at the time that we undertook this work. There was general agreement that the room temperature value was 350-400 meV. At low temperatures, magneto-optical studies and resonant tunneling data were used to support estimates of ΔE_v=0-40 meV [24,25], while transport measurements on HgTe/CdTe superlattices with near zero band gap suggested ΔE_v was at least 200 meV [2]. Our data contradict the suggestions of small ΔE_v at low temperatures. The separation between light and heavy hole transitions requires a value of ΔE_v of 350 - 400 meV at all temperatures [23]. An ultraviolet photoelectron spectroscopy measurement, performed at 50 K, gives a similar value ΔE_v=350 meV [26].

Quantitative information is extracted from the spectra by writing the absorption coefficient for each transition as being proportional to the product of the oscillator strength, the Coulomb enhancement (Sommerfeld factor), and the two dimensional density of states, phenomenologically broadened by homogeneous and inhomogeneous contributions [27]. This fit omits an excitonic contribution. Because of the small effective masses and large dielectric constant, the exciton has only ≈3 meV binding energy, and is therefore not expected to be observable. The fitting procedure, which is described in detail in Ref. [21], leads to excellent agreement between theoretical and experimental spectra, as shown in Fig. 4b.

The quantitative results of these calculations can be summarized as follows: (1) The predicted ratio of the heavy-to-light-hole absorption steps is 2.1:1, in excellent agreement with the measured value of 2.3:1. (2) Fig. 4 shows only n=1 transitions, but this sample has n=2 transitions as well. Another film, with 100 Å wells, has transitions up to n=3. For both films, the energy of each transition can be calculated to within 5 meV by using the isolated square well/envelope function model, provided that band nonparabolicity - which is important for the higher order transitions - is taken into account. (3) The low temperature linewidth yields an inhomogeneous broadening of 6 meV, comparable to the photoluminescence peak width that is seen in bulk alloys of the same composition. In view of the large confinement energy in these MQWs, we take the lack of additional broadening due to well width fluctuations as a sign of very good optical quality. (4) We expect that the thermal broadening is dominated by LO phonon collisions. The homogeneous broadening is normally fitted to the expression written as $\Gamma_{Hom}=\Gamma_{ph}N_{LO}$, where the population of LO phonons is given by $N_{LO}=[\exp(\hbar\Omega_{LO}/kT)-1]^{-1}$ [27]. With this expression, we find Γ_{ph} = 4.6 meV. Values of Γ_{ph}=5.5 and 7 meV have been reported for

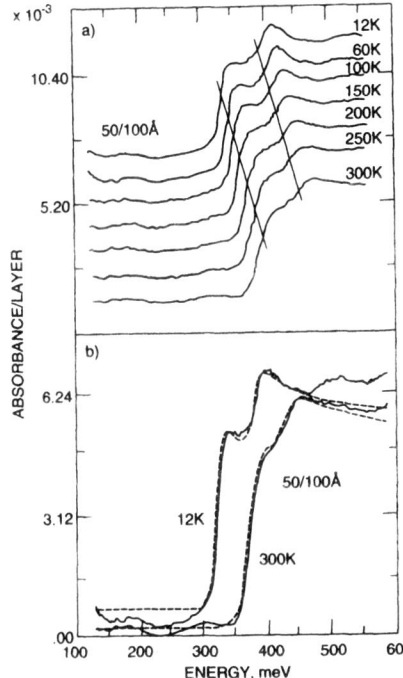

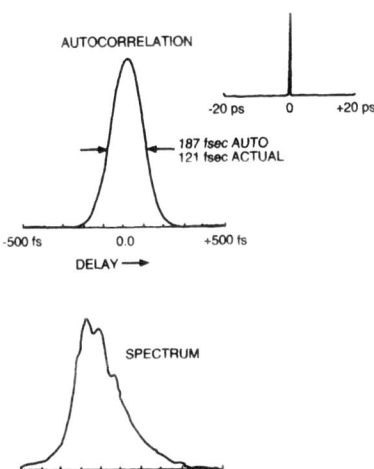

Figure 5. KCl:Li pulse autocorrelation (top) and spectrum (bottom).

Figure 4. (a) Absorption spectra from 12 K to room temperature for a 50 period multiple quantum well sample. (b) Comparison of theoretical and experimental spectra at two temperatures, using fitting parameters from Ref. 21.

excitons in GaAs/AlGaAs and InGaAs/InAlAs quantum wells, respectively [27,28].

The values of Γ_{ph} and Ω_{LO} predict a room temperature LO phonon collision time of about 150 fsec (150x10^{-15} sec) for HgCdTe/CdTe MQWs [21]. The comparable figures for GaAs and InGaAs quantum wells are 400 and 260 fsec, respectively [27,28]. The thermalization time is shorter in HgCdTe because the smaller LO phonon energy leads to a larger LO phonon population at room temperature.

We have taken advantage of the short thermalization time by using HgCdTe/CdTe MQWs as saturable absorbers to passively mode lock RbCl:Li and KCL:Li color center lasers (CCLs). The resulting pulses, an example of which is shown in Fig. 5, are as short as 120 fsec with a peak power of 470 W near 2.7 μm [29]. These are the first direct laser sources of subpicosecond pulses in the mid-infrared region, and, to our knowledge, are the shortest pulses at these wavelengths. This passive mode locking technique was previously demonstrated near 1.6 μm using a NaCl CCL and an InGaAs/InAlAs MQW to produce 200 fsec pulses [30].

In the mode locking experiment, a MQW is placed inside a laser cavity with the color center gain crystal. The total laser cavity has net loss when the MQW is absorbing, so no lasing is observed. Near the peak of the incident light pulse, however, the states that are available in the MQW to absorb the light become filled, and the MQW becomes transparent. During this period of transparency, the lasing cavity has net gain, and light is emitted. The mechanism for recovery of the absorption is thought to be the thermalization of the carriers in the MQW due to collisions with LO phonons. In support of this mechanism, we note that the observed pulse widths, both for HgCdTe and for InGaAs MQWs, are close to the LO phonon scattering times for each material. The fact that this time is shorter in HgCdTe than in InGaAs suggests that HgCdTe may be a good candidate for use at 1.6 μm as well.

CONCLUSION

We have demonstrated a multiple quantum well capability in the $Hg_{1-x}Cd_xTe$ system with a series of samples whose fundamental transitions occur between 2.5 and 3.5 μm. As a first step toward achieving this capability, we examined HgTe films deposited in two standard orientations, (111) and (100). The latter proved to be superior. The (111) orientation is difficult to nucleate without twins. In addition, growth of (111) quantum wells requires that conditions of growth rate, substrate temperature, and Hg flux be correct for both well and barrier layers. The window of conditions that yields single crystal, twin-free growth is so narrow for (111) films that growing within the optimum window for both layers may not be possible.

Optical studies show that (100) MQWs have inhomogeneous linewidths that are comparable to the alloy broadening in bulk material, which implies that the samples are very uniform in the planes and through the layers. Chemical lattice imaging confirm the uniformity of the layers, although there is interface roughness on a scale that is too fine to affect the optical properties. Thus interdiffusion is not a problem during growth. Above the growth temperature, interdiffusion was found to be highly nonlinear and strongly depth dependent.

We showed that a complete description of the MQW absorption data, including energy levels, line shapes, linewidths, and ratios of the oscillator strengths, is obtained by using the envelope function approach that has previously been applied to III-V quantum wells. From this analysis came a prediction of a 150 fsec thermalization time due to LO phonon collisions for carriers in HgCdTe at room temperature. We made use of this very fast thermalization time by developing a passively mode locked color center laser source with a HgCdTe saturable absorber operating near 2.8 μm.

This fast source may be useful in studies of optical nonlinearities at mid-infrared wavelengths. The HgCdTe/CdTe MQWs would themselves be good candidates for study, as the small effective masses and large band nonparabolicity in the well materials should enhance the nonlinearities.

Finally, extension of the HgCdTe/CdTe MQW technology to

shorter wavelengths, including 1.3-1.6 μm, would open up some new opportunities. Among these would be development of passively mode locked lasers that could yield shorter pulses than have been achieved with III-V saturable absorbers. It might also prove interesting to examine the effects of the large spin-orbit splitting on optical nonlinearities, since it comes into resonance with the quantum well transitions near 1.5 μm.

REFERENCES

1. J.P. Faurie, J. Reno, S. Sivananthan, I.K. Sou, X. Chu, M. Boukerche, and P.J. Wijewarnasuriya, J. Crystal Growth 79, 940 (1986).

2. J.R. Meyer, C.A. Hoffman, and F.J. Bartoli, in Narrow Gap Semiconductors and Related Materials, edited by D.G. Seiler and C.L. Littler (Adam Hilger, Bristol, England, 1990), pp. S90-S99.

3. J.N. Schulman and T.C. McGill, Appl. Phys. Lett. 34, 663 (1979).

4. J. Reno, R. Sporken, Y.J. Kim, C. Hsu, and J.P. Faurie, Appl. Phys. Lett. 51, 1545 (1987).

5. J.Y. Cheung, G. Nizawa, J. Moyle, N.P. Ong, B.M. Paine, and T. Vreeland, Jr., J. Vac. Sci. Technol. A4, 2086 (1986).

6. J.E. Hails, G.J. Russell, A.W. Brinkman, and J. Woods, J. Crystal Growth 79, 940 (1986).

7. J. Arias, S.H. Shin, J.T. Cheung, J.-S. Chen, S. Sivananthan, J. Reno, and J.P. Faurie, J. Vac. Sci. Technol. A5, 3133 (1987).

8. R.D. Feldman, M. Oron, R.F. Austin, and R.L. Opila, J. Appl. Phys. 63, 2872 (1988).

9. T. Okazaki and K. Shogenji, J. Phys. Chem. Solids 36, 439 (1975).

10. R.D. Hornig and J.-L. Staudenmann, Appl. Phys. Lett. 49, 1590 (1986).

11. R.D. Feldman, S. Nakahara, R.F. Austin, T. Boone, R.L. Opila, and A.S. Wynn, Appl. Phys. Lett. 51, 1239 (1987).

12. R.D. Feldman, S. Nakahara, R.L. Opila, R.F. Austin, and T. Boone, J. Crystal Growth 98, 581 (1989).

13. K.A. Harris, T.H. Myers, R.W. Yanka, L.M. Mohnkern, R.W. Green, and N. Otsuka, J. Vac. Sci. Technol. A8, 1013 (1990).

14. R.G. Benz II, B.K. Wagner, and C.J. Summers, J. Vac. Sci. Technol. A8, 1020 (1990).

15. A. Ourmazd, W.T. Tsang, J.A. Rentschler, and D.W. Taylor, Appl. Phys. Lett. 50, 1417 (1987).

16. R.D. Feldman, C.L. Cesar, M.N. Islam, R.F. Austin, A.E. DiGiovanni, J. Shah, R. Spitzer, and J. Orenstein, J. Vac. Sci.

Technol. $\underline{A7}$, 431 (1989).

17. Y. Kim, A. Ourmazd, R.D. Feldman, J.A. Rentschler, D.W. Taylor, and R.F. Austin, in **Advances in Materials, Processing and Devices in III-V Compound Semiconductors**, edited by D.K. Sadana, L.E. Eastman, and R. Dupuis (Mater. Res. Soc. Proc. $\underline{144}$, Pittsburgh, PA 1986) pp. 163-168.

18. A. Ourmazd, D.W. Taylor, J. Cunningham, and C.W. Tu, Phys. Rev. Lett. $\underline{62}$, 933 (1989).

19. Y. Kim, A. Ourmazd, M. Bode, and R.D. Feldman, Phys. Rev. Lett. $\underline{63}$, 636 (1989).

20. Y. Kim, A. Ourmazd, R.J. Malik, and J. A. Rentschler, in **Atomic Scale Structure of Interfaces**, edited by R.D. Briggins, R.M. Fenstra, and J.M. Gibson (Mater. Res. Soc. Proc. $\underline{159}$, Pittsburgh, PA 1990) pp. 351-355.

21. C.L. Cesar, M.N. Islam, R.D. Feldman, R.F. Austin, D.S. Chemla, L.C. West, and A.E. DiGiovanni, Appl. Phys. Lett. $\underline{56}$, 283 (1989).

22. G. Bastard and J. Brum, IEEE J. Quantum Electron. $\underline{QE-22}$, 1625 (1986).

23. C.L. Cesar, M.N. Islam, R.D. Feldman, R. Spitzer, R.F. Austin, A.E. DiGiovanni, J. Shah, and J. Orenstein, Appl. Phys. Lett. $\underline{54}$, 745 (1989).

24. Y. Guldner, G. Bastard, J.P. Vieren, M. Voos, J.P. Faurie, and A. Million, Phys. Rev. Lett. $\underline{51}$, 907 (1983).

25. D.H. Chow, J.O. McCaldin, A.R. Bonnefoi, T.C. McGill, I.K. Sou, and J.P. Faurie, Appl. Phys. Lett. $\underline{51}$, 2230 (1987).

26. R. Sporken, S. Sivananthan, J.P. Faurie, D.H. Ehlers, J. Fraxedas, L. Ley, J.J. Pireaux, and R. Caudano, J. Vac. Sci. Technol. $\underline{A7}$, 427 (1989).

27. D.S. Chemla, D.A.B. Miller, P.W. Smith, A.C. Gossard, and W. Wiegman, IEEE J. Quantum Electronics $\underline{QE-20}$, 265 (1984).

28. M. Wegener, I. Bar-Joseph, G. Sucha, M.N. Islam, N. Sauer, T.Y. Chang, and D.S. Chemla, Phys. Rev. B $\underline{39}$, 12794 (1989).

29. C.L. Cesar, M.N. Islam, C.E. Soccolich, R.D. Feldman, R.F. Austin, and K.R. German, Optics Letters $\underline{15}$, 1147 (1990).

30. M.N. Islam, E.R. Sunderman, C.E. Soccolich, I. Bar-Joseph, N. Sauer, T.Y. Chang, and B.I. Miller, IEEE J. Quantum Electronics $\underline{QE-25}$, 2454 (1989).

THERMAL TREATMENT OF HG CONTAINING II-VI SEMICONDUCTORS BY ANNEALING IN A MERCURY BATH (AMEBA)

R. KALISH AND C. UZAN-SAGUY

Technion-Israel Institute of Technology, Solid State Institute, Haifa, Israel 32 000.

ABSTRACT

An extremely simple and inexpensive technique (AMEBA) for the thermal treatment of Hg containing specimens which permits short or long time annealing in a Hg rich atmosphere is described. It is based on the immersion of the sample, properly protected by proximity caps in, or above, a hot mercury bath. The sample assembly is such that it permits Hg vapors, but not the liquid, to reach the specimen's surface.

The usefulness of AMEBA in improving the electrical and structural properties of as-grown $Hg_{1-x}Cd_xTe$ (x = 0.21) and in removing ion implantation related damage as well as electrically activating B implants in various p-type HgCdTe samples is demonstrated. All the data presented show that AMEBA treatment yields results which are comparable or superior to those obtainable by convensional annealing methods.

1. INTRODUCTION

The narrow band-gap semiconductor $Hg_{1-x}Cd_xTe$ (MCT) with the composition x = 0.2 is, up to date, the most widely used material for the fabrication of infrared (IR) detectors active at about 10vm. Even though most recently other Hg containing II-VI ternary semiconductors (like $Hg_{1-x}Zn_xTe$, x = 0.15) have also been considered as potential materials for IR detection , these have not yet reached the maturity for device realization[1,2]. The electrical properties of MCT, and to a certain extend also of other similar II-VI crystals, are determined, among others, by defects and deviations from stoichiometry and not only by real chemical doping. Hg vacancies are active acceptors while Hg interstitials are believed to be donors in MCT. As-grown HgCdTe crystals grown by the Solid State Recrystallization method or thin epitaxial layers grown on lattice matched substrates are usually Hg vacancy rich, and therefore exhibit high p-type carrier concentrations with poor mobilities. Post-growth thermal treatments are therefore often needed to improve the material and to give it the desired electrical properties. Damaged MCT, whether due to ion-implantation or to other surface treatment procedures (such as ion milling), always exhibit strong n-type features with very low mobilities [3-5]. To remove these, as is for example needed to achieve electrical activation of ion-implanted dopants, thermal annealing is required. Unfortunately MCT and to a lesser degree also other Hg containing crystals are extremely delicate, having the tendency to lose Hg when heated. This fact very much complicates their annealing as it must be done under such conditions that avoid Hg loss or, sometimes, even enhance Hg indiffusion . Post growth annealing of as-grown MCT under low Hg vapor pressure is known to improve the carrier mobility and to reduce the acceptor concentration, while, if performed under Hg overpressure the annealing turns the material n type with a high mobility [6]. Post implantation annealing is usually done either under Hg pressure or on encapsulated samples so as to avoid Hg loss. Both these methods are technically complicated since they require capping (i.e. sputter deposition of ZnS, of SiO_2, or growth of a Hg native oxide) or annealing under special enviromental conditions in closed ampoules (under excess Hg or noble gas pressure). Furthermore, rapid thermal treatment may sometimes be needed, in particular if undesirable diffusion processes need to be suppressed. Multistage annealing has proven to be useful in some cases, as for example for electrical activation of p type implants in MCT. It has been found that a relatively short, high temperature anneal (T = 400°, t = 4 hours) , followed by a longer, low temperature, (260°C 19 hours) anneal, leads to high electrical activation of implants in MCT.

The above mentioned requirements have led us to the development of an extremely simple, yet most affective, technique for the annealing of Hg containing materials[7-9]. The main idea behind this technique is to make use of the high Hg vapor pressure that exists in the vicintiy of a hot liquid mercury surface together with the excellent thermal conductivity of Hg to thermally treat delicate Hg containg crystals. In this, so called AMEBA (Annealing in a MErcury BAth) technique , the sample , mounted in a special arrangement that prevents its direct wetting, is immersed for a given time in a Hg bath held at the required temperature . Below we show that this simple and extremly inexpensive annealing method leads to defect removal in ion implanted HgCdTe, improves the electrical properties of as-grown bulk and epitaxial HgCdTe and leads to electrical activation of n type implants in this technologically important material.

2. EXPERIMENTAL

The annealing apparatus (fig. 1), placed in a well vented hood, is composed of a 3 neck standard pyrex flask filled with pure Hg. Any desired temperature, lower than the boiling point of Hg,(354°C) can be reached and kept constant by a heating element and a temperature controller. The bath temperature is measured by a thermocouple (A) embedded in quartz tube which in dipped in the liquid Hg. The sample to be immersion annealed is sandwiched between two clean Si plateletts and the whole assembly is placed in a small stainless-steel box, the dimensions of which fit tightly those of the two Si caps. A lid with a fine thread enables closing the box while gently pressing on the sample assembly in a way which prevents any wetting of the sample but allows the penetration of Hg vapors from a Hg vapor rich atmosphere . The temperature inside the annealing box is measured by a thermocouple glued onto the internal face of the lid. Fig. 2 shows the temperature-time profiles measured inside the box (using thermoconple B) and in the bath (with thermocouple A). As can be seen from the figure, the temperature of the bath drops slightly from its preset value immediately upon immersion (from 330°C to 315°C in the case shown) followed by a slow rise towards the preset value . The cooling down rate of the bath temperature, once the heating has been switched off is slow, and it may take up to 30 mintues to reach

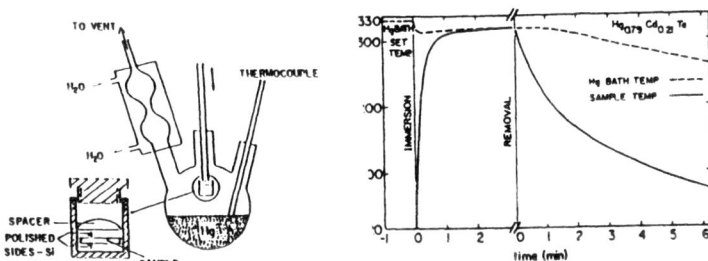

Fig. 1: Experimental setup used for the annealing of Hg(1-x)CdxTe by immersion in a hot Hg bath. Insert shows sample arrangment in the stainless-steel box.

Fig. 2: Temperature-time profiles of the Hg bath (dashed line) and sample-assembly (solid line) before immersion, during annealing and after removal of the sample assembly from the Hg bath.

room temperature (depending on the amount of liquid Hg in the bath). The actual sample temperature, as measured by thermocouple B (solid line in figure 2) rises sharply, and reaches the bath temperature within about 1 minute.(Much faster response-times have been obtained with an older version of a sample assembly [7] which had a much smaller heat capacity. The data shown below for annealing times of the order of 1min. were obtained with this arrangement). Upon removal of the sample assembly from the bath, it cools down rather rapidly, depending on how high above the hot mercury surface it is kept. The gradient in sample temperature as function of its distance from the liquid can actually be utilized to carry out , in situ, a multistage annealing sequence comprising of several steps by just changing the sample position. For example, by lifting the sample assembly 7cm above the Hg bath held at 320°C , a stable temperature of 200°C can be reached at the sample position. This lower temperature annealing which is now carried out above the hot bath is, of course, still in a Hg rich atmosphere.

The crystals used were mostly $Hg_{1-x}Cd_xTe$ (x = 0.21) grown by a variety of techniques. The results of AMEBA treatments have been evaluated by the following, complementary, experimental methods: The near surface crystal quality was evaluated before and after AMEBA by Rutherford Backscattering (RBS) combined with channeling and with Raman spectroscopy [10]. The atomic profiles of the constituents and dopants were measured by Secondary Ion Mass Spectrometry (SIMS Atomika 6500). The electrical properties of the samples were determined by Hall effect and differential Hall effect measurements at 77K using the Van der Pauw geometry at different magnetic fields varying from 0.5 to 8kG.

RESULTS

3.1 Material improvements by low temperature (260°C) AMEBA.

As mentioned above, most undoped HgCdTe crystals are as grown p-type with a poor mobility due to the presence of Hg vacancies. A standard procedure to improve the electrical properties of such crystals (whether bulk or epilayers) is to "recrystallize" the material for long times at fairly low temperatures in a Hg containing atmosphere. This is usually done by placing an ampoule containing the sample and a drop of Hg in a standard furnace. Below we show that AMEBA can lead to improvements of as-grown MCT which are at least comparable to those commonly obtained by standard furnace treatments.

3.1.1 Type conversion of as-recrystallized p type $Hg_{.79}Cd_{.21}Te$

Conversion of as-recrystallised p type samples to n type was performed by immersion annealing for increasing times at 260°C. This temperature was chosen as it is the temperature conventionally used when annealing is carried out in a sealed quartz ampoule with excess Hg[11]. Figures 3a and 3b show the changes in carrier concentration and mobility as a function of immersion time. In these annealing experiments (samples 1 and 2 solid and dashed lines in figure 3), the duration of the immersion was progressively increased to follow the improvement of the electrical characteristics. Following non interupted 50 hour immersion in the hot Hg, a clear type-conversion was achieved with a low donor concentration of $4 \times 10^{14} cm^{-3}$ and a high mobility of $10^5 cm^2 V^{-1} s$-1 at 2KG (sample 3, solid dashed lines). The results were found to exhibit some dependence on the value of the magnetic field, indicating mixed conductivity, possibly due to the perturbing effect of the thick underlying p type layer.

3.1.2 Improvement of n type $Hg_{0.79}Cd_{0.21}Te$.

Several n type $Hg_{0.79}Cd_{0.21}Te$ samples, annealed in the conventional way (under Hg pressure in an ampoule) could be further improved by the present immersion annealing method. The improvements of mobility as function of annealing temperatures for a fixed time of 1 min and as function of annealing times for a fixed temperature of 250°C are shown in figures 4 and 5 respectively. As can be seen in Fig. 4 the mobility improves with increasing temperature up to 250°C beyond which it starts to deteriorate. Fig. 5 shows that heating at 250°C for times up to 15 minutes improves the mobility, but further heating damages the crystal. Under optimal conditions (250°C for 1 min), an improvement in the mobility of nearly 50% (70 000 to 100 000 $cm^2 V^{-1} s^{-1}$) accompanied by a decrease in carrier concentration was obtained. Furthermore, ion channeling measurements have shown that the crystallinity of the above samples exposed to Hg immersion treatment has been improved. The results of the electrical properties of AMEBA treated samples are given in table 1 section A.

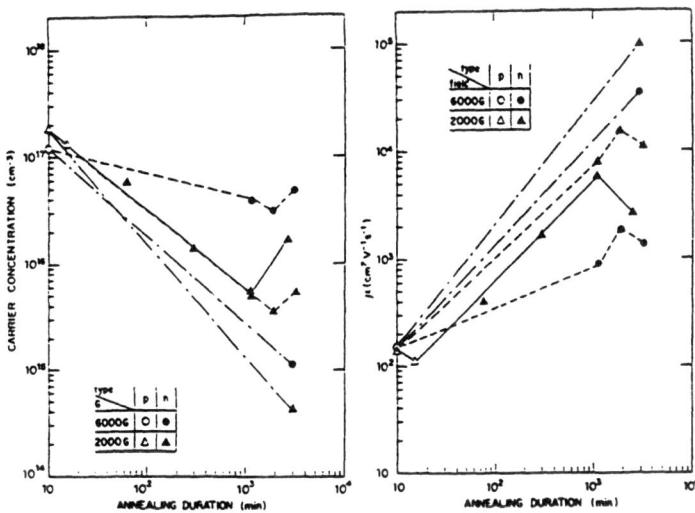

Fig. 3: Carrier concentration (a) and Hall mobility (b) of as-recrystallized p-type
$Hg_{.79}Cd_{.21}Te$ samples following AMEBA at 260°C as function of the
annealing duration; p-type conductivity is represented by open symbols while
the full symbols correspond to n-type conductivity, the circles and triangles
indicating Hall measurements in a magnetic field of 6 KG or 2 KG
respectively.

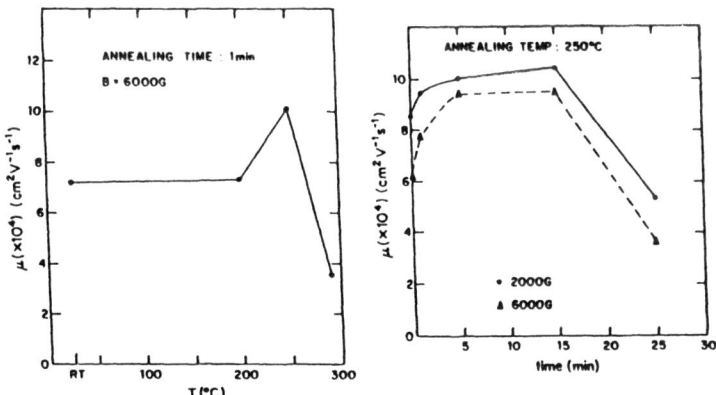

Fig. 4: Hall mobility of n-type $Hg_{.79}Cd_{.21}Te$ samples following AMEBA for 1 min.
as a function of anneal temperature.

Fig. 5: Hall mobility of n-type $Hg_{.79}Cd_{.21}Te$ samples following AMEBA at 250°C as
a function of time. The solid and dashed lines show the mobility obtained in
a magnetic field of 2 KG and 6 KG respectively.

SIMS measurements have shown that even following rather severe AMEBA (320°C, 10 min) only minimal changes in surface stoichiometry take place, as shown in figure 6. Raman scattering experiments (carried out at LN_2 temperatures using 2.41eV photons) on virgin and AMEBA treated samples did not reveal any observable difference in the spectra such as can be attributed to AMEBA induced changes in the near-surface properties of the samples.

3.2. Ion-implantation damage removal and electrical activation by high-temperature (320°C) AMEBA.

Ion implantation of HgCdTe, when carried out at room temperature, is accompanied by damage to the lattice. This damage is most severe for heavy ion implantations (i.e. In), for which it extends much beyond the ion range [12,13]. It is mainly comprised of dislocations or other extended defects and gives rise to n-type conduction with poor mobilities (typically $n = 2 \times 10^{18} cm^{-2}$, $m = 3000 cm^2 V^{-1}s^{-1}$). This electrical property of damage in MCT is actually used in practical applications for device realization . To achieve real chemical doping by ion-implantations this damage must be removed by proper annealing, and the implants must be driven to electrically active site.

Such anneals have been done most successfully by Bubulac et al [14] who have realized p type chemical doping by using a high temperature (defect removal) anneal (T = 400°C t = 4h) followed by a long low temperature recrystallization) stage (T = 260° t = 19h), both carried out in standard furnaces under Hg overpressure . Below we show the AMEBA is capable both to remove the

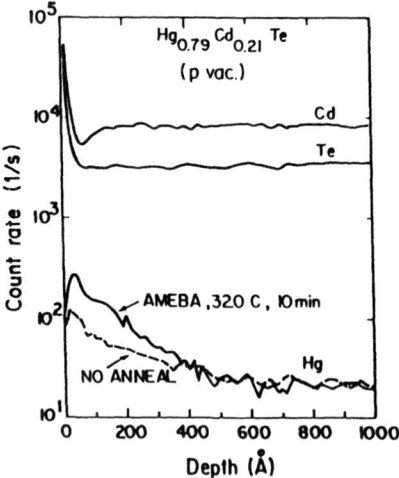

Fig. 6: *SIMS depth-profiles of Hg obtained in p-type vacancy $Hg_{.79}Cd_{.21}Te$ prior to annealing (no anneal) and following AMEBA at 320°C for 10 min.. The Cd and Te profiles are also displayed.*

implantation induced damage and to lead to electrical activation of the implants. Since damaged MCT is always n-types, pairs of samples one implanted with a dopant species and the other with a noble gas atom of comparable mass have been simultaniously AMEBA treated and their electrical properties compared, hence defect related effects could be eliminated.

3.2.1 Removal of implantation damage

The recovery of the near-surface crystallinity following ion implantation damage and AMEBA , has been monitored by RBS-Channeling (using 320keV protons) and by Raman scattering (using 2.41eV photons at LN_2) experiments. Because of the strong dependence of the density of the damage cascade on the implant mass, experiments were done for representative light (^{11}B, 200keV, $1x10^{14}cm^{-2}$), medium (^{75}As, 320keV, $5x10^{13}cm^{-2}$) and heavy (^{115}In and ^{107}Ag 320keV $1x10^{13}cm^{-2}$) implants. All implanted samples have been AMEBA treated at 320-330°C for times ranging from 8 min (B) to 4 hours (Ag). In all cases complete recovery of the crystallinity to at least its virgin value have been obtained (see figure 8). Raman scattering has proven to be extremely sensitive to the perfection of the mean surface structure of crystals. It is therefore an ideal probe to investigate the recovery of implantation related damage [15]. Raman measurements on In implanted MCT (340keV, $1x10^{13}cm^{-2}$) annealed under different AMEBA conditions have been performed at liquid nitrogen temperatures, using 2.41 eV photons which probe a depth in MCT of several hundred Angstroms. Figure 9 shows the Raman spectra of the as-grown, as-implanted and annealed samples. The effect of the implantation damage is evident through both changes in relative intensities of the TO_2, LO_2 and LO_1 lines as well as through the disappearance of the 2LO features in the spectra. However following 320°C AMEBA for 16 minutes the Raman spectrum has completely recovered being nearly indistinguishable from the spectrum of the virgin sample.

3.2.2 Electrical activation of the implants

Only data on the electrical activation of B (a donor in MCT) implants in p-type (due to vacancies or to As doping) $Hg_{1-x}Cd_xTe$ (x = 0.21) will be presented. Although we also have some data which indicate successful electrical activation of acceptor implants in MCT these are still somewhat unclear due to inherent difficulties in interpreting Hall data obtained from a sample which is composed of a thin p-type layer on a much thicker substrate (being either p or n) which dominates the Hall results.

3.2.2.1. B implant activation in p type (vacancy) MCT.

Boron and Neon ions have been implanted (200keV, 10^{14} B/cm^2 and 320keV $3x10^{13}$ Ne/cm^2) into two halves of a p-type (due to vacancies) $Hg_{.79}Cd_{.21}Te$ wafer. The implantation energies and doses have been chosen such as to roughly yield comparable damage profiles. Both implanted samples have been placed side by side in the sample assembly for AMEBA treatment, to assure

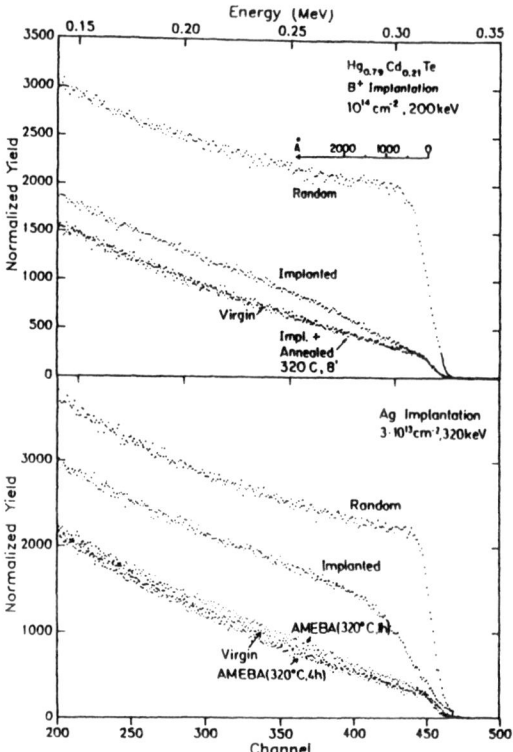

Fig. 7: *Channeling Rutherford backscattering spectra obtained in <111> Hg.79Cd.21Te with 320 keV protons showing the virgin, B implanted and immersion annealed spectra. A ramdom spectrum is also displayed.*

identical annealing conditions. Van-der-Pauw Hall measurement at LN_2 temperature in various magnetic fields (0.5-8KG) have been taken on the virgin samples, and on sample following implantation and following AMEBA. (320°C, 8 min.). The results are summarized in table 1, section B. By comparing the results for the Ne and the B implanted and annealed samples, it is obvious that chemical doping has indeed occured in the B implanted sample as a result of AMEBA; the B implanted sample exhibit an n-type carrier concentration of $7x10^{17}cm^{-3}$ with a mobility of $16000cm^2V^{-1}s^{-1}$, as compared to the poor p-type features of the control Ne implantation ($p=8x10^{16}cm^{-3}$, $\mu_p=65cm^2V^{-1}s^{-1}$).

Differential Hall data, obtained by consecutive layer removal and Hall measurements (figure 9, lines 2a and 3a) show that the n-type layer (line 2a) extends to a depth of about $2\mu m$, which is substantially deeper than the B implant profile as measured by SIMS (line 1 in fig. 9).

	sample	treatment impl.	AMEBA	cond. type	carrier concent.	mobility $cm^2V^{-1}s^{-1}$
A	p vac.	--	--	p	10^{17}	150
	bulk	--	260C, 50h	n	4×10^{14}	100 000
	n	--	--	n	2.5×10^{15}	70 000
	recrystal. bulk	--	250C, 15'	n	8.2×10^{14}	100 000
B	p vac.	--	--	p	1.5×10^{16}	580
		--	320C, 10'	p	1.8×10^{16}	300
	bulk	Ne	320C, 10'	p	6×10^{16}	65
		B	320C, 10'	n	7×10^{17}	16 000
C	p	--	--	p	6×10^{15}	600
	(As doped)		320C, 8'	p	4×10^{16}	200
	bulk	Ne	320C, 8'	p	2×10^{16}	70
		B	320C, 8'	n	10^{17}	24 000
D	epilayer	--	--	p	1.5×10^{16}	420
	p vac.	--	320C, 30'	n	2×10^{16}	1200
			+ 0.5µ etch	p	10^{17}	130
		Ne	320C, 30'	p	4×10^{17}	120
		B	320C, 10'	n	3×10^{17}	22 000

Table I

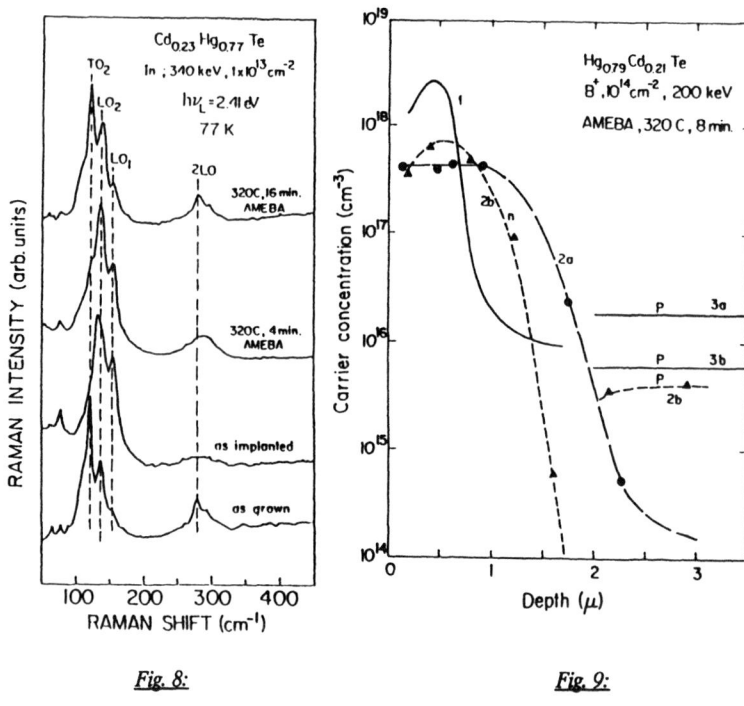

Fig. 8: Raman spectra obtained in $Hg_{.77}Cd_{.23}Te$ samples excited at 2.41 eV showing the as-grown, as-implanted and implanted and annealed spectra (AMEBA at 320°C for 4 min. or 16 min.).

Fig. 9: Atomic and electrical profiles as deduced from SIMS and differential Hall measurements on B implanted and annealed (AMEBA at 320°C for 8 min.) $Hg_{.79}Cd_{.21}Te$. The SIMS profile is represented by curve 1. Curves 2a and 2b display the carrier concentration in p-type vacancy and p-type As doped substrates respectively. Curves 3a and 3b show the p-type background carrier concentration for the p-vacancy and p-As-doped substates respectively.

3.2.2.2 B implant activation in p-type (As doped) MCT.

An identical experiment to that described in the previous section has been carried out on p-type $Hg_{.79}Cd_{.21}Te$ sample which have been intentionally doped with As during growth.

The result of this experiment are summarized in table 1 section C and in figure 9 lines 2b and 3b. Once more, the differences between B and Ne implantations followed by AMEBA are obvious, the first exhibiting good n-type features while the control shows poor p-type electrical properties. It should, however, be noted that in contrast to the differential Hall data obtained for the "vacancy doped" p-type material, which exhibits a rather thick n-type profile, the results of the similar experiment on As doped MCT show a fairly well defined n-type region beyond which the substrate p-type features(triangles in figure 9 line 2b) clearly appear. Hence, a rather sharp n-p junction can be obtained in the case.

3.2.2.3 B implant activation in a p-type (vacancy) epitaxial $Hg_{.76}Cd_{.24}Te$ layer.

The experiments described in 3.2.2.1 have been also carried out on a p-type (due to vacancies) epitaxial HgCdTe ($x = 0.23$) layer, 13μm thick, grown by the MOVPE technique on <111> CdTe.

The results of this experiment given in table 1 section D (320°C AMEBA for 10 , or 30 min.) show once more that AMEBA of B implanted samples leads to reasonable n-type features which, by comparison to the Ne data, can not be attributed to damage and must therefore be due to real chemical doping.

4. SUMMARY

A new method for the thermal treatment of Hg containing II-VI semiconductors has been described. It ist based on the immersion of the specimen in a hot Hg bath under conditions which permit Hg vapors to enter the vicinity of the sample. This, so called, AMEBA, (Annealing in a MErcuy BAth) technique has been shown in the present work to yield results comparable, or superior to those obtainable by the much more complicated commonly used annealing techniques. Both improvements of material properties of as-grown HgCdTe and electrical activation of B implants in HgCdTe by the use of AMEBA have been demonstrated.

ACKNOWLEDGEMENT

This research has been supported, in part by the US Army (Contract No. DAJA 45-86-C-0011) and by Semiconductor Devices,Israel.

REFERENCES

(1)   R. Triboulet, J. Cryst. Growth 86 (1986) 79.
(2)   A. Sher, A. Tsigelman, E. Weiss, and N. Mainzer, J. Vac. Sci. Technol. A8 (1990) 1093.
(3)   L.O. Bubulac, W.E. Tennant, R.A. Riedel, and T.J. Magae, J. Vac. Sci. Technol. 21 (1982) 251.
(4)   S. Margalit, Y. Nemirovsky, and I. Rotstein, J. Appl. Phys. 50 (1979) 6386.
(5)   G. Bahir and E. Finkman, J. Vac. Sci. Technol. A7 (1989) 348.
(6)   C.L. Jones, M.J.T. Quelch, P. Capper, and J.J. Gosney, J. Appl. Phys. 53 (1982) 9080.
(7)   R. Kalish, R. Fastow, V. Richter, and M. Shaanan, App. Phys. Lett. 51 (1987) 1158.
(8)   C. Uzan-Saguy and R. Kalish, Appl. Phys. Lett. 55 (1989) 1091.
(9)   C. Uzan-Saguy, D. Laser and R. Kalish, J. Crys. Growth 101 (1990) 864.
(10)   J. Wagner, P. Koidl, C. Saguy-Uzan and R. Kalish, to be published.
(11)   A.J. Sylliaos and M.J. Williams, J. Vac. Sci. Technol. 21 (1982) 201.
(12)   G. Bahir and R. Kalish, J. Appl. Phys. 54, (1983) 3129.
(13)   C. Saguy-Uzan, D.Comedi, V. Richter and R. Kalish J. Vac. Sci. Tech. A7 (1989) 2575.
(14)   L.O. Bubulac, D.S. Lo, W.E. Tennant, D.D. Edwall, J.C. Chen, J. Ratusnik, J.C. Robinson, and G. Bostrup, Appl. Phys. Lett. 50 (1987) 1586.
(15)   A. Lusson, J. Wagner, and M. Ramsteiner, Appl. Phys. Lett. 54 (1989) 1787.

STABILISATION OF $Hg_{1-x}Cd_xTe:Hg_{1-y}Cd_yTe$ ($x \neq y$) HETEROINTERFACES AND APPLICATIONS IN INFRA RED DEVICES

PAUL A. CLIFTON AND PAUL D. BROWN

School of Engineering and Applied Science, University of Durham, South Road, Durham, DH1 3LE, U.K.

ABSTRACT

The interface between $Hg_{1-x}Cd_xTe$ ($0 \leq x \leq 1$) and $Hg_{1-y}Cd_yTe$ ($0 \leq y \leq 1$) epitaxial layers of different composition ($x \neq y$) is unstable with regard to the intermixing of the Hg and Cd cations within the Group II sublattice. This phenomenon may give rise to long-term stability problems in HgTe-(Hg,Cd)Te superlattices and composition grading between (Hg,Cd)Te absorber layers and CdTe buffer or passivation layers in epitaxial infra red detectors. In this paper, a novel approach to the inhibition of interdiffusion in these systems is discussed. This involves the growth of an intervening ZnTe barrier layer at the heterointerface between two (Hg,Cd)Te layers. Initial results are presented which indicate the effectiveness of this technique in reducing interdiffusion in an experimental heterostructure grown by MOVPE. Some possible applications in a variety of HgTe-based long wavelength devices are discussed.

INTRODUCTION; INTERDIFFUSION IN (Hg,Cd)Te HETEROSTRUCTURES

Solid state interdiffusion of the metals mercury and cadmium may present a serious limitation to the long term prospects of optoelectronic devices incorporating cadmium mercury telluride heterojunctions. The high interdiffusivity between layers of CMT of different composition (in the extreme case, between HgTe and CdTe) gives rise to non-abrupt (compositionally graded) heterointerfaces at typical epitaxial growth temperatures.[1] Indeed, this property is exploited to some advantage in the interdiffused multilayer process (IMP)[2] for growth of CMT alloys by MOVPE. On the other hand, the same phenomenon gives rise to discernable interdiffusion in HgTe-CdTe superlattices at temperatures as low as 110°C [3] and possibly even near room temperature[4] and thus represents a real threat to the long-term viability of these materials. A serial vacancy-interstitial (interstitialcy) diffusion mechanism[5] may account for interdiffusion in (Hg,Cd)Te, at least at elevated temperatures, with interstitials becoming more important at lower temperatures. Kim et al[4] concluded that the rate of interdiffusion, $\tilde{D}$, in this system decreases more slowly than expected as the temperature is reduced and hence that extrapolation of values obtained at elevated temperatures to room temperature is highly optimistic. The 'residual' interdiffusion at lower temperatures is largely attributed by Shaw[6] to the presence of impurities. In epitaxial (Hg,Cd)Te grown on GaAs with a CdTe buffer layer, compositional (and hence bandgap) grading at the CMT-CdTe interface may also be undesirable and similarly for heterostructure passivation with CdTe.[7]

One approach to the reduction of interdiffusion in layers grown by MOVPE is to decrease the growth temperature by means of energy-assisted techniques[8] or through the use of tellurium precursors which pyrolyse at lower temperatures[9] [10]. However, there is inevitably a lower limit to the epitaxial growth temperature for both MOVPE and MBE and subsequent device fabrication necessarily involves further processing at elevated temperatures. As such, it would be advantageous to stabilise the heterointerfaces to the extent that interdiffusion is avoided even at MOVPE growth temperatures in excess of 250°C. An alternative approach to interface stabilisation is outlined here which involves the use of strained (ideally pseudomorphic) ZnTe layers as diffusion barriers, interposed at the CMT heterointerfaces.

Mat. Res. Soc. Symp. Proc. Vol. 216. ©1991 Materials Research Society

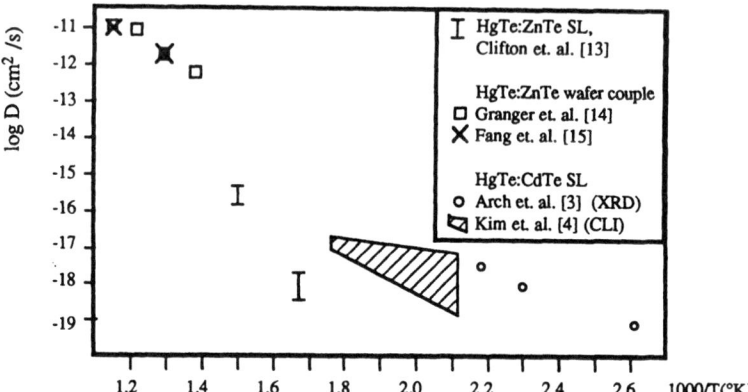

Figure 1; Arrhenius plot comparing interdiffusivities, $\tilde{D}$, determined experimentally by various authors for the HgTe:ZnTe and HgTe:CdTe systems.

ZnTe AS A DIFFUSION BARRIER.

It has been shown in earlier studies[11] [12] that HgTe-ZnTe superlattice structures may be realised by MOVPE at temperatures in excess of 300°C with apparently low rates of interdiffusion. $\tilde{D}$ values of $< 2 \times 10^{-18} cm^2 s^{-1}$ and $\approx 3 \times 10^{-16} cm^2 s^{-1}$ have been broadly estimated for this heteroepitaxial system by observation of layer contrast in cross-sectional transmission electron microscope (XTEM) images of such structures[13] . These values are at least two orders of magnitude lower than $\tilde{D}$ measured for the HgTe-CdTe system[6] in the same temperature range. They are also approximately two orders of magnitude lower than expected from extrapolation of measurements obtained at higher temperatures for the HgTe-ZnTe system using clamped HgTe and ZnTe wafer couples[14] [15] , as indicated in figure 1.

In considering the anomolously low interdiffusion rate in our heteroepitaxial structures, several factors are apparent;

(i) Interdiffusion in HgTe-based systems is to a large extent mediated by Hg vacancies[5] and epitaxial material may have lower vacancy concentrations. However, no measurements are available to confirm this in the present case.

(ii) The diffusion conditions in the heterostructure are distinct from those used in wafer couple studies. In the former, the zinc is present in a continuous layer of ZnTe, a compound with a quite strong chemical bond, whilst in the latter, the zinc is presented to the crystal surface in the vapour phase.

(iii) The continuous thin layers of ZnTe present in the heterostructures may consitute an impedance to the flux of metal atoms across the HgTe-ZnTe interfaces due to the great stability of the Zn-Te chemical bond. i.e. The harder and more stable ZnTe lattice wil be more resistant to the formation of the vacancies and interstitials required in the interstitialcy mechanism.

(iv) Although Hg may self diffuse rapidly within the metal sublattice, the rate of interdiffusion is limited by the slower diffusing species which is the zinc[5] .

(v) A further possible barrier to interdiffusion may arise from the tendency of the binaries HgTe and ZnTe to segregate[16] when the pseudobinary alloy (Hg,Zn)Te is formed. The corollary may be that ZnTe and HgTe layers deposited separately will not tend to interdiffuse to form the alloy.

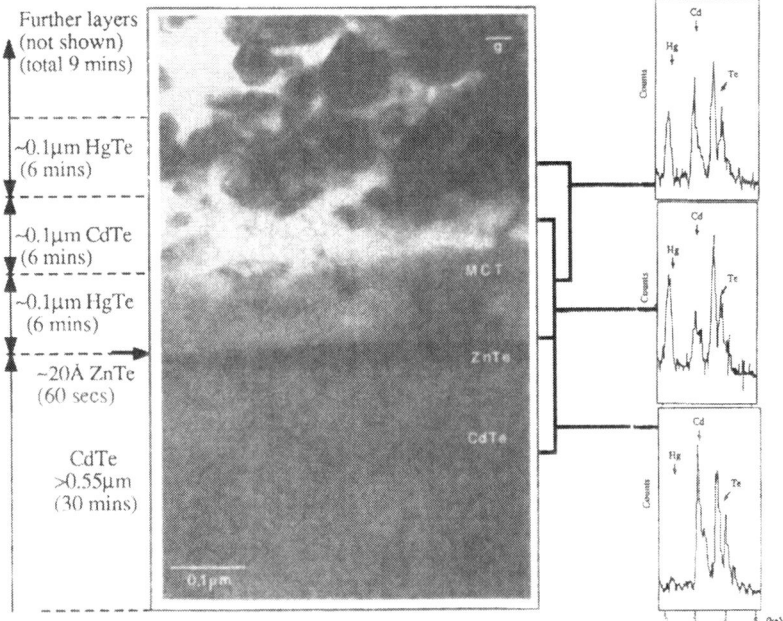

Figure 2; (a) Schematic of the intended heterostucture incorporating ZnTe barrier layers.
   (b) Cross-sectional TEM dark field image obtained from the structure grown by MOVPE at 325°C. g=220.
   (c) Electron probe microanalysis spectra shown with the approximate locations from which they were obtained.

GROWTH OF ZnTe BARRIER LAYERS AT (Hg,Cd)Te HETEROINTERFACES

In an initial experimental study, a heterostructure incorporating HgTe (x=0) - CdTe (x=1) interfaces both with and without an intervening ZnTe barrier layer has been grown by MOVPE at 325°C on a {100}GaAs substrate. A schematic diagram of the intended epitaxial structure is shown in figure 2(a), together with the respective growth periods. The growth was performed in an atmospheric pressure reactor using the metalorganic precursors dimethylzinc, dimethylcadmium and diisopropyltelluride (all at a nominal rate of 10^{-4} moles/min) and elemental mercury (heated *in situ* to 185°C). High hydrogen dilution flows (6.5 l/min) were admitted during the CdTe and ZnTe growth phases and a low total flow (1 l/min) was maintained for growth of HgTe. The growth sequence was established with the aim of producing the heterostructure shown schematically in figure 2(a). Conditions were based on prior experience of growth rates for very thin ZnTe layers and of the particular growth interruptions necessary for the realisation of HgTe-containing low-dimensional structures[11] . Unfortunately, during the course of the growth run, the dimethylcadmium bubbler became exhausted and had to be replaced subsequently.

CROSS SECTIONAL TEM; RESULTS AND DISCUSSION

Cross-sectional transmission electron micrograph (TEM) images and electron probe microanalysis data have been obtained from the heterostructure sample. Figure 2(b) shows a dark field TEM image of a region which includes the lower CdTe buffer layer and the structure up to the second CdTe layer. The image shows quite clearly that a band of precipitates had been incorporated during growth into the second CdTe layer. The six-fold symmetry of the precipitates is indicative of tellurium and their presence is attributed to the exhaustion of the Cd alkyl source with consequential growth under highly Cd-depleted vapour conditions. Given that such gross growth defects are present, the upper part of the heterostructure will not be discussed further. On inspection of the lower layers (in the sequence CdTe-ZnTe-HgTe-CdTe) however, two particular points may be noted;

(i) The ZnTe layer marks a distinct interface. As such, it may be inferred that the ZnTe layer has not interdiffused into the neighbouring CdTe or HgTe layers, even though it was maintained at the growth temperature for at least 1600 seconds. It should be noted that this image was obtained with the sample tilted away from the $< 110 >$ axis and that the dark band seen in the vicinity of the ZnTe layer is a contrast effect unrelated to its true thickness.

(ii) A significant reduction in the density of threading dislocations is apparent at the position of the barrier layer. The high dislocation content of the CdTe buffer layer is typical of the CdTe/GaAs heteroepitaxial system which is very poorly lattice matched. Possible reasons for the sudden decrease in dislocation density have been discussed previously[13] One possibility is that partial relaxation of the strained ZnTe layer has occurred. This would only need to be sufficient to generate an areal density of misfit dislocations equal to the density of threading dislocations incident from the underlying CdTe buffer layer according to the mechanism proposed by Petruzzello et al[17] . Alternatively, the dislocation filtering may be related in some way to the growth interruptions which occurred both before and after the growth of the ZnTe layer.

Electron probe microanalysis was also performed on the cross-sectional sample. A beam of approximate radius 150nm allowed composition to be determined at various points by energy dispersive X-ray analysis (EDX). The resulting EDX spectra are also shown in figure 2(c), together with the position within the epitaxial structure from which they were obtained. These results confirm several key points. The two layers (HgTe and CdTe) above the ZnTe barrier are considerably interdiffused, as indicated by the large amount of Cd in the HgTe layer and Hg in the CdTe layer (the two layers are jointly labelled as MCT in the figure). This observation is consistent with the expected interdiffusion length, $\mathcal{L}$, of $\approx 0.1\mu$m as estimated using the relation $\mathcal{L} = 2\sqrt{\tilde{D}}.t$ with $\tilde{D} = 1.84 \times 10^{-14}$cm^{2}s^{-1} at 325°C (obtained using an equation given by Shaw[6]) and $t = 1260$ secs. In contrast, the CdTe underneath the ZnTe barrier remains free of Hg, at least within the compositional resolution of the EDX and the spatial resolution of the TEM.

Thus it would appear that intermixing between the HgTe and CdTe layers on either side of the ZnTe barrier has indeed been effectively inhibited.

VALENCE BAND OFFSETS

Of particular relevance to the electrical and optical properties of heterostructures incorporating strained ZnTe layers, is the size of the valence band offsets at the respective HgTe-ZnTe and CdTe-ZnTe interfaces. In a common anion system, the bulk of the difference in bandgaps appears as a conduction band offset ΔE_C, which in the present case will be very large, approximately 2.2eV for the HgTe-ZnTe junction and 0.7eV for the CdTe-ZnTe junction. However, the valence band offsets are at present rather poorly determined and there is considerable disagreement in the literature over their values. It is acknowledged that strain has a role in determining the positions of the valence bands and in particular that the uniaxial component of strain (along the growth axis) has the effect of splitting the heavy hole (HH) and light hole (LH) subbands.[18] In the proposed application of ZnTe layers as pseudomorphic diffusion barriers, it is assumed that the misfit is taken up completely by elastic strain in the ZnTe layers (corresponding to biaxial tension and uniaxial compression), leaving the $Hg_{1-x}Cd_xTe$ layers on either side essentially unstrained. In this case, the LH subband should move up with respect to the HH subband and, in addition, the associated biaxial dilatation should cause a slight reduction in the bandgap. Overall, it is expected that the value of ΔE_V should be reduced by the strain in the ZnTe layer.

Hsu et al[19] have measured ΔE_V of for {100}HgTe-ZnTe , CdTe-ZnTe and HgTe-CdTe heterojunctions using the indirect method of X-ray photoelectron spectroscopy. Their study revealed noncommutativity of ΔE_V for the strained systems. For example, the valence band discontinuity for strained HgTe on a ZnTe substrate (311 meV) did not equal that of strained ZnTe grown on a HgTe substrate (401 meV). Nor was the magnitude of ΔE_V reduced by strain in the latter case but rather it was increased. According to the same set of results, nontransitivity of ΔE_V also arises in a three-component heterostructure such as HgTe-ZnTe-CdTe, as described by the expression;

$$\Delta E_V(HgTe - CdTe) \neq \Delta E_V(HgTe - ZnTe) + \Delta E_V(ZnTe - CdTe) \qquad (01)$$

Such a situation would not seem tolerable on the grounds of conservation of energy and the results may be questioned on certain points. Firstly, the XPS measurements were indirect in that they related to core levels, ΔE_{cl} on either side of the interface. Extraction of values for ΔE_V thus required knowledge of the electronic energy separations ($\Delta E_{cl} - \Delta E_V$) in the individual semiconductors. In Hsu's study, these were determined from bulk, and hence unstrained, samples and may therefore be inaccurate when applied to a strained system. Secondly, the epitaxial films were deposited to a depth of a few monomolecular layers on the substrate surface, leaving the upper surface free (i.e. forming a semiconductor-vacuum interface). This does not properly represent the usual situation in a heterostructure where the low dimensional layers are clad between heteroepitaxial layers.

Bertho et al have recently calculated ΔE_V for the same strained {100}heterojunctions using a tight-binding approach with the parameters modified by the change in interatomic separations resulting from tetragonal distortion. Their theoretical value for $\Delta E_V(HgTe-ZnTe)$ is also large (519 or 559 eV) but there is no indication of noncommutativity for either this or the CdTe-ZnTe junction. However, they do neglect splitting of the valence bands by the presence of biaxial strain, which is expected to be a sizeable effect. Furthermore, the results they present are for the 'average' of the top valence bands (heavy and light hole) which do not necessarily correspond to the valence bands probed by XPS. Therefore, considerable caution has to be exercised in comparing these theoretical and experimental values.

POSSIBLE APPLICATIONS OF ZnTe BARRIER LAYERS

As stated in the introduction, the main application foreseen for ZnTe layers as interdiffusion barriers is at the heterointerfaces in HgTe-(Hg,Cd)Te SLs and MQWs and in buffered or passivated epitaxial (Hg,Cd)Te . The first consideration in structural terms is the large misfit (6.5%;the ZnTe-(Hg,Cd)Te combination)which implies a critical thickness for the ZnTe layer of the order of 20Å [13] . Such a thickness has already been demonstrated by MOVPE and MBE. A secondary consideration is the effect of the thin ZnTe layers on the volume-averaged lattice parameter of the whole heterostructure. For example, if 10Å ZnTe layers were to be interposed at all the interfaces of a SL with 100Å HgTe and CdTe layers, the free-standing lattice parameter would be about 6.44Å. A further consideration is the effect of the ZnTe barrier on the optoelectronic properties of the interface. Certain key points may be noted;

(i) The large ΔE_C and ΔE_V should enhance the quantum confinement of both carriers in HgTe QWs.

(ii) Wide bandgap heterostructure passivation may yield improvements in performance and/or reproducibility of MCT photodiodes[7] as compared to current anodic sulphide passivation technology[20] . Although epitaxial CdTe introduces barriers in both the valence and conduction bands, these should be considerably enhanced by the presence of ZnTe layers.

(iii) The ΔE_V is probably too great for the ZnTe barrier to be used at a p(high x)-n(low x) $Hg_{1-x}Cd_xTe$ heterostructure for which quantum efficiency is impaired by hole potential barriers greater than 8meV and grading over at least 0.2μm is desirable.[21]

REFERENCES

1. M.A.Herman and M.Pessa, J. Appl. Phys. 57, 2671(1985)

2. J.Tunnicliffe, S.J.C.Irvine, O.D.Dosser and J.B.Mullin, J. Cryst. Growth 68, 245(198·

3. D.K.Arch, J.L.Staudenmann, J.P.Faurie, Appl. Phys. Letters 48(23), 1588 (1986).

4. Y.Kim, A.Ourmazd and R.D.Feldman, J. Vac. Sci. Technol. A8(2), 1116 (1990)

5. M.S.Tang and D.A.Stevenson, J. Vac. Sci. Technol. A6(4), 2650(1988)

6. D.Shaw, J. Cryst. Growth 86, 778(1988)

7. P.H.Zimmerman, M.B.Reine, K.Spignese, K.Maschoff and J.Schirripa, J. Vac. Sci. Technol. A8(2), 1182(1990)

8. S.J.C.Irvine, J.Giess, J.B.Mullin, G.W.Blackmore and O.D.Dosser, J. Vac. Sci. Technol. B3, 1450(1985).

9. F.Desjonqueres, A.Tromson-Carli, P.Cheuvart, R.Druilhe, C.Grattepain, A.Katty, Y.Marfaing and R.Triboulet, presented at the ICMOVPE-5, Aachen, Germany (1990) and to be published in J. Cryst. Growth .

10. I.B.Bhat, H.Ehsani and S.K.Ghandhi, J. Vac. Sci. Technol. A8(2), 1054(1990)

11. P.A.Clifton, J.T.Mullins, P.D.Brown, N.Lovergine, A.W.Brinkman and J.Woods, J. Cryst. Growth 99, 468(1990).

12. J.T.Mullins, P.A.Clifton, P.D.Brown, D.O.Hall and A.W.Brinkman in 'Properties of II-VI Semiconductors: Bulk Crystals, Epitaxial Films, Quantum Well Structures and Dilute Magnetic Systems', edited by F.J.Bartoli, H.F.Schaake and J.F.Schetzina (Mater. Res. Soc. Proc. 161, Pittsburgh, PA1990)pp.357 − 361.

13. P.A.Clifton and P.D.Brown, presented at the SPIE Conference on Advanced Materials Concepts for Optoelectronic Devices, Aachen, October 1990 (unpublished).

14. R.Granger, C.Pobla, S.Rolland and R.Triboulet, J. Cryst. Growth 101, 261(1990)

15. S.Fang, L.J.Farthing, M-F.S.Tang and D.A.Stevenson, J. Vac. Sci. Technol. A8(2),

16. A.Sher, A-B.Chen and M.van Schilfgaarde, J. Vac. Sci. Technol. A4(4), 1965(1986)

17. J.Petruzzello, D.Olego, X.Chu and J.P.Faurie, J. Appl. Phys. 66 (7), 2980(1989)

18. F.H.Pollack and M.Cardona, Phys. Rev. 172, 816(1968)

19. C.Hsu and J.P.Faurie, J. Vac. Sci. Technol. B6, 773(1988)

20. Y.Nemirovski and G.Bahir, J. Vac. Sci. Technol. A7, 450(1989)

21. K.Kosai and W.A.Radford, J. Vac. Sci. Technol. A8(2), 1254(1990)

HETEROEPITAXIAL HgCdTe/CdZnTe/GaAs/Si MATERIALS FOR INFRARED FOCAL PLANE ARRAYS

S.M. JOHNSON, J.B. JAMES, W.L. AHLGREN, W.J. HAMILTON, Jr., M. RAY, AND G.S. TOMPA[†]
Santa Barbara Research Center, 75 Coromar Drive, Goleta, CA 93117
† EMCORE Corporation, 35 Elizabeth Avenue, Somerset, NJ 08873

ABSTRACT

The structural properties of LPE-grown HgCdTe on heteroepitaxial MOCVD-grown CdZnTe/GaAs/Si substrates were evaluated using high-resolution x-ray diffraction techniques and TEM. Large tilts (up to 4°) between CdZnTe layers and GaAs/Si substrates are a general characteristic of this heteroepitaxial system and are attributed to the interaction of closely spaced misfit dislocations that arrange to form a tilt boundary. Either {112}CdTe or {552}CdTe can be grown on {112}GaAs/Si; the {552} was shown to result from a first-order twinning operation of {112}. Lamella {111} microtwins in {111}CdZnTe/{100}GaAs/Si substrates, measured by x-ray techniques, are not readily propagated into the LPE-grown HgCdTe layer. The x-ray FWHM of the LPE HgCdTe is typically at least a factor of two lower than that of the Si-based substrate from annealing and due to the increased thickness of the layer; both mechanisms promote dislocation interaction and annihilation. High performance MWIR and LWIR HgCdTe 128×128 hybrid focal plane arrays were fabricated on these Si-based substrates. An array average of $R_oA_j = 17.8$ ohm-cm^2 for a cutoff wavelength of 10.8 μm at 78K was demonstrated.

INTRODUCTION

Si-based alternative substrates are being actively developed as a replacement for bulk CdZnTe substrates for the growth of epitaxial HgCdTe for second-generation hybrid infrared focal plane arrays. This effort is motivated to seek improvements in substrate size, strength, cost, and the reliability of hybrid focal plane arrays. GaAs and sapphire are also candidates for substrates but both materials have thermal expansion coefficients close to that of CdZnTe and hence offer no advantages for thermal cycle reliability. Furthermore, backside illuminated arrays on sapphire substrates can only be used for MWIR applications since sapphire is not transparent in the LWIR region. The alternative substrate in this work is heteroepitaxial CdZnTe/GaAs/Si grown by metalorganic chemical vapor deposition (MOCVD).

Epitaxial growth of CdTe on Si has been achieved directly by MBE [1] and indirectly using CaF_2-BaF_2 buffer layers [2,3]. More commonly, CdTe or CdZnTe is grown on GaAs/Si substrates using congruent evaporation [4-6], MOCVD [7-12], and MBE [13]. HgCdTe has been grown on these substrates using close-spaced vapor phase epitaxy [4-6], MOCVD [9,10], LPE [11,12], and MBE [13]. The present work summarizes our recent results on MOCVD growth of CdTe on {112}GaAs/Si and {111}CdZnTe/{100}GaAs/Si and the fabrication of high-performance LWIR HgCdTe devices on these substrates. MOCVD-grown GaAs/Si was supplied by Kopin, Corp. CdTe and CdZnTe layers were grown using a vertical-flow, high-speed rotating disk reactor using dimethylcadmium (DMCd), diisopropyltellurium (DIPTe), and diethylzinc (DEZn) reactants; layers of 5-8 μm in thickness were grown at temperatures between 350-380° C. Controllably-doped HgCdTe layers were grown by vertical LPE from a Hg-rich solution [14].

STRUCTURAL ANALYSIS OF CdTe GROWN ON {112}GaAs/Si

X-ray rocking curve measurements were made utilizing a novel, compact Si four-crystal monochromator to produce $CuK\alpha_1$ x-rays. Lattice tilt and strain measurements were made using a high-resolution diffractometer that incorporates a Ge four-crystal monochromator and has an optically encoded angular readout of the diffractometer axis for precise angular measurements [15]. Layer tilt and strain ($\Delta d_l/d_s$) with respect to the Si substrate were determined by procedures described in detail elsewhere [16,17].

142

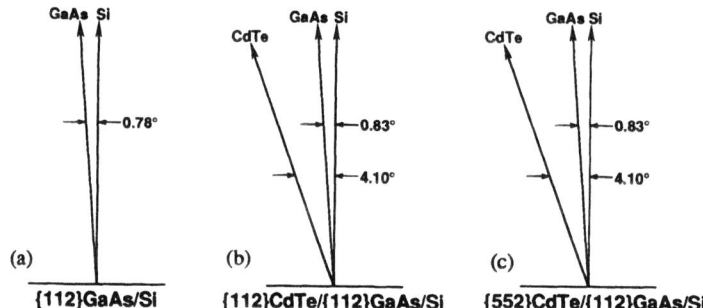

(a) {112}GaAs/Si   (b) {112}CdTe/{112}GaAs/Si   (c) {552}CdTe/{112}GaAs/Si

Fig. 1.   Schematic showing layer tilts relative to <112>Si surface normal.

Figures 1a, b, and c show a schematic representation of the relative layer tilts for multilayer structures of {112}GaAs/Si, {112}CdTe/{112}GaAs/Si, and {552}CdTe/{112}GaAs/Si, respectively, where the tilt angles are measured from the <112>Si surface normal. All layers were found to be tilted toward the nearest {111} about a <110> tilt axis. The GaAs layers are tilted approximately 0.8° while the CdTe layers are more substantially tilted, by 4.1°. Similar large layer tilts were previously reported for {100}CdZnTe/{100}GaAs/Si and {111}CdZnTe/{100}GaAs/Si [12] and for CdZnTe grown on bulk GaAs substrates [18]. These large layer tilts are attributed to the interaction of closely spaced misfit dislocations that arrange to form a tilt boundary.

We have observed that either {112}CdTe or {552}CdTe can be grown on a {112}GaAs/Si surface. Figure 2a shows a stereographic projection of a (112) surface. The (552) projection shown in Figure 2b can be obtained from the (112) by either a 180° rotation of the crystal about the [111] twist axis or by a 250°32' rotation about [011], [101], or [110] tilt axis. Both of these operations describe the relationship between a first order twin and its host lattice [19]. Narrow linear boundaries were observed on the {112}CdTe and {552}CdTe surfaces and were later found to originate in the {112}GaAs/Si. The relative orientations of these boundaries are shown in Figures 2a, 2b and correspond to the traces of {110} planes which are common to both orientations; their origin may be associated with the twinning process just described.

Cross-section TEM samples of {112}CdTe/{112}GaAs/Si were prepared using techniques described previously [20]. Figure 3a shows a low magnification image of the CdTe/GaAs interface. Twins are seen to extend from that interface and typically terminate approximately 1 μm into the CdTe layer. A similar termination was previously observed for lamella twins in {111}CdTe/{100}GaAs [21]. Threading dislocations decrease in density towards the surface. Figure 3b shows a high magnification image of the interface in area D in Figure 3a. The narrow twin lamella bounded by {111} are seen in this <110> projection. The narrow network of misfit dislocations is also visible at the CdTe/GaAs interface. None of the linear boundaries seen on the surface were included in the TEM specimen so no further information about their nature could be obtained.

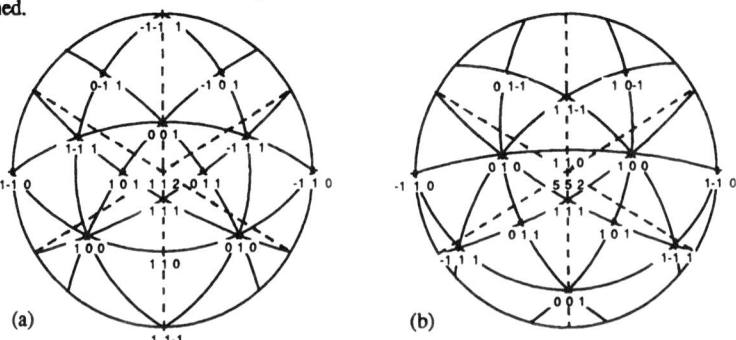

(a)   (b)

Fig. 2.   (a) Stereographic projection of a (112) surface and (b) (552) projection obtained from a first-order twinning operation of (112) as described in text.

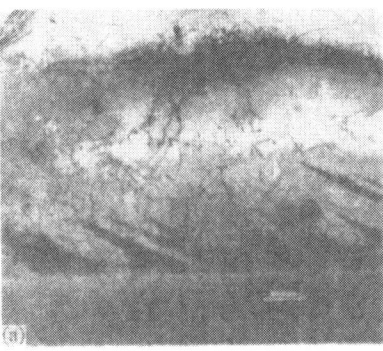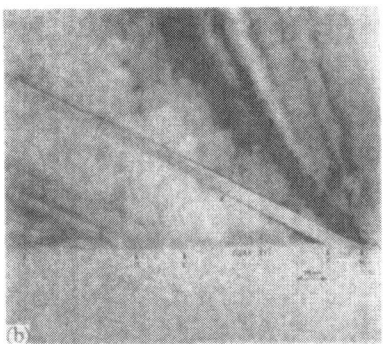

Fig. 3. TEM cross section of {112}CdTe/GaAs interface; (a) low magnification image of twins terminating in CdTe and (b) high magnification image of area D in (a) showing twins and misfit dislocations at the interface (observed along <110>).

Table I summarizes the tilt and strain measured with respect to the Si substrate for the layers discussed above. {224} symmetric reflections were used to determine the perpendicular strain for {224} CdTe and GaAs layers. Since {112} is not a principle plane and determination of strain parallel to the substrate requires multiple reflections, measurements using {333} reflections were used as an indication of strain in a non-perpendicular direction for comparison purposes. For {552}CdTe, the strain was calculated for two non-perpendicular planes using lattice constants determined from the Bond technique. Table I indicates that the GaAs layer is tetragonally distorted before and after CdTe growth; this is consistent with previous measurement made on {100} and {111} layers [12]. The CdTe layers have a slight tetragonal distortion but the majority of the mismatch is taken up by misfit dislocations at the CdTe/GaAs interface. Table I also shows that the CdTe layer tilt toward [111] for either the (112) or the (552) twinned orientation is exactly the same. This is expected since the [111] is a common axis in the first order twin relationship described above. As a relative indication of crystal quality {112}GaAs/Si has a typical rocking curve full-width at half-maximum (FWHM) of 150-200 arc-sec for 2.5 μm thick layers; the best {112}CdTe value was 90 arc-sec while the {552}CdTe is 220 arc-sec.

TWINS IN {111}CdZnTe/{100}GaAs/Si AND LPE-GROWN HgCdTe

A disadvantage of using {111} is that VPE-grown layers of CdTe and CdZnTe are usually heavily twinned, with the {111} twin planes lying parallel to the growth interface. This type of twinning can be detected using asymmetric {422} x-ray reflections; an untwinned crystal will have three such reflections (three-fold symmetry) accessible from a {111} surface while a twinned crystal will have six such reflections (six-fold symmetry) [22]. As a relative measure of twin volume fraction we have taken the ratio of {422} x-ray rocking curve intensities measured in the direction of a {422} and rotated 180° from that position [12]; this twin volume fraction is thus zero in an untwinned crystal and unity for a layer having 50% of its volume rotated 180° about the <111> surface normal. This value represents an average of the volume sampled by the x-rays.

Table I. Summary of heteroepitaxial layer tilt and strain relative to Si substrate.

Layer/Substrate	GaAs Tilt (deg)	GaAs (224) $\Delta d_{perp}/d_s$ (%)	GaAs (333) $\Delta d_l/d_s$ (%)	CdTe Tilt (deg)	CdTe (224) $\Delta d_{perp}/d_s$(%)	CdTe (333) $\Delta d_l/d_s$(%)	CdTe (335) $\Delta d_l/d_s$(%)
{112}GaAs/Si	0.781	3.994	4.030	-	-	-	-
{112}CdTe/GaAs/Si	0.835	3.999	4.058	4.098	19.306	19.355	-
{552}CdTe/{112} "	0.826	3.998	4.030	4.098		19.361	19.347

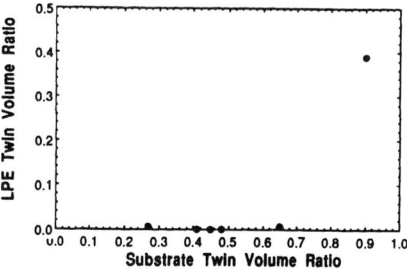

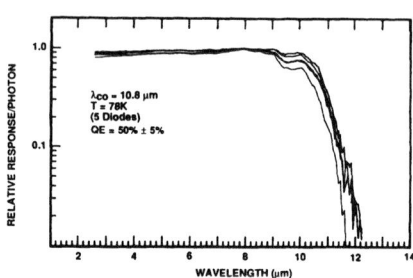

Fig. 4.   Twin volume fraction measured in LPE-grown HgCdTe layers versus twin volume fraction in CdZnTe/GaAs/Si substrate.

Fig. 5.   Relative 78K spectral reponse at zero bias of five diodes from HgCdTe 128×128 array fabricated on CdZnTe/GaAs/Si substrate

Figure 4 is a plot of the twin volume ratio measured in the LPE HgCdTe versus that of the {111}CdZnTe/{100}GaAs/Si substrate. Figure 4 shows that twins in the substrate do not readily propagate into the LPE layer unless their volume fraction is large. This seems to indicate that most of these twins in the CdZnTe layer are not close to the surface where they would easily propagate into the LPE HgCdTe layer. We have yet no experimental evidence supporting this explanation and further work is needed to understand this phenomena. In general, it is difficult to consistently reduce twinning in {111}CdZnTe layers although some success in doing this has been reported recently [23].

We have also observed that the x-ray FWHM of the LPE HgCdTe is typically at least a factor of two lower than that of the Si-based substrate [12]. This is due both to annealing, which increases dislocation mobility, and to the increased thickness of the layer; both mechanisms promote dislocation interaction and annihilation [12, 20]. This effect has also been observed for HgCdTe grown by MOCVD [24] and MBE [13] and for MOCVD-grown GaAs/Si [25].

LWIR HgCdTe FOCAL PLANE ARRAYS ON CdZnTe/GaAs/Si

LWIR HgCdTe 128×128 focal plane arrays with a p-on-n heterojunction structure were fabricated on {111}CdZnTe/{100}GaAs/Si substrates using a process described previously for demonstrating high-performance MWIR arrays [11]. An LWIR 128×128 array on this Si-based substrate was hybridized to a transparent readout (TRO) chip [26] that allowed all of the 16,384 detectors to be individually accessed for current-voltage (I-V) characterization.

Figure 5 shows the 78K spectral response at zero bias of five diodes measured near the four corners and at the center of the array. The cutoff wavelength was uniform with a mean value of 10.8 μm; the average quantum efficiency (8.2 μm spike filter) was 50% ± 5%. Figure 6 shows a histogram of the resistance-area product (R_0A_j) determined for each diode (diode junction area, $A_j = 1.22 \times 10^{-5}$ cm^2) measured at 78K, no FOV. The mean R_0A_j product was 17.8 ohm-cm^2 for a cutoff wavelength of 10.8 μm which demonstrates that high-performance LWIR arrays can be fabricated on a Si-based substrate. Not included in the histogram are 11% of the diodes having R_0A_j products less than 0.01 ohm-cm^2. These poor diodes were associated with either morphological defects or apparent linear twin boundaries, not visible on the surface, in the LPE layers. Both of these types of defects can be eliminated and do not appear to limit the technology.

To determine the mechanisms that limit the performance of these detectors, the R_0A_j product was measured as a function of temperature (no FOV) for a typical diode in the array and is shown in Figure 7. The results show that the R_0A_j product is diffusion-limited (proportional to n_i^{-2}, where n_i is the intrinsic carrier concentration) down to approximately 100K, generation-recombination-limited (g-r, proportional to n_i^{-1}) down to approximately 80K, and tunneling-limited at temperatures below 80K. The tunneling, which dominates the very low temperature I-V characteristics, is believed to be associated with dislocations [27]. Although further improvements

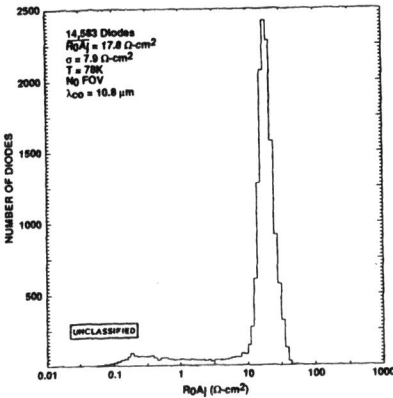

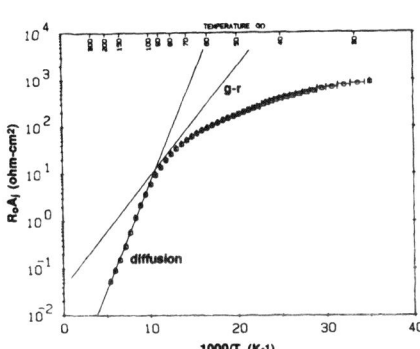

Fig. 6. Histogram of 78K R_oA_j product for HgCdTe diodes in 128×128 array fabricated on CdZnTe/GaAs/Si substrate

Fig. 7. R_oA_j product versus inverse temperature for one diode in 128×128 array fabricated on CdZnTe/GaAs/Si substrate

are needed in the heteroepitaxial growth technology to reduce the dislocation density below 10^6 cm^{-2} to improve very low temperature performance, high performance MWIR and LWIR arrays at temperatures of 80K and above can now be achieved.

CONCLUSIONS

The structural properties of LPE-grown HgCdTe on heteroepitaxial MOCVD-grown CdZnTe/GaAs/Si substrates were evaluated using high-resolution x-ray diffraction techniques and TEM. Large tilts (up to 4°) between CdZnTe layers and GaAs/Si substrates are a general characteristic of this heteroepitaxial system and are attributed to the interaction of closely spaced misfit dislocations that arrange to form a tilt boundary. Either {112}CdTe or {552}CdTe can be grown on {112}GaAs/Si; the {552} was shown to result from a first-order twinning operation of {112}. Lamella {111} microtwins in {111}CdZnTe/{100}GaAs/Si substrates, measured by x-ray techniques, are not readily propagated into the LPE-grown HgCdTe layer. The x-ray FWHM of the LPE HgCdTe is typically at least a factor of two lower than that of the Si-based substrate from annealing and due to the increased thickness of the layer; both mechanisms promote dislocation interaction and annihilation. High performance MWIR and LWIR HgCdTe 128×128 hybrid focal plane arrays were fabricated on these Si-based substrates. An array average of $R_oA_j = 17.8$ ohm-cm^2 for a cutoff wavelength of 10.8 μm at 78K was demonstrated.

ACKNOWLEDGEMENTS

The authors thank M.H. Kalisher, D.G. Voros, and R.F. Herald for LPE growth, V.L. Liguori for material characterization, P.S. Villa for hybridization, and S.R. Seay, J.J. Choquette-Ortega, and J.H. Deloo for device testing. The authors also thank K.T. Miller for his help with the high-resolution x-ray measurements.

REFERENCES

1. R. Sporken, M.D. Lange, C. Masset, and J.P. Faurie, Appl. Phys. Lett. 57, 1449 (1990).
2. H. Zogg and S. Blunier, Appl. Phys. Lett. 49, 1531 (1986).

146

3. A.N. Tiwari, W. Floeder, S. Blunier, H. Zogg, and H. Weibel, Appl. Phys. Lett. **57**, 1108 (1990).
4. K. Zanio, R. Bean, K. Hay, R. Fischer, and H. Morkoc in Heteroepitaxy on Silicon, edited by J.C.C. Fan and J.M. Poate (Mater. Res. Soc. Vol 67, Pittsburgh, PA, 1986), p. 141.
5. R. Kay, R. Bean, K. Zanio, C. Ito, and D. Mcintyre, Appl. Phys. Lett. **51**, 2211 (1987).
6. R. Bean, K. Zanio, and J. Ziegler, J. Vac. Sci. Technol. **A7**(2), 343 (1989).
7. A. Nouhi, G. Radhakrishnan, J. Katz, and K. Koliwad, Appl. Phys. Lett. **52**, 2028 (1988).
8. W.L. Ahlgren, S.M. Johnson, E.J. Smith, R.P. Ruth, B.C. Johnston, M.H. Kalisher, C.A. Cockrum, T.W. James, D.L. Arney, C.K. Ziegler, and W. Lick, J. Vac. Sci. Technol. **A7**(2), 331 (1989).
9. N.W. Cody, U. Sudarsan, and R. Solanki, J. Appl. Phys **66**, 449 (1989).
10. D.D. Edwall, J. Bajaj, and E.R. Gertner, J. Vac. Sci. Technol. **A.**, 1045 (1990).
11. S.M. Johnson, M.H. Kalisher, W.L. Ahlgren, J.B. James, and C.A. Cockrum, Appl. Phys. Lett. **56**, 946 (1990).
12. S.M. Johnson, W.L. Ahlgren, M.H. Kalisher, J.B. James, and W.J. Hamilton, Jr. in Properties of II-VI Semiconductors: Bulk Crystals, Epitaxial Films, Quantum Well Structures, and Dilute Magnetic Systems, edited by F.J. Bartolii, Jr., H.F. Schaake, and J.F. Schetzina (Mater. Res. Soc. Vol 161, Pittsburgh, PA 1990), p. 351.
13. J.M. Arias, S.H. Shin, M. Zandian, J.G. Pasko, and R.E. DeWames, 1990 U.S. Workshop on Physics and Chemistry of HgCdTe and Novel IR Detector Materials, submitted to J. Vac. Sci. Technol. A. (1991).
14. T. Tung, M.H. Kalisher, A.P. Stevens, and P.E. Herning, in Materials for Infrared Detectors and Sources, edited by R.F.C. Farrow, J.F. Schetzina, and J.T. Cheung (Mater. Res. Soc. Vol 90, Pittsburgh, PA 1987), p. 321.
15. K.T. Miller, Hughes Research Laboratories, unpublished.
16. W.J. Bartels and W. Nijman, J. Cryst. Growth **44**, 518 (1978).
17. T. Vreeland, Jr., A. Dommann, C.-J. Tsai, and M.-A. Nicolet, in Thin Films: Stresses and Mechanical Properties, edited by J.C. Bravman, W.D. Nix, D.M. Barnett, and D.A. Smith (Mater. Res. Soc. Vol 130, Pittsburgh, PA 1989) p. 3.
18. S.M. Johnson, W.L. Ahlgren, M.T. Smith, B.C. Johnston, and S. Sen, in Advances in Materials, Processing, and Devices in III-V Compound Semiconductors, edited by D.K. Sadana, L. Eastman, and R. Dupuis (Mater. Res. Soc. Vol. 144, Pittsburgh, PA 1989), p. 121.
19. K. Durose and G.J. Russell, J. Cryst. Growth **101**, 246 (1990).
20. H.-J. Kleebe, W.J. Hamilton, Jr., W.L. Ahlgren, S.M. Johnson, and M. Ruhle, in Properties of II-VI Semiconductors: Bulk Crystals, Epitaxial Films, Quantum Well Structures, and Dilute Magnetic Systems, edited by F.J. Bartolii, Jr., H.F. Schaake, and J.F. Schetzina (Mater. Res. Soc. Vol 161, Pittsburgh, PA 1990), p. 63.
21. P.D. Brown, J.E. Hails, G.J. Russell, and J. Woods, J. Cryst. Growth **86**, 511 (1988).
22. M. Oron, A. Raizman, H. Shtrikman, and G. Cinader, Appl. Phys. Lett. **52**, 1059 (1988).
23. E. Ligeon, C. Chami, R. Danielou, G. Feuillet, J. Fontenille, K. Saminadayar, A. Ponchet, J. Cibert, Y. Gobil, and S. Tatarenko, J. Appl. Phys. **67**, 2428 (1990).
24. A.M. Keir, A. Graham, S.J. Barnett, J. Giess, M.G. Astles, and S.J.C. Irvine, J. Cryst. Growth **101**, 572 (1990).
25. M. Yamaguchi, M. Tachikawa, Y. Itoh, M. Sugo, and S. Kondo, J. Appl. Phys **68**, 4518 (1990).
26. TRO chip was supplied by the Santa Barbara Research Center program on Manufacturing Technology (MANTECH) for HgCdTe Focal Plane Arrays, Wright Research and Development Center, Wright-Patterson Air Force Base contract No. F33615-86-C-5006.
27. S.M. Johnson, D.R. Rhiger, J.M. Peterson, and J.P. Rosbeck, unpublished.

NEW CLASS OF THE SEMICONDUCTOR MATERIALS FOR INFRARED PHOTODETECTION

BORIS A. AKIMOV, NIKOLAI B. BRANDT, SERGEI N. CHESNOKOV AND
DMITRIY R. KHOKHLOV
Physics Department, Moscow State University, Moscow 119899,
USSR

ABSTRACT

We present the new class of infrared photodetectors based
on the lead-tin tellurides doped with group III impurities. The
persistent photoconductivity effect appearing in these
materials provides the opportunity of internal signal
integration resulting in the considerable increase in signal-
to-noise ratio. The integration characteristic time may be
changed by means of the operating temperature or alloy
composition variation. Even if the integration time is higher
than the operation time required there exists an opportunity to
quench quickly ($\sim 10^{-5}$s) the persistent photoconductivity. In
some regime of quenching the effect of giant quantum efficiency
stimulation has been observed. Both the bulk crystal and thin
film technologies of the photodetector production are
developed.

Most of sensitive infrared photodetectors operating in the
wavelength region $\lambda \sim (3-30)$mcm are based on the semiconductor
materials. The detectivity of the most advanced systems is
close to theoretical limit. The restrictions on the
photodetector parameters are usually calculated in the
assumption that the photoresponce is proportional to the
infrared radiation intensity. The light flux integration may
lead to the strong increment in signal-to-noise ratio S/N. This
idea is realized, for example, in the charge coupled devices.
The fabrication technology of this kind of systems is, however,
rather sophisticated, and their dynamical range is not very
wide.

Another attractive opportunity is the application of
persistent photoconductivity phenomenon. The effect essence is
the following. If the permanent radiation flux illuminates the
sample the photoresponce increases linearly in time, and when
the light is switched off, the conductivity value remains
stable (the photomemory). This phenomenon is observed in a
range of materials, and usually it is due to the energy bands
modulation. Is this case the photogenerated free electrons and
holes are spatially separated in the semiconductor potential
field and therefore cannot recombine, their concentration
increases, and the light flux is "integrated". Is is rather
difficult, however, to reproduce the parameters of these
semiconductors, and their application in the real
photodetecting systems is problematic.

In some semiconductors the persistent photoconductivity
phenomenon is due to impurity center specifics, for example, in
$Ga_{1-x}Al_xAs$ [1]. In this case the energy bands are not

modulated, and therefore the crystal properties are much more reproducible. One of the most interesting materials of this kind is $Pb_{1-x}Sn_xTe(In)$.

Generally speaking, doping of the lead-tin tellurides with the group III impurities leads to the appearance of very interesting effects. PbTe(Tl) is the "high-temperature" superconductor among semiconductors with $T_c \sim 3K$ [2], in $Pb_{1-x}Sn_xTe(Ga)$ the impurity donor action is unstable and depends on temperature and external pressure [3].

When the lead-tin tellurides are doped with indium in concentration N_{In} exceeding other impurity concentration N_c, the Fermi level becomes pinned at some very definite position E_0 that practically doesn't depend on N_{In} [4]. E_0 doesn't shift with respect to the middle of the bandgap under the action of external factors: additional doping [4], temperature variation [5], external pressure application [6].

At the same time the E_0 position depends strongly on x. In PbTe the E_0-level lies at ~70meV above the conduction band bottom E_c. With the tin content x increase the level approaches E_c, then crosses the bandgap at $0.22 < x < 0.28$, and enters the valence band giving rise to the free hole concentration [7]. The conductivity of material being in the dielectric state ($0.22 < x < 0.28$) is defined by the charge carrier activation from the E_0-level, and therefore the free carrier concentration is very low $n, p < 10^8 cm^{-3}$ at the temperatures $T < 10K$. It should be noted that the n, p values in the undoped alloys are never lower than $10^{15} cm^{-3}$.

The second important consequence of the Fermi level pinning effect is the homogenization of material electrical properties. Indeed in PbTe(In) with the pinned Fermi level up to 50 periods of the Shubnikov-de Haas (SdH) oscillations are resolved [8] indicating very high degree of the sample homogenuity. It should be noted that no more than 10 periods of the SdH oscillations have ever been observed in undoped PbTe with the comparable electron concentration $n \approx 7 \cdot 10^{18} cm^{-3}$. So the electrical properties of the heavily doped $Pb_{1-x}Sn_xTe(In)$ system are much more reproducible and homogeneous than those of the undoped alloys.

The external infrared illumination leads to the substantial increment of material conductivity at the temperatures $T < 25K$ independently on the Fermi level pinning position [9]. Surely if the E_0-level lies within the bandgap the relative conductivity change is much higher, because there is practically no free carriers in the initial "darkness" state.

After the switch-off the light the photomemory effect is observed. It is due to the strong electron-lattice coupling in the system: when the electron is excited by the light quantum into the conduction band the impurity crystalline surrounding rearranges, and a barrier between the system states with the localized and the free electron appears [10]. The same reason causes the photoexcited free electron accumulation in the conduction band, and therefore the photoresponce is linear in time.

The maximal accumulation characteristic time is defined by the photoexcited electron lifetime τ. In $Pb_{1-x}Sn_xTe(In)$ $\tau \sim 10^4 s$ at $4.2 K < T < 10 K$, then it sharply decreases with the temperature rise, and $\tau \sim 10^{-2}s$ at $T \sim 20 K$. So one may chose the proper temperature in order to provide the photodetector operation time required τ_0. If the sample temperature is so that $T>T_0$ the photoresistor may operate, only if there exist a way to return the alloy conductivity to the initial "darkness" state, i.e. to quench quickly the persistent photoconductivity. Moreover, periodical accumulation and successive fast quenching of the photosignal leads to the substantial gain in the S/N ratio with respect to the case of ordinary single photodetectors.

One may transfer the photoexcited free electrons back into the localized states by means of the sample heating up to $T\approx25K$ and successive cooling down [9]. However this process is too slow.

Another possible way of the persistent photoconductivity quenching is the application of strong electric field on the sample contacts. In this case the system returns to the dielectric state due to the electrothermal breakdown [11]. The persistent photoconductivity quenching produced in this way is complete only in the photomemory regime, i.e. if the light is off. If the photogeneration is permanent, and the electric field pulses are applied periodically on the sample contacts, the photoconductivity quenching is unstable and not complete: the minimal conductivity value after the pulse end is higher than in the initial "darkness" state and changes from pulse to pulse. Moreover the periodical application of the strong electric field on the sample contacts leads to the irreversible photoresponce decrease and to its disappearance.

We worked out the new method of the persistent photoconductivity quenching by the UHF-pulses [12]. The sample was installed in the UHF-resonator loaded by the short rectangular UHF-pulses of $\Delta t \sim (10^{-6}-10^{-2})$s length. The UHF-frequency was ~ 400 MHz, the power P in the pulse – up to 38 W. We obtained that it is possible to transfer all of the free photogenerated electrons into the localized state for ~10^{-5}s without any substantial heating of the crystalline lattice. It means that the short UHF-pulse affects the electron system directly. This UHF-quenching is much more stable than the electrothermal one.

The possibility of the fast periodical persistent photoconductivity quenching permits to construct a photodetector with parameters depending on the integration time t_i. The t_i value variation allows to make photodetector dynamic range very wide.

The application of the short UHF-pulse leads not only to the persistent photoconductivity quenching, but affects also the character of the photoresponce after the pulse end. In the experiment to be described below the permanent flux of IR-radiation illuminated the sample. The persistent photoconductivity was quenched by the single UHF-pulse. We measured the photoconductivity signal increase dynamics after the pulse end.

The experiment results are shown in fig. 1. The photoresponce structure depends strongly on Δt. One can see that when Δt is close to Δt_0 – minimal pulse length necessary for the complete quenching – the initial photoconductivity increase rate R_0 is much higher than in the case $\Delta t \gg \Delta t_0$. The more is $(\Delta t - \Delta t_0)$ the less is R_0. This UHF-stimulation of the photoresponce acts only some time after the UHF-pulse end. After that the slope of the photosignal increment practically doesn't depend on Δt.

The effect described above may be due either to the strong reconstruction of the photoresponce spectrum (red shift of the

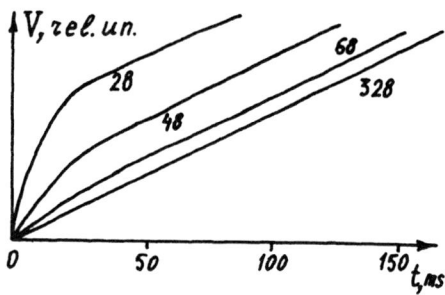

Fig. 1. The photosignal V dependence on the time t after the UHF-pulse end for the different pulse lengths Δt. P=38 W. Figures near the curves - Δt in μs. $\Delta t_0 = 20$ mcs

characteristic cutoff wavelength), or to the giant increment of the quantum efficiency η. Our preliminary measurements show that the second reason gives the main contribution to the effect. Moreover the evaluations of the quantum efficiency in the UHF-stimulation regime give η>10. As far as we know, this is the only case when the quantum efficiency exceeds 1 in the photoresistors.

To our opinion, the qualitative scheme of the processes involved is the following. There exist two types of the impurity localized states in $Pb_{1-x}Sn_xTe(In)$: ground two-electron ones $\mathcal{E}_2$ and metastable one-electron states $\mathcal{E}_1$ [13]. The $\mathcal{E}_2$-states that provide the Fermi level pinning are strongly localized, and their wavefunctions practically don't overlap [14]. The extended and two localized states are separated with each other by barriers in the configuration space due to the Yahn-Teller instability and strong electron-lattice coupling [10]. This suppresses the recombination of the photoexcited free carriers and results at the end in the persistent photoconductivity effect.

The one-electron states manifest themselves in fine effects: negative photoconductivity [15], giant negative magnetoresistance [16], singularities in the infrared reflection spectra [17]. The UHF-stimulation of the quantum efficiency is one more example of this kind of effects.

The application of the short UHF-pulse leads to electron transfer from conduction band to metastable one-electron state. The barriers separating these states from the extended and two-electron ones are rather small (W~1 meV [15]), that is why the longer UHF-pulse returns all the previously photoexcited electrons into the ground $\mathcal{E}_2$-state. So the short UHF-pulse results in the excess population of the one-electron states. The $\mathcal{E}_1$-states are much less localized than the 2-electron ones [16], and some of $\mathcal{E}_1$-centers may form a cluster with strong internal interaction. In these conditions excitation of one electron from this cluster to the extended state leads to reconstruction of the center crystalline surrounding and to the avalanche devastation of all of the centers in the cluster resulting in the quantum efficiency increase. This explanation is close to the ideas claimed in [18].

If the proposed mechanizm is valid the photorescponce spectrum must change in the UHF-stimulation regime. However measurements of the photoconductivity spectra of materials revealing the photomemory effect is a sophisticated problem. In fact the background radiation present in any spectrophotometer "pumps" the nonequilibrium free carriers, and one cannot begin the measurements from the ground "darkness" state. Moreover one can register only the spectral responce of the processes with the characteristic time τ_p lower than the time of spectral scanning τ_s. In $Pb_{1-x}Sn_xTe(In)$ $\tau_p \sim (10^4-10^5)s$ at T=4.2 K, and the ordinary spectral measurements are practically impossible.

We proposed the indirect technique of the red cutoff energy determination based on the analysis of the photoresponce increase dynamics under the action of the black-body illumination [19]. The advantage of this technique is the complete screening of the background radiation. Though the method accuracy is low ~(20-50)%, it allowed to estimate the cutoff energy $\mathcal{E}_{co} \approx 20$ meV in $Pb_{0.75}Sn_{0.25}Te(In)$ [19]. The UHF-stimulation effect should provide the red shift of $\mathcal{E}_{co}$. We are developing now the new set up that will allow the precise photoconductivity spectra measurements in any regime.

The impurity states peculiarities in the $Pb_{1-x}Sn_xTe(In)$ alloys give the opportunity to increase substantially the photoresponce in the rather low magnetic field H~0.5T. In fact the metastable one-electron states act as the traps for the injected free electrons, their energy changes in the magnetic field providing the traps devastation and producing the giant negative magnetoresistance effect [16]. The effect magnitude drastically rises under the action of the slight infrared illumination (fig.2). This fact may be due to the decrease of the capture process efficiency in the magnetic field.

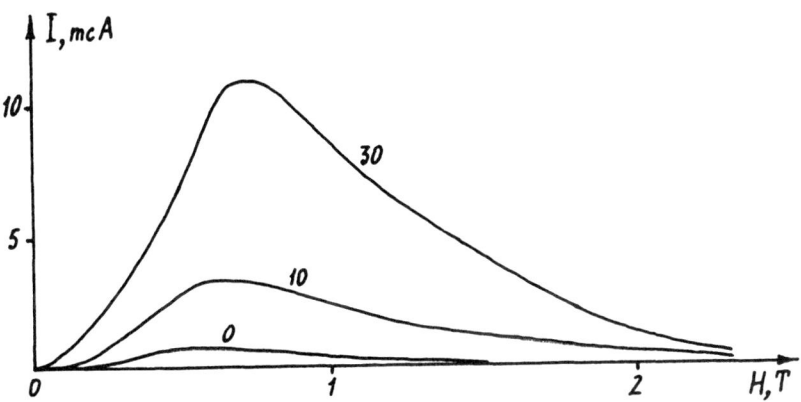

Fig. 2. The dependencies of the sample current I on magnetic field H under the action of the slight black-body infrared illumination. Figures near the curves - radiation exposition in s, black-body temperature is 12K. The sample size is $(0.5*0.5*5)mm^3$. The voltage on the sample V=4V. T=4.2K.

One more attracting feature of the $Pb_{1-x}Sn_xTe(In)$ alloys with respect to their practical application is the possibility of spatially unequilibrium state generation under the action of local irradiation. If some part of the sample is locally illuminated, the persistent photoconductivity effect is observed only in this part, and the photoexcitation doesn't propagate into the darkened regions. The situation remains stable even after the switch-off the light, i.e. in the photomemory regime. The excitation propagation characteristic time is at least more than 10^5s at T=4.2K. The spatial characteristic scale is ~10mcm [20].

Indium is not the only group III impurity that causes the Fermi level pinning and the persistent photoconductivity effect appearance. When lead telluride is doped with Ga, there exist some N_{Ga} interval in which the Fermi level is pinned near the middle of the bandgap. In contrary to the case of In doping the increment of Ga content leads to the sudden Fermi level jump high into the conduction band [21]. It is interesting to note that the Fermi level pinning effect has not been observed in $Pb_{1-x}Sn_xTe$(Ga) solid solutions even with very small tin content x.

The persistent photoconductivity effect appears in PbTe(Ga) at temperatures $T<T_c \approx 80$ K [21]. The respective cutoff wavelength $\lambda_{co} \approx 5.5$ mcm corresponds to the PbTe bandgap. The λ_{co} value may be changed by means of some additional doping (with Mn [22] or Ge[23]), but the T_c value remains stable and seems to depend only on the type of the group III impurity. The analysis of the infrared reflection spectra of lead-tin tellurides doped with In and Ga show that the temperature T_c of persistent photoconductivity effect appearance corresponds to the formation of one-electron metastable states [24].

Up to now we discussed the properties of bulk monocrystalline $Pb_{1-x}Sn_xTe$(In) and PbTe(Ga). Thin films of these materials deposited on BaF_2 substrate reveal some different properties. For example, it turned out to be very difficult to grow the film of $Pb_{1-x}Sn_xTe$(In) with the Fermi level pinned within the bandgap. Usually the "darkness" low-temperature resistivity of these films is much lower compared with the bulk crystals, and the photoresponce is sufficiently smaller [25]. Application of the sophisticated film growth technique allowed to obtain the films of $Pb_{1-x}Sn_xTe$(In) with the properties close to the bulk crystalline ones. However no progress in growing PbTe(Ga) films revealing the persistent photoconductivity effect is reached up to now.

The application of the lead-tin tellurides doped with the group III impurities as the base elements in photodetectors gives a challenging opportunity to produce universal and sensitive systems.

REFERENCES

1. Lang D. W. and Logan R. A. Phys. Rev. Lett. 39 635 (1977).
2. Kaidanov V. I., Nemov S. A., Parfen'ev R. V. and Chamchur D. V. Pis'ma v Zh. Eksp. Teor. Fiz. 32 517 (1982).
3. Akimov B. A., Brandt N. B., Ryabova L. I., Khokhlov D. R., Chudinov S. M. and Yatsenko O. B. Pis'ma v Zh. Eksp. Teor. Fiz. 31 304 (1980).
4. Averkin A. A., Kaidanov V. I. and Mel'nik R. B. Fiz. Tekh. Polupr. 5 91 (1971).
5. Akimov B. A., Brandt N. B., Ryabova L. I., Sokovishin V. V. and Chudinov S. M. J. Low Temp. Phys. 51 9 (1983).
6. Akimov B. A., Zhlomanov V. P., Ryabova L. I., Chudinov S. M. and Yatsenko O. B. Sov. Phys. Semicond. 13 759 (1979).

7. Akimov B. A., Ryabova L. I, Yatsenko O. B. and Chudinov S. M. Sov. Phys. Semicond. 13 441 (1979).

8. Akimov B. A., Brandt N. B., Kurbanov K. R., Ryabova L. I., Khasanov A. T. and Khokhlov D. R. Sov. Phys. Semicond. 17 1021 (1983).

9. Akimov B. A., Brandt N. B., Klimonskiy S. O., Ryabova L. I. and Khokhlov D. R. Phys. Lett. A 88A 483 (1982).

10. Volkov B. A. and Pankratov O. A. Sov. Phys. Dokl. 25 922 (1985).

11. Akimov B. A., Brandt N. B., Kerner B. S., Nikiforov V. N. and Chudinov S. M. Sol. St. Comm. 43 31 (1982).

12. Akimov B. A., Brandt N. B., Khokhlov D. R. and Chesnokov S. N. Sov. Tech. Phys. Lett. 14 325 (1988).

13. Zasavitskiy I. I., Matsonashvili B. N., Pankratov O. A. and Trofimov V. T. Sov. Phys. JETP Lett. 42 1 (1985).

14. Golubev V. G., Grechko N. I., Lykov S. N., Sabo E. P. and Chernik I. A. Fiz. Tekh. Polupr. 11 1704 (1977).

15. Zasavitskiy I. I., Matveenko A. V., Matsonashvili B. N. and Trofimov V. T. Sov. Phys. Semicond. 20 135 (1986).

16. Akimov B. A., Nikorich A. B., Khokhlov D. R. and Chesnokov S. N. Fiz. Tekh. Polupr. 23 668 (1989).

17. Romchevich N., Popovich Z. V. and Khokhlov D. R. Phys. Rev. B (in press).

18. Akimov B. A., Brandt N. B., Nikorich A. V., Ryabova L. I. and Sokovishin V. V. Sov. Phys. JETP Lett. 39 265 (1984).

19. Akimov B. A., Brandt N. B., Ryabova L. I. and Khokhlov D. R. Sov. Tech. Phys. Lett. 6 544 (1980).

20. Akimov B. A., Brandt N. B., Chesnokov S. N., Egorov K. N. and Khokhlov D. R. Sol. St. Comm. 66 811 (1988).

21. Akimov B. A., Brandt N. B., Gas'kov A. M., Zlomanov V. P., Ryabova L. I. and Khokhlov D. R. Sov. Phys. Semicond. 17 53 (1983).

22. Akimov B. A., Belokon' S. A., Dashevskiy Z. M., Egorov K. N., Lakeenkov V. M. and Ryabova L. I. Sov. Phys. Semicond. (in press).

23. Belogorokhov A. I. (private communication).

24. Belogorokhov A. I., Ivanchik I. I., Khokhlov D. R., Popovich Z. V. and Romchevich N., presented at the 1990 MRS Fall Meeting, Boston, MA, 1990 (unpublished).

25. S. W. McKnight and M. K. El-Rayess, in International Conference on Narrow-Gap Semiconductors and Related Materials Abstracts Gaithersburg, Maryland, USA, 1989.

OHMIC CONTACTS ON N-TYPE $Hg_{0.4}Cd_{0.6}Te$.

PATRICK W. LEECH[*], GEOFFREY K. REEVES[+] AND MARTYN H. KIBEL[*]
[*]Telecom Australia Research Laboratories, Melbourne, Victoria, Australia.
[+]Victoria University of Technology, Victoria, Australia.

ABSTRACT

The electrical characteristics of In, Sn, Au and Pt contacts on n-type $Hg_{0.4}Cd_{0.6}Te$ formed in the presence and absence of prior In^{2+} implantation have been examined. Measurements of specific contact resistance made using a Transmission Line Model have shown that the unimplanted $In/Hg_{0.4}Cd_{0.6}Te$ and $Sn/Hg_{0.4}Cd_{0.6}Te$ junctions gave values of $\rho_c = 3.0 \times 10^{-3}$ to 4.0×10^{-3} ohm.cm^2. Auger sputter profiles of the as-deposited $In/Hg_{0.4}Cd_{0.6}Te$ and $Sn/Hg_{0.4}Cd_{0.6}Te$ interfaces have shown a significant in-diffusion of the metal overlayer. The influence of shallow In^{2+} implantation prior to metallization was an increase in ρ_c which occurred above a dose of 10^{13} ions/cm^2. In contrast, Pt and Au formed Schottky barrier diodes on n-type $Hg_{0.4}Cd_{0.6}Te$ with $\varnothing_b = 0.69eV$ for Pt and $\varnothing_b = 0.79eV$ for Au. With prior In^{2+} implantation, both Pt and Au contacts exhibited an ohmic behaviour with $\rho_c = 2 \times 10^{-1}$ ohm.cm^2. These results have significance in the fabrication of devices for $1.0 - 2.5\mu m$ optical communications.

INTRODUCTION

The semiconductor $Hg_{1-x}Cd_xTe$ has important potential for application in optical communications devices such as sensors, emitters and field effect transistors operating in the near infrared spectral region ($1-2.5\mu m$) [1,2]. The realization of these device structures requires the formation of low resistivity ohmic contacts to n-type $Hg_{1-x}Cd_xTe$ in the corresponding range of stoichiometries, namely x=0.6 to 0.7. Based on theoretical models for Schottky barrier formation, Spicer et al.[3] have predicted that metal contacts on n-type $Hg_{1-x}Cd_xTe$ will have an intrinsic rectifying character for $x \geqslant 0.4$. Ohmic contacts to $n-Hg_{1-x}Cd_xTe$ in this range of x have been achieved by the use of a graded gap heterojunction which comprised metal/HgTe/$n-Hg_{1-x}Cd_xTe$ [4]. An alternative approach to ohmic contact formation not previously examined in $Hg_{1-x}Cd_xTe$, $x \geqslant 0.4$, is the generation of a highly doped n^+ surface layer either by diffusion of the contact metal or by shallow ion implantation. The possibility of impurity doping by the contact metal exists because of the extensive interaction between many metals and $Hg_{1-x}Cd_xTe$ at room temperature [5]; while low level implantation has previously been used in Si [6] and Ge [7] to reduce Schottky barrier height. The purpose of the present study is to examine the electrical characteristics of metal/$n-Hg_{0.4}Cd_{0.6}Te$ junctions using metals with substitutional donor (In, Sn), acceptor (Au) or unknown doping characteristics (Pt); both with and without the effects of prior In^{2+} implantation. In addition to the determination of the specific contact resistance, ρ_c using the transmission line model, metallurgical aspects of the $In/n-Hg_{0.4}Cd_{0.6}Te$ and $Sn/n-Hg_{0.4}Cd_{0.6}Te$ interfaces have been examined by Auger depth profiling.

Mat. Res. Soc. Symp. Proc. Vol. 216. ©1991 Materials Research Society

EXPERIMENTAL

The $Hg_{0.4}Cd_{0.6}Te$ layers were grown by metalorganic chemical vapour deposition (MOVCD) on 2° missoriented (100) GaAs substrates of $2x\ 10^{-3}$ ohm.cm resistivity. Formation of the $Hg_{0.4}Cd_{0.6}Te$ was by an interdiffused multilayer process (IMP) [8] in which alternate CdTe/HgTe layers were deposited from dimethyl cadmium, diethyl tellurium alkyls and an elemental Hg source. The HgTe/CdTe period thickness was 100nm with a subsequent in-situ anneal at 360°C. The resulting $Hg_{0.4}Cd_{0.6}Te$ was n-type, 1.5μm in thickness and with a carrier concentration of $3.5x10^{15}cm^{-3}$ as determined from Hall measurements. Alloy composition was in the range of x=0.59 to 0.62; with Selected area Electron Channelling and X-ray diffraction used to confirm the epitaxy of the $Hg_{0.4}Cd_{0.6}Te$ layers.

Contact pads in a pattern conforming to the transmission line model were then photolithographically defined on the surface of the $Hg_{0.4}Cd_{0.6}Te$/GaAs. Prior to metallization, selected samples were implanted with 50keV In^{2+} to doses of $6x10^{12}$, $1x10^{13}$ and $5x10^{13}$ ions/cm^2 at room temperature. During the implantation treatment, a low level of current density of the ion beam ($2\mu A.cm^{-2}$) eliminated any sample heating, with the photoresist pattern acting as a mask. Layers of 30nm thick Au, In, Pt or Sn followed by 200nm Au were deposited by either a resistively heated source (In, Sn) or by electron beam (Au, Pt) at a base pressure of $1x10^{-6}$ torr. Both unimplanted samples and samples with differing implant dosages were metallized. In a second series, unimplanted samples of $Au/In/n$-$Hg_{0.4}Cd_{0.6}Te$ were prepared with varying thickness of In (2.5nm to 30nm) overlaid by Au (200nm). Following the metal deposition, a lift-off technique was used to define the arrays of contact pads. Test structures were then isolated by chemical etching in I:KI:HBr [9], which selectively removed the $Hg_{0.4}Cd_{0.6}Te$ down to the semi-insulating GaAs substrate. Using a four point probing technique and based on the transmission line model, the specific contact resistance was calculated from ρ_c = Zw/a where w is the width of the contact pads. The constants a and Z were determined from measurements of contact resistance R_c and end contact resistance R_e where R_c = Z.coth(ad) and R_e=Z/sinh(ad) [10]. In addition, Auger depth profiles of $In/Hg_{0.4}Cd_{0.6}Te$ samples were performed in a Vacuum Generators HB100 Electron Spectrometer. For ion milling, a 5 keV Ar^+ beam was used at a current density of $40\mu A\ cm^{-2}$. The electron beam was rastered over an area of $45\mu m$x$30\mu m$ at a current density of $1.5mA\ cm^{-2}$ which was well below the limit for Hg desorption.

RESULTS AND DISCUSSION

The $In/Hg_{0.4}Cd_{0.6}Te$ and $Sn/Hg_{0.4}Cd_{0.6}Te$ junctions prepared on unimplanted surfaces were ohmic in character with measured specific contact resistances of ρ_c=$3x10^{-3}$ ohm.cm^2 and $4x10^{-3}$ ohm.cm^2 respectively; while the Au and Pt contacts were rectifying as described later. For $Au/In/n$-$Hg_{0.4}Cd_{0.6}Te$ contacts, Figure 1 shows the dependence of ρ_c on the thickness of the In layer. Two regimes may be defined here- above 15nm at which the value of ρ_c was approximately constant at $4x10^{-3}$ ohm.cm^2, and below 15nm at which ρc increases until frequently becoming non-ohmic in current-voltage response at 2.5nm. These characteristics may be attributed to an inward diffusion of In or Sn and the formation of a surface region more heavily n-type

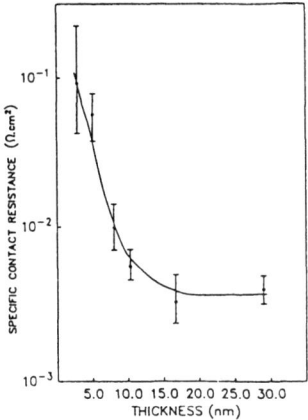

Figure 1. Specific contact resistance
as a function of In layer thickness
for Au/In/n-Hg$_{0.4}$Cd$_{0.6}$Te contacts.

doped than the bulk. Johnson and Schmitt [11] have shown that In is a sublattice substitutional donor which diffuses rapidly in Hg$_{1-x}$Cd$_x$Te, while Sn was designated as a donor atom which was electrically active at temperatures of 300°C. More generally, Group IVB elements exhibit strong donor properties in Hg$_{1-x}$Cd$_x$Te, with the behaviour of Sn dependent on the valence state adopted (Sn^{2+} or Sn^{4+}). Popovic [12] has considered the criteria for ohmic contact formation at a metal/n-type semiconductor contact with a thin, heavily doped layer. According to the above criteria, ohmic contacts occurred when the depth of the doped surface layer was at least equal to or greater than the equilibrium width of the depletion region. In this manner, a shallow n$^+$ layer on an n-type bulk can increase the interfacial electric field in the semiconductor, thereby allowing significant electron tunnelling. For the In/Hg$_{0.4}$Cd$_{0.6}$Te contacts, the depletion width was calculated as 0.1μm given ϵ = 13.2 at x=0.7 [13]. The diffusion of In in Hg$_{1-x}$Cd$_x$Te is known to be very high[13] despite wide variations in data. Using the experimental data from [14], a room temperature diffusion constant of D = 1.7x10$^{-4}\mu$m^{2}sec^{-1} is obtained. Thus even at room temperature In can diffuse across a depletion width of 0.1μm in a short period of time. Hence, for In/Hg$_{0.4}$Cd$_{0.6}$Te, the width of the In doped layer at the surface will probably exceed the equilibrium value of depletion width, which is consistent with the observed formation of ohmic contacts. The effect of initial thickness of In on ρ_c was thus the result of a decreasing availability of the In source. Figure 2 shows typical Auger depth profiles for In/Hg$_{0.4}$Cd$_{0.6}$Te and Sn/Hg$_{0.4}$Cd$_{0.6}$Te interfaces. For both metals, an extensive interdiffusion was evident to an estimated depth of 0.13-0.15μm for In and 0.12-0.16μm for Sn. Elemental Hg was present directly at the interfacial region while Hg, Cd and Te also appeared within the overlayer. In the case of the In coated material, a significantly lower level of Hg was indicated in the Hg$_{0.4}$Cd$_{0.6}$Te away from the interface than for the Sn coated material, despite the similar reactivity of In compared to Sn, (as determined from heat of telluride formation- Δ.H$_f$ = 15.3kcal/mole Te for In$_2$Te$_3$ and Δ.H$_f$ = 15.04kcal/mole Te for SnTe). In addition, Vydyanath has shown that for Hg$_{1-x}$Cd$_x$Te (x=0.2), most of the In is present as In$_2$Te$_3$ with only a small fraction present as donors on metal sites [15].

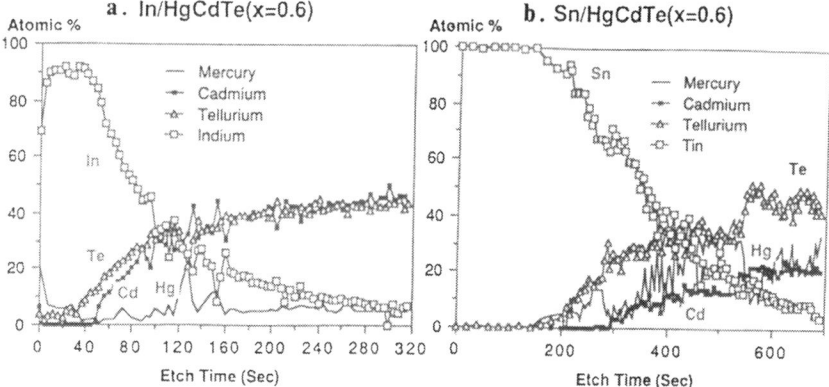

Figure 2. Auger sputter profiles for (a) In/n-$Hg_{0.4}Cd_{0.6}Te$ and (b) Sn/n-$Hg_{0.4}Cd_{0.6}Te$ interfaces.

The present results indicate that despite the probable formation of reacted telluride species, In and Sn acted as substitutional donors in the interfacial region. The effect of shallow ion implantation on specific contact resistance is shown in Figure 3.

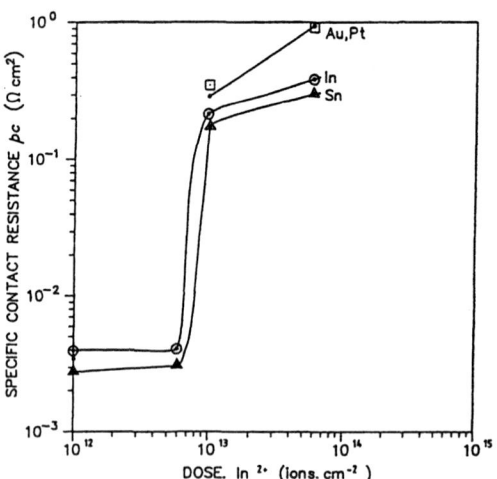

Figure 3. Specific contact resistance as a function of prior In^{2+} dosage for Au, Pt, In and Sn on $Hg_{0.4}Cd_{0.6}Te$.

At a dose of 6×10^{12} ions cm^{-2}, the In^{2+} implantation resulted in a 10^2 increase in ρ_c for contacts formed by both In and Sn. At larger doses, (10^{13}ions cm^{-2}), only a further minor increase in ρ_c was determined. This effect may be correlated with the reported minimum threshold dose for implantation damage to Hg$_{1-x}$Cd$_x$Te ($x=0.7$) which occurred at 10^{13}ions cm^{-2}[15]. Above this threshold, the damage density (measured by RBS) increased linearly with the ion dosage; with the damage at the lowest doses being comprised of point defects which agglomerated to form extended defects [15]. Subsequent annealing of the implanted and unimplanted samples at temperatures up to 200°C had little effect on the value of ρ_c.

In comparison, Au and Pt/n-Hg$_{0.4}$Cd$_{0.6}$Te contacts were rectifying at implant doses of 6×10^{12} ions cm^{-2} but exhibited ohmic behaviour at doses of 10^{13} ions cm^{-2} as shown in Figure 4. For doses of 10^{13} ions cm^{-2} the magnitude of the specific contact resistance was marginally greater for the metals (Au, Pt) than for (In, Sn).

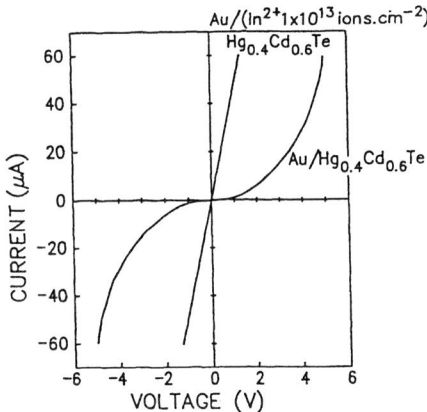

Figure 4. Current-voltage curves for Au/Hg$_{0.4}$Cd$_{0.6}$Te contacts with and without prior implantation with In^{2+} to a dose of 1×10^{13} ions.cm^{-2}.

This difference can probably be attributed to the doping effect of the indiffused In or Sn, although the extent of implantation damage appears the dominant factor controlling the magnitude of ρ_c. Au and Pt/Hg$_{0.4}$Cd$_{0.6}$Te contacts formed on unimplanted surfaces produced a Schottky barrier height of $\phi_b=0.70\pm0.01$V and ideality factor of $n=1.95$ for Pt and $\phi_b=0.79\pm0.03$V and $n=2.0$ for Au. These measurements were performed on Hg$_{0.4}$Cd$_{0.6}$Te grown on Si doped substrates of GaAs with an ohmic back contact. ϕ_b and n were determined from the intercept and slope of the lnJ versus V plot obtained from dot contacts of area 0.008cm^{-2}.

The permission of the Executive General Manager, Research, to publish this work is acknowledged.

REFERENCES

1.   B. Orsal, R. Alabedra, M. Valenza, G.P. LeCoy, J. Meslage and C.Y. Boisrobert, IEEE Trans. Electron Devices, ED-35, (1988), 101.

2   J. Thompson, P. Mackett, G.T. Jenkin, T. Nguyen Duy and P. Gori, J. Crystal Growth, 86, (1988), 917.

3.   W.E. Spicer, D.J. Friedman and G.P. Carey, J.Vac.Sci. Technol., A6, (1988), 2746.

4.   P.W. Leech, J. Appl. Phys., 68 (2), (1990), 1174.

5.   W.E. Spicer, J.Vac.Sci.Technol., A8 (2), (1990), 1174.

6.   J.M. Shannon, Solid State Electronics, 19, (1976), 537.

7.   C.C. Han, E.D. Marshall, F. Fang, L.C. Wang, S.S. Lau and D. Voreades, J. Vac. Sci. Technol. B6 (6), (1988), 1662.

8.   G. Pain, N. Bharatula, T.J. Elms, P. Gwynn, M. Kibel, M.S. Kwietniak, P. Leech, N. Petkovic, C. Sandford, J. Thompson, T. Warminiski, D. Gao, S.R. Glanvill, C.J. Rossouw, A.W. Stevenson, S.W.Wilkins and L. Wielunski, J. Vac. Sci. Technol., A8, (2), (1990), 1067.

9.   P.W. Leech, P.J. Gwynn and M.H. Kibel, App. Surf. Sci., 37, (1989), 291.

10.   G.K. Reeves and H.B. Harrison, IEEE Electron Device Lett. EDL-3, (1982), 111.

11.   E.S. Johnson and J.T. Schmitt, J. Electronic Mats., 6 (1), (1977), 25.

12.   R.S. Popovic, Solid State Electronics, 21, (1978) 1133.

13.   J. Brice and P. Capper (eds.), Properties of Mercury Cadmium Telluride, EMIS Datareviews Series No.3, The Institution of Electrical Engineers, London, (1987), Chap. 2.1, Chap. 3.2.

14.   S. Margalit and Y. Nemirovsky, J. Electrochem. Soc., 127 (6), (1980), 1406.

15.   H.R. Vydyanath, J. Electrochem. Soc., 128 (12), (1981), 2619.

16.   C. Uzan-Saquy, D. Comedi, V. Richter, R. Kalish and R. Triboulet, J.Vac.Sci. Technol., A7 (4), (1989), 2575.

PART III

Narrow Gap III-V and IV-VI Ternary and Quaternary Compounds

DEVELOPMENT OF INFRARED DETECTORS BASED ON TYPE II, InAsSb STRAINED-LAYER SUPERLATTICES

STEVEN R. KURTZ
Sandia National Laboratories, Albuquerque, New Mexico, 87185

Abstract

An overview is provided of long wavelength, photovoltaic detectors constructed with type II (also known as "staggered"), III-V superlattices. Specifically, the electronic properties of InAsSb strained-layer superlattices and prototype detectors utilizing these structures are described.

I. Introduction

Optical absorption in III-V semiconductors can be effectively extended to long wavelength ($\geq 10 \mu m$) for infrared detector applications by utilizing low energy, intersubband transitions [1] or by reducing the superlattice valence-conduction bandgap with a type II band offset.[2] (We define a type II superlattice as a structure having the electron and hole potentials wells located in different layers of the superlattice.) The type II superlattice detectors potentially could fill a useful niche in long wavelength detector technologies. Type II detectors offer advantages for focal-plane-array construction because unlike intersubband detectors [3], type II detectors can operate as photodiodes, responsive to normal incidence light. Furthermore, the detectivity of type II detectors may soon rival that of HgCdTe while offering advantages to HgCdTe in manufacturability, durability, and radiation hardness.[4,5]

In this overview, the electronic and optical properties of type II, InAsSb/InSb strained-layer superlattices (SLS) are briefly summarized and compared with alternative, InAs/GaInSb type II SLSs. A high detectivity InAsSb SLS photodiode is described which illustrates both the potential of type II photodiode technologies and device design considerations for SLS detectors. Lastly, we speculate on the future directions of InAsSb detector development .

II. Electronic Properties of InAsSb SLSs and Other Type II Infrared Structures

In the mid-1970's, the InAs/GaSb superlattice was investigated as a near-lattice-matched alternative to GaAs/AlAs type structures. Immediately, the type II band offset in InAs/GaSb was identified [6], and the potential of type II superlattices for long wavelength infrared detectors was recognized. However, long wavelength optical absorption in InAs/GaSb superlattices was considered to be too weak for use as an infrared detector. With the development of lattice mismatched SLSs, Osbourn examined InAsSb SLSs as potential III-V structures having significant optical absorption at long wavelength.[7] Subsequently, a type II offset was discovered in these structures.[8,9]

A comparison of optical absorption in InAsSb SLSs and InAs/GaSb superlattices is shown in Figure 1. Both structures exhibit photoresponses extending to approximately 9 μm. Improved long wavelength absorption in InAsSb SLSs is clearly demonstrated in Fig. 1 although the SLS layers are roughly twice as thick as those of the InAs/GaSb superlattice. Because of the spatial separation of electrons and holes, absorption coefficients of type II superlattices at energies near the bandgap are sensitive functions of the superlattice layer thickness. The wavefunction overlaps which determine the strength of the band edge transitions depend on the exponentially decaying (evanescent) tails of the

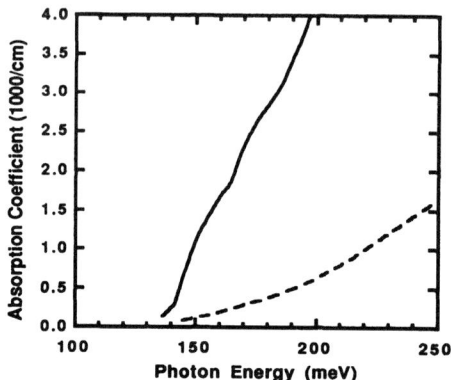

Figure 1 - Low temperature, optical absorption spectra for an $InAs_{0.13}Sb_{0.87}/InSb$ SLS (solid line, data from Ref. 8) with equal, 106 Å thick layers and an InAs/GaSb superlattice (dotted line, data from Ref. 10) with equal, 56 Å thick layers.

wavefunctions in the barrier layers. The decay lengths of these tails increase with decreasing barrier heights and decreasing effective masses. Useful absorption coefficients occur for decay lengths on the order of the barrier thickness, and band edge absorption in type II superlattices is increased by reducing layer thickness.

The relative bandgaps and barrier heights for InAs/GaSb and InAsSb SLSs are shown in Figure 2. The bandgaps of InAs and GaSb are much larger than that of InSb (See Fig. 2a), and the large barriers in InAs/GaSb superlattices result in the weak absorption seen in Fig. 1. Mailhoit and Smith proposed the use of InAs/GaInSb SLSs to increase the absorption coefficient from that of InAs/GaSb superlattices.[11] The bandgaps and barriers of a proposed InAs/GaInSb SLS are compared with that of an InSb/InAsSb SLS in Fig. 2b and 2c. The barriers in the InSb/InAsSb SLS remain significantly smaller than those of the InAs/GaAsSb SLS. This disadvantage is overcome by using very thin layers (< 40 Å) to increase absorption in InAs/GaAsSb SLSs. To obtain equal absorption coefficients and equal "cutoff" wavelengths in these two SLSs, the layers in the InAs/GaInSb SLS are thinner than those in the corresponding InSb/InAsSb SLS.

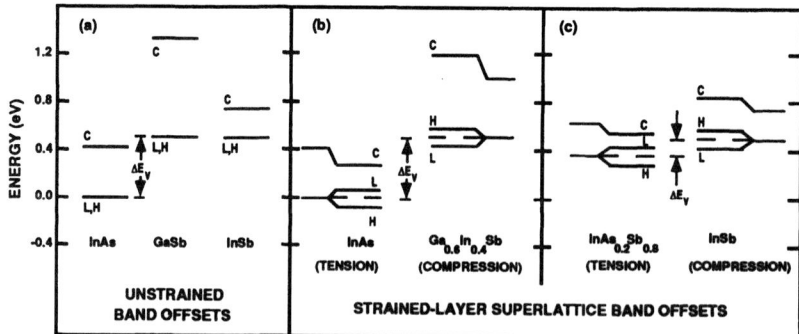

Figure 2 - (a) Relative energy levels of the conduction bands (C) and unstrained valence bands (L,H) of InSb, InAs, and GaSb. (b) Strained conduction band energies (C) and out--of-plane light-hole (L) and heavy-hole (H) energies for an $InAs/Ga_{0.6}In_{0.4}Sb$ SLS. (c) Strained conduction band (C) and valence band (L,H) energies for an $InAs_{0.2}Sb_{0.8}/InSb$ SLS. (Fig. 2a and 2b were obtained from Ref. 11.)

The energy levels of a typical InSb/InAsSb SLS are shown in Figure 3. (New magneto-optical and transport studies confirm this basic picture.[12]) Quantum size states have not been included in the figure. Characteristic of the type II, long wavelength SLSs, significant band edge absorption can result from electron minibands in the conduction band, and useful electron diffusion lengths, perpendicular to the SLS growth planes are easily obtained. Hole diffusion perpendicular to the growth planes is limited due to localization in the InSb layer of the lowest energy, out-of-plane heavy-hole (hh) state. Spatial separation of electrons and holes reduces band edge absorption in type II superlattices, but spatial separation also decreases electron-hole recombination rates. Long photocarrier lifetime increases the minority carrier diffusion length in photodiodes and can produce high gain in photoconductive detectors.

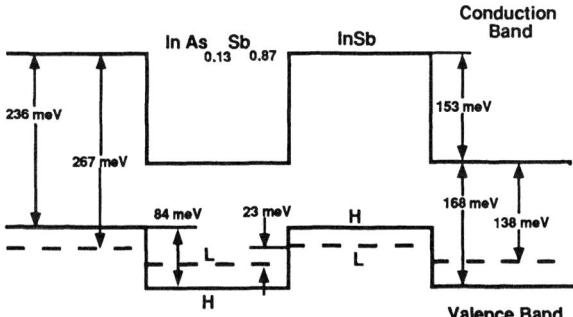

Figure 3 - Quantum well energies of an $InAs_{0.13}Sb_{0.87}$/InSb SLS.

III. A High Detectivity, InAsSb SLS Photodiode

A high detectivity, SLS photodiode (Figure 4) was fabricated from a p-p⁻-n junction embedded in an InSb/InAs$_{0.17}$Sb$_{0.83}$ SLS with equal, 150 Å thick layers. The p-n junction was formed during MBE growth; the p⁻ doping resulted from background doping in the growth chamber. The photodiode was mesa-isolated, with an area of 1.2×10^{-3} cm ^{2}. Operation of the diode as a photodetector requires diffusion of minority carriers perpendicular to the SLS layers. Other details about the device have been described previously.[4]

The SLS material in the detector was very high quality, essentially dislocation free. This was confirmed by TEM and x-ray diffraction analysis. Consistent with high sample quality, an infrared photoluminescence line was observed at 9.7 μm (15 K sample temperature), with a 12 meV linewidth. The zero-bias resistance of the device was very sensitive to surface treatment, thus demonstrating that the detector noise was limited by surface leakage, not SLS material quality. After passivating the surface of the mesa with a native oxide, the resistance-area product (R_0A) increased from 0.6 to 9 Ω·cm^2 (77 K).

The zero-bias, external current responsivity (77 K) is shown in Figure 5. The detector was illuminated at normal incidence. Both the weak increase in responsivity observed in reverse bias and the magnitudes of the responsivity and absorption indicate that the minority carrier diffusion length, perpendicular to the SLS layers, is 1-2 μm. The detectivity values shown in Fig. 5 are based on a measured noise current of 1.6×10^{-12} A/Hz$^{1/2}$ at 100 kHz. These detectivity values surpass the 300 K BLIP limit, and the noise measurements were made with the detector covered by a 77 K cold-shroud. It is noteworthy that detectivities $> 1 \times 10^{10}$ cm Hz$^{1/2}$/W at wavelengths ≤ 10 μm were achieved with this passivated, photovoltaic InAsSb SLS detector.

SLS: 150 Å InAs$_{0.15}$Sb$_{0.85}$/ 150 Å InSb

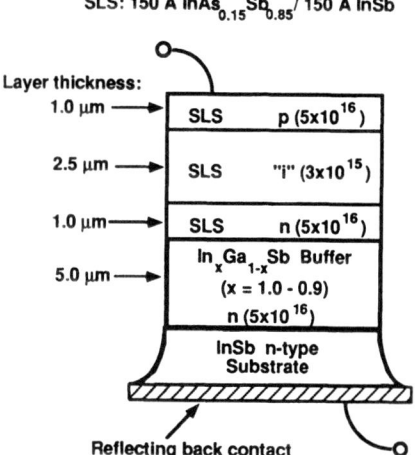

Figure 4 - Structure and composition of the InAsSb SLS photodiode.

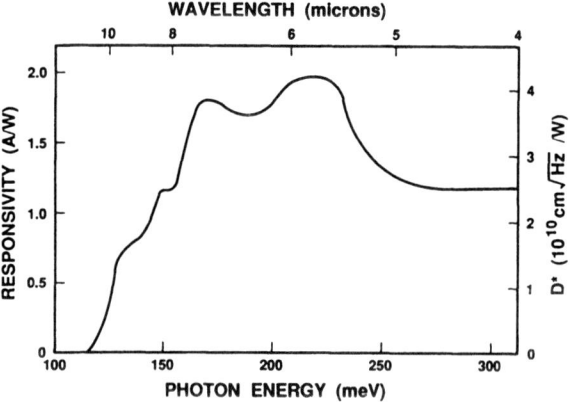

Figure 5 - External current responsivity of the photodiode at zero-bias, 77 K. Detectivity was determined from noise measurements made at 100 kHz, cold-shielded at 77 K.

IV. Future Directions of InAsSb Detector Development

The long wavelength, high detectivity photodiode in the previous section demonstrates the feasibility of an infrared detector technology based on InAsSb SLSs. Building on these results, it is conceptually straight-forward to improve the performance of these detectors. Detectivity can be increased by developing improved surface passivation techniques and by increasing the optically active volume while decreasing p-n junction area. A variety of novel device designs incorporating heterojunctions, lattice matched to the infrared active SLS, and lateral photodiode geometries may be used to increase detectivity. Many InSb microelectronic processing techniques can be transferred to InAsSb SLSs, and the spectral response of type II SLS detectors can be modified by changing SLS compositions and layer thicknesses.

The major obstacles to type II, SLS infrared detector development are associated with the immaturity of these III-V materials. We must demonstrate growth of high quality, infrared SLS wafers with excellent areal uniformity for focal-plane-arrays,and radiation-hard surface passivations must be developed. Like HgCdTe technology, the utilization of type II SLS detectors will be limited by material growth and microelectronic processing procedures and the manufacturability of these new technologies.

V. Acknowledgements

A large group has contributed to the infrared detector program at Sandia National Laboratories, and particularly, we thank R. M. Biefeld, L. R. Dawson, G. C. Osbourn, and R. E. Hibray for their enthusiasm and long-term collaboration on this project. This work was supported by DOE under contract No. DE-AC04-76P00789.

VI. References

1. L. C. West and S. J. Eglash, Appl. Phys. Lett. 46, 1156 (1985).
2. See for example, L. L. Chang and E. E. Mendez, in Synthetic Modulated Structures, edited by L. L. Chang and B. C. Giessen (Academic Press, New York, 1985) pp. 113-161.
3. B. F. Levine, C. G. Bethea, G. Hasnain V. O. Shen, E. Pelve, R. R. Abbott, and S. J. Hsieh, Appl. Phys. Lett. 56, 851 (1990).
4. S. R. Kurtz, L. R. Dawson, T. E. Zipperian, and R. D. Whaley, Jr., Elect. Dev. Lett. 11, 54 (1990).
5. S. R. Kurtz, J. G. Snyder, and L. R. Dawson, Proc. of the IRIS Specialty Group on Infrared Detectors, (in press) 1990.
6. G. A. Sai-Halasz, L. L. Chang, J. M. Wetter, C. A. Chang, and L. Esaki, Solid State Commun. 27, 935 (1978).
7. G. C. Osbourn, J. Vac. Sci. Technol. B2,176 (1984).
8. S. R. Kurtz, G. C. Osbourn, R. M. Biefeld, L. R. Dawson, and H. J. Stein, Appl. Phys. Lett. 52, 831 (1988).
9. S. R. Kurtz, G. C. Osbourn, R. M. Biefeld, and S. R. Lee, Appl. Phys. Lett. 53, 216 (1988).
10. D. K. Arch, G. Wicks, T. Tonaue, and J. Staudenmann, J. Appl. Phys. 58, 3933 (1985).
11. C. Mailhiot and D. L. Smith, J. Vac. Sci. Technol. A7, 445 (1989).
12. S. R. Kurtz and R. M. Biefeld, unpublished.

CHARACTERIZATION OF GaSb-BASED ALLOY SEMICONDUCTORS

H.UEKITA, N.KITAMURA*, M.ICHIMURA, A.USAMI and T.WADA

Nagoya Institute of Technology, Department of Electrical and Computer Engineering, Gokiso-cho, Showa-ku, Nagoya 466, Japan
*Suzuka College of Technology, Department of Electrical Engineering, Shiroko-cho, Suzuka 510-02, Japan

ABSTRACT

GaSb, $Al_xGa_{1-x}Sb$, and $In_xGa_{1-x}Sb$ epitaxial layers were grown by the liquid-phase epitaxy and characterized by photoluminescence, Raman spectroscopy, and double-crystal X-ray diffraction. The concentration of residual acceptors which are related to structural defects decreased with lowering growth temperature, but the GaSb epitaxial layer grown at an extremely low temperature of 270°C had poor crystalline quality. The $Al_xGa_{1-x}Sb$ (x≳0.15) and $In_xGa_{1-x}Sb$ (x=0.02) epitaxial layers grown at 270°C, however, had much better quality than the GaSb epitaxial layer grown at the same temperature.

INTRODUCTION

GaSb-based alloy semiconductors are interesting materials for optoelectronic devices in an infrared field; for example, $Al_xGa_{1-x}Sb$ is a promising material for an avalanche photodiode (APD) due to its large impact-ionization coefficient ratio of holes and electrons [1,2]. Epitaxial layers of the GaSb-based alloy semiconductors grown at a usual temperature, however, have high background acceptor concentrations because of structural defects which act as residual acceptors [3-5]. The concentrations of the residual acceptors in the epitaxial layers grown by the liquid-phase epitaxy (LPE) decrease with lowering growth temperature [6]. In this study, we grow the layers of GaSb, $Al_xGa_{1-x}Sb$, and $In_xGa_{1-x}Sb$ by LPE mainly at low temperatures (below 300°C) to reduce the acceptor concentration. The grown layers are characterized by photoluminescence (PL), Raman spectroscopy, and double-crystal X-ray diffraction.

EXPERIMENTS

The LPE growth was carried out at the temperatures of 270-500°C by the double-bin sliding boat method [7]. Epitaxial layers were grown without intentional doping on non-, Te-, and Zn-doped (100)-oriented GaSb substrates. The phase diagrams were calculated using the parameters of Cheng and Pearson [8] for $Al_xGa_{1-x}Sb$ and those of Blom and Plaskett [9] for $In_xGa_{1-x}Sb$. The Ga-rich melt was heat-treated at 700°C for two hours, but Al was added in the melt after the heat-treatment. Then, the substrate was placed in a recess of the boat. Again, the heat-treatment was carried out at 550°C for one hour to uniform the melt and clean the substrate surface. The furnace was cooled to a temperature 10-20°C higher than the growth temperature. The furnace was kept at that temperature for about one hour and then cooled with a rate of 0.3°C/min. The growth was started at the growth temperature and ended after a cooling of 5°C. In the case of the growth at temperatures lower than 350°C, the melt back was performed just before the growth and the temperature interval of growth was 15°C.

The PL spectra at 77K and the Raman spectra at a room temperature in a backscattering configuration were measured on the as-grown surfaces. The 514.5nm line of an Ar^+ laser was used in both measurements. The double-crystal X-ray rocking curve was measured on (400)-diffraction.

Mat. Res. Soc. Symp. Proc. Vol. 216. ©1991 Materials Research Society

EXPERIMENTAL RESULTS AND DISCUSSIONS

Figure 1 shows the PL spectra of the undoped GaSb epitaxial layers grown at temperatures of 270-500°C. Four emission peaks, labeled A, B, C, and D in decreasing order of the energy, are observed in the measured wavelength region at 77K. Emission A is a band-edge emission. Emission B has an energy about 30meV lower than that of emission A and is due to the transition between the conduction band and the residual acceptor level (Ev+30meV) [5,10-12]. The residual acceptor is considered to be related to a structural defect such as a Ga antisite [3,11]. As can be seen from the figure, the intensity of emission B decreases with lowering growth temperature. The origins of emissions C and D are not well identified, but they have been discussed in detail in our previous paper [6]. Emission A is not observed in the layer grown at 270°C, while the weak and broad emission is observed around the wavelength of 1700nm. The PL peak observed near 1580nm in the layer grown at 270°C is of the Zn-doped GaSb substrate.

The Raman spectra of the GaSb epitaxial layers grown at 270-500°C are shown in Fig. 2. Only the longitudinal optical (LO) phonon is observed in the layers grown at temperatures higher than 300°C. The transverse optical (TO) phonon, however, is also observed in the layer grown 270°C. The TO phonon is a forbidden mode on a (100)-oriented layer in a backscattering configuration.

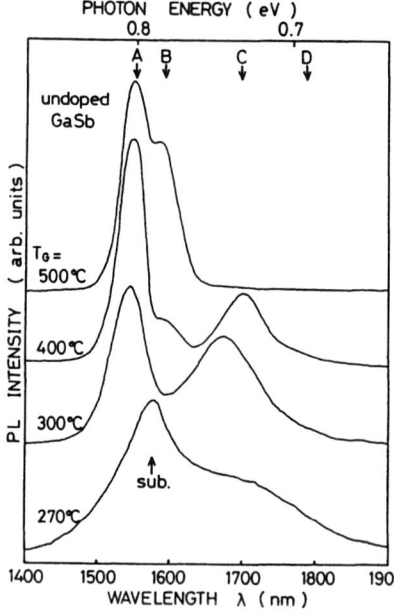

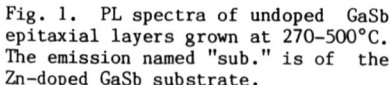

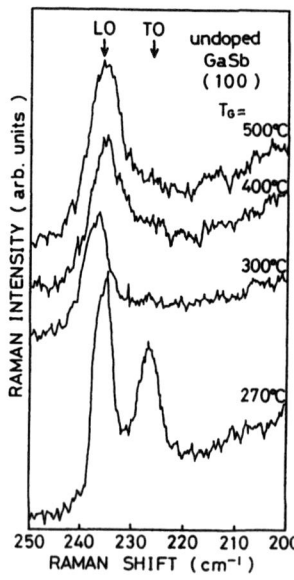

Fig. 1. PL spectra of undoped GaSb epitaxial layers grown at 270-500°C. The emission named "sub." is of the Zn-doped GaSb substrate.

Fig. 2. Raman spectra of undoped GaSb epitaxial layers grown at 270-500°C.

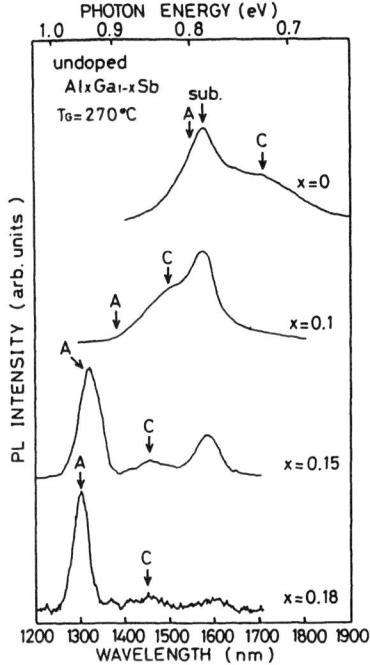

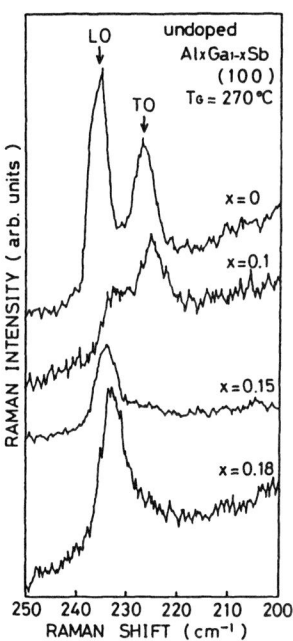

Fig. 3. PL spectra of undoped Al$_x$Ga$_{1-x}$Sb epitaxial layers grown at 270°C. The AlSb fraction x were estimated from the melt composition.

Fig. 4. Raman spectra of undoped Al$_x$Ga$_{1-x}$Sb epitaxial layers grown at 270°C. Only the GaSb-like modes are shown.

The double-crystal X-ray rocking-curves were measured for the GaSb epitaxial layers grown at 500°C, 400°C, and 270°C. Each of the layers grown at 500°C and 400°C has a peak with a narrow full-width at half maximum (FWHM) of 16sec. The layer grown at 270°C ,however, has a wide FWHM of 29sec.

These results show that the GaSb epitaxial layer has poor crystalline quality when the growth-temperature is extremely low, although its residual acceptor concentration becomes low.

Figures 3 and 4 show the PL and Raman spectra, respectively, of the undoped Al$_x$Ga$_{1-x}$Sb epitaxial layers grown at 270°C. The weak PL spectrum without emission A (band-edge emission) and the Raman spectrum with the TO phonon are observed for the Al$_{0.1}$Ga$_{0.9}$Sb layer, as for the GaSb layer grown at the same temperature. However, for the Al$_x$Ga$_{1-x}$Sb layer whose AlSb composition x is greater than 0.15, emission A is observed in its PL spectrum and only the allowed LO-phonon appears in its Raman spectrum. In addition, FWHMs of the double-crystal X-ray rocking-curves of the Al$_{0.15}$Ga$_{0.85}$Sb layers grown at 270°C and 400°C are 40sec and 33sec, respectively. The FWHM ratio of the 270°C-grown Al$_{0.15}$Ga$_{0.85}$Sb layer to the 400°C-grown one is smaller than that of GaSb.

The similar results are obtained for the $In_{0.02}Ga_{0.98}Sb$ layer. Figures 5 and 6 show the PL and Raman spectra, respectively, of the undoped $In_{0.02}Ga_{0.98}Sb$ epitaxial layers grown at 270–500°C. The band-edge emission (emission A) is observed in the PL spectrum of the 270°C-grown layer and only the allowed LO-phonon appears in the Raman spectrum of the same layer. Emission A in the PL spectrum is broader for the 270°C-grown layer than for the 400°C-grown one. We have not clarified the origin of the broadening. However, we may conclude that the $In_{0.02}Ga_{0.98}Sb$ layer grown at 270°C has much better quality than the GaSb layer grown at the same temperature, since emission A is completely unobservable for the GaSb layer and the forbidden TO-phonon disappears in the $In_{0.02}Ga_{0.98}Sb$ layer.

Emission B is not observed in the PL spectrum of the $In_{0.02}Ga_{0.98}Sb$ layer grown at 400°C, as shown in Fig.5, while emission B is observed for the GaSb layer grown at the same temperature. Emission B of the $In_{0.02}Ga_{0.98}Sb$ layer grown at 500°C is also weaker than that of the GaSb layer grown at the same temperature. These show that the residual acceptor concentration decreases by the In doping.

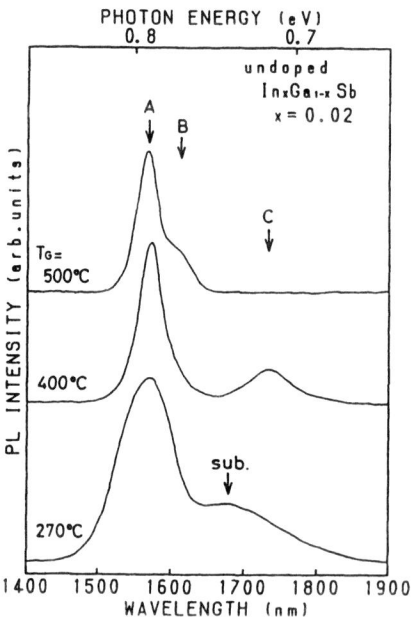

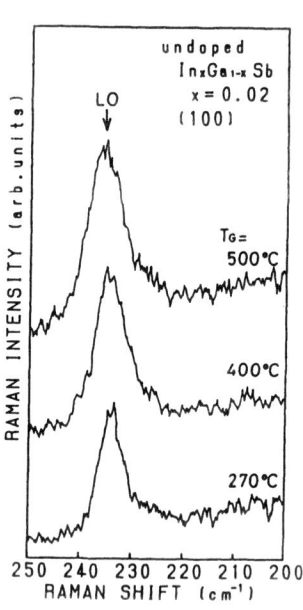

Fig. 5. PL spectra of undoped $In_{0.02}Ga_{0.98}Sb$ epitaxial layers grown at 270–500°C.

Fig. 6. Raman spectra of undoped $In_{0.02}Ga_{0.98}Sb$ epitaxial layers grown at 270–500°C.

These results show that the crystalline quality of the GaSb-based alloy semiconductors is much better than that of GaSb binary compound. We reported that the residual acceptor concentration in the $Al_xGa_{1-x}Sb$ epitaxial layers grown at 500°C decreased with increasing AlSb composition [6]. It is expected that the residual acceptor concentration in the layers grown at lower temperatures (<500°C) has the similar compositional dependence according to the thermodynamic calculation of defect concentrations by Higuchi et al. [13]. In addition, the present results show that the $In_{0.02}Ga_{0.98}Sb$ epitaxial layers also have lower residual acceptor concentrations than the GaSb epitaxial layers grown at the same temperatures (see Figs. 1 and 5). Accordingly, it is suggested that the concentration of the residual acceptor in the GaSb epitaxial layer is decreased by doping Al or In. The effects of the Al doping have been discussed in Ref. [13], but those of the In doping have not been understood. Tanaka et al. [14] reported the decrease of native defects in GaP by In doping and discussed it using a thermodynamic model. However, it is not certain that the mechanism of the reduction is the same, because In composition in the melt is very different between their and our cases. On the other hand, Yang et al. [15] and Beneking et al. [16] reported that the quality of LPE-GaAs layers was improved by doping In, which pinned dislocations. However, the alloying-effects observed in this study are more drastic than their observations: we observed the forbidden TO-phonon in the Raman spectrum and no band-edge emission in the PL spectrum for the GaSb layer grown at 270°C, but those anomalies disappear in the alloys. These results cannot be interpreted considering only point defects such as the residual acceptor or dislocations. We therefore suppose that the quality-improvement in the GaSb-based alloy semiconductors is related to larger scale defects than point defects or dislocations. Although the defects have not been identified yet, we consider that Al and In atoms suppress the extension of the defects through a mechanism similar to the dislocation pinning by In.

SUMMARY

LPE-GaSb, $Al_xGa_{1-x}Sb$, and $In_xGa_{1-x}Sb$ layers were grown at 270-500°C and characterized. The experimental results showed that the crystalline quality of the GaSb-based alloy semiconductors $Al_xGa_{1-x}Sb$ (x≳0.15) and $In_{0.02}Ga_{0.98}Sb$ was much better than that of the GaSb binary compound when the growth temperature is very low. Further studies should be conducted to clarify the mechanism of the quality-improvement in the alloy semiconductors.

ACKNOWLEDGEMENTS

We would like to thank Messrs. A.Kobayashi, N. Hoshikawa, K. Kato and A. Tanahashi of Nagoya Institute of Technology for experimental assistance, and Prof. A. Sasaki and Dr. Y. Takeda of Kyoto University for the double-crystal X-ray diffraction studies.

REFERENCES

1. H.D. Law, R. Chin, K. Nakano and R.A. Milano, IEEE J. Quantum Electron. QE-17, 275 (1981)
2. O. Hildebrand, W. Kuebart, K.W. Benz and M.H. Pilkuhm, IEEE J. Quantum Electron. QE-17, 284 (1981)
3. R.N. Hall and J.H. Racette, J. Appl. Phys. 32, 856 (1961)
4. M.H. van Maaren, J. Phys. Chem. Solid. 27, 472 (1966)
5. K. Nakashima, Jpn. J. Appl. Phys. 20, 1085 (1981)
6. N. Kitamura, K. Higuchi, H. Uekita, M. Ichimura, A. Usami and T. Wada, Jpn. J. Appl. Phys. 29, 1403 (1990)
7. A. Sasaki, A. Onishi, E. Sogawa, S. Mizugaki, Y. Takeda and S. Fujita, Inst. Phys. Conf. Ser. No. 63, (The Institute of Physics, Bristol, 1982), p. 83.
8. K.Y. Cheng and G.L. Pearson, J. Electrochem. Soc. 124, 753 (1977)
9. G.M. Blom and T.S. Plaskett, J. Electrochem. Soc. 118, 1831 (1971)
10. E.J. Johnson and H.Y. Fan, Phys. Rev. 139, A1991 (1965)
11. R.D. Baxter, R.T.Bate and F.J. Reid, J. Phys. Chem. Solids 26, 41 (1965)
12. C. Benoit a la Guillaume and P. Lavallard, Phys. Rev. B 5, 4900 (1972)
13. K. Higuchi, N. Kitamura, Y. Hattori, M. Ichimura and T. Wada, to be published in Proc. Int. Conf. Sci. & Tech. of Defect Control in Semicond.
14. A. Tanaka, T. Sugiura and T. Sukegawa, Oyo Buturi 52, 326 (1983), (in Japanese)
15. B.H. Yang, Z.G. Wang, H.J. He and L.Y. Lin, J. Crystal Growth 103, 371 (1990)
16. H. Beneking, P. Narozy and N. Emeis, Inst. Phys. Conf. Ser. No. 79, (The Institute of Physics, Bristol, 1986), p. 403

THE PREPARATION OF InAsSb/InSb SLS AND InSb PHOTODIODES BY MOCVD

R. M. Biefeld, B. T. Cunningham, S. R. Kurtz, and J. R. Wendt
Sandia National Laboratories, Albuquerque, NM

ABSTRACT

Infrared absorption and photoluminescence have been demonstrated for $InAs_{1-x}Sb_x/InSb$ strained-layer superlattices (SLS's) in the 8-15 μm region for As content less than 20%. This extended infrared activity is due to the type II heterojunction band offset in these SLS's. The preparation of the first MOCVD grown p-n junction diode was achieved by using dimethyltellurium as an n-type dopant. Several factors, such as background doping and dopant profiles affect the performance of this device. InSb diodes have been prepared using tetraethyltin. The resulting current-voltage characteristics are improved over those of diodes grown previously using dimethyltellurium. Doping levels of 8×10^{15} to 5×10^{18} cm^{-3} and mobilities of 6.7×10^4 to 1.1×10^4 cm^2/Vs have been measured for Sn doped InSb. SLS diode structures have been prepared using Sn and Cd as the dopants. Structures prepared with p-type buffer layers are more reproducible.

INTRODUCTION

$InAs_{1-x}Sb_x/InSb$ strained-layer superlattices (SLS's) have been proposed for use as long wavelength detectors in the 8-15 micron range. The preparation of high quality materials was achieved by the minimization of cracks and dislocations in these $InAs_{1-x}Sb_x/InSb$ SLS's by using 2-3 μm thick, compositionally graded $InAs_{1-x}Sb_x$ buffer layers [1]. Infrared absorption and photoluminescence were determined for these high quality SLS's in the 8-12 μm region for As content less than 20% [2]. This extended infrared activity is due to the type II heterojunction band offset in these SLS's.

Fabrication of a photodiode is an important step in the development of a new infrared material. The preparation of the first MOCVD grown p-n junction diode was achieved by using dimethyltellurium as an n-type dopant [3]. The structure of the SLS used in this photodiode consisted of $InAs_{0.18}Sb_{0.82}/InSb$ layers with equal layer thicknesses, 13.0 nm. Several factors, such as background doping and dopant profiles, are believed to affect the performance of this device. Recently, a high detectivity, > 1×10^{10} $cmHz^{1/2}/W$ at 10 μm, InAsSb SLS photodiode was prepared by MBE [4]. The MBE InSb has a lower background carrier concentration than the MOCVD InSb and Se was the n-type dopant. This paper discusses the use of tetratethyltin as an n-type dopant and Cd as a p-type dopant. Current-voltage characteristics of an InSb diode prepared using tetraethyltin are presented and compared to the previously reported diode results. These measurements indicate that Sn is the preferred n-type dopant for InSb. The results of the growth of step-graded buffer layers using Cd or Sn as the dopant are also discussed.

EXPERIMENTAL

Studies investigating the use of tetraethyltin (TESn) and dimethylcadmium (DMCd) were carried out in a previously described horizontal, atmospheric pressure system [5]. The sources of In, Sb and As were trimethylindium (TMIn), trimethylantimony (TMSb) or triethylantimony (TESb) and arsine (AsH3). Tetraethyltin was used in its pure state in a bubbler at a variety of different flow rates and temperatures. Purified hydrogen was used as the carrier gas. The layers were grown at 470 C on (100) InSb substrates. The optimum growth conditions have been previously described [5]. At 470 C and a pressure of 630 torr, a V/III ratio of 2.4 and a growth rate of 0.5 μm/h were used. At 410 C and 630 torr, a V/IIIratio of 3.0 and a growth rate of 0.1 μm/h were necessary to obtain the same surface morphology. The buffer layers discussed in this paper were grown at low pressure, 200 torr, a V/III ratio that varied between 15 and 60, and 470 C. The InSb substrates were cleaned by degreasing in hot solvents and deionized water. They were then etched for two minutes in a 10 to 1 mixture of lactic acid and nitric acid, rinsed with deionized water and blown dry with filtered nitrogen.

Structures for Hall measurements were grown on compensated, Cd doped InSb with measured hole densities at 77 K of 10^{12}-10^{13} cm^{-3}. The general structure which was used for the Hall measurements consisted of a single layer of InSb 2 to 4 μm thick grown directly on the substrate. Hall measurements were made by standard van der Pauw techniques. The reported mobilities were determined with a magnetic field of 2.0 or 3.0 kG. The epitaxial layers were uniformly doped. The samples were examined by optical microscopy and a lapping technique to determine layer thicknesses.

The structure of the diode reported on in this paper consisted of an n-type substrate, N_D-N_A = 3 x 10^{17} cm^{-3}, a 1.6 μm Sn doped layer with N_D-N_A = 1 x 10^{16} cm^{-3}, and a final layer of undoped InSb with N_A-N_D = 1 x 10^{16} cm^{-3}. The p-doping in the final layer is the present background level of the InSb grown using TMIn and TMSb at 470 C. The diodes were mesa isolated with an area of 1.2 x 10^{-3} cm^2.

RESULTS AND DISSCUSSION

One of the major difficulties that has been encountered in the growth of InAsSb SLS photodiodes by MOCVD is the high background carrier concentration [3]. One approach to lower this background that has been investigated is the use of low pressure MOCVD. The use of low pressure MOCVD has resulted in a small improvement in the background carrier concentrations for the best samples and a slight improvement in mobilities. The carrier concentration improved from 7 x 10^{15} cm^{-3} to 5 x 10^{15} cm^{-3} and the mobility changed from 5000 to 7400 cm^2/Vs. However, in order to prepare a photodiode, a graded buffer layer and doped layers still need to be grown. The results of doping experiments which used TESn at both atmospheric and reduced pressure are illustrated in Fig. 1. The low pressure values for the mobility versus carrier concentration are very similar to the atmospheric pressure results.

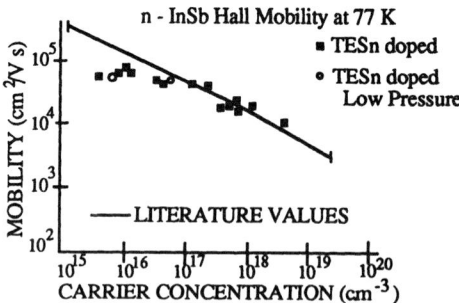

Figure 1. Mobility versus net carrier concentration at 77 K for uniformly doped epitaxial layers of InSb. The open circles are the data for InSb grown at 200 torr and the filled squares are for the samples grown at 630 torr.

The current-voltage characteristics of the photodiode grown using TESn and described above are shown in Fig. 2. The diode characteristic is typical of a narrow bandgap semiconductor [3]. This device exhibits considerably better diode behavior than the first diode reported which was grown using Te as the n-type dopant. The doping levels are also somewhat different than those of the Te diode. Also, the Te diode was grown in an SLS structure with a bandgap of about 10 μm compared to the bandgap of InSb of 5.5 μm. This could explain some of the difference in the I-V curve. The preparation of SLS diodes using Sn will determine if the use of Sn improves the device parameters for infrared detection. The background carrier concentration is still the same for the undoped InSb. Further improvements in this number are needed to enhance the photodiode behavior.

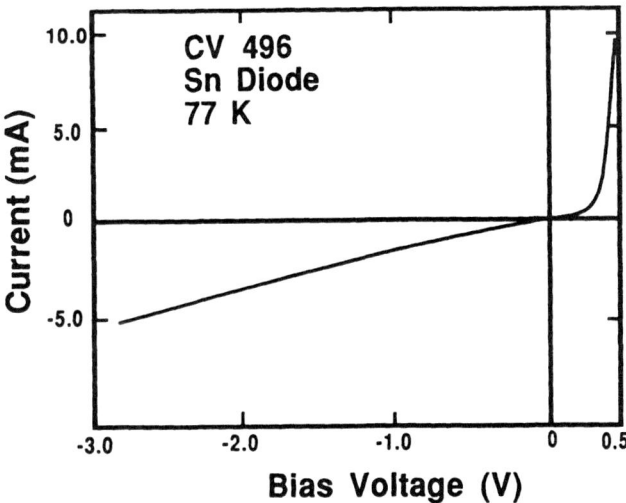

Figure 2. The current-voltage characteristics at 77 K for an unpassivated grown junction, InSb diode prepared using Sn as the n-type dopant and undoped InSb as the p-type material at 630 torr.

Initial attempts to grow the step-graded InAsSb buffer layer using Sn doping resulted in poor morphology as illustrated in the micrograph in the upper right hand corner of Fig. 3. The growth of an InAsSb/InSb SLS on top of these buffers resulted in poor morphology and cracks as illustrated by the micrograph in the upper left hand corner of Fig. 3. Attempts to prepare diodes from the Sn doped InAsSb step-graded buffer layers resulted in devices that exhibited no rectification. Since the initial photodiodes grown at atmospheric pressure were grown on p-type buffer layers, this structure was reproduced at low pressure using diethylcadmium as a source for Cd. The micrograph in the lower left hand corner of Fig. 3 shows the morphology of the Cd doped buffer layers. These p-type buffer layers have improved morphology and improved reproducibility over that of the n-type buffer layers. The result of the first attempt to grow an Sn doped SLS on top of the Cd-doped buffer layers resulted in very poor morphology. This is illustrated in the micrograph in the lower right hand corner of Fig. 3. A possible reason for this type of surface is the growth of an SLS with a large lattice mismatch to the substrate. However, the surface of this Sn doped layer is much worse than the surface of the SLS grown on the Sn doped buffer layers which was grown under identical conditions except for the dopants. This experiment will be reproduced in an attempt to confirm the observed effect.

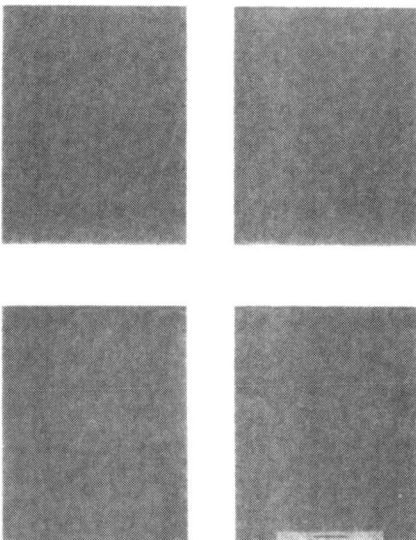

Figure 3. The surface morphology of InAsSb step-graded buffer layers grown using Sn and Cd as dopants and similar structures with InAsSb/InSb SLS's grown on top of them. See text for a detailed description.

The reproducibility of several MOCVD InAsSb growths is illustrated in Fig. 4 where the As composition of InAsSb as determined by x-ray diffraction is plotted against the arsine flow rate. The variation from run to run is approximately one percent. This type of mismatch is not sufficient to explain the extremely rough surface morphology that was observed for the Sn doped SLS grown on top of the Cd doped InAsSb step-graded buffer layer. Until an InAsSb/InSb SLS photodiode can be grown using Sn at low pressures, there is no way to know if the low pressure growth by MOCVD will yield improved electrical characteristics.

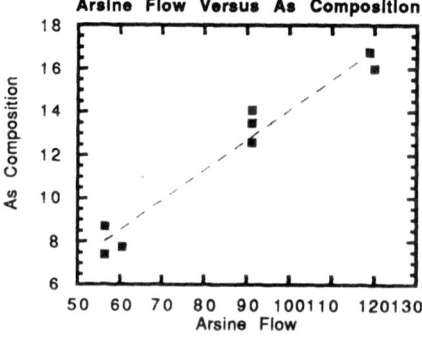

Figure 4. Arsine flow versus As composition in InAsSb grown at 200 torr and 470 C.

REFERENCES

1. R. M. Biefeld, C. R. Hills and S. R. Lee, J. Crystal Growth 91 (1988) 515.
2. S. R. Kurtz, G. C. Osbourn, R. M. Biefeld, and S. R. Lee, Appl. Phys. Letters 53 (1988) 216.
3. S. R. Kurtz, R. M. Biefeld, and T. E. Zipperian, Semicond. Sci. Technol., 5 (1990) S24 .
4. S. R. Kurtz, L. R. Dawson, T. E. Zipperian, and R. D. Whaley, Jr., IEEE Elec. Dev. Letters 11 (1990) 54.
5. R. M. Biefeld, S. R. Kurtz, and I. J. Fritz, J. Electron. Mater. 18 (1989) 775.

InAsSb PHOTODIODES GROWN ON InAs, GaAs AND Si SUBSTRATES BY MOLECULAR BEAM EPITAXY

W. DOBBELAERE, J. DE BOECK, W. DE RAEDT, J. VANHELLEMONT, G. ZOU, M. VAN HOVE, B. BRIJS, R. MERTENS and G. BORGHS
Interuniversity Micro-Electronics Center (IMEC vzw), 75 Kapeldreef, B-3001 Leuven, BELGIUM

ABSTRACT

InAs and $InAs_{0.85}Sb_{0.15}$ p-"i"-n structures were grown on InAs, GaAs and Si substrates by Molecular Beam Epitaxy. The structural quality of the layers is discussed using Transmission Electron Microscopy and Rutherford Back Scattering. The influence of the material quality on the 77 K current-voltage measurements is explained. The spectral response of the devices was measured demonstrating peak responsivities of 2.1 A/W at 3.5 μm wavelength for $InAs_{0.85}Sb_{0.15}$ detectors with a 4.3 μm cut-off wavelength.

INTRODUCTION

The growth of InAsSb on GaAs and Si substrates is very attractive for the monolithic integration of infrared detectors with electronic devices. The Molecular Beam Epitaxial (MBE) growth of InAsSb epilayers on these highly mismatched substrates has already been investigated by different groups [1]. Photoconductive InAsSb detectors on GaAs and Si substrates [2] have a very low resistance. Since these devices can not be operated at zero-bias, large leakage currents and a significant power dissipation limit their applications. These disadvantages can be overcome by using photovoltaic detectors. InAsSb photodiodes have already been demonstrated on nearly lattice matched InAs, GaSb and InSb substrates by various growth techniques [3].

In this paper we demonstrate the first InAsSb photovoltaic detectors, grown on GaAs and Si substrates by MBE. The devices were optimized for operation in the 3 to 5 μm atmospheric window, requiring an Sb content of about 15 %. $InAs_{1-x}Sb_x$ n(Si)-p(Be) structures were grown simultaneously on InAs (Lattice mismatch $\Delta a/a=1$ % with respect to $InAs_{0.85}Sb_{0.15}$), GaAs ($\Delta a/a=8.4$ %) and Si ($\Delta a/a=12.8$ %) substrates. The GaAs and Si substrates were precoated with an InAs buffer layer. We first describe the experimental conditions. Then we discuss the material quality using Nomarsky contrast microscopy, Transmission Electron Microscopy and Rutherford Back Scattering. In the next section we demonstrate the impact of the material quality on the current-voltage measurements and on the zero bias Resistance.Area ($R_0.A$) product. Finally we present infrared photocurrent measurements.

EXPERIMENTAL

The epilayers used for this study were grown using a Riber 2300 MBE equipped with an As_2 cracker cell and a conventional Sb_4 cell. Two different procedures were followed to fabricate diodes on Si substrates. In a first run an InGaAs buffer layer was deposited on the Si substrate which was cleaned as described in [2]. We first deposited a 0.1 μm thick GaAs nucleation layer at a growth temperature of 350 °C and a growth rate of 0.5 μm/h, followed by a 0.5 μm thick InGaAs graded buffer and a 1 μm InAs layer grown at 480 °C and 1 μm/h. In a second run, the Si substrates were precoated with a 1 μm GaAs buffer using the growth conditions described in [4]. The GaAs and GaAs-coated Si wafers were degassed for 15 min. at 585 °C. Then the substrate temperature was lowered to 400 °C under As_2 pressure and a 0.1 μm thick InAs nucleation layer was grown at a growth rate of about 0.5 μm/h. The In temperature was then increased to reach the standard growth rate of 1 μm/h and a 0.5 μm thick film was grown at a substrate temperature of 480 °C.

The InAs-coated wafers were unloaded and indium-glued on a molybdenum block together with an InAs wafer which we use as a reference. The wafers were degassed for 30 minutes at 480 °C and the InAs diode structure was grown at 1 μm/h and 460 °C. The diode structure consists of a 1.5 μm n-type buffer layer, followed by a 1 μm thick lowly doped p-type layer and a 0.5 μm thick p-type cap layer. For the InAsSb diodes the growth was started with a 0.75 μm compositionally graded buffer followed by the same 3 μm thick diode structure. Several growths were performed with different compositions and doping levels. The growth parameters of some diodes that will be further considered in this paper are summarized in table I. The fluxes of the different cells were measured with a Bayard-Alpert gauge by closing the cell shutter after a stabilisation period. We found that in this growth mode (low Sb_4 fluxes), the composition is nearly linearly proportional to the Sb_4/As_2 flux ratio. The Sb-content x of the $InAs_{1-x}Sb_x$ layers was measured using Wavelength Dispersive Spectroscopy and is also given in table I. The p and n doping are estimated from the Be and Si cell temperatures. The Be cell was calibrated using GaAs reference samples. For the Si cell we used both GaAs reference samples and optical measurements on Si-doped InAsSb layers (the Burstein-Moss effect) [5,6]. The properties of the (p-) region are determined by the back-ground doping level which is probably significantly larger than the Be-doping level ($>3x10^{15}cm^{-3}$).

Photodiodes with an area of $1.25x10^{-4}$ cm^2 were formed by chemical etching in a solution of (5 : 1) (Lactic Acid : Nitric Acid). The p and n regions were contacted with a 60 nm Ti / 150 nm Au metallisation.

TABLE I : Cell fluxes, Sb-content, and device structure (doping levels) of InAsSb diodes.

Growth	ϕ_{In} (10^{-7} Torr)	ϕ_{Sb} (10^{-7} Torr)	ϕ_{As} (10^{-7} Torr)	x (%)	n (cm^{-3}) d= 1.5 μm	p- (cm^{-3}) d= 1.0 μm	p (cm^{-3}) d= 0.5 μm
#1	7.1	0	16	0	$2x10^{17}$	$3x10^{15}$	$3x10^{17}$
#2	7.0	2.7	14	15	$2x10^{17}$	$3x10^{15}$	$3x10^{17}$
#3	7.1	0	16	0	$2x10^{16}$	"i"	$4x10^{16}$
#4	5.6	0	23	0	$1x10^{18}$	$3x10^{15}$	$2x10^{18}$

MATERIAL CHARACTERISTICS

In this section we compare the morphology and the crystal quality of InAs and InAsSb epilayers in relation with the substrate mismatch. Fig. 1 shows Nomarsky Contrast images of InAs layers (#1) grown on InAs, GaAs and Si substrates and InAsSb (#2) on the same substrates. The InAs diodes result in a perfectly specular morphology on InAs (A) and an increasingly rough morphology on GaAs(B) and Si (C). The Si wafer was not precoated with a thick GaAs layer

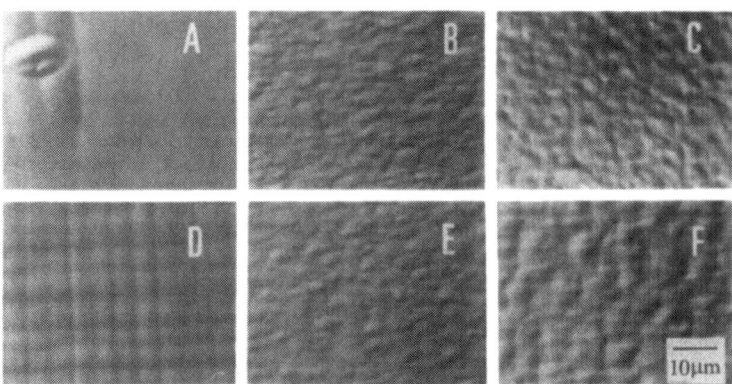

Fig. 1 Morphology of InAs (#1) photodiodes grown on InAs (A), GaAs (B) and Si (C) and InAsSb (#2) photodiodes grown on InAs (D), GaAs(E) and Si (F). The surface roughness clearly increases with the misfit.

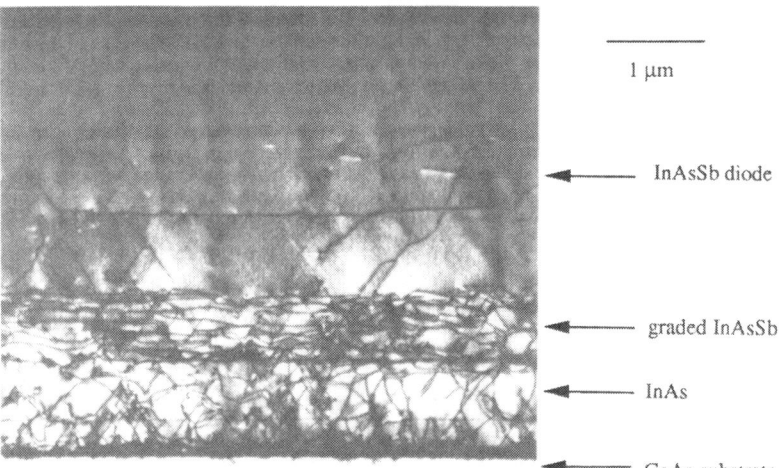

1 μm

InAsSb diode

graded InAsSb

InAs

GaAs substrate

Fig. 2. TEM cross section of InAsSb (#2) grown on GaAs.

which resulted in a rather bad morphology. By precoating the Si-substrate with a 1 μm thick GaAs layer [4] equal morphologies could be obtained on GaAs and Si. The InAsSb diodes (#2) exhibit a very flat morphology with a typical cross-hatched pattern when grown on slightly mismatched InAs substrates (D). The morphology of the same layers on GaAs (E) and Si (F) is much rougher.

The defect structure was studied using Transmission Electron Microscopy (TEM). Fig. 2 shows a TEM cross-section of the InAsSb diodes grown on GaAs. The micrograph shows a high density of dislocations near the InAs/GaAs interface, decreasing toward the surface. The beneficial influence of using the graded InAsSb buffer growth can be clearly observed. In the graded buffer a network of horizontal dislocations is formed. This network effectively reduces the penetration of dislocations toward the surface by annihilation mechanisms. For the growth on Si substrates, a similar defect structure was obtained if the Si-wafers were precoated with a thick GaAs buffer layer.

The crystallinity of the epilayers was also investigated using Rutherford Back Scattering (Ion Channeling). Fig. 3 shows the aligned yield vs. channel # (energy) for InAsSb (#2) grown

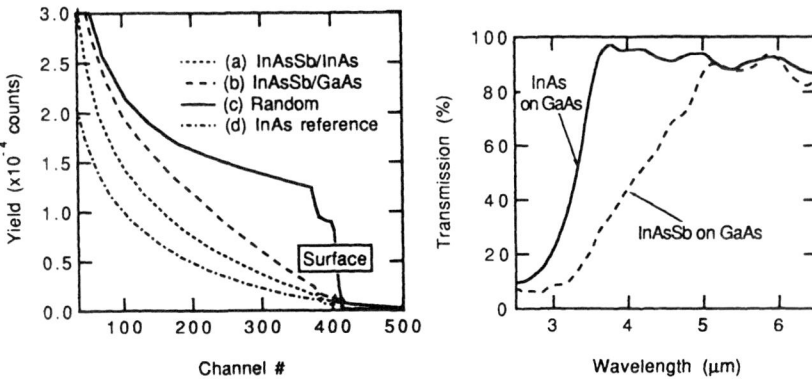

Fig. 3. Aligned yield for InAsSb layers grown on InAs and GaAs as measured with Rutherford Back Scattering (Ion channeling).

Fig.4. Room temperature infrared transmission of InAs and InAsSb photodiodes on GaAs.

on InAs (a) and GaAs (b). Curve (c) shows the random yield and curve (d) represents a reference measurement on a dislocation-free InAs layer. The effect of the mismatch is reflected in an increasing yield toward the heterointerface. Near the surface the InAsSb layers and the reference sample exhibit a comparable yield, indicating that high quality crystals can be obtained by increasing the layer thickness. The ratio of the aligned to random yield near the surface is about 4 %.

The incorporation of Sb results in a lowering of the bandgap so that devices can be made with cut-off wavelengths around 5 μm, making them very well suited for operation in the atmospheric window. This is demonstrated in Fig. 4 which shows the room temperature infrared transmission spectrum of the InAs (a) and InAsSb (b) diodes grown on GaAs. The Sb content of about 15 % shifts the room temperature cut-off wavelength from 3.6 μm for InAs to 5 μm for InAsSb.

CURRENT-VOLTAGE MEASUREMENTS

In this paragraph we present current-voltage measurements of InAs and InAsSb p-"i"-n structures grown on InAs, GaAs and Si substrates. Fig. 5 shows the 77 K current voltage measurements of growths #1 and #2 on InAs (a), GaAs (b) and Si (c). The homoepitaxial InAs/InAs diodes exhibit good diode characteristics. The 77 K reverse bias characteristics of these diodes are most likely determined by the generation in the depletion layer. Since the devices were not passivated, a significant contribution of surface leakage can be expected. The same diodes grown on GaAs and Si substrates exhibit significantly larger reverse leakage currents. The presence of defects reduces the carrier lifetime so that the generation-recombination currents become increasingly important. In addition there will be a shunt leakage current along 1 or 2-dimensional defects that cross the junction. The result is a significant lowering of the Resistance.Area ($R_0.A$) product for diodes on GaAs and Si with respect to InAs. The $R_0.A$ products are given in table II. For growth #2, an Sb content of about 15 % was incorporated. This results in a serious increase of the reverse leakage current at higher bias voltages for the three substrates. The InAsSb/InAs diodes are influenced by the 1 % lattice mismatch, resulting in a lowering of $R_0.A$ in comparison with InAs/InAs. For the diodes on GaAs and Si the incorporation of Sb only slightly increases the lattice mismatch.

TABLE II. Zero-bias Resistance.Area product of InAs (#1) and InAsSb (#2) at 77 K.

Growth	$R_0.A$ ($\Omega.cm^2$)		
	InAs	GaAs	Si
#1	1500	225	4.4
#2	49	25	1.9

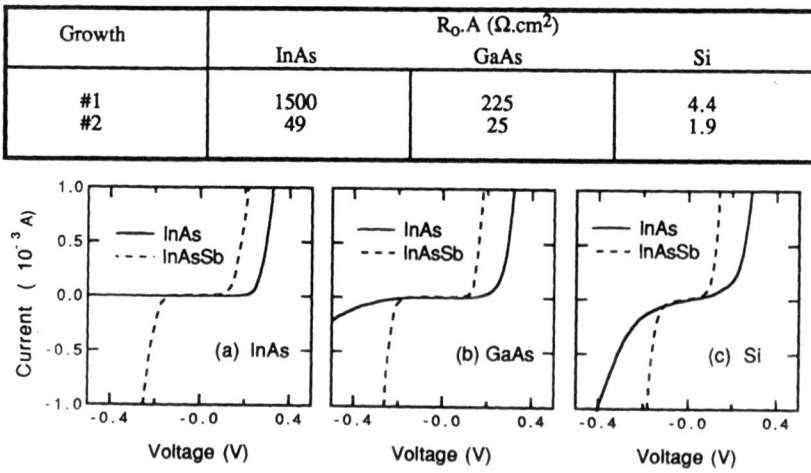

Fig 5. Current-voltage characteristics at 77 K of InAs (#1) and InAsSb (#2) photodiodes grown on InAs (a), GaAs (b) and Si (c).

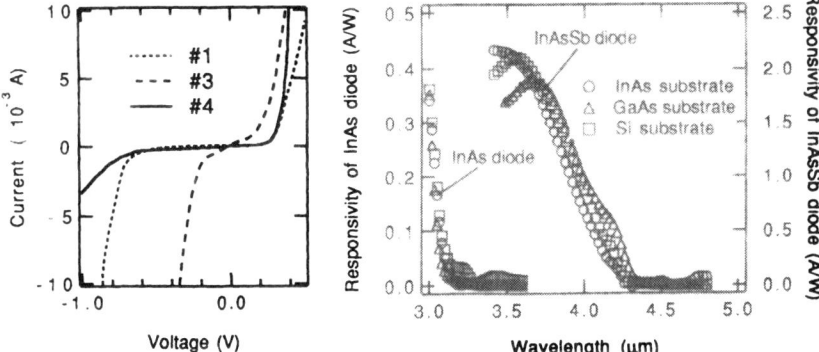

Fig 6. Current-voltage characteristics at 77 K of InAs potodiodes with different doping levels (#1,#3 and #4) grown on GaAs.

Fig. 7 Current responsivity of InAs and InAsSb photodiodes grown on InAs, GaAs and Si substrates.

The lowering of the bandgap from 400 meV for InAs to 300 meV for InAsSb may enhance several leakage mechanisms, typical for narrow bandgap diodes. These mechanisms have been investigated by several authors for InSb which has a 77 K bandgap of 230 meV. [7] First of all interband tunneling will drastically increase. Secondly tunneling processes at the surface are expected to become important especially since the devices were not passivated. [7] Appropriate passivation techniques are necessary to reduce these leakage currents. The diode characteristics of the layers grown on Si are slightly worse than those grown on GaAs. By pre-coating the Si wafer with a thick GaAs buffer we were able to improve the diodes on Si so that they became nearly as good as those on GaAs. It is evident that the number of dislocations that is introduced at the InAs/GaAs interface is much larger than the residual dislocation density at the surface of a GaAs/Si buffer. Therefore we can state that using the GaAs-coated Si technology it is possible to obtain equal quality InAsSb diodes on Si as on GaAs.

We investigated the influence of the doping levels on the characteristics of the InAs/GaAs diodes (Growths #1,#3,#4) which is shown in Fig. 6. We found that an increasing doping level resulted in an improvement of the reverse characteristics. This can be explained by the reduction of the depletion layer so that generation currents become smaller. A further increase of the doping levels however causes tunneling leakage currents.

We also fabricated InAs diodes in wells, pre-etched in the Si or GaAs substrates in order to obtain a coplanar surface. The material quality of those embedded structures did not differ from the planar layers. However a peculiar problem arose from the fact that the n-side of the embedded diodes could not be contacted from the top. Therefore we used the n-type GaAs substrate to contact the photodiodes. We found that for an n-doping level of 5×10^{17} the GaAs-InAs interface resulted in a parasitic heterostructure diode. This problem can be solved by using higher doping levels near the interface.

SPECTRAL RESPONSE

The zero-bias current responsivity of the devices was measured at 77 K. The infrared radiation of a 1 inch diameter 1200 °C black body source was focused with a ZnSe lens on the entrance slit of a 65 cm focal length monochromator and chopped at 1 kHz. The dispersed light was focused with a second lens on the InAsSb photodiodes which were positioned in a cryostat equipped with ZnSe windows. The photocurrent was measured using a lock-in amplifier with a trans-impedance preamplifier. The power incident on the detectors was measured with a pyro-electric radiometer. Fig. 7 shows the results for InAs diodes and InAsSb diodes. The incorporation of Sb results in a significant shift of the spectral response to longer wavelengths. The current responsivity was comparable ($R_{max}=2.1$ A/W at 3.5 μm) for the three different substrates because it merely depends on the quantum efficiency which is not much influenced by the material quality. While the $R_0.A$ product gives an important indication of the material quality, the responsivity is a good measure for the detector design. Closely related to the $R_0.A$ product is the noise performance. We did not measure the detector noise, however since the $R_0.A$ product is quite low, we believe that Johnson Noise will dominate. Under this assumption, we find a peak detectivity D* at 3.5 μm wavelength of 2.3×10^{11} cm.Hz$^{1/2}$W^{-1} for InAsSb/InAs, 1.6×10^{11} for InAsSb/GaAs and 4.5×10^{10} for InAsSb/Si. The 300 K background limited detectivity at this wavelength is about 6×10^{11}. [8] We believe that by optimising the buffer layer it will be possible to significantly improve the detectivity of the detectors grown on GaAs and Si such that background limited operation may be possible.

CONCLUSIONS

We have demonstrated the first $InAs_{0.85}Sb_{0.15}$ photodiodes on GaAs and Si substrates. The performance of the devices is strongly influenced by the very high lattice mismatch that exists between the substrate and the diode layer. We obtained $R_0.A$ products of 25 Ωcm^2 at 77 K and current responsivities of 2.1 A/W at 3.5 μm wavelength for detectors with a 4.3 μm cut-off wavelength on GaAs substrates. The GaAs-coated Si technology allows to obtain the same material quality on Si substrates. Our results clearly demonstrate the feasibility of the monolithic integration of InAsSb-based infrared photodiodes and GaAs or Si read-out electronics.

ACKNOWLEDGMENTS

The authors would like to thank H. Bender, P. Heremans, W. Vandervorst and J. Vermeiren for their contributions to this work. W. Dobbelaere acknowledges financial support from the "Instituut tot aanmoediging van het Wetenschappelijk Onderzoek in Nijverheid en Landbouw (IWONL)".

REFERENCES

[1] See e.g. G.S. Lee, Y. Lo, Y.F. Lin, S.M. Bedair, and W. D. Laidig, Appl. Phys. Lett., 47(11), 1219-1221 (1985); J.-I. Chyi, S. Kalem, N.S. Kumar, C.W. Litton, and H. Morkoç, Appl. Phys. Lett. 53(12), 1092-1094 (1988); M.Y. Yen, J. Appl. Phys. 64(6), 3306-3309 (1988).

[2] W. Dobbelaere, J. De Boeck, M. Van Hove, K. Deneffe, W. De Raedt, R. Mertens and G. Borghs, IEDM Techn. Dig. 717-720 (1989), Electron. Lett. 26(11), 259-261 (1990).

[3] L.O. Bubulac, A. M. Andrews, E.R. Gertner, and D. T. Cheung, Appl. Phys. Lett. 36(9), 734-736 (1980); P.K. Chiang, and S. M. Bedair, Appl. Phys. Lett., 46(4), 383-385 (1985); K. Mohammed, F. Capasso, R.A. Logan, J.P. Van Der Ziel, and A. L. Hutchinson, Electron. Lett. 22(4), 215-216 (1986); S.R. Kurtz, L.R. Dawson, Thomas E. Zipperian, and R.D. Whaley, IEEE, Electron Device Lett., 11(1), 54-56 (1990).

[4] W. Dobbelaere, J. De Boeck and G. Borghs, Appl. Phys. Lett. 55(18), 1856-1858 (1989).

[5] J. De Boeck, W. Dobbelaere, M. Van Hove, J. Vanhellemont, W. De Raedt, W. Vandervorst, R. Mertens and G. borghs, Proceedings of MRS Spring Meeting Symposium V (Materials Research Symposium V (Materials Research Society, Pittsburgh, 1990), in press.

[6] W. Dobbelaere, J. De Boeck, P. Van Mieghem, R. Mertens and G. Borghs, to be published in the Journal of Applied Physics, (February 1991).

[7] See e.g. Tai-Ping Sun, Si-Chen Lee and Sheng-Jenn Yang, J. Appl. Phys. 67(11), 7092-7097 (1990).

[8] P.N.J. Dennis, Photodetectors, (Plenum Press, New York, 1986), p.22.

SEMICONDUCTOR DEVICES AND MATERIALS FOR OPTICAL
COMMUNICATION AT 2-4 μm WAVELENGTHS RANGE

GAN FUXI AND WANG HAILONG
Shanghai Institute of Optics and Fine Mechanics, Academia Sinica
P.O.Box 800-216 Shanghai 201800 China

ABSTRACT

In this paper we reported most attractive semiconductor
materials for optical sources and detectors with wavelength in
the range of 2-4 μm, such as GaInAsSb/GaAlAsSb, InAsPSb/InAs
heterostructure and PbCdSSe HgCdTe. The Lasers and detectors have
been made using these materials. The performance of the devices
were discussed.

INTRODUCTION

Extremely low losses $(10^{-2}-10^{-3} dB/km)$ have been predicted
for novel fiber materials like fluoride glasses at 2-4μm
wavelenth range(1,2). It has stimulated considerable interest
in researching on radiation sources and detectors operating in
this region for future mid-infrared optical communication
Among the compound semiconductor materials potentially
useful for sources and detectors at these wavelengths are III-V
solid solutions based on GaSb or InAs substrates(3),IV-VI Pb-
salt compounds(4) and II-VI narrow bandgap compounds(5), such as
$Ga_{1-x}In_xAs_ySb_{1-y}$ lattice-matched to GaSb $(y/x \sim 0.9)$, $InAs_{1-x-y}P_xSb_y$ lattice-matched InAs, $Pb_{1-x}Cd_xS_{1-y}Se_y$ and $Hg_{1-x}Cd_xTe$ etc.
In this report, these materials and their application in
light sources and detectors have been discussed.

GaAlAsSb/GaInAsSb/GaAlAsSb/GaSb Double Heterostructure

C.Caneau et al.(6) reported GaInAsSb/GaAlAsSb double
heterostructure lasers cw operating up to 235K. We studied some
problems in growing this heterostructure and fabricated broad
area diode lasers by using these materials.
The solid composition can be chosen according to design
emission wavelength of lasers. The variation of energy gap of
active layer materials $Ga_{1-x}In_xAs_ySb_{1-y}$ with composition are
approximated by

$$Eg(X,Y)=0.726+3.009X-4.389Y-3.565X^2+5.084Y^2-0.504XY \qquad (1)$$

Liquid phase epitaxial (LPE) growth was carried out in a
conventional sliding boat, horizontal reactor in an ambient of
palladium-purified hydrogen. Details of the growth procedure and
liquid solution composition were taken from Ref.7.
Fig.1 shows the photograph of cleave surface of typical DH
of GaAlAsSb/GaInAsSb/GaAlAsSb grown on (100) GaSb substrate
etched by $K_3Fe(CN)_6$+KOH etchant. The 2μm thick active layer was
grown in 12 second, whereas the 5μm cladding layers were grown
in 3 minutes.
We found that in GaAlAsSb/GaInAsSb/GaAlAsSb/GaSb double
heterostructures, it is easy for GaAlAsSb grown on GaSb and
GaInAsSb grown on GaAlAsSb, as long as to choose GaSb substrate
with good crystal perfection (double crystal x-ray diffraction
rocking curve like Fig.2a) and appropriate etching method.

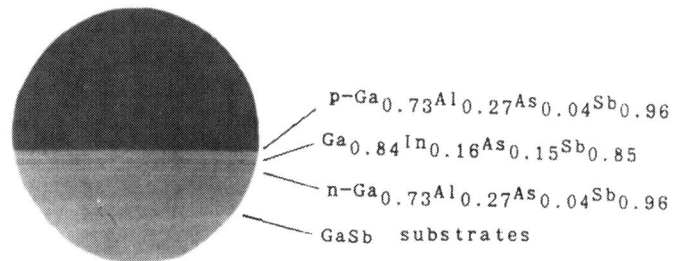

p-$Ga_{0.73}Al_{0.27}As_{0.04}Sb_{0.96}$
$Ga_{0.84}In_{0.16}As_{0.15}Sb_{0.85}$
n-$Ga_{0.73}Al_{0.27}As_{0.04}Sb_{0.96}$
GaSb substrates

Fig.1 Photograph of cleave surface of typical DH
of GaAlAsSb/GaInAsSb/GaAlAsSb grown on (100) GaSb

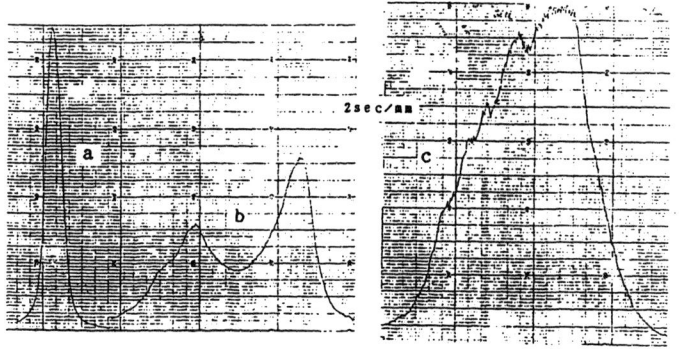

Fig.2 Double crystal X-ray diffraction rocking curves
of GaSb substrates ,best crystallinity FWHM ~12″.

However, there is troublesome problem for GaAlAsSb grown on
GaInAsSb, such as rough interface and surface, island growth
polynucleation, large area residual melt to cover on the grown
wafers etc. This is essentially connected with the surface
quality of an active layer and consequently with degree of
lattice-matching. For exact lattice-matched structure and growth
condition including precise weigh of melt liquid phase
compositions, these problem can be solved.

The wafers were processed into broad area lasers. The
contacts were evaporated and alloyed Au-Zn on the p side and Au-
Ge-Ni on the n side. A chips were soldered p side down on copper
studs electroplated In of special package.

The broad area lasers(stripe width s=240μm cavity length
L=500um) were measured under pulsed operation, with current
pulses 2μs duration, at a repetition rate 5KHz. The detector was

a HgCdTe photodiode operating at room temperature in 1-3μm spectral region. The threshold current densities is of 4.8KA/cm² at
operating temperature of 82K. Pulse operation was achieved up to
212K. Figure 3 shows the spectrum of a broad area laser at a
temperature of 16K, recorded using a grating monochromator and a
liquid nitrogen cooled InSb detector. The emission wavelength
is 2.063μ. Fig.4 shows threshold current desities vs temperature
curve for one of the broad area lasers. The characteristic
temperature To was 44K between 16-82K. The low value of To shows
heating characteristic of broad area lasers were very poor.
Further improvement could be obtained by decreasing the optimal
active layer thicknesses, to reduce the series resistance of
chips, increasing the Al content of the confinement layers and
using stripe geometry structure which should decrease the
threshold current.

InAsPSb/InAs Heterostructures Lasers and Detectors

$InAs_{1-x-y}P_xSb_y$ is one of the most attractive III-V compound
for optical source and detector with wavelength in the range of
2-4μm. A recent investigation(8) of losses in infrared fiber has
shown that there exists a low-loss narrow window in 2.5-2.6um
wavelength range. In InGaAsSb quanternary system the miscibility
gap(MG) just covers the window, which cause troublesome problem
for LPE. However the InAsPSb alloy can be grown outside the MG
on InAs by LPE.

The quanternary alloy $InAs_{1-x-y}P_xSb_y$ with y=0.47x (x+y<1)
are lattice matched to InAs. To a first-order approximation its
energy band gap Eg can be expressed as

$$Eg(InAs_{1-x-y}P_xSb_y)=(1-X-Y)Eg(InAs)+XEg(InP)+YEg(InSb)$$
$$-X(1-X-Y)C_1-Y(1-X-Y)C_2-XYC_3 \tag{2}$$

in which the $C_1=0.10ev$ $C_2=0.61ev$ and $C_3=1.6ev$ are the bowing

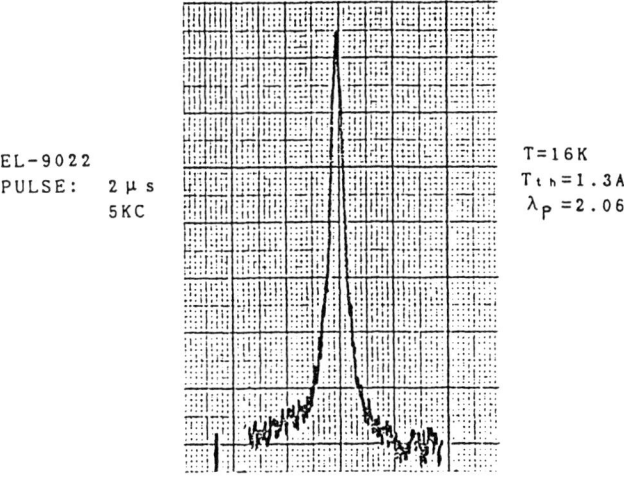

EL-9022
PULSE: 2 μs
5KC

T=16K
$T_{th}=1.3A$
$\lambda_P=2.063\mu$

Fig.3 Emission spectrum of $Ga_{0.73}Al_{0.27}As_{0.04}Sb_{0.96}/$
$Ga_{0.84}In_{0.16}As_{0.15}Sb_{0.85}/Ga_{0.73}Al_{0.27}As_{0.04}Sb_{0.96}/GaSb$
DH laser, recorded at 16K at current slightly above
threshold.

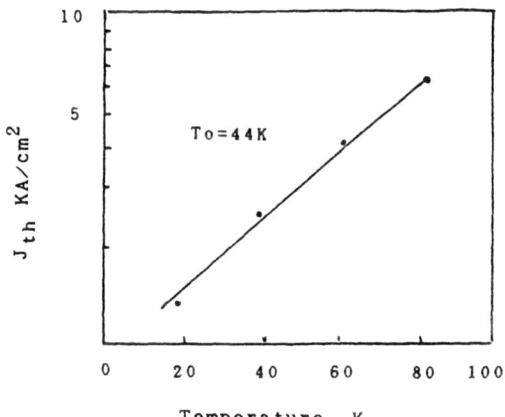

Fig.4 Threshold current density agaist
temperature for a broad laser.

parameter of the ternary systems InAsP, InAsSb and InPSb(9). So
its band gap could be changed from 0.36 to 0.62ev at 300K. But
the lattice matched quanternary alloy with Eg>0.56ev are
unavailable because of the miscibility gap. So the available
wavelength is from 2.2 to 3.5μm in this system which just
includes the 2.5um low loss window
   The lattice-matched heterostructure wafers were grown by
gradual temperature lowering and step-cooling methods using a
conventional sliding graphite boat in LPE
   In InAsPSb/InAs system there is a determinate relation
between the surface morphology and the lattice mismatch of the
epilayer, using which we can easily control the melt composition
to grow high quality heterostructures. We found that the best
morphology occurs at positive lattice mismatch about 0.1%, all
wafers with negative lattice mismatch shows poor morphology(10).
It could be seen from Fig.5 and Table I that the FWHM of the
DCXD rocking curves also related to the lattice mismatch and
surface morphology, best FWHM for the substrate and epilayer are
about 10sec and 46sec. The large lattice mismatch causes evident
broading on the DCXD rocking curves because of the strain
between the substrate and epilayer.
   The solid composition of epilayer was determined by the
EPMA analysis. Fig.6 shows a typical result for the composition
variation in the epilayer along the growth direction(11). The
phosphorous content has maximum value at interface and decreases
gradually alog the growth direction. On the contrary, Sb content
increases gradually alog the growth direction. But the
phosphorous and antimony content have almost a constant value
alog the growth direction for epilayer grown by step-cooling
method.

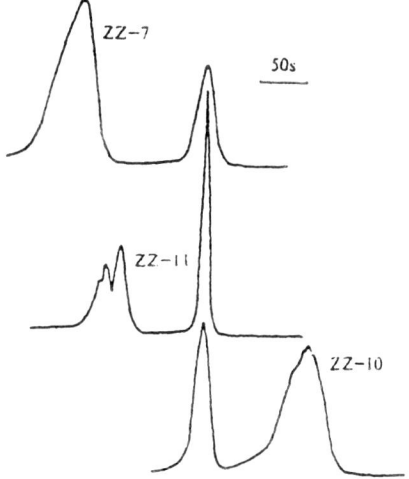

Fig.5 Typical DCXD rocking curves of
InAsPSb/InAs wafers.

Table I. Measured data of different InAsPSb / InAs epi-wafers

Sample No.	x(P) × 10⁻⁴	Δa / a %	Surface morphology	FWHM of Sub.	DCXD epi.
ZZ-6	6.447	+0.362	Cross wave	70"	92"
ZZ-7	7.513	+0.158	Shiny ripple	28"	64"
ZZ-11	8.046	+0.110	Shiny uniform	10"	46"
ZZ-26	8.143	+0.033	Sparse pit	25" *	60" *
ZZ-9	8.240	−0.046	Dense pit	26"	62" *
ZZ-10	9.064	−0.137	Very dense pit	27"	68"

Estimated value

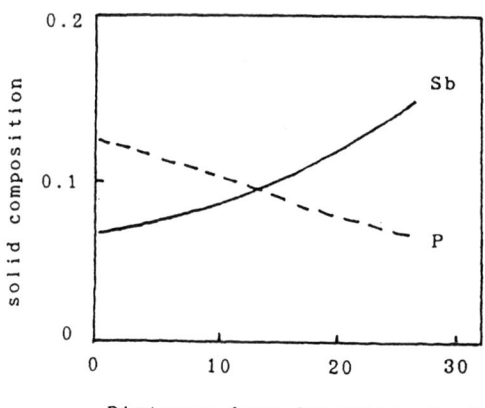

Fig.6 Typical result of composition variation
in epilayer along the growth direction.

P-InAsPSb (Ze-doped) was successively grown on the (100) n-
InAs substrate at temperature of 530°C. Fig.7 shows the I-V
characteristic of P-InAs$_{0.82}$P$_{0.12}$Sb$_{0.06}$/N-InAs(sub) epitaxial
wafers at 77K. Broad area(200umX350um) diode chips were
fabricated using these wafers. The diodes were driven by current
pulses 10us duration and 5kHz repetition rate. The optical
output was detected by InSb detector. The threshold current
densities is of 950A/cm^2 at operating of 14K. Fig.8 presents
emission spectrum of a InAs$_{0.82}$P$_{0.12}$Sb$_{0.06}$/InAs single
heterostructure diode laser at 14K. The emission wavelength is
3.09um.

InAs$_{0.76}$P$_{0.16}$Sb$_{0.08}$/InAs$_{0.94}$P$_{0.04}$Sb$_{0.02}$/InAs$_{0.76}$P$_{0.16}$Sb$_{0.08}$
/InAs DH epitaxial wafers have been grown. We belive that
performance of lasers made by using these DH epitaxial wafers
would be improved notably.

The InAs$_{0.71}$P$_{0.2}$Sb$_{0.09}$/InAs pin mesa photodetectors were
made(12). Fig.9 shows schematic diagram of the InAsPSb/InAs mid-
infrared photodetector structure. The response characteristic of
the detector was measured at room temperature, Peak response
wavelength is 2.6μm, detectivity D*>9X10^{8}cmHz$^{1/2}$W^{-1} (D*~
10^{12}cmHz$^{1/2}$W^{-1} at 77K) reponse time is 1.2ns (100μm mesa) and
spectra responsivities is ranging from 1.0-3.2μm. The response
spectrum of the detector as shown in Fig.10.

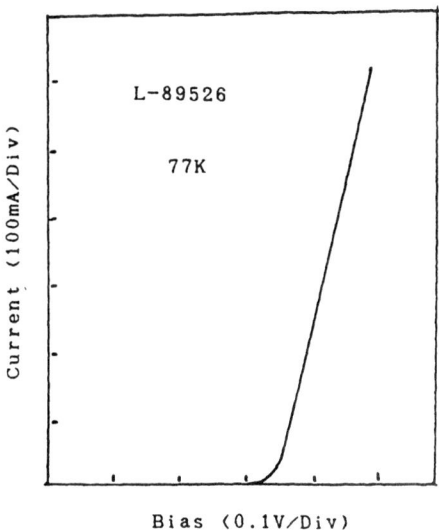

Fig.7 The I-V characteristic of a p-InAs$_{0.82}$P$_{0.12}$Sb$_{0.06}$/n-InAs at 77K

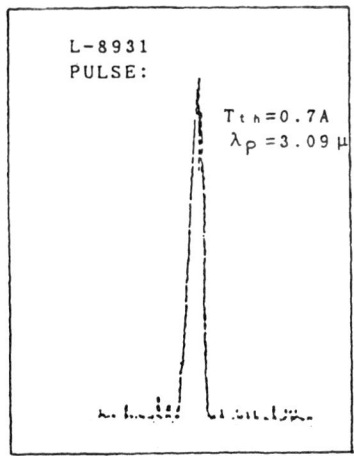

Fig.8 Emissin spectrum of In$_{0.82}$P$_{0.12}$Sb$_{0.06}$/InAs diode laser at operating temperature of 14K.

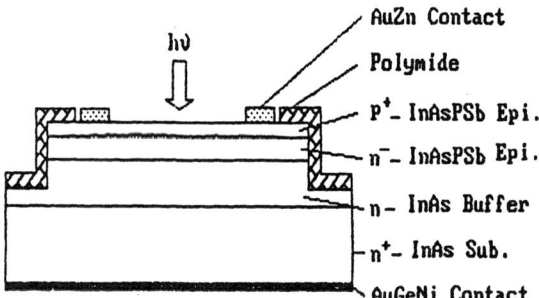

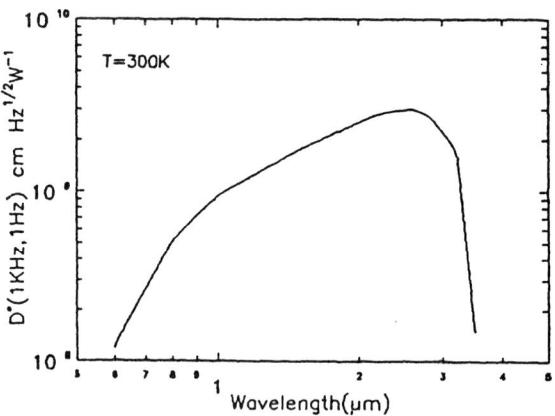

Fig.9 Schematic diagram of the InAsPSb/InAs
mid-infrared photodetector structure.

Fig.10 Response spectrum of the InAsPSb/InAs
photodetector.

$Pb_{1-x}Cd_xS_{1-y}Se_y$ Diode Laser for Mid-infrared

The IV-VI compound semiconductor is one of a potentially useful materials for mid-IR laser sources. PbS has an energy gap of 0.42ev at 300K. It is a potentially useful material for such laser source. The lattice-matched double heterostructure lasers consising of $Pb_{0.95}Cd_{0.05}S_{0.89}Se_{0.11}$ confinement layers and PbS active layer grown by molecular beem epitaxy(MBE) operated up to 200K at 3.27um(13).
The energy gap of $Pb_{1-x}Cd_xS_{1-y}Se_y$ can be approximately expressed as

$$Eg(X,Y,T)=264.5+3600X-140Y+(400+0.265T^2)^{1/2} \text{ (mev)} \quad (3)$$

where T is operating temperature.
We have firstly got bulk single crystal of PbCdSSe by using horizontal unseeded vopor growth(HUVG). The dislocation densities of the crystal lies in the range of 10^3-10^4/cm^2 and carrier concentration N lies in the range of $5X10^{17}-5X10^{18}$/cm^3. The bulk crystal has a very nice (100) grown facet. It is convenient to fabricate diode lasers.
The crystal block together with a few of free selenium were seeled in a quartz ampoule evacuated to $5X10^{-6}$ Torr. A p-n junction with a depth of 20um was obtained by annealing at 450°C for one hour. The crystal block with p-n junction was lapped into wafers of 200um thick. The contacts were evaporated Au and electroplated In on p-side and n-side. Fig.11 is schematic diagram of the mesa-stripe chip fabricated by standard photolithography. Finally, the chips were mounted in special packages.
The $Pb_{0.97}Cd_{0.03}S_{0.90}Se_{0.10}$ homojunction laser was driven by current pulse 4us long, 10KHz repetition rate. The threshold current is of 400mA and the emission wavelength is 3.13um at 15K Pulse operation was achieved up to 90K.

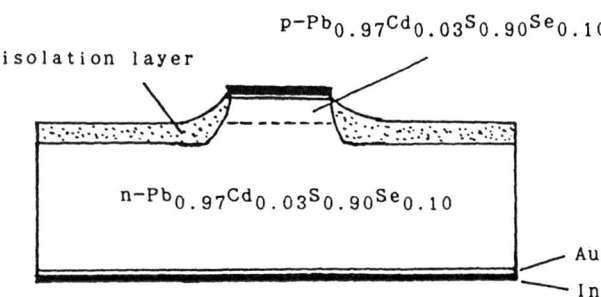

p-$Pb_{0.97}Cd_{0.03}S_{0.90}Se_{0.10}$

isolation layer

n-$Pb_{0.97}Cd_{0.03}S_{0.90}Se_{0.10}$

Au

In

Fig.11 Schematic diagram of mesa PbCdSSe homojuction laser.

HgCdTe Detector for the 2-4μm region at Room Temperature

The binary alloy semiconductor (Hg,Cd)Te has been selected as a detector materials for fabricating sensitive high speed laser receivers that operate at room temperature in 2-4 spectral region(5).

It is completely miscible alloy, its energy band gap Eg can be expressed as

$$Eg(X,T)=-0.295+1.87X-0.28X^2+(6-14X+3X^2)10^{-4}T+0.35X^4 \quad (4)$$

The relative detector cut-off wavelength can be adjusted from roughly zero to 1.6ev by varying the composition. At operating of room temperature and selecting X=0.3-0.5, the response spectral of the detector can cover 2-4um range.

The detector with a peak spectral response at 2.5 and operating near room temperature have been made(14). Its peak specific detectivities is about $1X10^{10}cmHz^{1/2}/W$ and response time<6ns.

Reference

(1)D.C.Tran, G.H.Sigel,Jr., and B.Bendow, J. Lightwave Technol. LT-2, 566 (1984).
(2)M.E.Lines, Science 226, 663 (1984).
(3)A.Joullie, F.Jia Hua, F.Karouta, H.Mani and C.Alibert, SPIE. 587, 46 (1985).
(4)Nobuyuki Koguchi, Satoshi Takahashi and Teruo Kiyosawa, JAP. J. of Appl Phys 27, L2376 (1988).
(5)D.A.Soderman and W.H.Pinkston, Appl Optics 11, 2162 (1972).
(6)C.Caneau, A.K.Srivastava, A.G.Dentai, J.L.Zyskind, C.A.Burrus, and M.A.Pollack, Electron. Lett 22, 992 (1986).
(7)C.Caneau, A.K.Srivastava, J.L.Zyskind, J.W.Sulhoff, A.G.Dentai and M.A.Pollack, Appl.Phys.Lett. 49, 55 (1986).
(8)B.W.France, S.F.Carter, M.W.Moore, and J.R.Williams, SPIE. 618, 51 (1986).
(9)Fukui T. and Horikoushi Y., Jpn.J.Appl.Phys. 20,2301 (1981).
(10)Zhang Yonggang, Zhou Ping, Chen Huiying and Pan Huizhen, Rare Metals 9, 46 (1990).
(11)Jin Changchun, Wang Yongzhen, Lu Guijin, Wang Tifeng, Jiang Jinxin and Wang Hailong, "Liquid phase epitaxial growth of InAsPSb for laser emission in 2-4um wavelength region", to be published.
(12)Zhang Yonggang, Zhou Ping, Zhou Ying and Pan Huizhen, "Investigation of InAsPSb/InAs high speed mid-infrared photodetector", to be published.
(13)N.Koguchi, T.Kiyosawa and S.Takahashi, J. Crystal Growth 81, 400 (1987).
(14)Tong Feiming, Yang Xiuzhen, Chen Siyuan, "HgCdTe detector at operating of room temperature", private communication.

PHOTOLUMINESCENCE OF InAs/InAsPSb AND InAs/InAlAsSb HETEROJUNCTIONS

A. N. BARANOV, M. S. BRESLER, O. B. GUSEV, K. D. MOISEEV,*
V. V. SHERSTNEV, YU. P. YAKOVLEV AND I. N. YASSIEVICH*
A. F. Ioffe Physico-Technical Institute, Leningrad, U.S.S.R.
*also Theoretical Physics Institute, Minnesota University,
Minneapolis, U.S.A.

ABSTRACT

    Photoluminescence of p-P InAs/InAsPSb and InAs/InAlAsSb
heterojunctions grown by LPE method was studied at liquid helium temperature. The recombination spectra contained a new
broad band lying between the substrate and the layer lines
which was identified as an emission from the interface. This
line is characterized by a strong blue shift when the excitation intensity increases. The intensities of bulk and interface lines show an unusual dependence on the pumping power.
On the basis of experimental findings the interface line is
attributed to emission from electrons confined at the interface due to reflection of electrons moving above the barrier.

    We report here results of photoluminescence (PL) studies
in p-InAs/P-InAsPSb heterostructures and also p-InAs/P-InAlAsSb
heterostructures grown for the first time. A new PL interface
line was found in both systems interpreted as an emission from
electron quasilocal states due to electron reflection from the
boundary. This fact suggests a sharp potential discontinuity
on the interface demonstrating high quality of heterojunction.
It is found that comparatively strong electron reflection from
the interface takes place in heterojunctions studied.

    1. p-InAs/P-InAs$_{1-x-y}$P$_x$Sb$_y$ (x=0.25,y=0.12) and p-InAs/
P-In$_{1-x}$Al$_x$As$_{1-y}$Sb$_y$ (x=0.058,y=0.06) heterojunctions were grown
by LPE method n-type indium arsenide [100] substrates. Doping
to p-type was achieved by zinc diffusion during growth.

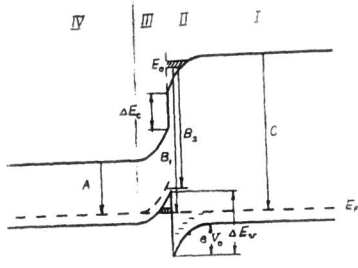

Fig.1

Structures InAs/InAsPSb
with doping level from 4.10^{16}
cm^{-3} to 7.10^{17} cm^{-3} at the
layer surface were studied in
detail. EBIC measurements have
shown that p-n junction was
displaced by 10-30 microns into indium arsenide, the epitaxial layer thickness not exceeding 2-3 microns.
    To reveal the heterojunc-
tion type we estimated the bands
offsets using the empirical data
on valence band edge positions
quoted in [1]. The valence band
shift in the solid solution in-
duced by addition of phosphorus and antimony to indium arsenide
was calculated in a linear approximation. In this case the valence band offset is 80 meV, the valence band edge of solid solution being lower than that of indium arsenide. From the lumi-

nescence data we deduced the forbidden gap difference of 105 meV. Thus the heterojunction studied must be attributed to the first type p-P one. The energy band diagram is shown in Fig.1.

2. Photoluminescence in heterostructures was excited by pumping with neodymium cw laser. The excitation power (up to 500 mW) was focused onto samples immersed in liquid helium at 1.8 K. Measurements of PL were made in reflection geometry, illumination being done from the epitaxial layer side.

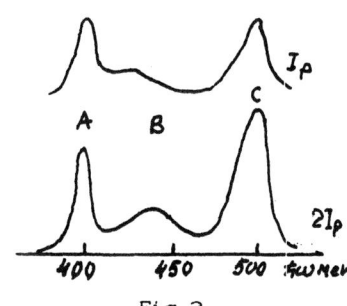

Fig.2

In Fig.2 photoluminescence spectra are given for p-InAs/ P-InAsPSb heterojunction and two excitation intensities. The luminescence spectrum consists of two narrow lines (A and C) and a broad band B, situated between them. The interpretation of the narrow lines is straightforward: line A corresponds to a well known transition in p-InAs – conduction band – acceptor and line C can be identified with the same transition for P-InAsPSb. In this case one can deduce the sum of band offsets from the PL data: $\Delta E_c + \Delta E_v = 105$ meV. The band B has a halfwidth of 30 meV which is independent of the pumping power. However, the peak position of band B suffers rather strong blue shift with the excitation intensity increase. For the lowest pumping level the distance of B band from the narrow-gap bulk line does not exceed 20 meV, and for the highest excitation band B merges into bulk line C. It should be stressed that this type of behaviour is characteristic for interface lines. The interface origin of B band is confirmed by its behaviour in magnetic field: in a magnetic field parallel to the interface the B band intensity decreases (and C line intensity simultaneously increases) demonstrating a reduction in diffusion flow of photoelectrons to the interface.

In first type heterojunctions the localization of holes in quantum well at the interface leads to emission with transition energy less than that of narrow-gap material. Such line was observed in GaAs/GaAlAs heterojunction [2].

The interface line B observed is characterized by its unusual position between the bulk lines of narrow-gap and wide-gap materials. This fact suggests the accumulation of charge carriers at the interface in the energy range where no localization occurs according to conventional reasoning.

The luminescence lines intensities as functions of excitation power were also measured.

3. We shall start the theoretical discussion from the construction of the energy band diagram for the heterojunction in question. Basing on the experimental results we adopted the following scheme: the heterojunction is the first type p-P one with equal doping level on both sides of the interface, the Fermi level of holes coincides with the acceptor level at T=0 K.

The potential profile at the interface was calculated according to [3]. Here we assumed the potential profile to be triangular in the space charge region and parabolic-shaped in the depletion region.

Taking into account the position of the hole Fermi level
being located on the acceptor level, i.e. $E_F = \Delta E_v - E_A - eV_o$
we obtain the following equation for band-bending potential
drop

$$\frac{m_h}{\pi^2 \hbar}\left\{\Delta E_v - E_A - eV_o - \left[\frac{\hbar^2}{2m_h}\right]^{1/3}\left[\frac{9\pi}{4}e\left(\frac{2\pi N_A}{\varkappa}eV_o\right)^{1/2}\right]^{2/3}\right\} = N_A\left[\frac{eV_o\varkappa}{2\pi e^2 N_A}\right] \quad (1)$$

where eV_o is the band-bending potential drop, N_A the acceptor
concentration, E_A the acceptor energy, $\varkappa$ the dielectric constant assumed equal on both sides of the interface, m_h the
heavy hole effective mass, m_e the electron effective mass.
The potential profile near the interface calculated according
to (1) is shown in Fig.1. We assumed $m_h = 0.4\,m_o$, $m_e = 0.03\,m_o$,
$\varkappa = 14.3$, $\Delta E_v = 80$ meV, $E_A = 20$ meV. Table I contains the
band-bending potential drops for various doping levels N_A.

Table I

Band-bending potential drop eV_o, the zeroth level energy
of heavy holes E_{oh}, the zeroth and the first level energy for electrons in a quasiwell E_{oe}, E_{1e}, the width
of space charge l_S and depletion l_D region as a function of the acceptor concentration N_A. Values of E_{oe}
given in parenthesis are calculated for triangular potential.

N_A 10^{16} cm^{-3}	eV_o meV	E_{oh} meV	E_{oe} meV	E_{1e} meV	l_S Å	l_D Å
1	47.8	11.3	24.6 (26.7)	39.7	110.8	869
2	45.0	13.9	29.4 (33.1)	44.4	99.4	596
3	43.1	15.7	32.3 (37.3)	–	93.2	477
4	42.2	17.2	34.5 (40.8)	–	86.2	408
5	40.3	18.2	35.7 (43.3)	–	87.2	357
6	39.4	19.2	36.8 –	–	84.3	322
7	38.4	20.1	37.4 –	–	82.6	295
8	37.5	20.8	– –	–	81.7	272
9	37.5	21.7	– –	–	77.0	257
10	36.6	22.2	– –	–	77.0	240

4. Now we want to discuss the nature of the B band. The
spectral position of this broad band could be explained neither
by surface states nor by hot luminescence of non-thermalized
electrons driven by diffusion to narrow-gap material.

However, the phenomena observed can be effectively treated
using the conception of electron confinement due to their reflection from the interface.

The occurence of sufficiently strong reflection from the
interface for electrons moving above the potential barrier generates electron quasi-local states at the interface with complex energy eigenvalues $\varepsilon = E - i\,\Gamma/2$, the level broadening
$\Gamma/2$ being small for strong reflection. If $\Gamma/2 \ll E$, accumulation of photoelectrons in a sort of 'quasi-well' occurs. In
this case a significant blue shift of the interface line is possible with the excitation intensity increase because the photoelectron accumulation leads to reduction of the quasiwell width,

the electron level being simultaneously pushed upward.

Comparatively large linewidth of the B band ~ 30 meV implies the existence of photoelectron transitions from the quasiwell states not only to the holes located in the potential notch but also to the acceptors in the space charge region (transitions B_1, B_2 in Fig. 1).

Quasilocal level energies (in semiclassical approximation) can be found for parabolic-shape potential from the equation

$$k - \left[1 - k^2 \right] \ln \left[\frac{1+k}{(1-k^2)^{1/2}} \right] = \left[\frac{\hbar^2}{2m_e} 8\pi^3 \frac{e^2 N_A}{eV_0 \varkappa} \left(n + \frac{3}{4} \right)^2 \right]^{1/3} \qquad (2)$$

where $k = (E/eV)^{1/2}$, n=0,1,2... The calculation results are shown in Table I, where data for triangular-shape potential are also presented (In the last case an exact solution of the Schroedinger equation was found). For low doping levels the results for triangular potential coincide with those for parabolic-shape one as expected.

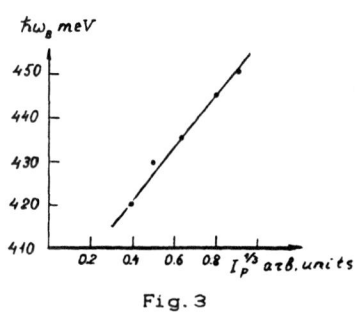

Fig. 3

The large blue shift of the B band with the excitation intensity increase can be easily treated with use of previous arguments. In the first approximation we shall assume that the effect of electron accumulation is equivalent to increase of acceptor concentration. In this case one finds for triangular potential, i.e. for k << 1, that the zeroth level position E_{oe} is related to the pumping intensity I_p as $E_{oe} \sim I_p^{1/3}$. If we neglect the change of band-bending potential drop eV_0 with I_p (which is actually weaker than the corresponding change in the zeroth level energy) then the interface line position would shift with the excitation power according to the same rule. In Fig. 3 we present the B line energy plotted against $I_p^{1/3}$. It is clear that $I_p^{1/3}$ dependence is valid, demonstrating the agreement of the theory and the experiment.

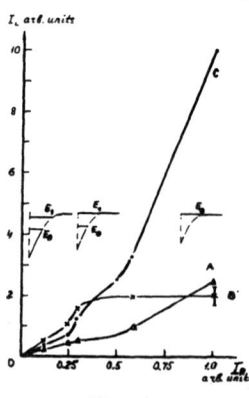

Fig. 4

In Fig. 4 the intensity dependence of A,B and C lines on the pumping power I_p is shown. We can connect this unusual behaviour of the PL lines with the displacement of electron energy levels in a quasiwell. According to our calculations the quasiwell accomodates two electron states (the zeroth and the first level) at low pumping levels. With the excitation power increase the first (upper) level would be pushed out of the well. When the level comes out into the continuous energy spectrum the electron reflection from the boundary suffers a resonant increase which is accompanied by a resonant increase in the electron capture to a quasiwell [4]. We believe that this mechanism is responsible for a sharp increase of B and

C lines intensities (and a simultaneous decrease in the slope
of A line dependence) observed at $I_p \approx 0.25$. At higher
excitation powers the resonance in Preflection and capture
disappears. The second sharp increase of A and C intensities
may be due to the zeroth level displacement being pushed out
of the well at $I_p \approx 0.6$.

5. Consider the space distribution of excess electrons
near the heterojunction which determines the relative intensities of PL lines (The excitation is being done from the
epitaxial layer side, region I).

For estimates we shall assume the quartenary-alloy parameters to be close to their values for indium arsenide, i.e.
the absorption coefficient being $\alpha \sim 3.10^4$ cm^{-1}, the electron mobility $\mu \sim 2.10^4$ cm^2/V.s (for electron Fermi energy $E_F \sim 5$ meV) and the diffusion coefficient $D \sim 60$ cm^2/s. The
characteristic lifetime of nonequilibrium electrons for indium arsenide is $\tau \lesssim 10^{-9}$ s at helium temperatures. As the
quantum efficiency of indium arsenide is small, we may assume
that the radiation lifetime is significantly greater than the
quoted value. The same is true for the four-component alloy.

In the experimental conditions the electron diffusion
length is $L = (D \tau)^{1/2} \sim 2.5$ microns to be compared to the
epitaxial layer thickness d of 2 microns and the depletion
depth l (being 869 Å for $N_A = 10^{16}$ cm^{-3} and 240 Å for $N_A = 10^{17}$
cm^{-3}). Thus, $L \gtrsim d \gg 1$. Significant part of photoelectrons excited is driven out of the epitaxial layer by diffusion, the
corresponding lifetime in region I $\tau_D \sim d^2/2 D \sim 6.10^{-10}$ s
being of the order of nonradiative lifetime.

Photoelectrons coming to the boundary of region II (depletion region) are effectively captured in quasilocal states.
To estimate their lifetime in a quasiwell (which is governed
by transfer across the interface) we used semiclassical approximation. In this case for arbitrary potential profile we obtain $\tau = T / \mathfrak{D}$, where T is the classical period of electron
oscillation in the bound state, $\mathfrak{D}$ is the particle transmission coefficient for over-barrier reflection. The transmission
coefficient $\mathfrak{D}$ cannot be estimated on theoretical grounds with
reasonable accuracy and we shall treat it as a physical parameter to be determined from experiment.

For parabolic-shape potential we have

$$T = (m\varkappa / \pi e^2 N_A)^{1/2} \ln(1 + \sqrt{E/eV_o}) \qquad (3)$$

In simple case of quadratic dispersion relation and for
electron reflection being caused only by difference in electron kinetic energy on both sides of the interface

$$\mathfrak{D} = \frac{4\sqrt{E(E+\Delta)}}{(\sqrt{E} + \sqrt{E+\Delta})^2} \qquad (4)$$

where E is the electron kinetic energy (above the barrier), Δ
is the kinetic energy drop at the boundary. In our experimental conditions $\Delta \approx \Delta E$, and E is the electron energy measured from the quasiwell bottom. For our set of parameters the
transmission coefficient estimated from (4) does not differ
very much from unity. However, this value puts only the upper
limit on the transmission coefficient as it may be significantly less due to wave function mixing in the Kane model and/or
the effect of inconsistency of the envelope wave function approximation for the heterojunction in question (the last point
may be connected with different nature of materials on both
sides of the heterojunction). Both these factors may lead to

substantial increase of the electron lifetime in a quasiwell.
Nonetheless one cannot expect the transmission coefficient
to be less than 10^{-2}, i.e. the electron lifetime to exceed
10^{-10} s.
On the other hand the B line intensity is unexpectedly strong
being of the same order of magnitude as that for line A (taking the integral intensity value) or being lower by a factor
of 10 if one compares the peak values. This fact implies a
great osillator strength for the transition involved. We
believe that it is a sequence of carrier confinement at the
interface.

6. Similar results were obtained for InAs/InAlAsSb heterojunctions. The sum of band offsets was found to be 75 meV,
the valence band offset being estimated as 20 meV. A broad
interface line B of the same nature was also observed which
moved from 405 to 480 meV as the excitation power increased.

7. In conclusion, we have studied photoluminescence in
p-InAs/P-InAsPSb and p-InAs/P-InAlAsSb heterojunctions demonstrating the existence of electron confinement at the interface due to over-barrier electron reflection. Energy band diagrams were constructed from the experimental data. The over-
barrier electron reflection is found to be unexpectedly
strong.
We believe that the quasilocal states observed can be of
importance for the properties of single heterojunctions and
also of multilayer semiconductor laser structures with charge
carrier collection. In particular, the heterostructures with
strong over-barrier reflection provide a possibility to rea-
lize the electron Fabry-Perot interferometer with a finesse
$Q \sim \Gamma /E$.
We are indebted to Prof. B. P. Zakharchenya, Prof. V. I. Perel'
and Prof. A. A. Rogachev for support of this work and discussion
of results.

References

1. G. Margaritondo, in Electronic Structure of Semiconductor
   Heterojunctions, edited by G. Margaritondo (Kluwer, Dor-
   drecht, 1988), p. 20.

2. Y. R. Yuan, K. Mohammed, M. A. A. Pudensi and J. L. Merz, Appl.
   Phys. Lett. 45, 739 (1984).

3. H. Morkoc, in Molecular Beam Epitaxy and Heterostructures,
   edited by L. L. Chang and K. Ploog (Martinus Nijhoff, Dor-
   drecht, 1983), p. 631.

4. S. V. Kozyrev, A. Ya. Shik, Fiz. Tekhn. Polupr. 19, 1667
   (1985).

LPE GROWTH AND CHARACTERIZATION OF InAsSbP/In$_{1-x}$Ga$_x$As$_{1-y}$Sb$_y$ /InAsSbP (X≥0,Y≥0) HETEROSTRUCTURES FOR LONG WAVELENGTH (λ>3µm) LEDS AND LASERS.

M.AYDARALIEV,T.S.ARGUNOVA,N.V.ZOTOVA,S.A.KARANDASHOV,R.N.KUTT, B.A.MATVEEV,S.S.RUVIMOV,L.M.SOROKIN,N.M.STUS',G.N.TALALAKIN

A.F.IOFFE PHYSICO-TECHNICAL INSTITUTE, ACADEMY OF SCIENCES OF THE USSR, 194021, LENINGRAD

The band gap of InAsSbP and InGaAsSb alloys enriched with InAs correspods to the spectral range 2.5 - 5 µm which make possible to manufacture LEDs and detectors for the second atmosphere window. Elevated hardness (see Fig.1a) and small plasticity of the alloys results in an inversed plastic deformation process during LPE growth of InAsSbP (InGaAsSb) on InAs substrate.That is when growing graded(grad a ≈ 5*10^{-8}) epilayers at elevated temperatures (680 - 720° C) misfit dislocations are formed throughout the entire substrate thickness, dislocation densities in epilayer (curves 2,4 on Fig.1b) and substrate (curves 1,3) are dependent on the initial substrate

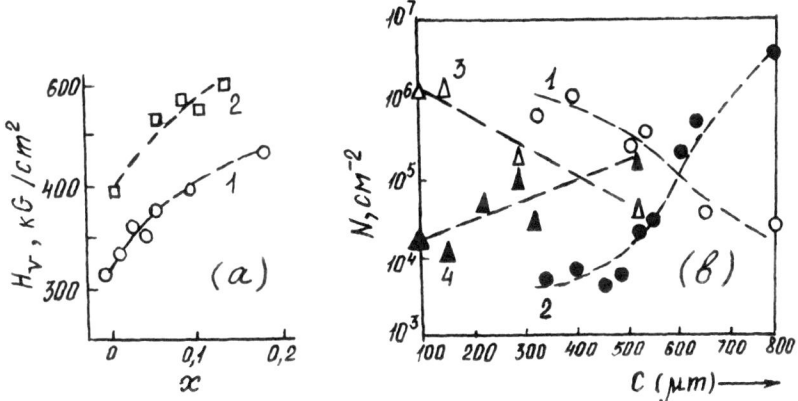

Fig.1. Microhardness as a function of composition x of In$_{1-x}$Ga$_x$As(1) and In$_{1-x}$Ga$_x$As$_{0.9}$Sb$_{0.1}$(2) - a ; InAs substrate(1,3), InGaAsSb(4),InAsSbP(2) dislocation density as a function of the initial substrate thickness - b.

thickness (C). Simultaneously with increasing dislocation density in InAs (curves 1,3), the curvature of the structure increased within $\ae = 0.02-0.2$ cm^{-1} . This fact enable us to describe the above dependence by applying Nye-correlation: $N=\frac{1}{R*a}$, where a is the lattice constant and R is the curvature radius. The dislocation density in the graded layer (curves 2,4) declines with decreased substrate thickness, a fact assumed to be due to the specific feature of the crystal cell of the graded crystal(see Fig.2) whose dislocation density at

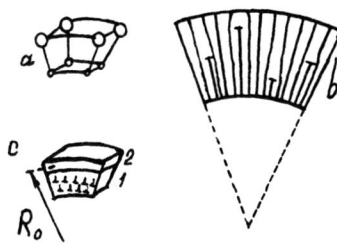

Fig.2. Lattice schemes of
a) graded crystal unit cell,
b) dislocated graded crystal,
c) heterostructure containing
the plastically deformed
substrate(1) and the strained
epilayer(2) having R_0=a/grad a

full strain relaxation and at larger curvature declines as $N=\frac{1}{a}(\frac{\text{grad } a}{a} - \frac{1}{R})$, where a is the lattice constant, R -the radius of curvature [1].

Optical absorption was measured in graded InAs$_{1-z}$Sb$_z$/ InAs$_{1-x-y}$Sb$_x$P$_y$ (0.15 $\leqslant$ z $\leqslant$ 0.54) samples grown on thin (< 350 μm) InAs (III) substrates (see Fig.3).

The absorption edge (optical density D) can be described using the semiempirical expressions [2]: D = D$_0$ exp((hν − E$_{g \text{ min}}$)/ε) for hν<E$_{g \text{ min}}$ and D=D$_0$+(hν−E$_{g \text{ min}}$)α$_0$/∇ for hν>E$_{gmin}$, where E$_{g \text{ min}}$ − band gap at the surface, α$_0$ − absorption coefficient at hν = E$_{g \text{ min}}$, ∇ − band gap gradient, D$_0$,ε − empirical parameters. Employing abovementioned expressions and transmission measurements on samples of various compositions band gaps from 0.175 (α$_0$ = 1400 cm^{-1}) to 0.09 (α$_0$ = 644 cm^{-1}) eV and ε ~ 7 meV were obtained.

High crystalline quality of layers grown on thin substrates and resonance disturbances (E$_g$≠Δ) in p-InAsSbP(Zn) and p-InGaAs (Mn) permit to create ambient operating single heterostructure diodes in the spectral range of 2.5-4.6 μm (see Fig.3) used in optoelectronic sensors [3].

Using the elaborated technology we have been able to obtain DH with 5-15 μm cladding layers of n- or p(Zn)-InAsSbP and

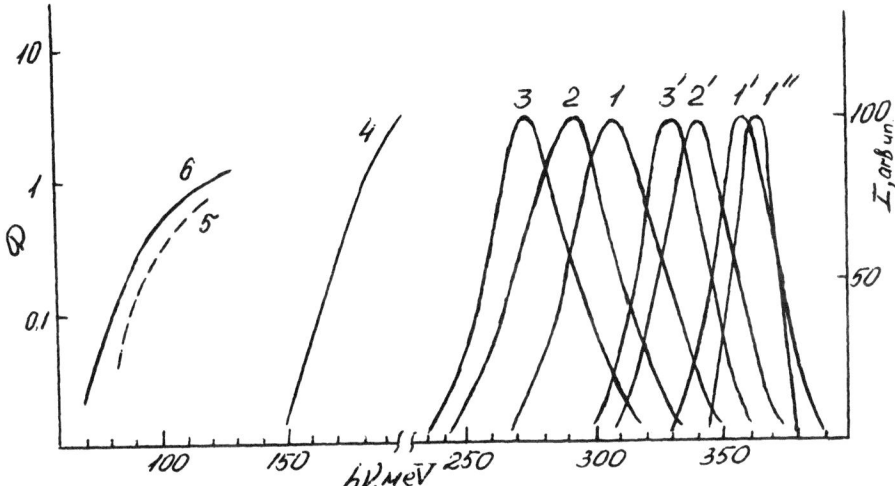

Fig.3. Emission (1,2,3) and absorption (4,5,6) spectra for
InAs$_{1-x-y}$Sb$_x$P$_y$ diodes and epilayers InAs$_{1-z}$Sb$_z$ with
different compositions: 4 - z=0.15, 5,6 - z=0.54.
1,2,3,4,5,6 - 300 K, 1',2',3' - 77 K, 1" - 4.2 K,
6 - spectra of epilayer without substrate.

1-3 μm narrow gap undoped n-type layers of InAs ,
InGaAs, InAsSb or InGaAsSb. TEM, x-ray diffractometry and topography showed the presence of 60°-misfit dislocations(MD).Most
of MD were at the interface close to InAs substrate. The latter
result was obtained both by TEM cross section method and triple-
crystal x-ray diffractometry. The analysis was made by means of
constructing a two - coordinate presentation of intensity-along
the vector reciprocal lattice H - I(q_H), q_H|H, and perpendicu-
larlly to it.

MD networks were far enough from emitting p-n junctions so
it was possible to obtain low threshold (I_{th}<100 A/cm^2, 77 K)
DH broad contact (250x300 μm^2) lasers 3.0 -3.56 μm operating
at 4.2 - 160 K, emitting CW and pulse power (see Table I). For
laser fabrication ohmic contacts were formed by thermal evaporation under vacuum of Au-Zn on the p-InAsSbP cap layer and
Au-Ag on the n-InAs (III) substrate [4,5].

Simultaneous generation of 2 wavelengths of stimulated
emission was perfomed in the DH-structures with p-n junction

located in the InAsSbP wide-band emitter having high quantum efficiency. Two wavelength generation takes place when distance from p-n junction to heterojunction InAsSbP / InAs $W \leqslant 3 \div 4\ L_p$ (diffusion length of holes in n - type InAsSbP emitter). Two peaks of stimulated emission were observed, with energy 407 meV (recombination in the narrow band region) and 427 meV (recombination in p-n junction) when W=10 μm, $W/L_p \sim 4$. Threshold currents of 370 and 800 A/cm^2 were observed respectively .

In summary we have fabricated InAsSbP layers with energy band gaps down to 0.09 eV (300 K) and InAsSbP/In$_{1-x}$Ga$_x$As$_{1-y}$Sb$_y$ double heterostructure lasers continiously operating from 3.0 to 3.6 μm at cryogenic temperatures. These are the longest wavelength reported to date for DH injection laser made from III-V system and III-V layers grown by LPE.

Table I.

Struc-ture	Composition of the active layer		Wavelength of genera-tion	Threshold current at 77 K, A/cm^2		Temperature range, K
	In$_{1-x}$Ga$_x$As$_{1-y}$Sb$_y$		μm	pulsed τ=5 μs	cw	
	x	y				
HS	0	0.13	3.90	200(4.2K)	–	4.2
SH	0	0.08	3.60	4000	–	4.2 - 140
DH	0	0	3.05	64	103	4.2 - 140
DH	0.01	0	3.04	100	130	4.2 - 140
DH	0.07	0.065	3.23	39	60	4.2 - 160
DH	0	0.07	3.55	87	130	4.2 - 140

REFERENCES

1. B.A.Matveev, N.M.Stus', G.N.Talalakin, Crystallografia, 32, 216 (1988) (Sov. Crystallography).

2. B.M.Morozov, Yu.B.Bolkohovitianov, R.S.Gabaraev, A.F.Kravohen-ko,V.I.Yudaev, Fiz. Tekh. Polyprovodn. 14, 1486 (1980) (Sov. Phys. Semicond.).

3. N.P.Esina, N.V.Zotova et al.,Jurnal Prikladnoi Spektroskopii, 42, 692 (1985) (Sov. J. Appl. Spectroskopy).

4. M.Aidaraliev, N.V.Zotova et al., Pis'ma v Jurnal Techniches-koy Fiziki, 15, 49 (1989) (Sov. Tech. Phys. Lett.).

5. M.A i daraliev, N.V.Zotova, S.A.Karandashov, B.A.Matveev N.M.Stus',G.N.Talalakin, Phys. Stat. Sol.(a),115, K117 (1989).

MBE GROWTH OF GaInAsSb/AlGaAsSb DOUBLE HETEROSTRUCTURES FOR DIODE LASERS EMITTING BEYOND 2 μm

S.J. EGLASH, H.K. CHOI, G.W. TURNER, and M.C. FINN
Lincoln Laboratory, Massachusetts Institute of Technology, Lexington, MA 02173-9108

ABSTRACT

Molecular beam epitaxy has been used to grow GaInAsSb/AlGaAsSb double heterostructures, lattice matched to GaSb substrates, for diode laser emission at 2.3 μm. Double-crystal x-ray diffraction measurements were used to determine alloy lattice constants, and photoluminescence and infrared absorption spectroscopies were used to determine the bandgaps of the GaInAsSb layers. Alloy compositions measured by Auger electron spectroscopy were consistent with measured lattice constants and bandgaps. Diode lasers fabricated from the double heterostructures were operated in the pulsed mode at room temperature with threshold current densities as low as 1.5 kA cm^{-2}, differential quantum efficiencies as high as 50 percent, and output power as high as 900 mW per facet.

INTRODUCTION

Double-heterostructure diode lasers consisting of GaInAsSb active layers and AlGaAsSb confining layers lattice-matched to GaSb substrates can potentially provide room-temperature emission from 1.7 to 4.4 μm. Such lasers would be useful for a variety of applications, including optical fiber communications employing low-loss fluoride-based fibers, laser radar exploiting atmospheric transmission windows, remote sensing of atmospheric gases, and molecular spectroscopy. Room-temperature cw operation at wavelengths from 2.0 to 2.3 μm has been reported [1,2] for GaInAsSb/AlGaAsSb double-heterostructure lasers grown by liquid phase epitaxy on GaSb substrates, and pulsed room-temperature operation has been demonstrated for such lasers grown by molecular beam epitaxy (MBE) [3,4]. We have used MBE to grow lasers of this type emitting at 2.3 μm, and have recently reported [5] pulsed threshold current density J_{th} as low as 1.5 kA cm^{-2}, differential quantum efficiency η_d as high as 50 percent, and output power as high as 900 mW per facet. This J_{th} value is equal to the lowest reported previously for GaInAsSb/AlGaAsSb diode lasers [1], and the η_d and output power values are the highest reported for the room-temperature operation of any diode laser emitting beyond 2 μm. In this paper we report on the growth and properties of the constituent layers of our GaInAsSb/AlGaAsSb double heterostructures.

EXPERIMENTAL PROCEDURE

MBE Growth

The sources used for MBE growth were the group III and group V elements, which yielded beams of Al, Ga, and In atoms and of As$_4$ and Sb$_4$ molecules. The impurity used for n-type doping was Te provided by the sublimation of GaTe, and the p-type dopant was Be. Layers were grown on commercial Te-doped n-GaSb and semi-insulating GaAs (100) substrates. The GaAs substrates were used for growing test layers for van der Pauw measurements, while the GaSb substrates were used for the other test layers as well as for the laser structures. The surface structure was monitored during growth by observing the reflection high-energy electron diffraction (RHEED) pattern formed by a glancing-incidence 8-keV beam. RHEED intensity oscillations were measured and analyzed to obtain beam fluxes by using a digital data acquisition system described previously [6]. Further details of the growth procedure will be reported elsewhere [7].

Materials Characterization

The approximate bandgap of GaInAsSb test layers was obtained from the peak wavelength of the photoluminescence (PL) spectrum measured at 4.2 K by using the 647-nm line of a Kr-ion laser with a power density on the sample of approximately 90 or 600 W cm^{-2}, a 0.5-m spectrometer, and a PbS detector cooled to 77 K. The absorption edge was obtained from the transmission spectrum measured at 300 K with a Fourier transform infrared spectrometer. The lattice mismatch between the GaInAsSb and AlGaAsSb layers and the GaSb substrate was obtained from x-ray diffraction measurements of the (400) reflection made with a double-crystal diffractometer equipped with a GaSb first crystal. The alloy composition was measured by Auger electron spectroscopy, as calibrated by Auger analysis of AlSb, GaSb, and GaAs layers and also of a thick GaInAsSb test layer whose composition had been determined by electron microprobe analysis.

Laser Fabrication and Characterization

The laser structure used for 2.3-μm emission consisted of the following layers: 0.2-μm-thick n^+-GaSb buffer, 2-μm-thick n-Al$_{0.50}$Ga$_{0.50}$As$_{0.04}$Sb$_{0.96}$ cladding, 0.4-μm-thick nominally undoped Ga$_{0.84}$In$_{0.16}$As$_{0.14}$Sb$_{0.86}$ active, 3-μm-thick p-Al$_{0.50}$Ga$_{0.50}$As$_{0.04}$Sb$_{0.96}$ cladding, and 0.05-μm-thick p^+-GaSb cap. The composition of the cladding layers was selected to provide good carrier and optical confinement. All layers were nominally lattice matched to the n-GaSb substrate. The carrier concentrations in the n- and p-AlGaAsSb cladding layers were 1×10^{17} and 6×10^{16} cm^{-3}, respectively. Following MBE growth, broad-stripe lasers 300 μm wide with cavity lengths ranging from 300 to 700 μm were fabricated utilizing Au/Sn/Au and Ti/Au metallizations on the n- and p^+-GaSb surfaces, respectively. Further details of the fabrication procedure have been reported previously [4]. Lasers were operated in the pulsed mode at room temperature. Laser emission spectra were measured with a 0.5-m spectrometer and a PbS detector cooled to 77 K. The output power was measured by using an ellipsoidal reflector to focus the emission onto a pyroelectric detector.

RESULTS

RHEED Intensity Oscillations

Accurate control of alloy compositions is critical to obtaining high-performance lasers with the desired emission wavelength. To minimize the formation of misfit dislocations, the lattice mismatch $\Delta a/a_{sub}$ should be less than 10^{-3} [8], where $\Delta a = a_{epi} - a_{sub}$, and a_{epi} and a_{sub} are the lattice constants of the epitaxial layer and substrate, respectively. For lattice-matched Ga$_x$In$_{1-x}$As$_y$Sb$_{1-y}$ active layers with a 300 K emission wavelength of 2.3 μm, the design composition is given by $x = 0.84$ and $y = 0.14$ [9,10]. The parameters x and y must be maintained to within ± 0.01 of these values to achieve $\Delta a/a_{sub} < 10^{-3}$ and $\Delta E_g < 0.01$ eV, where ΔE_g is the difference in bandgap from the design value of 0.54 eV. For the Al$_{0.50}$Ga$_{0.50}$As$_{0.04}$Sb$_{0.96}$ cladding layers, the Sb content is the most critical and must be maintained at 0.96 ± 0.01 to achieve $\Delta a/a_{sub} < 10^{-3}$. The growth conditions leading to lattice-matched layers with the desired bandgaps were determined by growing and characterizing test layers. The reproducible growth of layers is facilitated by the careful calibration of beam fluxes and substrate temperatures.

The group III fluxes were calibrated by measuring the frequency of the RHEED intensity oscillations observed during the growth of test layers of AlSb, GaSb, and InAs. As an example, a plot of RHEED intensity vs time for a GaSb layer grown on a GaSb substrate is shown in Fig. 1(a). The corresponding power spectrum, which was obtained by using fast Fourier transform techniques [6], is shown in Fig. 1(b). The fundamental oscillation frequency, which is given by the peak in the power spectrum, is the reciprocal of the time required for the deposition of one monolayer. Another example of RHEED oscillations is shown in Fig. 2 for the growth of Al$_{0.50}$Ga$_{0.50}$As$_{0.04}$Sb$_{0.96}$ on GaSb. The persistence of the oscillations is indicative of continuing layer-by-layer growth proceeding by the nucleation and growth of two-dimensional islands.

Selection of the appropriate group III fluxes for alloy growth is straightforward, because the efficiency of incorporation is essentially unity for these elements. Selecting the appropriate group V fluxes is more difficult because the incorporation of As and Sb is influenced by the As:Sb and V:III flux ratios and the substrate temperature. In this work the group V fluxes were chosen by iteratively growing and characterizing test layers until the desired properties were obtained. Although the alloy layers contain much more Sb than As, it was necessary to use a large excess As flux because Sb is incorporated more readily than As [7,11]. The Sb:III flux ratio was near unity.

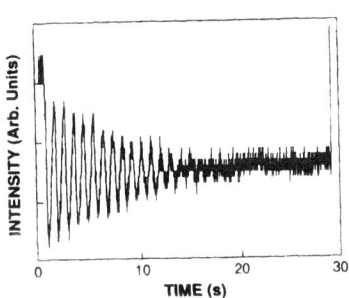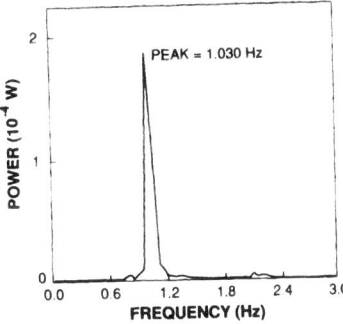

Fig. 1. (a) RHEED intensity vs time observed during the growth of GaSb.
(b) Power spectrum of the data.

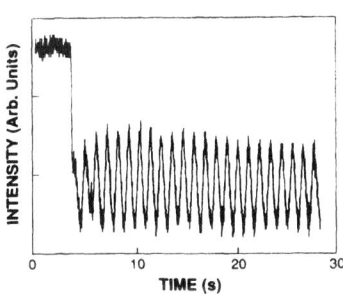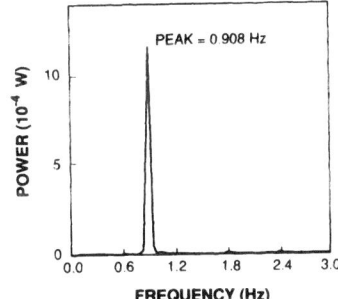

Fig. 2. (a) RHEED intensity vs time for the growth of AlGaAsSb.
(b) Power spectrum of the data.

GaInAsSb Layers

The design composition of the active layer, $Ga_{0.84}In_{0.16}As_{0.14}Sb_{0.86}$, was chosen to give room-temperature emission at 2.3 μm and lattice matching to the GaSb substrate. Table I summarizes the properties of four GaInAsSb test layers, which were grown sequentially in a series of runs that led to the successful growth of our most recent laser. Figure 3 shows PL spectra from sample D measured at 4.2 K and at incident powers of 90 and 600 W cm^{-2}. The PL intensity is approximately proportional to excitation power in this range. The peak is centered at 2.00 μm and has a full width at half-maximum (FWHM) of 16 meV. Figure 4 shows the infrared transmission spectrum of sample D, measured at 300 K. The absorption edge is located at 2.30 μm. The energy difference of 0.08 eV between the absorption edge and the PL peak is equal to the change in

TABLE I. Properties of $Ga_xIn_{1-x}As_ySb_{1-y}$ layers.

Sample	PL Peak* (μm)	Absorption Edge† (μm)	x-ray $\Delta a/a_{sub}$	PL and x-ray x	PL and x-ray y	Auger x	Auger y
A	2.14	2.42	-7.1×10^{-3}	0.81	0.26	0.83	0.23
B	2.03	2.33	-3.5×10^{-3}	0.85	0.18	0.85	0.19
C	1.94	2.22	-3.5×10^{-3}	0.88	0.16	0.83	0.15
D	2.00	2.30	$+1.5 \times 10^{-3}$	0.84	0.12	0.81	0.16

*Photoluminescence measured at 4.2 K
†Infrared transmission measured at 300 K

bandgap between 300 and 4.2 K calculated for $Ga_{0.84}In_{0.16}As_{0.14}Sb_{0.86}$ from the values for the four binary compounds [9]. Double-crystal x-ray diffraction measurements on sample D show that $\Delta a/a_{sub} = +1.5 \times 10^{-3}$ for the GaInAsSb layer.

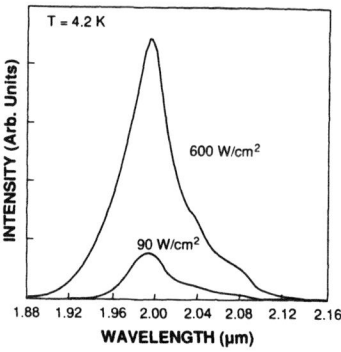

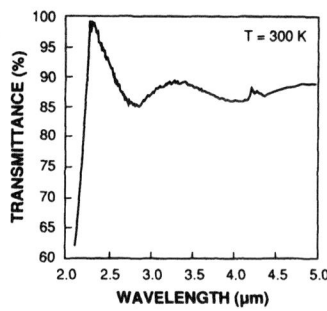

Fig. 3. Photoluminescence spectra from a 1.2-μm-thick GaInAsSb test structure (sample D in Table I).

Fig. 4 Transmission spectrum of the GaInAsSb sample of Fig. 3. The substrate and background were subtracted to give a peak transmittance of 100 percent. The oscillations below the band edge are due to reflections at the epitaxial layer surfaces.

The composition of the GaInAsSb layer of sample D measured by Auger spectroscopy is $Ga_{0.81}In_{0.19}As_{0.16}Sb_{0.84}$. Using available expressions [9,10], the composition calculated from the measured bandgap and lattice constant is $Ga_{0.84}In_{0.16}As_{0.12}Sb_{0.88}$. The differences between the measured and calculated compositions listed in Table I are dominated by random errors arising from uncertainties in Auger peak heights and in the determination of the bandgap. The measured values of x listed in Table I are in good agreement with the values calculated from the Ga and In beam fluxes as calibrated by RHEED oscillation measurements. The Sb content exhibited an approximately linear dependence on Sb flux, and the Sb incorporation efficiency was slightly less than unity [7].

AlGaAsSb Layers

The design composition of the cladding layers, $Al_{0.50}Ga_{0.50}As_{0.04}Sb_{0.96}$, was chosen to provide adequate electrical and optical confinement together with lattice matching to the GaSb substrate. The most critical property of these layers is the lattice constant, which is determined almost entirely by the As:Sb ratio. Figure 5 shows the double-crystal x-ray diffraction spectrum from a sample with a 2-μm-thick AlGaAsSb test layer. The peak at 0 s is due to the GaSb substrate, while the peak near + 350 s is due to the AlGaAsSb layer. For this layer, $\Delta a/a_{sub} = -2.8 \times 10^{-3}$. For AlGaAsSb growth the Sb content exhibited a sublinear dependence on Sb flux [7], indicating that the Sb incorporation efficiency is reduced significantly below unity when the As flux is large and the As content of the alloy is very small.

For Te-doped $Al_{0.50}Ga_{0.50}As_{0.04}Sb_{0.96}$, we have obtained values of electron concentration n as high as 1×10^{17} cm^{-3}. The highest electron mobilities were 1300 and 4150 cm^{2}V^{-1}s^{-1} at 300 and 77 K, respectively, for a sample with $n = 2 \times 10^{16}$ cm^{-3}. (These values were obtained directly from van der Pauw measurements without correction for multiband conduction.) Even higher n values and mobilities have been obtained for GaSb and AlAsSb. For GaSb the values of n range from 8×10^{15} to 1×10^{18} cm^{-3}, with mobilities as high as 4000 and 5850 cm^{2}V^{-1}s^{-1} at 300 and 77 K, respectively. The results for Be doping of p-type GaSb and AlGaAsSb are straightforward. Values of hole concentration up to 2×10^{18} cm^{-3} have been obtained. Further details of the doping of these materials will be reported elsewhere [7].

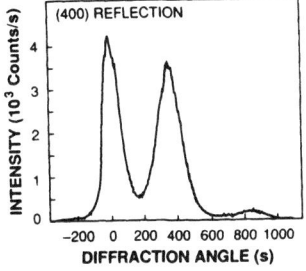

Fig. 5 Double-crystal x-ray diffraction spectrum from a 2-μm-thick AlGaAsSb test layer. The peak at 0 s is due to the GaSb substrate and the peak near + 350 s is due to the AlGaAsSb layer.

Laser Characteristics

Figure 6 shows the room-temperature emission spectrum of a representative GaInAsSb/AlGaAsSb laser, which was fabricated from a wafer grown after the GaInAsSb test samples listed in Table I. The spectrum is composed of multiple longitudinal modes near 2.27 μm. The PL spectrum measured for this wafer at 4.2 K is centered at 2.02 μm and has a FWHM of 21 meV. (The lowest FWHM we have observed for a laser wafer is 7.5 meV.) The lattice mismatch was determined by x-ray diffraction to be within $\pm 1.6 \times 10^{-3}$. The surface was specular, but a slight cross-hatch pattern was visible. The lasers fabricated from the wafer exhibited excellent characteristics for pulsed operation at room temperature, including $J_{th} = 1.5$ kA cm^{-2} for devices with a cavity length of 700 μm. Values of η_d up to 50 percent and output power up to 900 mW per facet were obtained for devices with a cavity length of 300 μm. The near-field pattern, which was imaged with a PbS camera, was fairly uniform, indicating the good spatial uniformity of the epitaxial layers. Values of ~ 100 percent for the internal quantum efficiency and 43 cm^{-1} for the internal loss coefficient were determined from the measured dependence of differential quantum efficiency on cavity length.

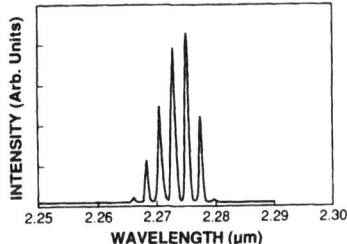

Fig. 6. Room-temperature emission spectrum of a GaInAsSb/AlGaAsSb diode laser having a cavity length of 300 μm.

SUMMARY

The growth and characterization of GaInAsSb/AlGaAsSb double heterostructures have been reported. RHEED oscillation measurements were used to provide calibrated beam fluxes, which facilitated accurate and reproducible control of alloy compositions and growth rates. X-ray diffraction, PL, and infrared absorption measurements were employed for characterizing GaInAsSb and AlGaAsSb test layers as well as GaInAsSb/AlGaAsSb double heterostructures. Diode lasers operating at 2.3 μm with excellent pulsed room-temperature characteristics have been demonstrated. By varying the GaInAsSb active layer composition it may be possible to obtain diode laser emission from 1.7 to 4.4 μm in lattice-matched structures.

ACKNOWLEDGEMENTS

We are grateful to J.V. Pantano and L. Krohn, Jr. for expert technical assistance in the MBE laboratory, and to D.R. Calawa, J.W. Chludzinski, and K.A. McIntosh for materials characterization. We are particularly grateful to A.J. Strauss for his many useful insights. This work was sponsored by the Department of the Air Force.

REFERENCES

1. A.E. Bochkarev, L.M. Dolginov, A.E. Drakin, P.G. Eliseev, and B.N. Sverdlov, Sov. J. Quantum Electron. **18**, 1362 (1988).

2. A.N. Baranov, T.N. Danilova, B.E. Dzhurtanov, A.N. Imenkov, S.G. Konnikov, A.M. Litvak, V.E. Usmanskii, and Yu.P. Yakovlev, Sov. Tech. Phys. Lett. **14**, 727 (1988); A.N. Baranov, A.N. Imenkov, M.P. Mikhailova, A.A. Rogachev, and Yu.P. Yakovlev, Proc. SPIE **1048**, 188 (1989).

3. T.H. Chiu, W.T. Tsang, J.A. Ditzenberger, and J.P. van der Ziel, Appl. Phys. Lett. **49**, 1051 (1986).

4. S.J. Eglash and H.K. Choi, Appl. Phys. Lett. **57**, 1292 (1990).

5. H.K. Choi and S.J. Eglash, IEEE J. Quantum Electron., June 1991 (to be published).

6. G.W. Turner, B.A. Nechay, and S.J. Eglash, J. Vac. Sci. Technol. B **8**, 283 (1990).

7. S.J. Eglash, H.K. Choi, and G.W. Turner, J. Cryst. Growth (to be published).

8. See, for example, B.W. Dodson and J.Y. Tsao, Annu. Rev. Mater. Sci. **19**, 419 (1989).

9. H.C. Casey, Jr., and M.B. Panish, Heterostructure Lasers, Part B (Academic Press, New York, 1978), pp. 1-48.

10. J.C. DeWinter, M.A. Pollack, A.K. Srivastava, and J.L. Zyskind, J. Electron. Mater. **14**, 729 (1985).

11. C-A. Chang, R. Ludeke, L.L. Chang, and L. Esaki, Appl. Phys. Lett. **31**, 759 (1977).

OPTICAL PROPERTIES OF MBE-GROWN GaInAsSb

WENGANG BI, AIZHEN LI, AND SONGSHENG TAN
Department of Semiconductor Materials
Shanghai Institute of Metallurgy, Academia Sinica
Shanghai 200050, China

ABSTRACT

The infrared optical absorption properties near and above the fundamental absorption edge of MBE grown undoped GaInAsSb quaternary semiconductor alloy deposited on GaSb and GaAs substrates have been measured and analyzed at room temperature by means of a Fourier Transform Infrared Spectrometer, and were found to be fully characterized by the interband transition theory and Urbach's rule. The optical band gap of MBE-GaInAsSb has been determined using a linear extrapolation of $(\alpha h\nu)^2$ as a function of the photon energy $h\nu$, and the refractive index n deduced from the interference pattern, which shows good agreement with the theory of Sadao Adachi's.

INTRODUCTION

Recently, much attention has been focused on the research of light sources and detectors operating in the 2-4µm wavelength region for applications to present silica fiber, possible ultralow loss new generation fluoride glass fiber system, infrared image technique, optical fiber sensor, heterodyne optical spectroscopy and so on. GaInAsSb quaternary alloy lattice matched to GaSb, InAs and InP can cover this entire wavelength range of interest by changing its alloy compositions, thus becoming the most promising and attracting 2-4µm optoelectronic materials which will lead to high performance detectors and emitters. Up to date, epitaxial growth of GaInAsSb has been reported by the methods of MBE [1], OMVPE [2] etc. and injection lasers based on this alloy system have also been fabricated [3,4]. However, little work has been done concerning the optical properties of this material, which are very important for device design applications. As we know, the absorption coefficient determines the peak response or cutoff wavelength of detectors, the band-gap energy responds for the operating wavelength of the optoelectronic devices [5], and the refractive index step between the active and cladding layer confines and guides lasing light through the heterostructure dielectric waveguide, so knowledge of the properties of these optical constants forms an important part in the design of heterostructure lasers and other waveguide devices [5].
    It is our purpose in this paper to study these optical properties of GaInAsSb material in order to provide some useful information for the device design applications based on this material.

EXPERIMENTAL

The samples used in the investigation were grown by MBE technique in a chinese made model FS-2M MBE system. A two step MBE growth procedure described elsewhere [6] was applied for

the growth of GaInAsSb on GaAs substrate. The sample compositions were determined by an electron beam microprobe. For the
optical measurements, the samples were prepared by lapping and
polishing the back side of the substrate, during which special
care was taken to have a uniform thickness, and by etching the
final specimen in $3H_2SO_4:1H_2O_2:1H_2O$ for GaInAsSb/GaAs and in
$0.125Br_2:100CH_3OH$ for GaInAsSb/GaSb to reduce the surface
damage. Room temperature IR transmission was performed on a
NICOLET NIC-7199C FTIR spectrometer. The spectral range is
$5000-400cm^{-1}$, and the resolution is $0.06cm^{-1}$

RESULTS AND DISCUSSION

A. Absorption coefficient

    Room temperature infrared transmission spectrum of the
grown samples is shown in Fig.1, in which the strong interference oscillation indicates relatively smooth interfaces in
the sample, and the steep reduction of the transmission value
near the absorption edge shows the excellent compositional
uniformity and crystalline perfection.

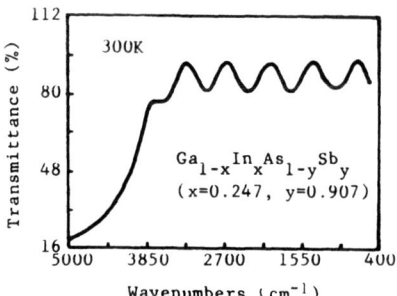

Fig.1   Room temperature infrared transmission
spectra for GaInAsSb materials.

    To study the optical absorption properties of GaInAsSb
materials, we have calculated their absorption coefficient α
from these measured IR spectroscopies according to the non-
destructive technique proposed by Swanepoel [7], and the results are shown in Fig.2. It is clearly seen that an exponential dependence of α on $h\nu$ exists in the range of energies near
the absorption edge,i.e. $\alpha=\alpha_0\exp[(h\nu-h\nu_0)/kT_{eff}]$. Here $h\nu$ is
the photon energy and α_0, ν_0, and T_{eff} are fitting parameters.
This behavior is known to be Urbach's tail and has been found
in an extensive number of ionic materials, semiconductors, and
organic and amorphous systems. The origin of which might results from interactions other than band-to-band transition, such
as electron hole, electron phonon, and electron impurity interactions. As the energy increases above the band gaps, α deviates
from the exponential dependence on $h\nu$, and can be described by
$\alpha=A(h\nu-Eg)^{\frac{1}{2}}/h\nu$ (Fig.2(b)). This consists with the theory of
interband optical absorption, in which the absorption coeffi-

cient α at the absorption edge varies with the photon energy $h\nu$ in terms of the following expression:

$$\alpha h\nu = A(h\nu - Eg)^n \tag{1}$$

Here A is a constant, Eg is band gap, and n is a number that characterizes the optical transition process, i.e. n takes the value 1/2 for direct allowed transition, 3/2 for direct forbidden, 2 for indirect allowed, and 3 for indirect forbidden. A noticeable feature of Eq.(1) is that $(\alpha h\nu)^{1/n}$ varies linearly with $h\nu$, and that a plot of $(\alpha h\nu)^{1/n}$ against $h\nu$ must be a straight line if the appropriate value of n has been used. The quite satisfactory description of $\alpha = A(h\nu - Eg)^{\frac{1}{2}}/h\nu$ for our experimental data indicates that the optical transitions in GaInAsSb are direct allowed, which is consistent with the notation that GaInAsSb being direct gap material.

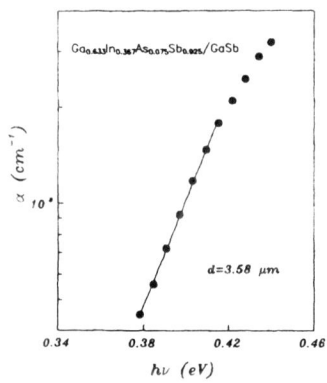 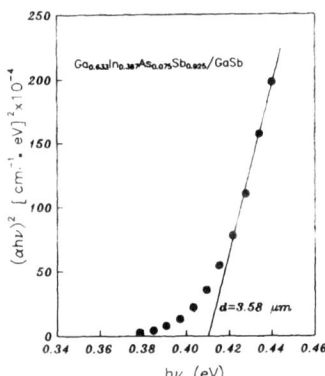

Fig.2 (a) Room-temperature absorption coefficient as a function of photon energy for a typical alloy sample (x=0.266, y=0.891); (b) Direct transition analysis on the same GaInAsSb sample. Threshold energy is 0.41eV.

B. Optical band gap

The band-gap energy Eg is known to be one of the most important device parameters which is strongly connected with the operating wavelength of the optoelectronic devices. To determine this quantity precisely for GaInAsSb materials, two methods were employed: one is directly from the measured transmission spectra, in which it is assumed that the point on the spectrum corresponding to the optical energy gap Eg is at half of the maximum transmission, the other is from the intercept of the graph $(\alpha h\nu)^2$ versus $h\nu$ on the $h\nu$ axis. Fig.3 shows the results of Eg thus obtained as a function of the theoretical band gap value EgO. Here EgO is deduced from

$$E_{go}(x,y) = (1-x)E_{ACD}(y) + xE_{BCD}(y) - \Delta E \tag{2}$$

where $\Delta E = x(1-x)[yC_{ABD} + (1-y)C_{ABC}] + y(1-y)[xC_{BCD} + (1-x)C_{ACD}]$, E_{ijk}

and C_{ijk} are the band gap energy and bowing parameters of the corresponding ternaries, respectively.

It is seen that, although the two experimental values of Eg are different for each sample, the difference is quite small, indicating that, within the experimental error, the two methods indeed result in the same value of Eg, and that the particular criterion used will have little effect on the results obtained. It is also worth noting that the experimental data are located closely near the hypothetical line, which shows good experiment-theory agreement, and demonstrates the validity of Eq.(2) being able to predict band gaps for GaInAsSb materials to a first order approximation.

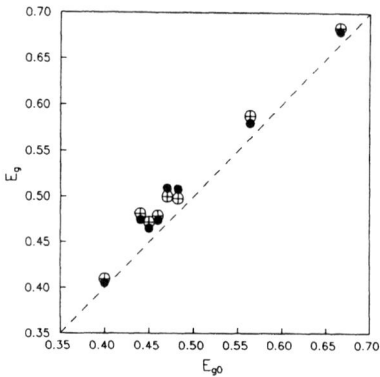

Fig.3 Experimental results of the optical energy gap Eg for each sample plotted as a function of the calculated band gap Ego: ● Half the maximum transmission; ⊕ intercept of $(\alpha h\nu)^2$ versus hν graph.

C. Refractive index

In light of the interference patterns in the IR transmission spectrum, the refractive indices of GaInAsSb were obtained and the results are shown in Fig.4. It should be noted that the refractive index n of SM941, which has smaller band gap than that of SM9015, is smaller than that of SM9015. This is contrary to most of the III-V compound alloys, like $Al_xGa_{1-x}As$, $In_{1-x}Ga_xAs_{1-y}P_y$ and $Al_{1-x}Ga_xAs_{1-y}Sb_y$, for which the smaller Eg-gap material has a larger value of refractive index. Also shown in Fig.4 are the theoretical values of n derived from the theory of Sadao Adachi's [8], and quite satisfactory agreement between the experimental results and the theory can be seen. As is known, the knowledge of refractive index is very important for device design, e.g. for a injection laser, the refractive index step Δn between the active layer and cladding layer defines the optical confinement of phonons, i.e., if n in the active region is larger than that of the cladding layer on both sides, the cladding layer will not confine radiation to the neighborhood of the active region, and thus offers an optical loss in the waveguide which might lead to a large threshold current. So Δn is a very important parameter for device design. For this purpose of device design applications, we have calculated the refractive index step Δn between GaInAsSb active layer (lattice matched to GaSb and InAs) and $Al_{1-x}Ga_xAs_{1-y}Sb_y$ cladding layer as a function of x at lasing wavelength λ=2.55μm, and the results are shown in Fig.5, which might be useful for device design of 2-4μm light sources.

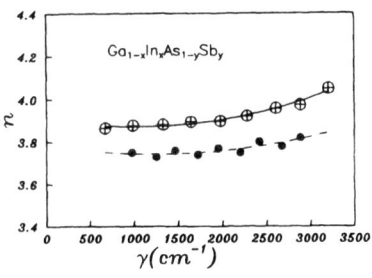

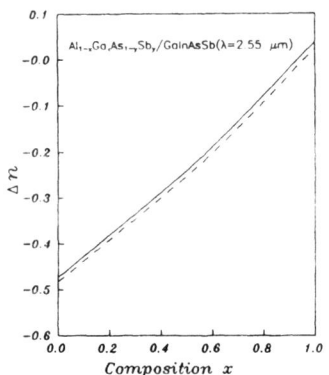

Fig.4 Refractive index of two $Ga_{1-x}In_xAs_{1-y}Sb_y$ samples:
⊕ SM9015 (x=0.247, y=0.907);
● SM941 (x=0.266, y=0.891);
— theoretical curve derived from Sadao Adachi's theory.

Fig.5 Refractive index steps between $Al_{1-x}Ga_xAs_{1-y}Sb_y$ cladding layer and GaInAsSb active layer lattice matched to GaSb (solid line) and InAs (dashed line) at operating wavelength of λ=2.55μm.

CONCLUSION

Room temperature optical absorption properties of MBE-grown GaInAsSb have been studied by employing Fourier Transform Infrared Spectrometer technique. For all samples investigated, absorption coefficient above the intrinsic edge obeys satisfactorily the relation $α=A(hν-Eg)^{\frac{1}{2}}/hν$, indicating that the optical absorption process is direct allowed, while absorption coefficient less than $1.7×10^3 cm^{-1}$ exhibits an exponential tail which can be described by $α=α_o exp[(hν-hν_o)/kT_{eff}]$. The optical band gap of MBE-GaInAsSb has been determined using a linear extrapolation of $(αhν)^2$ as a function of the photon energy hν, and the refractive index n deduced from the interference pattern. Good agreement is obtained between the experimental data n and the theoretical curve derived from the theory of Sadao Adachi's, and the refractive index step between GaInAsSb active layer and $Al_{1-x}Ga_xAs_{1-y}Sb_y$ waveguiding layer (lattice matched to InAs and GaSb) as a function of the waveguiding-layer composition x at lasing wavelength λ=2.55μm is presented for the purpose of achieving the best-structure device design applications.

ACKNOWLEDGEMENTS

This work was supported by a Chinese Advanced Technology Program. The authors are grateful to Ms. Y.L. Zheng, Mr. J.X. Wang, Mr. P.X. Cai and Ms. W.F. Gu for help in sample preparation and FTIR measurements.

REFERENCES

[1] T.H. Chiu and W.T. Tsang, J. Appl. Phys. 57, 4572 (1987).
[2] M.J. Cherng, H.R. Jen, C.A. Larsen, G.B. Stringfellow, H. Lundt, and P.C. Tayler, J. Cryst. Growth 77, 408 (1985).
[3] N. Kobayashi, Y. Horikoshi, and C. Uemura, Jpn. J. Appl. Phys. 19, L30 (1980).
[4] C. Caneau, A.K. Strivastava, J.L. Zyskind, J.W. Sulhoff, A.G. Dentai, and M.A. Pollack, Appl. Phys. Lett. 49, 55 (1986).
[5] H.C. Casey, Jr. and M..B. Panish, Heterostructure Lasers (Academic, New York, 1978), Parts A and B.
[6] A.Z. Li, Y.L. Zheng, J.H. Qiu, J.X. Wang and F.Y. Hu in 1989 16th Int. Symp. on GaAs and Related Compounds Proc.106, edited by T. Ikoma and H. Watanabe, pp.159-163.
[7] R. Swanepol, J. Phys. E 16, 1214 (1983); ibid., 17, 896 (1984); J. Opt. Soc. Am. 2, 1339 (1985).
[8] Sadao Adachi, J. Appl. Phys. 53, 5863 (1982).

PART IV

Narrow-Gap (In, Ga) Sb
III-V Binary Compounds

LOW TEMPERATURE p- AND n- TYPE DOPING OF InSb GROWN ON GaAs USING MOLECULAR BEAM EPITAXY.

PHILLIP E. THOMPSON,* JOHN L. DAVIS,* and DAVID S. SIMONS**
*Naval Research Laboratory, Washington, DC 20375
**National Institute of Standards and Technology, Gaithersburg, MD, 20899

ABSTRACT

30 nm thick layers of Be-doped InSb were grown by MBE, sandwiched between undoped layers of InSb on semi-insulating GaAs substrates. Using secondary ion mass spectroscopy (SIMS), it has been demonstrated that the p-type dopant, Be, undergoes an anomalous migration towards the top surface for substrate growth temperatures in excess of 340°C. A thermal diffusion process can be eliminated as an explanation since there was not an associated diffusion towards the substrate. At a substrate growth temperature of 340°C, the Be-doped layer had backside and frontside doping concentration gradients of 0.067 and 0.147 decade/nm, respectively. A sample grown at 420°C had a comparable backside doping concentration gradient, but the frontside gradient was essentially zero, i. e., uniform doping to the surface. Electrical activation of the Be dopant, when grown at 340°C, compared to activation in GaAs grown at 580°C, was 38% and 47% for growth rates of 1.0 and 0.5 µm/h. While the Si dopant does not undergo a migration comparable to that of Be, it has been demonstrated that n-type doping of InSb with Si must be done at growth temperatures less than or equal to 340°C to obtain high carrier concentrations. The maximum n-type carrier concentration obtained was $6.2 \times 10^{18}/cm^3$, for a 0.5 µm/h growth rate at 340°C.

INTRODUCTION

There has been renewed interest [1-6] in the use of InSb and InAsSb epitaxial layers for the detection of infrared radiation in the region of 8 to 12 µm. While these materials would not have the problem of the instability of Hg found in HgCdTe, they present a new set of difficulties. One major hurdle is to extend the operational regime of photodetectors fabricated from an InSb-based layer to 12 µm, which requires device structures such as strained layer superlattices or nipi structures. Another problem is the lack of a lattice-matched, electrically insulating substrate, which would permit characterization of epitaxial layers by Hall measurements or the use of photoconductors as photodetectors. Recently, there have been reports [5,6] of epitaxial growth of InSb on GaAs, where, in spite of a 15% lattice mismatch, the electron mobilities at 77 K are close to those of bulk InSb. To make electron devices in these materials requires techniques to form n- and p-type regions. In this paper we report on the formation of n- and p-type regions using Si and Be during the molecular beam epitaxy (MBE) growth process. The focus is on the positional stability of these layers and on the doping efficiency. In associated papers we report on the doping of InSb/GaAs using ion implantation [7] and on the growth and characterization of Si delta-doped InSb [8].

EXPERIMENTAL PROCEDURES

Both InSb ($n_{77} = 2 \times 10^{14}/cm^3$, $\mu_{77} = 4.5 \times 10^5$ cm²/Vs) and semi-insulating GaAs substrates were used for the epitaxial growth. The GaAs wafers underwent an organic degreasing procedure and were etched in 5:1:1 H_2SO_4:H_2O_2:H_2O to remove 4 µm from the surface. The InSb wafers were chemically-mechanically polished using bromine-methanol, degreased with organic solvents, and etched in 25:4:1 lactic acid:HNO_3:HF to remove 2 µm from the surface. After being blown dry with N_2, both types of substrates were baked at 300°C in the preparation chamber to further remove volatiles. The growth was performed using elemental sources of Ga, As, In, Sb, Be, and Si. Prior to growth, the GaAs substrates were heated to 625°C in an arsenic flux to remove the surface oxide. Similarly, the InSb wafers were heated to 440°C in an antimony flux. The fluences of the As, Sb, In, and Ga cells were monitored with a nude ionization gauge. The growth rate was determined from the cell temperatures and the measured pressure using kinetic theory. This growth rate was confirmed by measuring the thickness of the epitaxial film using surface profilometry. The Si and Be molecular fluxes were too small to be measured with the ionization gauge. Thus, throughout this paper, the Si and Be fluxes are characterized solely by their cell temperatures during growth. After growth, the samples were characterized with Hall measurements for electrical properties, x-ray rocking curves for structural properties, and secondary ion mass spectroscopy (SIMS) for dopant distributions. All SIMS profiles were made with a 300 nA O_2^+ primary ion beam at 5.5 keV impact energy and positive secondary ion detection. The relative sensitivity factors (RSFs) for Be and Si in InSb, measured from ion implants, were supplied by another laboratory (9). The accuracy of concentrations determined by this method is unknown. However, in this study, we are concerned with the relative position of dopants, which is not affected by the RSF value.

RESULTS AND DISCUSSION

Be Doping of InSb

We have studied two features of the p-type dopant profiles produced in InSb with Be. The first feature is the positional stability of the profiles with selected substrate growth temperatures. The second feature is the relative doping efficiency of Be in InSb compared to Be in GaAs.

To study the positional stability of Be we grew the following structure on n-type InSb substrates: 100 nm undoped InSb, 30 nm Be-doped InSb, and 100 nm undoped InSb. InSb substrates were chosen for this study to avoid dislocation-enhanced diffusion which might occur if the epitaxial layers were grown on GaAs. The Be cell temperature was set to 945°C, which would correspond to a doping level of $2.3 \times 10^{18}/cm^3$ in GaAs. The substrate temperature during growth was kept at 420, 380, or 340°C. The growth rate was 1 μm/h when the substrate temperature was either 420 or 380°C and 0.5 μm/h for the growth at 340°C. The atomic Be profiles were measured with SIMS, Fig. 1. It is observed that the Be underwent an anomalous migration towards the front surface for substrate growth temperatures in excess of 340°C. With a substrate growth temperature of 340°C, the Be-doped layer had backside and frontside doping concentration gradients of 0.067 and 0.147 decade/nm, respectively. The sample grown at 420°C had a comparable backside doping concentration gradient, but possessed a frontside gradient of zero, i. e., uniform doping to the surface. A thermal diffusion process can be eliminated as an explanation since there is not an associated diffusion toward the substrate. If one integrates the atomic concentration profiles with depth from 10 nm to 200 nm, the result is the Be/cm² incorporated into the sample. The amount of Be taken into the lattice at 340°C compared to 380°C and 420°C is approximately two and three times higher, respectively. At the higher substrate growth temperatures, much of the Be rides the growth interface and accumulates onto the surface, which accounts for the large Be signal at the surface.

Since the Be-doped samples above were grown on n-type InSb, no electrical characterization could be performed. To perform Hall measurements to obtain the carrier density and mobility, thick (1.5 - 3 μm), Be-doped InSb layers were grown on semi-insulating GaAs. The InSb layers were grown at a substrate temperature of 340°C for maximum dopant incorporation. The hole concentrations measured at 77 K for two Be-doped InSb layers, grown with the Be source set at 980°C but at different growth rates (1 and 0.5 μm/h), are presented in Fig. 2 along with the Be-doping calibration curve, obtained with the same Be source, for GaAs, also measured at 77 K. The growth-rate of the GaAs was 1 μm/h. At a given Be cell temperature, there is available at the growth interface a fixed number of Be atoms per cm² per second. If one assumes that the Be incorporation efficiency is 100% for GaAs, then one can obtain the doping efficiency for InSb. For the

InSb sample grown at 1 μm/h, doping efficiency was 38%. The InSb sample grown at 0.5 μm/h had a hole concentration equal to 94% of that of the GaAs sample grown at the identical Be cell temperature, but it was exposed to the Be molecular beam for twice the time for an equivalent thickness. Therefore, for the 0.5 μm/h growth-rate, the InSb Be incorporation efficiency was 47%.

The results of the Be doping study suggest several experiments to understand the activation process better. While we have shown that at a 340°C substrate

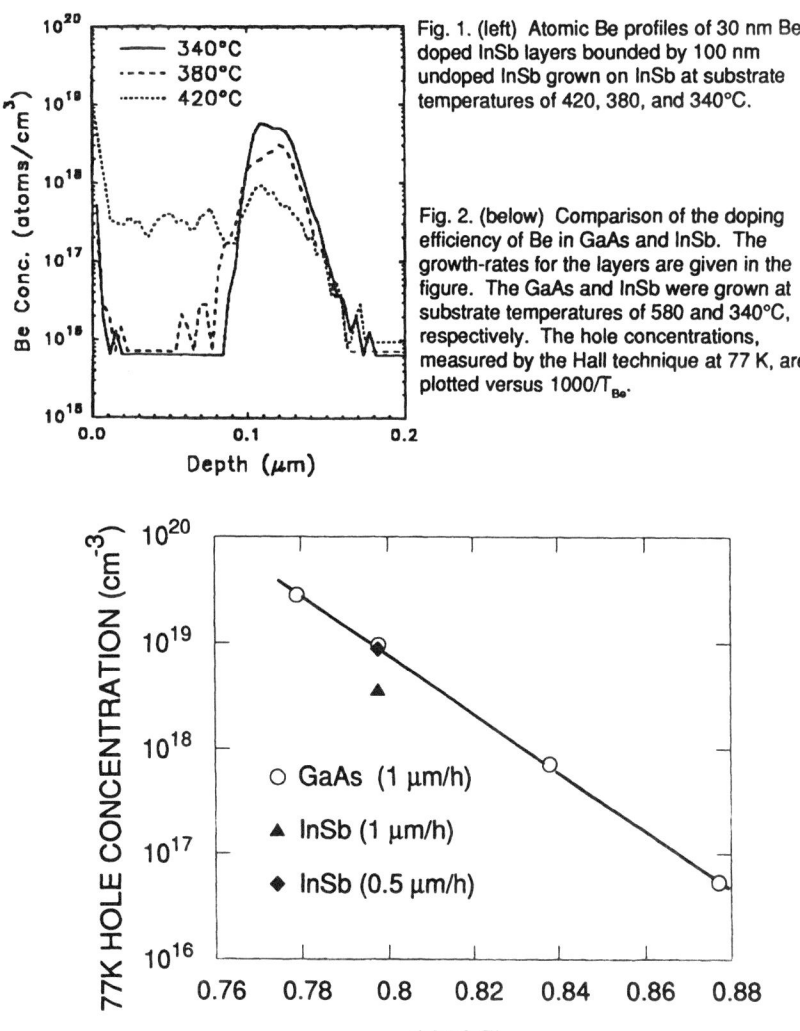

Fig. 1. (left) Atomic Be profiles of 30 nm Be-doped InSb layers bounded by 100 nm undoped InSb grown on InSb at substrate temperatures of 420, 380, and 340°C.

Fig. 2. (below) Comparison of the doping efficiency of Be in GaAs and InSb. The growth-rates for the layers are given in the figure. The GaAs and InSb were grown at substrate temperatures of 580 and 340°C, respectively. The hole concentrations, measured by the Hall technique at 77 K, are plotted versus $1000/T_{Be}$.

temperature, the Be was stable, we must explore lower temperatures, which may improve Be incorporation into electrically active sites. We must use SIMS to investigate Be-doped slabs of InSb embedded in InSb grown on GaAs to look for defect-enhanced diffusion. Finally we should explore lower Be doping concentrations to produce a Be doping calibration curve for InSb. These experiments are currently being performed.

Si Doping of InSb

The profile stability of Si-doped InSb layers, 50 nm wide buried between two 100 nm undoped InSb layers on n-type InSb grown at substrate temperatures of 420 and 360°C, were measured with SIMS. The atomic profiles for Si and As of the sample grown at 360°C are presented in Fig. 3. The Si atomic profile shows the doped layer at the proper position. The background level for Si of $3 \times 10^{17}/cm^3$ is probably caused by a mass interference with N_2^+. A better detection limit for Si could be achieved by Cs^+ bombardment with negative ion detection. However, this resulted in much poorer depth resolution. Also it is observed that these samples have an As background doping of $4 \times 10^{19}/cm^3$. This nonintentional doping comes from residual As in the growth chamber. The sample grown at 420°C has these same artifacts, but the

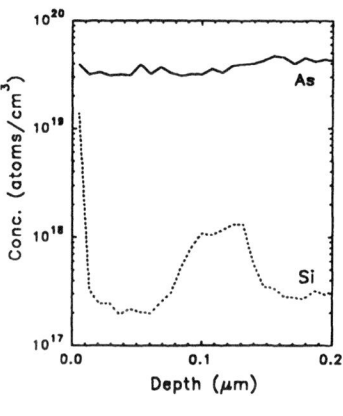

Fig. 3. (left) Atomic Si and As profiles of a 50 nm Si-doped layer, 100 nm below the surface in InSb grown at a substrate temperature of 360°C.

Fig. 4. (below) Net electron carrier concentration measured by the Hall technique at 77 K versus $1000/T_{Si}$ of Si-doped InSb grown on GaAs substrates at 420, 380, and 340°C.

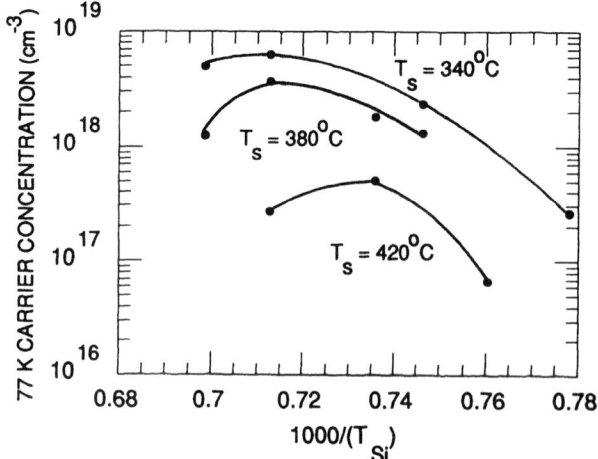

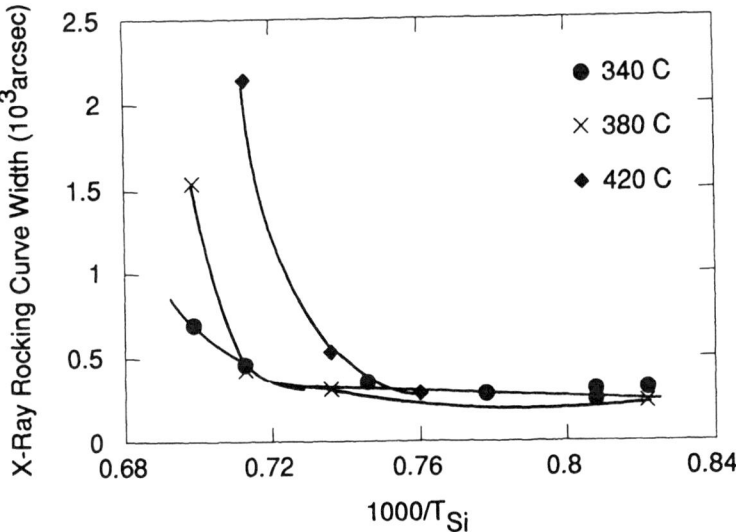

Fig. 5. X-ray rocking curve width versus $1000/T_{Si}$ of Si-doped InSb grown on GaAs substrates at 420, 380, and 340°C.

doped layer was slightly narrower (30 nm) and had a higher peak atomic Si concentration (4×10^{18}/cm³). From this study, it is concluded that the Si does not undergo the same anomalous diffusion process at substrate temperatures of 420°C or less experienced by Be.

The electrical activation of thick (~2 µm), Si-doped InSb epitaxial layers grown on GaAs at substrate temperatures of 420, 380, and 340°C was obtained using Hall measurements. The growth-rate was 1 µm/h for the samples grown at 420°C and 380°C and 0.5 µm/h for the growth at 340°C. The results of this study are presented in Fig. 4, where the 77 K carrier concentration has been plotted versus the reciprocal of the temperature of the Si cell. It is seen that the maximum net n-type carrier concentration is obtained with the lowest substrate temperature. A maximum 77 K n-type carrier concentration of 6.2×10^{18}/cm³ was obtained with growth at a substrate temperature of 340°C. For each substrate temperature there is a Si cell temperature, i. e., Si flux, at which a maximum carrier concentration is achieved. At higher Si cell temperatures, the net n-type carrier concentration and the electron mobility decreased, along with a large increase in the x-ray rocking curve width, Fig. 5. The reduction in the net n-type carrier concentration and electron mobility might imply a compensation mechanism due to the amphoteric doping nature of Si. However, these reductions, when considered along with the dramatic increase in the x-ray rocking curve width, can be better explained by a mechanism in which Si aggregates are formed when the Si flux is too high for a given substrate temperature. When compared to the activation of Si in GaAs, the best activation, using a substrate temperature of 340°C, of Si in InSb/GaAs was 65% at the same Si flux, normalized to the same growth-rate.

The improved activation of Si in InSb with lower substrate temperature has been reported by others [10], however this is the first report of a relative maximum in the n-type carrier concentration as a function of Si cell temperature at different substrate temperatures. This effect of substrate temperature on Si activation will be investigated further to see if substrate temperatures less than 340°C improve the n-type activation.

CONCLUSIONS

It has been shown that p-type doping during MBE growth must be done at substrate temperatures less than or equal to 340°C to prevent the Be from being swept to the surface. Be dopant incorporation in InSb compared to GaAs at the same dopant flux is 38% at 1 μm/h and 47% at 0.5 μm/h. While Si does not undergo a migration comparable to that of Be, it has been demonstrated that n-type doping with Si must also be done at temperatures less than or equal to 340°C to obtain high n-type carrier concentrations. Si activation in InSb at 340°C was on average 35% less than that of Si in GaAs at the same Si flux. The maximum n-type carrier concentration obtained was $6.2 \times 10^{18}/cm^3$.

REFERENCES

1. M. Y. Yen, J. Appl. Phys. **64**(6), 3306 (1988).

2. J. -I. Chyi, S. Kalem, N. S. Kumar, C. W. Litton, and H. Morkoc, Appl. Phys. Lett. **53**(12), 1092 (1988).

3. C. F. McConnville, C. R. Whitehouse, G. M. Williams, A. G. Cullis, T. Ashley, M. S. Skolnick, G. T. Brown, and S. J. Courtney, J. Crystal Growth **95**, 228 (1989).

4. W. Dobbelaere, J. De Boeck, and G. Borghs, Appl. Phys. Lett. **55**(18), 1856 (1989).

5. J. L. Davis and P. E. Thompson, Appl. Phys. Lett. 54(22), 2235 (1989).

6. J. E. Oh, P. K. Bhattacharya, Y. C. Chen, and S. Tsukamoto, J. Appl. Phys. 66(8), 3618 (1989).

7. M. V. Rao, R. Echard, P. E. Thompson, A. K. Berry, S. Mulpuri, and F. Moore to be published in Long-Wavelength Semiconductor Devices, Materials, and Processes, edited by A. Katz, B. M. Biefeld, R. J. Malik, and R. L. Gunshor (Mat. Res. Soc. Proc.).

8. M. -J. Yang, J. R. Waterman, R. J. Wagner, P. E. Thompson, and J. L. Davis to be published in Electronic, Optical, and Device Properties of Layered Structures, edited by J. R. Hayes, M. S. Hybertsen, and E. R. Weber (Mat. Res. Soc. Proc.)

9. R. G. Wilson (private communication).

10. S. D. Parker, R. L. Williams, R. Droopad, R. A. Stradling, K. W. J. Barnham, S. N. Holmes, J. Laverty, C. C. Phillips, E. Skuras, R. Thomas, X. Zhang, A. Staton-Bevan, and D. W. Pashley, Semicond. Sci. and Tech. 4(9), 663 (1989).

INTERFACE PROPERTIES of
HETEROVALENT InSb MULTILAYER STRUCTURES

M. Kobayashi, R. L. Gunshor, J. L. Glenn Jr., Sungki O, J. Han,
D.R. Menke, and L. A. Kolodziejski*
School of Elec. Eng., Purdue Univ., W. Lafayette, IN 47907 USA
D. Li and N. Otsuka
School of Mat. Eng., Purdue Univ., W. Lafayette, IN 47907 USA

ABSTRACT

Multilayer InSb/CdTe and InSb/MnTe heterostructures are grown
by molecular beam epitaxy. The analysis of multilayer structures
by transmission electron microscopy (TEM) and x-ray rocking curves
confirms the presence of periodic heterointerfaces. TEM, x-ray,
and x-ray photoelectron spectroscopy (XPS) measurements are found
to provide a sensitive indicator for the presence of ultra-thin
interfacial layers. Double barrier structures in the
MnTe/InSb/MnTe configuration exhibit the negative differential
resistance which is characteristic of resonant tunneling diodes.

Introduction

InSb is expected to have unique and interesting properties as
a quantum well layer. The small electron effective mass results
in easily achievable quantum well dimensions providing access to
the 2-5.5µm wavelength range. The high carrier mobility and band
nonparabolicity suggest the possibility of interesting optical and
electronic device characteristics. InSb has not heretofore been
exploited in heterostructure configurations as the large lattice
constant of InSb makes it incompatible with other III-V materials
for lattice-matched superlattice and quantum well structures. One
of the few zincblende semiconductors available with appropriate
lattice parameter and bandgap to serve as a barrier to InSb
quantum wells is the II-VI compound, CdTe [1,2]; in this case the
lattice match (0.05%) is comparable to that of GaAs and AlAs. In
an effort to provide an even wider bandgap alternative to CdTe as
a barrier material for InSb quantum structures, we have explored
the use of zincblende MnTe for this purpose. MnTe behaves as a
II-VI compound when the Mn 2s electrons bond with Te. The
realization of the proposed InSb/CdTe heterostructures has been
hampered by the materials problems associated with the growth of
heterostructures. The optimum temperatures for the growth of high
quality InSb and CdTe are quite different (approximately 400°C [3]
and 200°C [4], respectively). Our approach has been to i) keep
the growth temperature for the two materials in the vicinity of
300°C, ii) improve the low temperature growth of InSb by using an
antimony cracker as a source of Sb_2, and iii) use very low growth
rates (from 0.2 to 0.4Å/s) to minimize the formation of defects.
Previously, a Raman and x-ray photoelectron spectroscopy
(XPS) study of CdTe nucleated on an ion milled and annealed InSb
substrate provided evidence for the formation of interfacial
compounds (such as In_2Te_3) with thickness ranging to 50 Å [5]. It
was suggested by Golding et al [6] that the interface compound
formation was the result of an excess in Te due to the relatively

high vapor pressure of Cd.

Recently, the as-grown heterointerfaces between high resistivity ZnSe and p-GaAs, another II-VI/III-V heterostructure of interest, have been evaluated by C-V measurements. It was shown that the ZnSe/GaAs interface state density can be reduced by orders of magnitude when an appropriate GaAs epilayer surface stoichiometry is chosen prior to nucleation of a ZnSe epilayer [7]. Recent transmission electron microscopy (TEM) studies combining image simulation with the experimental examination of cross-sectional samples tended to indicate the presence of an average of 2 monolayers of a strained interfacial compound at the interface [8]. *In situ* XPS was used to analyze the interfacial layer, and confirmed the interfacial layer as Ga_2Se_3 [9]. It is becoming evident in a general sense that a III-VI coherent interfacial layer tends to appear when forming II-VI-on-III-V heterovalent junctions by molecular beam epitaxy (MBE).

In this paper, x-ray rocking curve and TEM measurements are used to evaluate the microstructure of InSb/CdTe multilayer structures. In addition, we briefly describe a study of the chemical bonding at the interface between InSb and CdTe for epilayer/epilayer structures using an *in situ* XPS.

As an alternative to CdTe, a series of strained single quantum wells were fabricated with MnTe forming widegap barrier layers for quantum wells of InSb. The MBE growth technique enabled the metastable growth of the zincblende phase of MnTe at 300°C, whereas bulk-grown crystals of MnTe exhibit a hexagonal NiAs crystal structure [10]. The NiAs phase has an optical bandgap of 1.3 eV while the bandgap of the zincblende phase is 3.18 eV at 10 K [11,12]. The lattice constant of zincblende MnTe was determined from x-ray diffraction and TEM electron diffraction to be 6.33 Å [11].

InSb/CdTe multilayer structures

Single quantum well, multiple quantum well (MQW), and InSb/CdTe superlattice structures have been grown on InSb buffer layers at calibrated InSb substrate temperatures between 280 and 310°C. (The CdTe growth rate was 0.3Å/s, while the InSb was grown at 0.4Å/s.) The x-ray rocking curve for a 15 period superlattice is shown in Fig. 1. Of the two main peaks in the diffraction spectrum, the peak at the higher angle side (FWHM = 22 arcsec) is attributed to the zero order diffraction peak of the multilayer structure. Satellite peaks, having a separation of approximately 208 arcsec, are observed in a symmetrical angular distribution about the zero order peak. The spacing between satellite peaks corresponds to a periodicity of about 870Å, in good agreement with the dimensions (883±10Å) determined from TEM images. The other high intensity peak seen in the figure, having a FWHM value of 11 arcsec, is attributed to the (004) reflection from the InSb buffer/substrate. Since the lattice spacing in the growth direction for the InSb substrate/buffer should be smaller than the average lattice plane spacing associated with the periodic structure, the zero order diffraction peak from the superlattice would be expected to lie at a *lower* angle than the peak corresponding to InSb. The observation that the positions of the peaks were reversed can be explained by assuming that an ultra-thin, and highly strained layer of some interfacial compound, such as zincblende In_2Te_3, was incorporated at the interfaces [5,6].

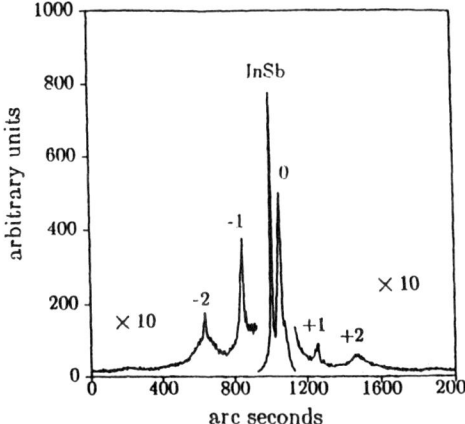

Figure 1. X-ray rocking curve of a 15 period superlattice (period = 870Å) grown at 310°C.

(Zincblende In_2Te_3 has a lattice constant of 6.14Å, a dimension significantly smaller than that of InSb/CdTe.) The calculation predicted the measured angular positions of the diffraction peaks, provided several (~5) monolayers of In_2Te_3 was distributed in some fashion between the two interfaces associated with each period of the structure.

The possible existence of the In_2Te_3 layers is further examined by analyzing changes of spacing and orientation of lattice fringes in HREM images. If thin pseudomorphic layers of stoichiometric In_2Te_3 exist at the interfaces, the spacing of the (200) lattice fringes in the In_2Te_3 layer are expected to be smaller than those of InSb and CdTe by 0.27Å; also the {111} type lattice fringes of the layer should change their orientations from those of InSb and CdTe by 2.8° in [011] HREM images.

HREM images of two MQW structures, taken at ten different areas for each sample, were examined. Due to the extremely small difference in the spacing and orientation of lattice planes, the changes in lattice fringes were detected only from areas which appeared to have more than two In_2Te_3 monolayers. Figure 2 is a [011] HREM image of one such CdTe-on-InSb interface area. In the image, (200) lattice fringes in the range between the two arrows had slightly smaller spaces than those in the CdTe and InSb layers. Also both (111) and (111) lattice fringes showed small shifts of their positions at the interface, as a result of the change of their orientations in the narrow range.

Another form of evidence for the existence of interfacial In_2Te_3 layers was obtained from the observation of dark field images. In a recent study of the ZnSe/GaAs interface [13], we have shown the existence of an interfacial Ga_2Se_3 layer by examining a set of dark field images which were taken by using different reflections. Similarly to Ga_2Se_3, In_2Te_3 forms a zincblende structure in which one third of cation sites are left as vacancies [14]. Because of the presence of vacancies, crystal structure factors of 200 type reflections of In_2Te_3 become larger than those of CdTe and InSb, while crystal structure factors of

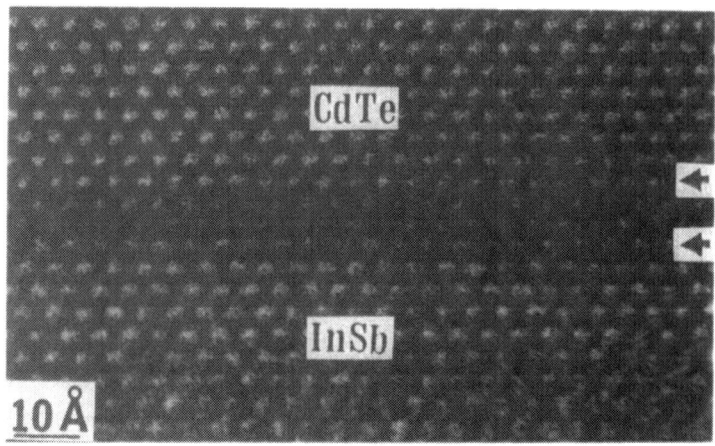

Figure 2. Magnified [110] HREM image of a CdTe on InSb interface.

400 type reflections have significantly smaller values than those of CdTe and InSb. Observed dark field images of CdTe/InSb multilayer structures confirm the existence of thin In_2Te_3 layers, showing a bright interfacial line in 200 type dark field images and a dark interfacial line in 400 type dark field images as expected from their crystal structure factors [8]. Simulated dark field images based on the model of an In_2Te_3 interfacial layer is in an excellent agreement with observed images.

In situ XPS experiments were performed (using a Perkin Elmer hemispherical analyzer and monochromatic x-ray source) to observe the formation of the interfacial layer when a CdTe epilayer is nucleated on an InSb epilayer, as well as for the inverse structure. The conditions used for the growth of the multilayer structures were duplicated in each case. First an XPS scan was taken from an InSb epilayer grown on an InSb substrate. The InSb growth was then resumed until interrupted for the nucleation of CdTe. The CdTe compound source shutter was opened for 6 seconds corresponding to a layer thickness of the order of two monolayers. The sample was again transferred under UHV to the XPS chamber. The process was repeated for the inverse interface. For CdTe-on-InSb, the XPS measurements show distinct changes in the In, Te, and Sb spectra which can be interpreted as the formation of additional chemical species when the heterointerface is formed. In this case, the single In peak associated with the InSb epilayer broadens into two clear peaks. One of the peaks has a binding energy relative to Sb within 0.1eV of that of InSb. The second, additional In peak emerging upon nucleation of CdTe has the same (within 0.1eV) binding energy relation with Te as is measured in the case of a deliberately grown In_2Te_3 epilayer.

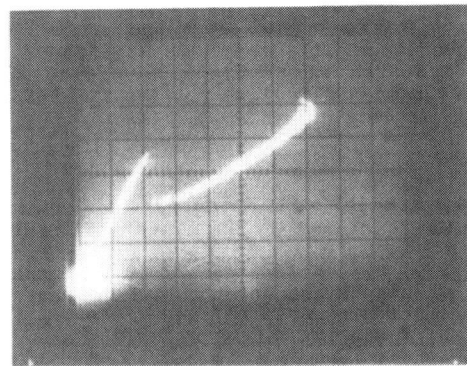

Current (280 A/cm**2/div)

Figure 3. I-V characteristic of the InSb/MnTe resonant tunneling structure. The measurement is carried out at 77 K.

Voltage (0.15 V/div)

InSb/MnTe Resonant Tunneling Structures

The Study of InSb/MnTe heterovalent tunneling structures was motivated by the potential application of resonant tunneling at high frequencies (InSb has the highest electron mobility among conventional semiconductors.), with improved device performance at room temperature. Computer simulations of InSb/MnTe double barrier resonant tunneling have been performed [15-17] and predict peak-to-valley ratios of the order of several thousand; peak current densities of 5×10^3 A/cm^2 are expected corresponding to barrier and well dimensions of 20 Å and 50 Å, respectively. The film growth was started in a III-V growth chamber where an InSb buffer layer was grown on a (100) InSb substrate at a temperature of 420 °C. The sample was subsequently transferred under UHV to another growth chamber (used for the InSb/CdTe multilayer structures described above) in which the InSb/MnTe resonant tunneling structures were grown using elemental sources at a substrate temperature of 300 °C [18]. Preliminary I-V results employing pulse measurement (to avoid sample heating) are shown in Fig. 3. A peak-to-valley ratio of 1.7:1 is observed at 77 K with a peak current density of 980 A/cm^2. Asymmetry about the origin in the I-V characteristics is observed, with the resonant peak appearing at a higher voltage for the reverse bias condition as compared to the forward bias condition. The origin of the deviation of the I-V characteristics from theoretical predictions is not yet known, and is still under investigation. One could speculate about such factors as interface scattering (MnTe/InSb interfaces have not been previously reported.), leakage current paths along the side wall of the etched mesas, etc. To our knowledge, the observed resonant tunneling represents the first report of a quantized state in InSb.

232

Acknowledgments

The authors thank R. Venkatasubramanian for aid in the interpretation of the x-ray features, D. Lubelski for his participation in the operation of the MBE machines, and T. Golding for sharing his unpublished results involving the growth and evaluation of CdTe/InSb multilayers. Research support was provided by National Science Foundation-MRG Grant DMR-8913706, Air Force Office of Scientific Research AFOSR-89-0438, Office of Naval Research N00014-89-J-1604, and NSF Equipment grant ECS-8606241.

References

* Present address: Department of Electrical Engineering and Computer Science, Massachusetts Institute of Technology, Cambridge, MA 02139

1. J.L.Glenn, Jr., Sungki O, L.A. Kolodziejeski, R.L. Gunshor, M. Kobayashi, D. Li, N. Otsuka, M. Haggerott, N. Pelekanos, and A.V. Nurmikko, J. Vac. Sci. Technol. B7, 249 (1986)

2. G.M. Williams, C.R. Whitehouse, A.G. Cullis, N.G. Chew, and G.W. Blackmore, Appl. Phys. Lett. 53, 1847 (1988)

3. A.J.Noreika, J.Greggi, Jr., W.J.Takei, and M.H.Francombe, J.Vac. Sci. Technol. A1, 558 (1983).

4. Z.C.Feng, A.Mascarenhas, W.J.Choyke, R.F.C.Farrow, F.A.Shirland, and W.J.Takei, Appl. Phys. Lett. 47, 24 (1985).

5. K. J. Mackey, D.R.T. Zahn, P.M.G.Allen, R.H.Williams, W.Richter, and R.S.Williams, J.Vac. Sci. Technol. B5, 1233 (1987)

6. T.D. Golding, M. Martinka, and J.H. Dinan J. Appl. Phys. 64, 1873 (1988)

7 J. Qiu, Q.-D. Qian, R.L. Gunshor, M. Kobayashi, D.R. Menke, D. Li, and N. Otsuka, Appl. Phys. Lett., 56, 1272 (1990)

8 D. Li, N. Otsuka, J. Qiu, J. Glenn, Jr., M. Kobayashi and R.L. Gunshor, Mat. Res. Soc. Symp. Proc., 161, 127 (1990)

9 J. Qiu, D.R. Menke, M. Kobayashi, R.L. Gunshor, Q.-D. Qian, D. Li, and N. Otsuka, to appear in J. Crystal Growth

10 J.W. Allen, G. Lucovsky, and J.C. Mikkelsen, Jr., Solid State Commun. 24, 367 (1977)

11 S.M. Durbin, J. Han, Sungki O, M. Kobayashi, D.R. Menke, R.L. Gunshor, Q. Fu, N. Pelekanos, A.V. Nurmikko, D. Li, J. Gonsalves, and N. Otsuka, Appl. Phys. Lett. 55, 2087 (1989)

12 Y. Lee and A.K. Ramdas, Phys. Rev. B38, 10600 (1988)

13 D. Li, J.M. Gonsalves, N. Otsuka, J. Qiu, M. Kobayashi, and R.L. Gunshor, Appl. Phys. Lett., 57, 449 (1990)

14 B. Greta-Plekovic, S. Popovic, B. Celustka, Z. Ruzic-toros, B. Santic, and D. Sold, J. Appl. Cryst., 16, 415 (1983)

15 M.J. McLennan and S. Datta, SEQUAL 2.1 User's Manual, Purdue University, (TR-EE 89-17 1989).

16 E.T. Yu and T.C. McGill, Appl. Phys. Lett., 53, 60 (1988)

17 R.G. van Welzenis and B.K. Ridley, Solid State Electronics, 27, 113 (1984)

18 J.Han, S.M. Durbin, R.L.Gunshor, M. Kobayashi, D.R. Menke, N. Pelekanos, M. Hagerott, A.V. Nurmikko, Y. Nakamura, and N. Otsuka, to appear in J. Crystal Growth

LOW PRESSURE MOCVD GROWTH OF InSb

B.T. CUNNINGHAM, R.P. SCHNEIDER, JR., AND R.M. BIEFELD
Sandia National Laboratory, Albuquerque, NM 87185

ABSTRACT

Low pressure (200 Torr) metalorganic chemical vapor deposition (MOCVD) of InSb has been examined through variation of the Column III (TMIn) and Column V (TMSb or TESb) precursor partial pressures. The use of lower growth pressure significantly enhanced the range of allowable Column III and Column V partial pressures in which specular morphology InSb could be obtained without the formation of In droplets or Sb crystals. In addition, a 70% improvement in the average hole mobility was obtained, compared to InSb grown in the same reactor at atmospheric pressure. SIMS analysis revealed that Si at the substrate/epitaxial layer interface is an important impurity that may contribute to degradation of the mobility. Substitution of TESb for TMSb did not result in any improvement in the purity of the InSb.

Introduction

There is considerable interest in InSb/InAsSb strained-layer superlattices (SLS) due to their potential use as efficient long-wavelength detectors in the 8-12 um spectral region [1,2]. There is also a great deal of recent interest in epitaxial InSb, generated by its potential use for infrared detectors, for high speed circuit elements because of its narrow bandgap (0.17 eV at 300K) and its low effective mass ($m^*/m = 0.0145$), and for magnetic Hall devices for its large mobilities ($u_n > 1 \times 10^6$ cm^2/Vsec). High residual background impurity concentrations in the intrinsic region of InSb/InAsSb SLS p-i-n photodiodes grown by metalorganic chemical vapor deposition (MOCVD) currently limit the detectivity through large reverse bias dark currents caused by tunneling through impurity states [3]. In addition, the InSb/InAsSb heterointerface quality may suffer degradation due to the long (30 sec) growth pause required to purge growth precursors from the reactor at atmospheric pressure [4]. One problem particular to the atmospheric pressure MOCVD growth of InSb is the narrow window of V/III ratios which are acceptable for the growth of specular epitaxial layers without the formation of In droplets (In rich condition) or Sb crystals (Sb rich condition) [5]. This narrow range places severe restrictions upon the choice of process parameters and upon process reproducibility.

The use of reduced MOCVD growth pressure has been successfully demonstrated in several material systems as a means of reducing the background impurity concentration and reactant switching times, but has not yet been applied to InSb. In this work, we demonstrate the use of reduced pressure (200 Torr) MOCVD for the growth of InSb using TMIn as a Column III precursor, and TMSb or TESb as the Column V precursor. The material purity, as measured by the Hall effect, has been improved significantly with respect to InSb grown in the same reactor at atmospheric pressure. In addition, a very wide range of Column III and Column V partial pressures were found to yield InSb with specular morphology. Sb crystals did not form under any conditions, while In droplets formed only when the growth rate exceeded 0.6 um/hr. Secondary ion mass spectroscopy (SIMS) analysis of the InSb revealed the presence of Si at the substrate/epitaxial layer interface of every sample. Samples with the lowest Si concentration demonstrated the best mobilities. Substitution of TESb for TMSb did not result in any improvement in the purity of the InSb.

Experimental

The horizontal quartz reactor chamber used in this work was not equipped with a glove box or a load lock, so the growth chamber was exposed to air prior to each run. The p-InSb substrates were packaged in paraffin by the vendor (Cominco) for shipping. The wax was melted and removed with several rinses in hot TCE. Prior to growth, the substrate was degreased in TCE again, and rinsed in organic solvents. The substrate was then etched for 2 min in 10:1 lactic:nitric acid, followed by a DI rinse and N_2 drying just prior to loading. All growths were performed at a pressure of 200 Torr, a temperature of 470 C, and a total H_2 flow rate of 7 slm. All undoped InSb samples had a p-type background carrier concentration. All the samples reported in this work had a smooth, specular morphology without In droplets or Sb crystals. The appearance and electrical properties of the layers were consistent over the 2-inch susceptor area.

When comparing Hall mobility data of InSb, it is important to consider the mobility and carrier concentration dependence on the Hall magnetic field. The variable field Hall data in Figure 1 is typical for p-type InSb. The hole mobility decreases sharply with increased magnetic field for 0.0 < B < 2.0 kG, but remains fairly constant for B > 2.0 kG. This effect is caused by an increase in the Hall scattering factor via increased impurity scattering at higher magnetic fields [6]. For this work, a Hall magnetic field of 3.4 kG was chosen for all Hall measurements in order to avoid measurement error in the mobility due to small fluctuations in the magnetic field.

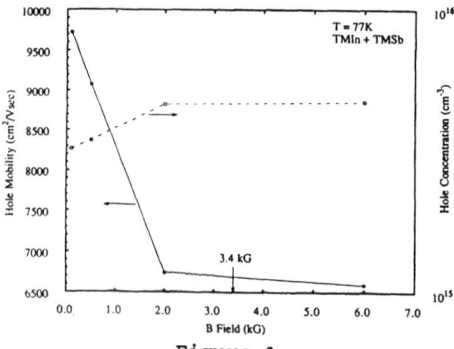

Figure 1.

Results

The variation of hole mobility with TMSb flow rate for InSb grown with TMIn and TMSb is shown in Figure 2. Sb crystals did not form for any TMSb flow rate investigated. The largest Sb flow rate investigated, TMSb = 1×10^5 moles/min corresponds to a V/III ratio of 74. Atmospheric pressure growth required a V/III ratio of 3 for specular growth. The hole mobility first increases with increased TMSb flow rate, but then declines after reaching a maximum.

The variation of hole mobility with TMIn flow rate for InSb grown using TMIn and TMSb is shown in Figure 3. The TMIn flow rate is directly proportional to the growth rate, and a TMIn flow rate of 2.7×10^{-6} moles/min corresponds to roughly 1/2 um/hr growth rate. At a growth rate above 0.6 um/hr, the surface took on a rough, textured appearance, and In droplets were observed. The sample with u_p = 5300 cm^2/Vsec was grown at a very high TMSb flow rate (TMSb = 1.9 $\times 10^{-4}$). With this exception, Figure 3 demonstrates a trend towards increased mobility with increased TMIn flow rate.

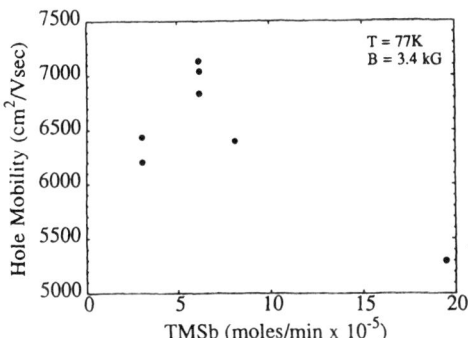

Figure 2.

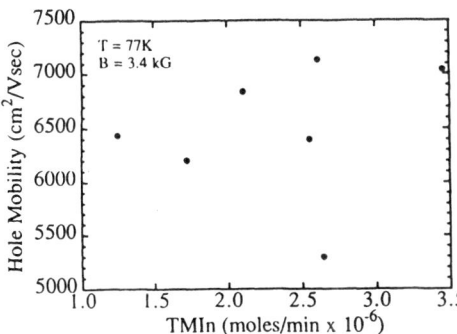

Figure 3.

SIMS analysis was performed for several of the most common donor and acceptor impurities using an oxygen primary ion beam. The donors Sn, Ge, and Te were not detected in any sample with a background detection sensitivity limit of approximately 1×10^{15} cm^{-3}. Likewise, the acceptors Zn, Be, Mg, Cd, and Hg were not detected with a background detection sensitivity limit of 1×10^{15} cm^{-3}. The detection sensitivity using an oxygen primary ion beam was not sufficient to draw any conclusion as to the presence or absence of carbon. However, Si was detected in every sample at the substrate/epitaxial layer interface. Both ^{28}Si and ^{29}Si were detected, confirming the presence of Si, rather than a combination of carbon and oxygen. A typical Si "spike" is shown in the SIMS profile of Figure 4. The SIMS Si secondary ion count was converted to Si concentration using a Si implanted InSb standard. The sample shown in Figure 6 had a hole mobility of 6401 cm^2/Vsec, and a peak Si concentration of 4.7×10^{15} cm^{-3}. Samples with a peak Si concentration as high as 1×10^{18} cm^{-3} had mobilities near 5000 cm^2/Vsec, while a sample with a peak Si concentration of 1×10^{16} cm^{-3} had a mobility of 7000 cm^2/Vsec. Thus, there appears to be some correlation between mobility and the amount of Si incorporated at the substrate/epitaxial layer interface. Because Si is only observed near the substrate surface, we suggest that it originates from the surface preparation or pregrowth procedure, rather than from incorporation from the metalorganic precursors. One possible source of Si contamination is the paraffin wax material that the substrates are packaged in. Research is in progress to investigate surface preparation techniques which will remove this material more effectively.

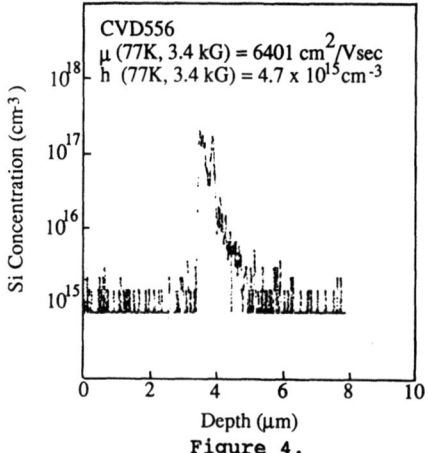

Figure 4.

One disadvantage in the use of TMSb as an Sb precursor is that is does not dissociate easily at the growth temperature. As the growth pressure is reduced, the cracking efficiency of TMSb declines significantly. To optimize the InSb mobility, approximately seven times more TMSb is required at 200 Torr than is needed at atmospheric pressure (650 Torr). Because TESb dissociates much more easily than TMSb, the substitution of TESb would be advantageous if InSb of equal or improved purity could be grown. Ethylated metalorganic precursors for Ga, Al, and In have demonstrated greatly reduced levels of carbon incorporation than methylated precursors due to the absence of reactive methyl radicals. The hole mobility and hole concentration as a function of TESb flow rate at a fixed TMIn flow rate is shown in Figure 5. The maximum mobility obtained was 6500 cm^2/Vsec, and the minimum hole concentration obtained was 1×10^{16} cm^{-3}. The InSb grown using TMSb had consistently better purity than that grown using TESb. This trend probably indicates that the purity of our TESb source was inferior to that of our TMSb. More research must be done to determine precisely the extent that carbon will incorporate in InSb, and which lattice sites it will occupy.

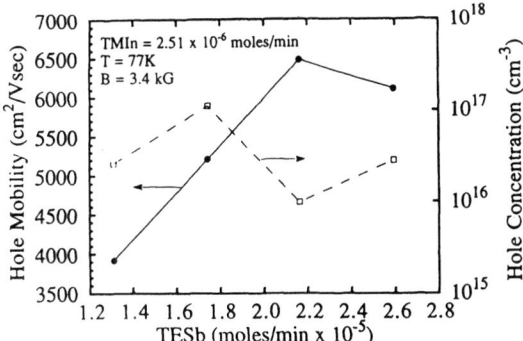

Figure 5

Table I summarizes the comparison between InSb grown at 200 Torr, and that grown at 650 Torr in the same reactor. The best mobility obtained thus far by either technique has been improved by 48% at 200 Torr. However, when averaged over many samples, the mobility has been improved by 70% at low pressure, indicating the high degree of reliability the growth process has achieved with the large window of precursor flow rates to obtain specular morphology material. However, the average hole concentration at 200 Torr is higher than that at 650 Torr, indicating that the low pressure InSb is less heavily compensated than that grown at atmospheric pressure.

Table I

	200 T (B = 3.4 kG)	650 T (B = 2.0 kG)	% Improved
Best u_h (cm^2/Vsec)	7374	4987	48%
Avg. u_h (cm^2/Vsec)	6592	3872	70%
Avg. h (cm^{-3})	5.2×10^{16}	3.16×10^{16}	

Conclusions

In summary, low pressure MOCVD growth of InSb has been investigated as a means of reducing the level of background impurities. On the average, the mobility of the InSb has been improved by 70% as compared to InSb grown by atmospheric pressure in the same reactor. SIMS analysis revealed that Si at the substrate is an important impurity species which may be limiting the mobility. Substitution of TESb for TMSb did not enhance the purity of the material. We have also demonstrated that a wide range of Column III and Column V flow rates are available for growing specular morphology films at low pressure, whereas one is limited to a very narrow range of flows at atmospheric pressure.

This work was supported by the Department of Energy under contract DE-AC0476P00789

References

1.   P.K. Chiang and S.M. Bedair. J. Electrochem. Soc. 131, 2422 (1984).
2.   S.R. Kurtz, L.R. Dawson, R.M. Biefeld, I.J. Fritz, and T.E. Zipperian, IEEE Electron Device Lett. 10, 150 (1989).
3.   G.C. Osbourn, Semicond. Sci. Technol. 5, S5 (1990).
4.   G.A. Hebner, R.M. Biefeld (unpublished).
5.   R.M. Biefeld, J. Cryst. Growth 91, 515 (1988).
6.   D.L. Leslie-Pelecky, D.G. Seiler, M.R. Loloee, and C.L. Littler, Appl. Phys. Lett. 51, 1916 (1987).

OMVPE GROWTH OF EPITAXIAL InSb THIN FILMS USING A NOVEL GROUP V SOURCE COMPOUND

Gregory T. Stauf,* D.K. Gaskill,* N. Bottka* and R.W. Gedridge, Jr.**
*Naval Research Laboratory, Washington, DC 20375-5000
**Naval Weapons Center, China Lake, CA 93555

ABSTRACT

To date, OMVPE has failed to grow epitaxial InSb at temperatures below about 400 °C. Therefore, we have studied InSb deposition using the novel group V source triisopropylantimony (TIPSb) with trimethylindium (TMIn), comparing this with results in the same reactor using trimethylantimony (TMSb). We have grown InSb on both GaAs and on high-resistivity p-type InSb substrates, exploring the effects of temperature, substrate, V/III ratio, and thickness on film properties. Our best results with TMSb and TMIn were obtained using a V/III ratio of 8 and growth temperature of 450 °C. The unintentionally doped films exhibited good crystallinity and were n-type about 1×10^{15} cm^{-3} down to 5 K, with 77 K mobilities of 90,000 and 253,000 cm^2/V-sec for InSb on GaAs and on InSb, respectively. These results are comparable to the best published MBE data. While InSb films could not be grown well at temperatures below 425 °C using TMSb, specular epilayers were obtained with TIPSb at temperatures as low as 300 °C, though with a low deposition rate. Our best conditions with TIPSb were with a V/III ratio of about 12 and a substrate temperature of 350 °C. The results of variable temperature and variable magnetic field Hall, double crystal x-ray diffractometry, Nomarski microscopy, Auger electron spectroscopy and scanning electron microscopy will be presented.

INTRODUCTION

The compound semiconductor InSb holds great promise as an infrared radiation detector, due to its low bandgap (0.18 eV at room temperature) [1]. A potential method of extending its response to the energy region of 8-12 µm (0.1 eV) where few other candidate materials exist is by alloying InSb with As and Bi. An obstacle has been the solid solubility limit of Bi in InSb (about 2%). To reach the higher incorporation needed to make a detector [2], growth of this alloy must be carried out at low temperatures to produce a metastable material, perhaps in the range of 300-350 °C. Until now, OMVPE of InSb has only been successful at temperatures of 400 °C or higher [3]. This is because the typically used antimony source, trimethylantimony (TMSb) does not decompose below 400 °C. Thus, alternative antimony source compounds which decompose at lower temperatures must be explored.

Another potential obstacle to fabrication of useful detectors with InSb has been residual background impurities. Until recently, OMVPE of InSb has not produced material with impurity levels much under 10^{16}, which has limited 77 K mobilities to about 70,000 cm^2/V-sec [4]. This cast doubt on whether OMVPE with non-hydride group V sources could compete with the quality of films produced by MBE.

The objective of our work was to establish two points. First, that non-hydride organometallic precursors do not limit InSb film properties; high purity sources, careful choice of reactor geometry and strict control of the V/III ratios used can result in high film quality. Second, an alternative antimony source, triisopropylantimony, (TIPSb), should allow lower film growth temperatures, eventually permitting higher Bi incorporation. Results will therefore be presented on InSb grown using trimethylindium (TMIn) and TIPSb, and compared with InSb growth in the same reactor using TMSb.

Mat. Res. Soc. Symp. Proc. Vol. 216. ©1991 Materials Research Society

EXPERIMENTAL

The reactor used in this work has been described previously [5]. The vertical 3" nominal ID reactor was operated at atmospheric pressure with palladium-diffused hydrogen as a carrier gas. A rotating 2" graphite susceptor inside was heated with a RF coil. The TMIn, TMSb and TIPSb bubblers were kept at 17.0 $^\circ$C, -12.0 $^\circ$C and 16.5 $^\circ$C, yielding vapor pressures of 1 torr, 15.1 torr and 0.3 torr, respectively. Total gas flow through the reactor was 4 l/min, with individual bubbler flows in the range 6-20 sccm for TMIn, 30-45 sccm for TMSb and 800-1600 sccm for TIPSb. Deposition temperatures examined were from 300 $^\circ$C to 400 $^\circ$C during TIPSb work and 400-450 $^\circ$C using TMSb. The TIPSb was synthesized by previously described methods from reagent grade chemicals [6].

InSb epilayers were typically grown on (100) semi-insulating GaAs in order to facilitate electrical measurements. Unfortunately, the 14% lattice mismatch in this system results in the generation of defects, which adversely affect morphology and electrical properties. For this reason, some growths were also done on p-type "high resistivity" (~100 Ω-cm) InSb substrates so that lattice-matched properties could be examined.

RESULTS AND DISCUSSION

Growth using TMSb

During growth experiments using TMSb as a Sb source, the best results were found at a growth temperature of 450 $^\circ$C. Under optimum conditions, the central 2/3 of the substrate could be covered with near-specular InSb, while the outer third was Sb rich and non-specular. Electrical transport property results from these films and some from less optimized 425 $^\circ$C depositions are shown in Table 1. Experiments at 400 $^\circ$C.resulted in poor quality films.

In Table 1 iit is apparent that under similar growth conditions, much better film properties are obtained on a lattice-matched substrate, as would be expected. X-ray rocking curves support this, with full width at half maximums (FWHM) of about 180 arc-sec for the best growth on GaAs using TMSb, vs. 13 arc-sec for a similar InSb/InSb growth.

TABLE 1. A summary of 77 K carrier concentrations and mobilities for InSb epilayers.

Sb source	Growth temp. $^\circ$C	V/III ratio	Thickness (μm)	Substrate	mobility (cm^2/V-s)	Carrier conc. (cm^{-3})
TMSb	450	7.4	4.5	GaAs	90,000	-1.4 E15
TMSb	450	8.5	4.8	InSb	253,000	-1.4 E15
TMSb	425	10.0	2.3	GaAs	9,940	-2.4 E15
TMSb	425	10.0	2.3	InSb	160,000	-3.0 E15
TMSb	450	7.2	0.9	GaAs	500	-4.8 E16
TIPSb	375	12.0	1.1	GaAs	9,300	-3.4 E16
TIPSb	350	13.0	1.0	GaAs	20,000	-1.8 E16
TIPSb	350	13.0	1.0	InSb	63,200	-3.7 E16
TIPSb	325	20.0	0.6	GaAs	8,000	-1.8 E17
TIPSb	300	40.0	0.5	GaAs	1,600	-1.1 E17

The major result from this series of experiments is that low impurity levels have been achieved in these films. Carrier concentrations of around 1×10^{15} cm^{-3} have allowed near-bulk like mobilities of 250,000 cm^2/V-sec to be achieved by homo-epitaxial growth. Electron cyclotron resonance using 70.6 μm radiation on a 5 mm spot gave a local mobility of 180,000 cm^2/V-sec even at a temperature of 4 K, confirming high purity. When thin layers of InSb are grown at around 450 °C on GaAs, though, lattice mismatch causes tremendous degradation in electrical properties. Thus, mobility drops to just 500 cm^2/V-sec for the 0.9 μm thick film grown with TMSb in Table 1.

Variable temperature (5 K up to 180 K) and variable magnetic field Van der Pauw Hall measurements have been reported in more detail elsewhere [5]. Peak mobilities were found at 65 K, 253,000 and 220,000 cm^2/V-sec at magnetic fields of 500 and 2000 G respectively, for InSb/InSb films grown at 450 °C. Based on low temperature mobility results, phonon scattering is believed to be the dominant mobility limiting mechanism.

Growth using TIPSb

The best growth temperatures using TIPSb as a Group V source were seen to be in the range 300-375 °C. The V/III ratio needed for smooth morphology was found to be very narrow with this system, more so than in the TMIn/TMSb experiments. Preferential depletion of one of the two organometallic reactants caused different morphology regions on the substrate during a single run, as shown in Figure 1. The locations of these varying morphologies depended on the V/III ratio used, while the sizes of the substrate area covered by them depended mostly on the growth temperature. Normally a black material was deposited in the center of the susceptor, while a white or tan Sb-rich material formed on the outside, with a smooth, shiny region between. Lowering the V/III ratio of the input gases (more TMIn) increased the size of the center haze relative to the outer haze, while increasing the V/III ratio caused the opposite. Higher growth temperatures caused larger total hazy areas relative to smooth regions.

FIGURE 1. (a) SEM micrograph of a center region exhibiting needle-like crystals, from InSb growth on GaAs at 350 °C using TIPSb. (b) Nomarski micrograph of Sb rich area on outer edge in similar growth.

(a)  (b)

The Sb-rich regions looked similar to those grown with TMSb, with "plateaus" and small flat crystals, as can be seen in Fig. 1(a) on the previous page. The regions in the center, however, looked very different, exhibiting the clusters of needle-like crystals in Fig. 1(b). AES measurements showed these areas to be between 5% and 20% Sb-rich, making their exact cause and nature unknown at this time.

Locations that achieved specular growth of InSb on GaAs substrates were in general smoother than similar regions from TMSb experiments. This can be seen in Figure 2, below.

At a growth temperature of 300 °C, a V/III ratio of 40, and a growth rate of 0.077 μm/hr, almost the entire 2" substrate could be covered with very smooth InSb. The V/III ratio could be varied over the widest range at this low temperature and still produce large specular areas, although the growth rate was quite low. Figure 2(b) shows a Nomarski micrograph of one of these growths.

FIGURE 2. (a) Nomarski micrograph of InSb grown on GaAs at 450 °C using TMSb, 0.9 μm thick. (b) InSb/GaAs grown at 300 °C using TIPSb, 0.5 μm thick. (c) InSb/GaAs grown at 325 °C using TIPSb, 0.6 μm thick. (d) InSb/GaAs grown at 350 °C using TIPSb, 1.3 μm thick.

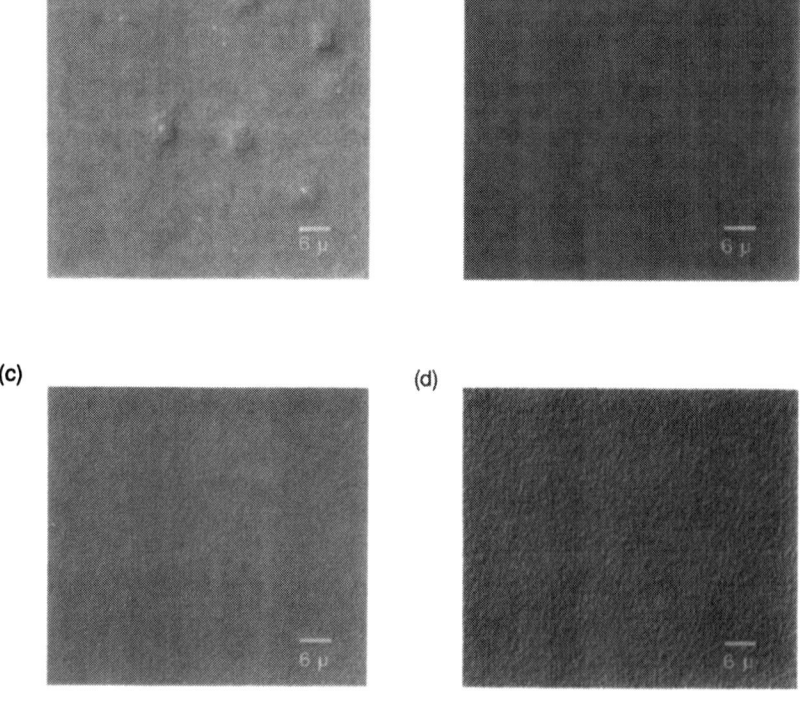

At 325 °C, smooth morphologies at the edge of the substrate (outer~1/2 inch) and a growth rate of 0.1 μm/hr were seen with a V/III ratio of 20. Figure 2(c) shows one of these regions. Variation of the V/III ratio by 10% from these conditions shifted the shiny region substantially, either nearer to or farther from the edge, and 20% or more eliminated it. Larger smooth areas in the center were produced when TIPSb flow was raised from 800 to 1600 sccm, for a *nominal* V/III ratio of 40, though we do not believe that the TIPSb bubbler flow is saturated at 1600 sccm.

Deposition at 350 °C produced smooth epilayers at 0.24 μm/hr with a V/III of 12, as seen in Fig. 2(d). These conditions covered about 50% of the substrate area at the outer edge with shiny InSb. If the V/III ratio was changed by more than about 10%, morphology became much worse. No substantial increase in size of the smooth InSb area came from increased TIPSb flow, contrary to results at 325 °C.

Finally, at 375 °C, a V/III ratio of 12 was used, producing a growth rate of 0.6 μm/hr. Locations and appearances of smooth morphologies were similar to those at 350 °C, as was sensitivity to V/III ratio. Since the V/III ratio for growth was unchanged from that at 350 °C, this indicates that even at 350 °C both organometallics are decomposing well.

To briefly summarize, morphologies for InSb deposition using TIPSb were seen to improve at lower growth temperatures, as is apparent in Figure 2. Deposition rates comparable to the 1 μm/hr achieved using TMSb could not be attained in these experiments. As higher deposition temperatures were used, the size of the smooth InSb areas grown became smaller, indicating that either the V/III ratio range for good epitaxy narrows or depletion effects are more severe.

Mobilities and carrier concentrations for heteroepitaxial films grown at different temperatures are shown in Table 1. Electrical properties of these films are seen to be at their best at 350 C, though even here they are not as good as the TMSb results. We believe this to be a result of unintentional n-type doping from impurities in the reagent grade starting materials used for the TIPSb synthesis. The 63,000 cm^2/V-s homoepitaxial mobility is in fair agreement with bulk InSb mobility when doped to a similar carrier concentration [7], indicating low levels of compensation. Variable tempearture Hall measurements made on these films have supported this view [8].

A key point, though, is that transport properties are much better for *thin* films on GaAs (under 2 μm) than they were using TMSb. The low growth temperatures achievable using TIPSb may be helping to isolate defects to positions near the mismatched interface, so that they do not interfere as strongly with mobility. Good crystalline quality was attained even at these low temperatures, though, as demonstrated by the 14 arc-sec x-ray FWHM of a 1.2 μm thick homoepitaxial film grown at 350 °C.

CONCLUSIONS

OMVPE of InSb using non-hydride group V sources such as TMSb can produce state-of-the-art quality material. Carrier concentrations of $1x10^{15}$ cm^{-3} and mobility of 253,000 cm^2/V-s are both comparable to the best homoepitaxial MBE films reported to date. While transport properties of films grown on GaAs are still somewhat less than the best MBE results, this may be due to lack of any GaAs or ALE type buffer layers in our work. In addition,the novel group V source triisopropylantimony has been used to grow InSb at deposition temperatures as low as 300 °C. The V/III ratio is very critical with this system, with variations of only 10% causing major changes in morphology. Electrical measurements on thin (~1 μm thick) layers grown on GaAs at 350 °C yielded a mobility of 20,000 cm^2/V-s, indicating that these low temperatures play a major role in overcoming defects from lattice mismatch. Electrical properties in homoepitaxial films are consistent with low levels of compensation, indicating that with more careful purification, TlPSb would be a very promising source compound for OMVPE growth of InSb at lower temperatures than heretofore possible. This capability could allow growth of InBiSb alloys without segregation of the Bi, making available a III/V semi-conducting material for use in infrared detection devices.

ACKNOWLEDGEMENTS

We would like to thank M. Fatemi, M.-J. Yang and J. Mittereder for the SEM and AES data, as well as R. Gorman for technical support. We are also grateful for many helpful discussions with P. Thompson, J. Davis, R. Sillmon and J. Pazik. G.T Stauf would like to acknowledge support by the NRC/NRL Research Associateship Program.

REFERENCES

(1) Zwerdling, S. et. al., Phys. Rev. 108 (6), 1402 (1967).

(2) M.A. Berding, A. Sher, A.-B. Chen and W.E. Miller, J. Appl. Phys. 63, 107 (1988).

(3) K.Y. Ma, D.H. Jaw, Z.M. Fang, R.M. Cohen and G.B. Stringfellow, Appl. Phys. Lett. 55, 2420 (1989).

(4) R.M. Biefeld, S.R. Kurtz and I.J. Fritz, J. Electr. Mater. 18, 775 (1989).

(5) D.K. Gaskill, G.T. Stauf and N. Bottka, submitted to Appl. Phys. Lett..

(6) H.J. Breunig and W. Kanig, J. Organomet. Chem. C5, 186 (1980).

(7) A.J. Strauss, J. Appl. Phys.30, 559 (1959).

(8) G.T. Stauf, D.K. Gaskill, N. Bottka and R.W. Gedridge, Jr., submitted to Appl. Phys. Lett.

RAMAN AND SECONDARY ION MASS SPECTROSCOPY OF EPITAXIAL CdTe/InSb INTERFACES GROWN BY LOW-ENERGY BIAS SPUTTERING

SUHIT R. DAS, DAVID J. LOCKWOOD, STEPHEN J. ROLFE, JOHN P. McCAFFREY, AND JOHN G. COOK
Institute For Microstructural Sciences, National Research Council, Ottawa, Ontario, Canada K1A 0R6

ABSTRACT

Heteroepitaxial (100)CdTe ∥ (100)InSb structures have been fabricated by growing CdTe epilayers, at growth temperatures below 200°C, on single crystal InSb substrates by low-energy bias sputtering. Controlled low-energy ion bombardment at the substrate was employed to clean the growth surface in-situ just prior to film deposition and to modify the growth kinetics and enhance adatom mobility during deposition. Raman spectroscopy of the interface revealed no evidence of In_2Te_3 and secondary ion mass spectroscopy showed the interface to be chemically abrupt.

INTRODUCTION

The mixed II-VI/III-V CdTe/InSb heterostructure with a near perfect lattice match ($\Delta a/a < 0.05\%$) and widely different bandgaps (CdTe 1.44 eV, InSb 0.18 eV) is an attractive system for the investigation of quantum effects [1-3] and for fabrication of quantum-well based opto-electronic devices. A high quality interface is obviously the key to achieving good device performance. Fabrication of CdTe/InSb heterostructures of high structural quality by molecular beam epitaxy (MBE) usually involves an in-situ cleaning of the InSb surface by 500 eV Ar^+ ion bombardment followed by annealing at about 200°C prior to growth of the CdTe layer at temperatures below 200°C [1-5]. Chew et al. [5] have shown the existence of Te precipitates at the interface and polycrystalline growth for CdTe growth temperatures of ≤ 150°C. MBE growth of CdTe on InSb at growth temperatures above 240°C [6] is observed to lead to structural imperfections in the epilayer due to the formation of In droplets at the interface and interdiffusion of the different elements across the interface, although chemically abrupt (as revealed by secondary ion mass spectroscopy) interfaces of good structural quality have been reported at high growth temperatures (≥ 240°C) using a modified two-step growth technique [6,7] or a Cd/Te flux ratio 3/1 during growth.

Although the CdTe epilayers are well-ordered when grown on InSb by MBE under the conditions mentioned above, Raman scattering studies [8] and high resolution electron microscopy (HREM) investigations [9] have shown the existence of an interfacial layer of In_2Te_3. We have previously reported [10] the growth of epitaxial CdTe/InSb heterostructures by low-energy bias sputtering (LEBS). In this technique, Ar^+ ions are drawn from an rf magnetron discharge by a small bias voltage on the substrate, thus causing low-energy ion bombardment at the substrate which can be utilised to clean the substrate surface just prior to film deposition or to enhance adatom mobility during deposition. In this paper we report the results of Raman scattering and secondary ion mass spectroscopy (SIMS) studies of the interface of CdTe/InSb heterostructures

Mat. Res. Soc. Symp. Proc. Vol. 216. ©1991 Materials Research Society

fabricated by LEBS.

EXPERIMENTAL DETAILS

The (LEBS) apparatus employed a cryopumped stainless steel chamber with a Meissner trap and a base pressure of 2×10^{-7} Torr. The target was a hot-pressed 15-cm-diameter disc of 99.999% pure CdTe bonded to a magnetron cathode of commercial design (MRC, Orangeburg, N.Y.). Power was fed to the target and to the substrate heater block independently from separate rf power supplies (Advanced Energy RFX-600) through Mercator Control Systems ATN-500 tuning networks. A dc power supply, connected to the substrate via a low-pass filter, was employed to adjust the substrate dc bias.

Single crystal (100)InSb substrates were etched for 15 seconds in a 10% HF solution, rinsed in deionised water, blow-dried with filtered nitrogen, and bonded to the heater block with In-Ga eutectic. The deposition conditions for CdTe were 100 SCCM argon flow, sputter pressure of 4.1 mTorr, and rf power of 50 W and a dc self bias of -58 V on the target. Samples of thicknesses ranging from 1900 to 2500 A° were grown at temperatures between 180-240°C. Growth rates of about 3 A°/s were obtained under the above conditions. Prior to CdTe growth, an in situ surface preparation procedure, consisting of low-energy bias sputtering of the InSb substrate, was carried out for 30 minutes with the substrate held at 300°C. The rf power to the substrate during this procedure was 15 W and the dc bias on the substrate was varied between -20 V to -50 V.

High resolution transmission electron microscopy of cross-sectional samples of the CdTe/InSb heterostructures, prepared by conventional argon atom milling, was performed in a Philips EM 430T microscope operating at 250 keV. SIMS depth profiling was performed using a CAMECA IMS 4f secondary ion microprobe, equipped with a duoplasmatron ion source and a primary mass filter. The samples were sputtered with 2.75 keV O_2^+ primary ions rastered over 350 microns. Positive secondary ions (^{114}Cd, ^{115}In, ^{121}Sb, and ^{130}Te) were analyzed from a 62 micron area. Depths were obtained with a Tencor Alphastep 200 and sputtering rates were about 5 A°/s.

The Raman scattering measurements were carried out at room temperature in a quasi-backscattering geometry [11]. The spectra were excited with 30 mW of 6471 A° Krypton laser light, analyzed with a Spex 14018 double monochromator, and detected with a cooled RCA 31034A photomultiplier. The laser light was incident at Brewster's angle (19.3° at 6471 A°) for CdTe. The polarization of the incident light was contained within the scattering plane, while the unanalyzed scattered light was directed through a polarization scrambler into the spectrometer. The sample was contained in a helium gas atmosphere to eliminate Raman scattering from air.

RESULTS AND DISCUSSION

Although electron diffraction and high resolution lattice images of the CdTe/InSb heterostructures showed that epitaxial growth with (100)CdTe ∥ (100)InSb had occurred under all growth conditions, a distinct improvement of the interfacial morphology and a decrease in defect density at the interface was noticed with decreasing growth temperature, a result in conformity with studies of MBE grown structures [4,6]. A higher substrate bias during ion cleaning also produced better interfaces. However, no evidence was

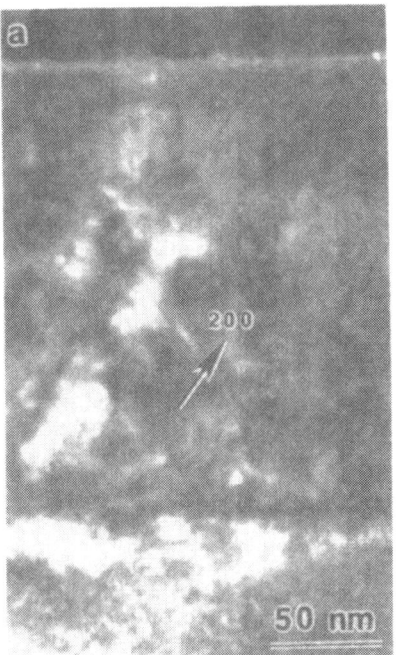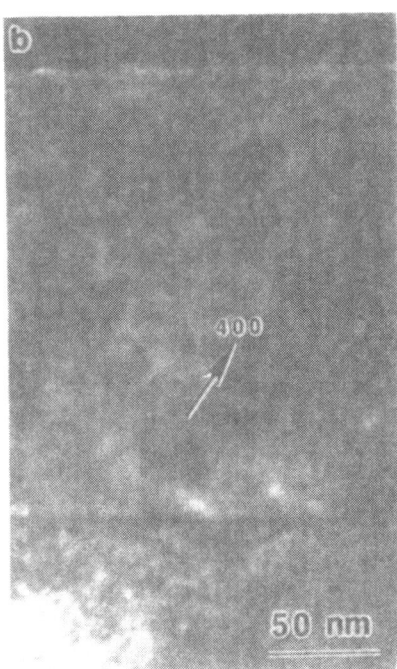

Fig. 1: Dark field images of a CdTe epilayer and the CdTe/InSb interface region: (a) 200 reflection and (b) 400 reflection.

found of In_2Te_3 precipitates. Figure 1 shows two cross-sectional dark field images (foil normal = [011]) of an epitaxial sample grown at 180°C with a substrate bias of -40 V during ion cleaning for (a) g = 200 DF and (b) g = 400 DF. A dark band is visible at the interface which is also visible in all bright field [011] images except in very thin regions near the edge of the sample. According to Li et al. [9], an In_2Te_3 layer should appear as a bright line in 200 dark field images and as a dark line in 400 dark field images. As can be seen in Figure 1, there is no evidence of a bright interface region in the 200 dark field image, suggesting that the dark band is caused by something else, probably structural defects at the interface.

The electron microscopy results were supported by SIMS depth profiles. Figure 2 shows the SIMS depth profiles for two samples grown at the same temperature but with different substrate bias during ion cleaning and for two other samples grown at different temperatures but with the same substrate bias during ion cleaning. The sharpest interfaces were obtained for samples grown at the lowest temperature and for samples grown with the highest substrate bias during ion cleaning.

A typical Raman spectrum of a CdTe/InSb heterostructure grown by LEBS is shown in Figure 3. The first-order Raman spectrum of InSb (see Fig. 3) is comprised of a strong peak at 190 cm^{-1} due to

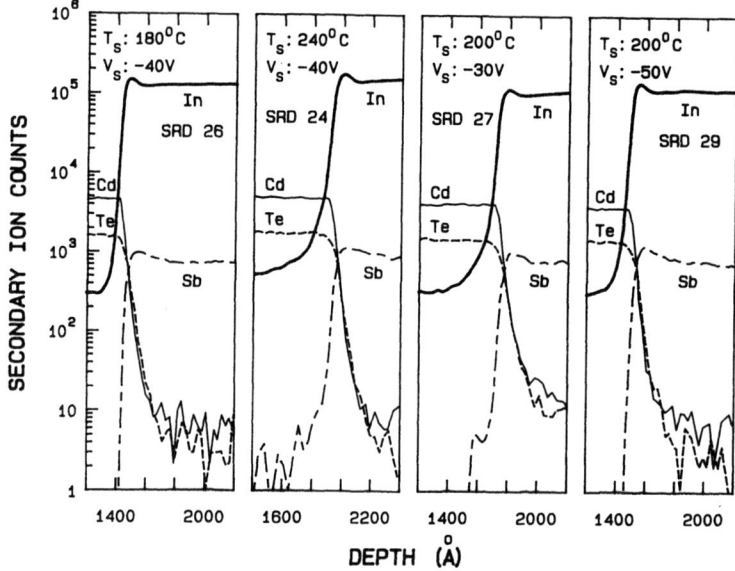

Fig. 2: SIMS depth profiles of CdTe/InSb heterostructures grown under different conditions by LEBS.

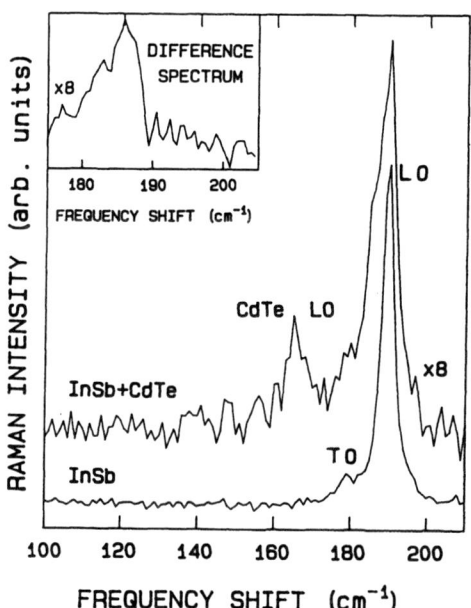

Fig. 3: Raman spectra of the InSb substrate and substrate plus epitaxial layer of CdTe recorded at a resolution of 1.6 cm^{-1}. The inset shows the difference between the two spectra when recorded at 1.0 cm^{-1} resolution.

the longitudinal optical (LO) phonon and a much weaker transverse optical (TO) phonon peak at 179 cm^{-1} (this scattering is forbidden in true backscattering). At lower frequency, a weaker peak is seen at 166 cm^{-1} due to scattering from LO phonons in CdTe. Scaling and subtraction of the InSb substrate spectrum from the CdTe-on-InSb spectrum yielded the difference spectrum shown in the inset to Figure 3. The subtraction process revealed a weak, broad, and asymmetric feature peaked at 185 cm^{-1} just below the InSb mode. No other Raman peaks were found in these samples grown under the range of deposition conditions mentioned above. In other studies of the CdTe/InSb system [8,12] Raman lines at 105, 126, and 143 cm^{-1} were observed and attributed to In$_2$Te$_3$ formation at the interface. The absence of such lines in the Raman spectrum of all our samples indicates little, if any, In$_2$Te$_3$ formation at the epilayer-substrate interface, thus supporting the electron microscopy results mentioned above. The Raman scattering studies, however, indicate that the 185 cm^{-1} line arises from some other chemical species or some other effect at the CdTe/InSb interface. The origin of this line is under investigation.

In conclusion, it has been demonstrated that low-energy bias sputtering is capable of producing epitaxial and chemically abrupt CdTe/InSb heterostructures without the formation of an In$_2$Te$_3$ interfacial layer. The results presented indicate that growth conditions could be further optimized probably towards lower growth temperatures and higher substrate bias during ion cleaning. The electronic quality of the heterostructures need to be assessed by transport measurements. Such measurements are currently being undertaken.

ACKNOWLEDGEMENT

The authors wish to thank H. Labbe for technical assistance and Marie-C. Leonard for help with manuscript preparation.

REFERENCES

1.   Y-D. Zheng, Y.H. Chang, B.D. McCombe, R.F.C. Farrow, T. Temofonte, and F. Shirland, Appl. Phys. Lett. 49, 1187 (1986).

2.   M. Alikacem, M.L. Leadbeater, D.K. Maude, M. Davies, L. Eaves, M. Heath, L. Dmowski, J.C. Portal, D. Ashenford, and B. Lunn, Surf. Sci. 229, 428 (1990).

3.   T.D. Golding, S.K. Greene, M. Pepper, J.H. Dinan, A.G. Cullis, G.M. Williams, and C.R. Whitehouse, Semicond. Sci. Technol. 5, S311, (1990).

4.   R.F.C. Farrow, S. Wood, J.C. Greggi, Jr., W.J. Takei, F.A. Shirland, and J. Furneaux, J. Vac. Sci. Technol. B 3, 681 (1985).

5.   N.G. Chew, A.G. Cullis, and G.M. Williams, Appl. Phys. Lett. 45, 1090 (1984).

6.   G.M. Williams, C.R. Whitehouse, N.G. Chew, G.W. Blackmore, and A.G. Cullis, J. Vac. Sci. Technol. B 3, 704 (1985).

7.   J.L. Glenn, Jr., O. Sungki, L.A. Kolodziejski, R.L. Gunshor,

M. Kobayashi, D. Li, N. Otsuka, M. Haggerott, N. Pelekanos, and A.V. Nurmikko, J. Vac. Sci. Technol. B 7, 249 (1989).

8. D.R.T. Zahn, K.J. Mackey, R.H. Williams, H. Munder, J. Geurts, and W. Richter, Appl. Phys. Lett. 50, 742 (1987).

9. D. Li, N. Otsuka, J. Qiu, J. Glenn Jr., M. Kobayashi, and R.L. Gunshor, Mat. Res. Soc. Symp. Proc. 161, 127 (1990).

10. S.R. Das, J.P. McCaffrey, J.G. Cook, and J.B. Webb, Semicond. Sci. Technol. 5, S315 (1990).

11. D.J. Lockwood, M.W.C. Dharma-wardana, J.-M. Baribeau, and D.C. Houghton, Phys. Rev. B 35, 2243 (1987).

12. K.J. Mackey, D.R.T. Zahn, P.M.G. Allen, R.H. Williams, W. Richter, and R.S. Williams, J. Vac. Sci. Technol. B 5, 1233 (1987).

REACTIVE ION ETCHING OF GaSb, (Al,Ga)Sb, AND InAs FOR NOVEL DEVICE APPLICATIONS.

D.C. La Tulipe, D.J. Frank, and H. Munekata
IBM Research Division, T.J.Watson Research Center
P.O. Box 218, Yorktown Heights, NY. 10598

Abstract-Although a variety of novel device proposals for GaSb/(Al,Ga)Sb/InAs heterostructures have been made, relatively little is known about processing these materials. We have studied the reactive ion etching characteristics of GaSb, (Al,Ga)Sb, and InAs in both methane/ hydrogen and chlorine gas chemistries. At conditions similar to those reported elsewhere for RIE of InP and GaAs in CH_4/H_2, the etch rate of (Al,Ga)Sb was found to be near zero, while GaSb and InAs etched at $200\text{Å}/$minute. Under conditions where the etch mechanism is primarily physical sputtering, the three compounds etch at similar rates. Etching in Cl_2 was found to yield anisotropic profiles, with the etch rate of (Al,Ga)Sb increasing with Al mole fraction, while InAs remains unetched. Damage to an InAs "stop layer" was investigated by sheet resistance and mobility measurements. These etching techniques were used to fabricate a novel InAs-channel FET composed of these materials. Several scanning electron micrographs of etching results are shown along with preliminary electrical characteristics.

As III-V device geometries continue to shrink and material systems become more complex, dry etch processes must be used to define patterns and obtain material selectivity. Historically, III-V compounds have been dry etched using either pure Cl_2 or a gas which has a chlorine component.[1-6] More recently, a dry etch process which uses methane and hydrogen in what has been described as a reverse MOCVD process has gained popularity for its ability to etch indium containing III-V's as well as GaAs and AlGaAs. This was a necessary development since the evolution of III-V devices has progressed to include lattice matched In compounds on InP substrates. The next step in this evolutionary path might include heterostructure FET's based on GaSb, (Al,Ga)Sb, and InAs combinations. These devices would offer the high carrier velocities and charge carrying capabilities of InAs, coupled with very large conduction band offsets between InAs and the antimonides, which enhance charge confinement. Indeed, a number of recent device reports have made use of these materials.[7,8]

The purpose of this paper is twofold. First we discuss the RIE characteristics of GaSb, (Al,Ga)Sb, and InAs in pure Cl_2 and CH_4/H_2 gas chemistries. Second, we present electrical results of a new FET based on a combination of these materials, formed using the reported etching techniques.

Experimental Technique-Experiments were carried out in a Leybold Heraeus Z401 load-locked, planer diode etching system. A 13.56MHz rf generator was used to power the 20 cm diameter electrode. Typical base pressure of the main chamber is $1.5\text{x}10^{-7}$ Torr; a load lock pressure of at least $1\text{x}10^{-5}$ Torr is obtained before sample transfer. Samples were prepared using a variety of masking materials, including photoresist, PECVD silicon nitride, and sputtered tungston nitride. Patterns were defined in the two nitride materials using photoresist and standard fluorine based RIE processes. In all cases, the photoresist was left on the sample for the CH_4 and Cl_2 based etches to determine erosion characteristics of the polymer at different RIE conditions. A typical etch cycle would involve an initial one minute argon sputter oxide preclean at 60mT and a dc self-bias voltage of -250V, immediately followed by a two minute purge with Cl_2 or CH_4 gas mixture and a five minute etch in the same gas.

The semiconductor material consisted of nominally undoped samples of GaSb, (Al,Ga)Sb with several different Al mole fractions, and stacks of $GaSb/Al_{0.35}Ga_{0.65}Sb$ /InAs /$Al_{0.35}Ga_{0.65}Sb$ all of which were grown by MBE on (100) semi-insulating GaAs substrates. Bulk InAs substrates were also etched.

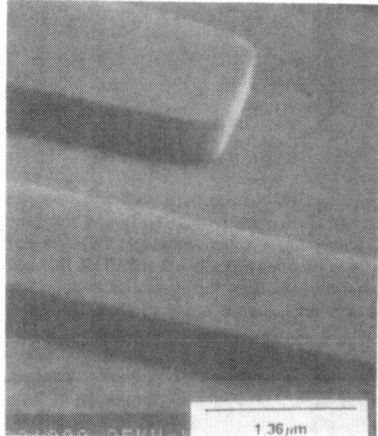 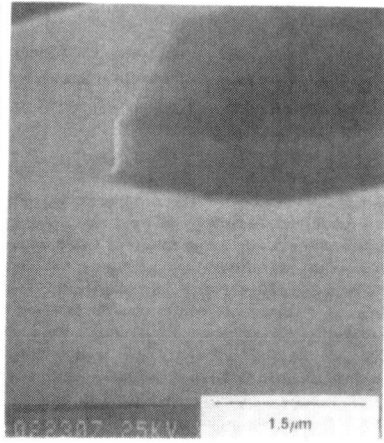

Figure 1a. Sidewall profile of GaAs etched in Cl$_2$. Photoresist mask has been removed.

Figure 1b. Sidewall profile of GaSb etched in Cl$_2$. Photoresist mask still in place.

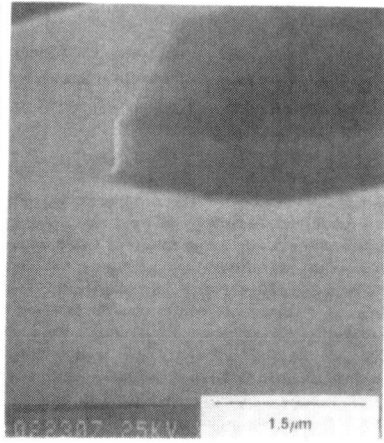

Figure 2. Etch rate of (Al,Ga)Sb at several Al mole fractions, in Cl$_2$.

Etching Results-Chlorine was used for the anisotropic etching of GaSb and (Al,Ga)Sb with Al mole fractions between 0.15 and 1.00. Samples were etched at 4mT, and 0.042 W/cm^2 (-170Vdc), a process characterized by Lee, et al. for GaAs in this vacuum system.[2] Figure 1 shows scanning electron micrographs of GaAs and GaSb etched with Cl$_2$ at these conditions. The etch rate of (Al,Ga)Sb with an Al mole fraction of 0 - 0.48 is plotted in figure 2. Etch rate was found to increase as a function of increasing Al concentration. In figure 3, the sidewall profile of a GaSb/Al$_{0.35}$Ga$_{0.65}$Sb bi-layer is shown. In this case a 300Å undoped InAs layer was used as an etch-stop. This layer was measured by Hall and Van der Paw at 300K and 77K to determine the amount of damage introduced by this process. Values of approximately 9000cm^2/V-sec. mobility and a sheet resistance of about 300Ω/□ have been measured on as-grown thin InAs layers, in this case, values of 9000 cm^2/V-sec. mobility, 210Ω/□ sheet resistance at 300K, and 12000 cm^2/V-sec. mobility, 200Ω/□ sheet resistance at 77K were measured. These results indicate that any damage

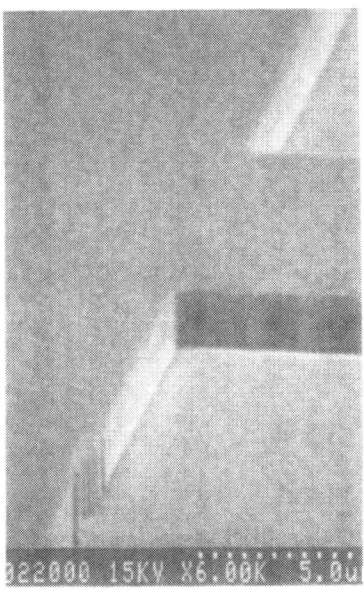

Figure 3. Sidewall profile of a GaSb/(Al,Ga)Sb/InAs heterostructure etched by Cl$_2$ RIE. Etching was stopped by the InAs etch-stop.

done to this InAs layer, chemical or physical, was not sufficient to degrade its ability to provide low resistance regions to contact a channel of an FET.

A CH$_4$ and H$_2$ process was also developed for the Z401 system based on processes reported in the literature.[9-12] Gas flows of 100sccm H$_2$, 30sccm CH$_4$, and 15sccm Ar and a process pressure of 60mT were used. A power of 0.15 W/cm^2 (-250Vdc) was found to yield anisotropic profiles in GaAs at an etch rate of 150Å/minute. These same conditions were used to study the etch characteristics of GaSb, (Al,Ga)Sb, AlSb, and InAs. The etch rate of both InAs and GaSb appeared to be equal and about 200Å/minute. Figure 4 shows SEM's of GaAs, GaSb and InAs etched with this process. All samples which contained Al at any concentration were not etched by this chemistry at these conditions. Selectivity of GaSb over (Al,Ga)Sb was found to be independent of surface oxide removal by etching (Al,Ga)Sb layers capped with 50Å of GaSb. Again, only the GaSb cap was etched. This type of selectivity has been reported for the GaAs/AlGaAs system and is explained in terms of the difference

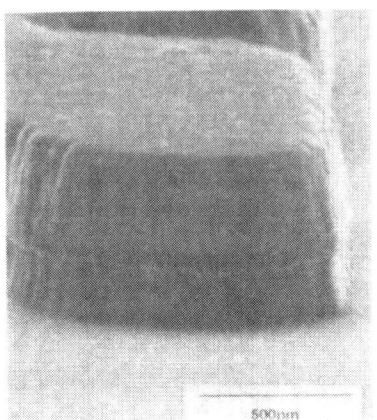

Figure 4a. Sidewall profile of GaAs etched in CH$_4$ + H$_2$ + Ar. WN$_x$ mask in place.

Figure 4b. Sidewall profile of GaSb etched in CH$_4$ + H$_2$ + Ar. WN$_x$ mask in place.

Figure 4c. Sidewall profile of InAs etched in $CH_4 + H_2 + Ar$. Masking material has been removed.

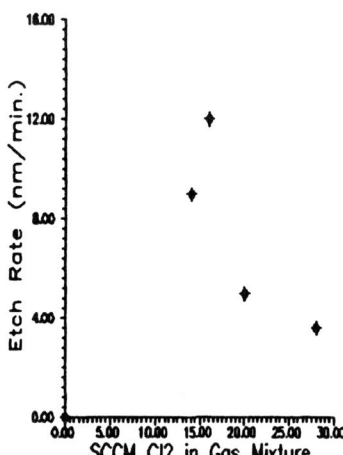

Figure 5. Etch rate of $Al_{0.28}Ga_{0.72}Sb$ as a function of Cl_2 concentration in $CH_4 + H_2$ etch mixture.

in the bonding energies; the heat of formation for AlAs is 35.8kJ/mole compared to 30.6kJ/mole for GaAs. This slows the etch rate to a value below the polymer deposition rate.[13] At conditions where the physical etch rate was large enough to overcome the polymer formation rate, equirate etching seems to be possible. Conditions at which a useful etch rate could be achieved were at a dc bias of approximately -450V for this RIE system. The etch rate due to physical sputtering by Ar alone was determined to be 50Å/minute out of the 200Å/minute rate measured with the reactive gases present. Etching GaSb/(Al,Ga)Sb/InAs devices at such high dc bias however, appears to present several disadvantages. Etch damage due to ion bombardment of the crystal lattice may be a serious problem, since the heterostructure may not withstand damage healing anneal cycles. Mask erosion and surface contamination due to sputtering of cathode material are additional problems.

Device fabrication steps such as mesa isolation require the development of an RIE process that is capable of etching both antimonides and InAs at equal rates. The CH_4 and H_2 process described above can etch InAs and GaSb, but when run at conditions where physical etch damage is minimized, does not etch (Al,Ga)Sb at several Al mole fractions from 22% to 100%. We have shown that Cl_2 can be used to etch the antimonides anisotropically at fairly high rates, but InAs is not. Gas mixtures composed of the standard CH_4, H_2, and Ar added with various amounts of Cl_2 were studied as a way of combining the advantages of the two processes. It can be seen from Figure 5 that the etch rate of $Al_{0.28}Ga_{0.72}Sb$ increases with Cl_2 concentration until reaching a maximum at 16sccm in the mixture, it then drops back towards zero as the fraction of Cl_2 increases. Etching is limited by the formation of polymer at higher Cl_2 concentrations. Thus it is possible to etch GaSb/(Al,Ga)Sb/InAs at approximately equal rates. In the past, others have reported using a mix of the two processes to etch both GaAs and InP as a way of realizing better surface morphology and increased etch rates.[14,15]

Device Fabrication-The etching processes described above were used to fabricate a new FET based on the antimonide/arsenide system. Unlike other heterostructure III-V FETs, this device is novel in that no insulating dielectric is needed between the gate and the active channel, thus its name, ZFET (Zero-insulator-thickness FET). The idea for such a FET was first proposed by Solomon in the (Al,In)As/InP system,[16] and has recently been proposed for the present system by Luo, et al..[8] In our device, a p⁺ (Al,Ga)Sb gate layer is placed in direct contact with the undoped InAs channel layer. The carriers remain separated because of the staggered bandgap alignment of (Al,Ga)Sb and InAs.[17] The holes from the gate layer cannot flow into the channel because they are blocked by the InAs valence band offset, while the electrons in the InAs channel cannot enter the (Al,Ga)Sb gate because of the conduction band offset. The threshold voltage of this FET can be adjusted by varying the Al mole fraction, but there must be enough Al to keep the heterojunction out of the broken band gap alignment regime since such a junction would not be able to adequately block gate-to-channel current flow. The motivation for making a FET in which the gate insulator has been scaled all the way to zero is that it should enable one to obtain extremely high transconductance, since the ability of the gate to modulate the channel has been maximized. More details about the operation of the ZFET can be found in Refs. 16 and 18.

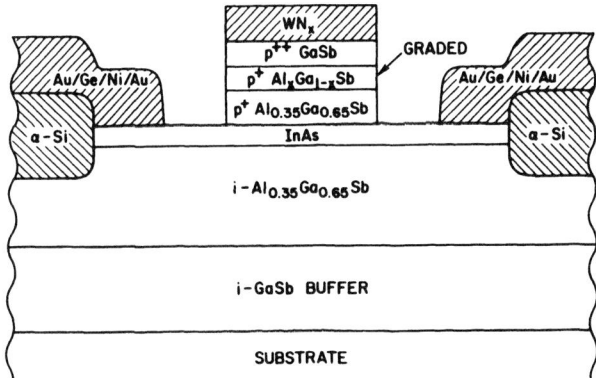

Figure 6. Cross sectional diagram of the GaSb/(Al,Ga)Sb/InAs ZFET showing epitaxial layer structure as well as process related structure.

A typical device structure was grown by MBE consisting of the layer structure shown in figure 6. A 1.5um nominally undoped GaSb buffer was grown on (100) semi-insulating GaAs followed by a 2200Å undoped layer of (Al,Ga)Sb and a 300Å layer of undoped InAs. Gate layers include a 50Å thick undoped (Al,Ga)Sb spacer layer followed by a 450Å (Al,Ga)Sb layer doped p⁺; with Be, graded to p⁺ GaSb over 200Å and finally 200Å of p⁺ GaSb. An Al mole fraction of 35% was used throughout the structure. Typically, buffer conduction due to background p-type impurities is a problem for this material system, so a process was developed in which a dielectric could be deposited in a self aligned manner to isolation mesas. These mesas were etched with a photoresist stencil using the $CH_4/H_2/Cl_2$ process described earlier to a depth of approximately 1500Å. A 1500Å film of α-Si was then deposited by e-beam evaporation to refill the etched areas with dielectric. Figure 7 shows an SEM of the isolation mesa patterned and etched to a photoresist lift-off stencil prior to Si deposition. After the Si film is lifted off to expose the GaSb mesa surface, a blanket film of WN_x was sputter deposited on the sample and patterned with photoresist to define the gate. The metal film was anisitropically etched with CF_4 and O_2, and the resist stripped. The Cl_2 process described earlier was used to etch the semiconductor gate, using the WN_x as a mask, and stop on InAs. Source/drain contacts and device interconnects were patterned and Au/12%Ge,

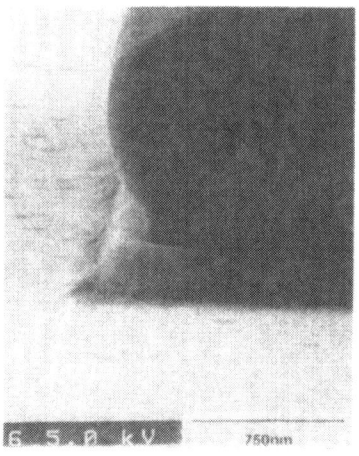

Figure 7. SEM showing the formation of a device mesa structure using the $CH_4 + H_2 + Cl_2$ process described earlier. The structure is ready for α-Si deposition.

Ni, Au metallurgy was evaporated on to the wafer and lifted off. Figure 8 shows an SEM of the finished device structure.

FET performance was measured at 77K and a maximum transconductance, g_m, of 500 mS/mm was found for an FET with a 1μm gate length. The device exhibited both current and voltage gain in the region where g_m reached its maximum. Figure 9 shows an I-V curve and a plot of g_m and current gain versus V_g for this particular device. InAs sheet resistance was found to be 200Ω/□, and a S/D metal to InAs contact resistance of 0.2 Ω-mm was measured. An unexpectedly large gate voltage was required to modulate the transistor. We have concluded that this is due to high contact resistance between the metal gate contact and the p⁺ GaSb.[18] We believe that with further process modifications, transconductances exceeding 1000 mS/mm should be attainable.

Figure 8. SEM showing the finished 1μm gate length ZFET, the performance of which is reported here.

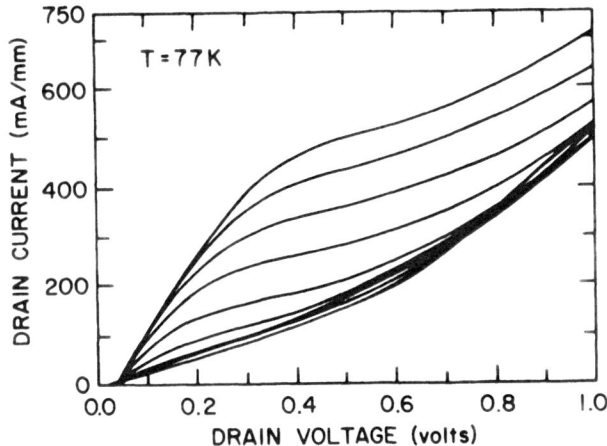

Figure 9a. Drain I-V curves for a 1 μm gate length device. The gate voltage steps from 0.3 V to 1.8 V in 0.15 V steps.

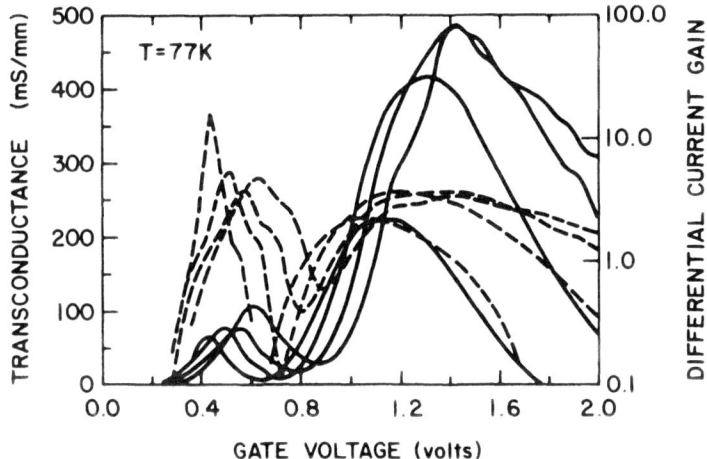

Figure 9b. Transconductance (solid lines) and differential current gain (dashed lines) versus gate voltage for the same device. The drain voltage steps from 0.15 V to 0.75 V in 0.15 V steps.

Conclusion-In conclusion, we have investigated dry plasma etching of GaSb, (Al,Ga)Sb, and InAs in Cl_2 and CH_4/H_2 gas mixtures. We have found that both GaSb and (Al,Ga)Sb etch in pure Cl_2 at conditions where etch damage is kept to a minimum, yet anisotropic etch profiles are maintained. Selectivity of the antimonides over InAs was also demonstrated. CH_4 and H_2 were used to selectively etch InAs and GaSb relative to (Al,Ga)Sb at several Al mole fractions. A $CH_4/H_2/Cl_2$ process was described, we believe for the first time, to achieve relatively equirate etching of both the antimonides and InAs. The etch processes were used to produce a p^+ (Al,Ga)Sb / n^- InAs heterojunction FET in which the separation of the channel from the gate can be reduced to zero. This FET exhibited an extrinsic transconductance of 500mS/mm at 77K for a gate length of 1μm.

Acknowledgements We would like to acknowledge P. Solomon and J. Magerlein for helpful discussions, and C. Knoedler and A. Ginzberg for preliminary etch studies.

References

1) E.L. Hu, R.E. Howard, J. Vac. Sci. Technol.B 2(1), Jan.-Mar. 1984.
2) B.S. Lee, H. Baratte, J.Electrochem. Soc., Vol. 137, No. 3, March 1990.
3) G.A. Vawter, L.A. Coldren, J.L. Merz, and E.L. Hu, Appl. Phys. Lett., 51, 719, 1987.
4) V.M. Donnelly, D.L. Flamm, and D.E. Ibbotson, J. Vac. Sci. Tech., A 1(2), Apr-June 1983.
5) S.W. Pang, J. Vac. Sci. Tech., 133, 784, 1986.
6) V.M. Donnelly, D.L. Flamm, C.W. Tu, and D.E. Ibbotson,J. Electrochm. Soc.: Solid State Science and Tech., vol.29, no.11, Nov. 1982.
7) K. Yoh, T. Moriuchi, and M. Inoue, 48th Annual Device Research Conference, June 1990.
8) L.F. Luo, R. Beresford, W.I. Wang, and H. Munekata, Appl. Phys. Lett., 55, 789,1989.
9) R. Cheung, S. Thoms, S.P. Beamont G. Doughty, V. Law, C.D.W. Wilkinson, Electronics Letters, Vol. 23, No. 16, 30 July 1987.
10) L. Henry, C. Vaudry, P. Granjoux, Electron. Letters, Vol 23, 19 Nov. 1987.
11) T.R. Hayes, M.A. Dreisbach, P.M. Thomas, W.C. Dautremont-Smith, and L.A. Heimbrook, J. Vac. Sci. Tech., B 7(5), Sep.-Oct. 1989.
12) T.R. Hayes, U.K. Chakrabarti, F.A. Baiocchi, A.B. Emerson, H.S. Luuftman, and W.C. Dautremont-Smith, J. Appl. Phys., 68(2), 15 July 1990.
13) V.J. Law, G.A.C. Jones, D.A. Ritchie, D.C. Peacock, and J.E.F. Frost, J. Vac. Sci. Tech. B, 7(6), Nov.-Dec. 1989.
14) G.J. Van Gurp, J.M. Jacobs, Phillips Journal of Research, Vol. 44, Nos. 2/3, 1989.
15) N. Vodjdani, P. Parrens, J. Vac. Sci. Tech., 5(6), Nov.-Dec. 1987.
16) P. M. Solomon, 'Staggered Bandgap Gate Field Effect Transistor,' U.S. Patent 4962409, Oct. 9, 1990.
17)H. Munekata, T.P. Smith,III, and L.L. Chang, J. Vac. Sci. Tech. B 7(2), Mar.-April 1989.
18) D.J. Frank, D.C. La Tulipe, H.Munekata, 'Novel InAs/(Al,Ga)Sb FET with Direct Gate-to-Channel Contact,' to be published.

ION IMPLANTATION IN InSb GROWN ON GaAs

M. V. RAO[†], R. ECHARD[†], P. E. THOMPSON[‡], A. K. BERRY[†],
S. MULPURI[†], and F. G. MOORE[*]
†Department of Electrical & Computer Engineering,
George Mason University, Fairfax, VA 22030.
‡Naval Research Laboratory, Washington, D.C. 20375.
*NRC/NRL Associate, Washington, D.C. 20375.

ABSTRACT

Be, S, Si, and Ne implantations were performed at room temperature into InSb layers grown on undoped semi-insulating GaAs substrates. The implant damage in InSb is of n-type behavior. The implanted material was subjected to both isochronal and isothermal annealing schemes using a molybdenum strip heater. A maximum p-type activation of 90 % and n-type activation of 16 % was achieved for Be and S implants, respectively. Si implant has an amphoteric doping behavior.

INTRODUCTION

InSb is an attractive material for long-wavelength photodetector application. Due to its low bandgap, to obtain interdevice isolation InSb needs to be grown on semi-insulating (SI) substrate of a large bandgap material. Undoped SI GaAs substrates seems to be a suitable choice for this purpose. The InSb interdevice isolation can be obtained by mesa etching down to the GaAs substrate. Due to the maturity of the GaAs technology the electronic circuitry made using GaAs can be monolithically integrated with InSb detectors by using InSb grown on GaAs (InSb/GaAs here after). Recently there have been some reports on the MBE growth of good quality InSb layers on GaAs substrates [1,2]. Ion implantation is an attractive method to define selective area ohmic contacts and p-n junctions required for InSb devices. Though there are several reports on implantation doping of bulk InSb [3-9] to date there are no reports on ion implantation doping in InSb/GaAs. The ion implantation results in InSb/GaAs can be significantly different from those of bulk InSb due to 14.6 % lattice mismatch between InSb and GaAs. Due to this reason we have investigated ion implantation doping of Be, S, Si, and Ne into InSb/GaAs.

EXPERIMENT

InSb layers of 1 μm thick grown by MBE on (100) oriented undoped SI GaAs substrates were used in this study. A 250 nm thick undoped GaAs buffer layer was grown on the substrate before initiating the InSb growth. The typical liquid nitrogen temperature (LNT) resistivity, net carrier concentration, and mobility in these layers are 5×10^{-3} Ω-cm, 3×10^{16} cm^{-3}, and 40,000 cm^2/V s, respectively. Multiple energy Be, S, Si, and Ne ion implants were performed at room temperature with 7 ° tilt so that the implant atomic concentration is almost uniform over the entire thickness of the InSb layer. The highest energy of the implant was selected so that its $R_p + \Delta R_p$ value is roughly equal to the InSb layer thickness. This was done to make accurate dopant electrical activation measurements without parallel conduction from the high mobility undamaged InSb layer. All multiple energy implant schedules used for various impurity species in this study are given in Table I. To protect the InSb surface during annealing a 50 nm thick Si_3N_4 layer was deposited on InSb by plasma enhanced chemical vapor deposition at 100 °C. For additional protection during annealing the samples were placed face down on a silicon wafer. All anneals in this study were performed using a molybdenum strip heater. After annealing the samples were electrically characterized by using van der Pauw Hall technique at LNT. Some of the samples were also characterized by SIMS and x-ray rocking curve measurements to study the redistribution of the impurity profile and crystalline lattice quality of the material, respectively.

Table I. Multiple energy implant schedules used in this study

Species	Implant schedule (keV/cm^{-2})
Be	$190/1.15x10^{14}$, $70/2.5x10^{13}$, $40/1.8x10^{13}$, and $10/1x10^{13}$
S	$850/8.8x10^{13}$, $550/4x10^{13}$, $220/3x10^{13}$, $60/7x10^{12}$, and $10/1.5x10^{12}$
Ne	$480/6x10^{13}$, $200/2.4x10^{13}$, $70/1x10^{13}$, and $10/2.2x10^{12}$
Si	$580/4x10^{13}$, $300/1.5x10^{13}$, $140/1x10^{13}$, and $30/3x10^{12}$

RESULTS AND DISCUSSION

A sample of the multiple energy Be implanted material was subjected to 30 s duration isochronal anneals in the temperature range 200 - 500 °C. The as-implanted material is n-type with a sheet carrier concentration of $1.9x10^{14}$ cm^{-2}. The plots of measured sheet carrier concentration and mobility as a function of annealing temperature are shown in Fig. 1. The material remained n-type up to an annealing temperature of 250 °C, with a decrease in the sheet carrier concentration as the annealing temperature increases thereafter. This is due to annealing of the implant lattice damage with increasing temperature. The material has mixed type conduction after a 300 °C annealing. But after 350 °C anneal the material turned p-type with increasing electrical activation and mobility with an increase in the annealing temperature. The electrical activation and hole mobility measured after 500 °C anneal are 89 % and 570 cm^2/V s, respectively. We have also performed isothermal anneals on two different samples at 400 and 500 °C. In both samples the electrical activation of the dopant and hole mobility almost remained constant with increasing annealing time at a given temperature. This indicates that the electrical activation od Be in InSb depends primarily on the annealing temperature and is not a sensitive function of the annealing time. The electrical activation of Be measured in this study is higher than the maximum electrical activation (50 %) reported earlier[4,9] in bulk InSb.

Smoothed SIMS Be atomic density depth profiiles measured in a single energy 60 $keV/1x10^{14}$ cm^{-2} Be implanted material before and after annealing are shown in Fig.

2. No redistribution of the dopant was observed for 15 s annealing at either 400 or 500 °C. The x-ray rocking curves obtained at different stages of the implantation/annealing process are shown in Fig. 3. The full width at half maximum (FWHM) which gives an indication of the crystal quality of the material is given beside each rocking curve. There is an increase in FWHM after Be implantation indicating a lattice damage in the material. But after a 500 °C annealing for 15 s the FWHM is similar to the value in as-grown crystal indicating an almost complete lattice recovery.

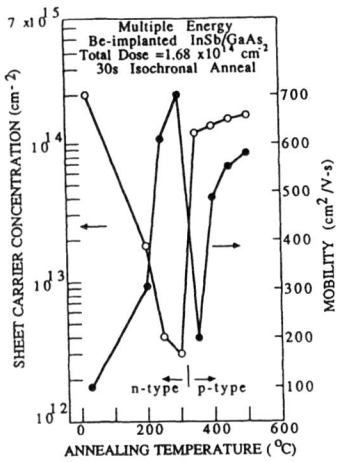

Fig. 1. Variation of sheet carrier concentration and hole mobility with annealing temperature for 30 s isochronal annealing on multiple energy Be implanted InSb/GaAs.

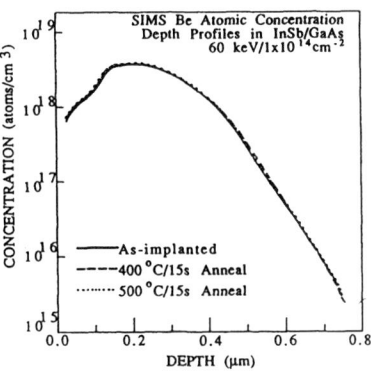

Fig. 2. SIMS Be atomic density depth profiles in 60 keV/1×10^{14} cm^{-2} Be implanted InSb/GaAs before and after annealing.

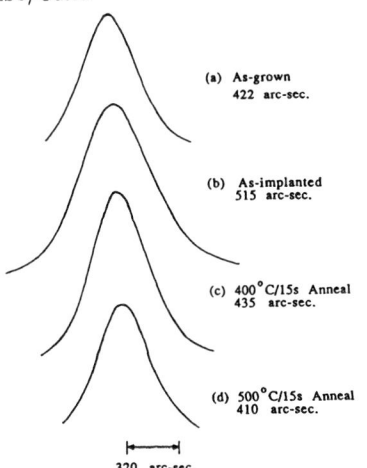

Fig. 3. X-ray rocking curves at various stages of Be implantation/annealing process in InSb/GaAs.

For comparison we have also performed multiple Be implants into a 0.6 μm thick InSb layer which also gives a uniform 2×10^{18} cm^{-3} Be concentration as the previously mentioned multiple energy Be implant in 1 μm thick InSb. The general trend in the variation of electrical dopant activation and carrier mobility in this sample is similar to what has been observed in Fig. 1 for 1 μm thick InSb layer. But the maximum dopant electrical activation and mobility measured in this layer are 58 % and 535 cm^2/V s, respectively, compared to 89 % and 600 cm^2/V s in the 1 μm thick layer. This reduced activation and mobility is believed to be due to the effect of more dislocations in the thinner layer.

The multiple energy S implantation schedule used in this study is given in Table I. The S as-implanted layer is n-type with a sheet carrier concentration of 4.7×10^{14} cm^{-2}. This value is more than what has been measured for Be due to a higher atomic mass of S compared to Be. A sample of this material was subjected to 30 s isochronal anneals. As in case of Be initially the damage related sheet carrier concentration decreased with increasing annealing temperature. Dopant activation is initiated at a temperature of > 450 °C. The results of isothermal anneals at 460 and 510 °C are shown in Fig. 4. For 460 °C anneal the electrical activation has increased with an increasing annealing time

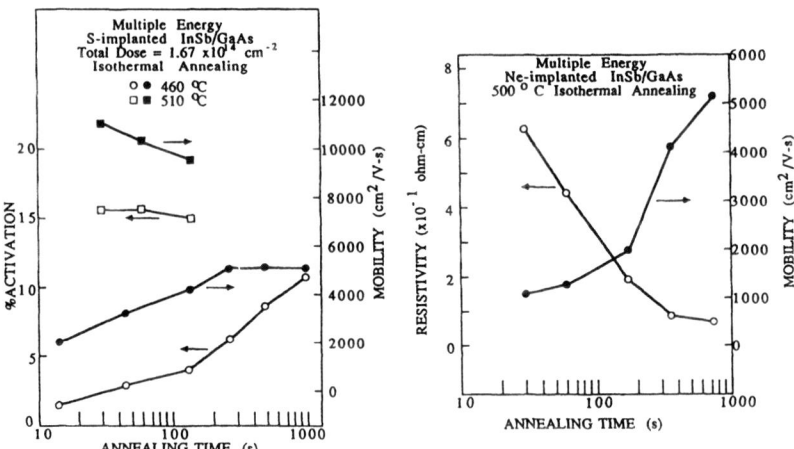

Fig. 4. Variation of percentage of dopant activation and carrier mobility with annealing time for 460 and 510 °C isothermal annealings on multiple energy S implanted InSb/GaAs.

Fig. 5. Variation of resistivity and carrier mobility with annealing time for 500 °C isothermal annealing on multiple energy Ne implanted InSb/GaAs.

giving a maximum activation of 10 % and a carrier mobility of 4,200 cm²/V s after a 16 min. total anneal time. Whereas for 510 °C anneal the maximum dopant activation and mobility are obtained after 30 s anneal and the values are 16 % and 11,200 cm²/V s, respectively. Surface morphology started deteriorating for longer duration anneals at 510 °C. For InSb the annealing temperature can not be increased above 510 °C due to its low melting temperature ($\approx$ 520 °C). The low S donor activation is believed to be due to a greater degree of lattice damage which can not be repaired satisfactorily even after 510 °C anneal. Higher activations probably can be achieved if the implant is performed at an elevated temperature instead of the room temperature used in this study.

To evaluate the effect of residual lattice damage left after annealing on the electrical properties of the InSb material we have performed multiple energy Ne implants using the schedule given in Table I. The as-implanted material has a resistivity of 7×10^{-2} Ω-cm. The variation of resistivity and electron mobility with annealing time is shown in Fig. 5 for 500 °C isothermal annealings. The sheet carrier (electron) concentration has remained almost constant at $\approx 1.5 \times 10^{12}$ cm^{-2} for all anneal times. The electron mobility of the material has increased from 1090 to 5200 cm²/V s for an increase in annealing time from 30 s to 12 min. causing a decrease in the resistivity. This is due to a reduction in the implant lattice damage with increasing annealing time. But a mobility of 5,200 cm²/V s is far less compared to the as-grown material mobility of 42,500 cm²/V s. This indicates a large degree of residual implant lattice damage left in InSb/GaAs even after 500 °C/12 min. annealing. Based on these results it can be stated that the poor dopant activation in S implanted InSb/GaAs is due to a large amount of residual implant lattice damage left in the material after annealing.

Multiple energy Si implantation schedule used in this study is given in Table I. Annealing was performed on these samples at 500 °C for various time durations. In all samples the material has p-type conduction with a sheet carrier concentration of 3×10^{13} cm^{-2} and carrier mobility of 550 cm²/V s. This indicates amphoteric nature of Si in InSb/GaAs. A similar nature has been reported earlier in the implantation doping studies in bulk InSb [7,8] and MBE growth studies of InSb [10-12]. In the growth studies it was found that the nature of the electrical activation of Si depends on the growth temperature showing a n-type behavior at low growth temperatures and a p-type behavior at high growth temperatures. Low annealing temperatures can not be used to activate Si implanted InSb due to a large degree of implant damage in the material as in the case of S.

CONCLUSIONS

Be, S, Si and Ne implantations were performed in 1 μm thick InSb layers grown on semi-insulating GaAs substrates. The as-implanted material is n-type for all implant species used in this study. An electrical activation of $\approx$ 90 % was observed for 15 s anneal at 500 °C on Be implanted material. No Be redistribution was observed after annealing. A maximum of only 16 % activation was observed in S implanted material due to a large degree of residual implant damage in the material. This was confirmed by performing Ne implants in InSb/GaAs. Si implant has an amphoteric conduction behavior resulting in a net acceptor behavior after 500 °C annealing.

ACKNOWLEDGMENTS

The authors thank John Davis, Larry Ardis, and Mark Goldenberg of NRL and P. Chi and D. S. Simons of NIST for their help during the course of this study.

REFERENCES

1. J. L. Davis and P. E. Thompson, Appl. Phys. Lett. 54, 2235 (1989).

2. J. E. Oh, P. K. Bhattacharya, Y. C. Chen, and S. Tsukamoto, J. Appl. Phys. 66, 3618 (1989).

3. P. J. McNally, Radiat. Effects 6, 149 (1970).

4. C. E. Hurwitz and J. P. Donnelly, Solid-state Electron. 18, 753 (1975).

5. K. H. Wiedeburg, H. Berz, and H. Kranz, Phys. Status Solidi A 31, K69 (1975).

6. M. I. Guseva, A. N. Mansurova, V. G. Tikhonov, and S. N. Khorvat, Sov. Phys. Semicond. 10, 872 (1976).

7. V. A. Bogotyrev and G. A. Kachurin, Sov. Phys. Semicond. 11, 798 (1977).

8. I. Fujisawa, Jpn. J. Appl. Phys. 19, 2137 (1980).

9. S. J. Pearton, S. Nakahara, A. R. Von Neida, K. T. Short, and L. J. Oster, J. Appl. Phys. 66, 1942 (1989).

10. S. D. Parker, R. L. Williams, R. Droopad, R. A. Stradling, K. W. J. Barnham, S. N. Holmes, J. Laverty, C. C. Phillips, E. Skuras, R. Thomas, X. Zhang, A. Staton-Bevon, and D. W. Pashey, Semicond. Sci. Technol. 4, 663 (1989).

11. R. L. Williams, E. Skuras, R. A. Stradling, R. Droopad, S. N. Holmes, and S. D. Parker, Semicond. Sci. Technol. 5, S 338 (1990).

12. V. M. Glazov and E. B. Smirnova, Sov. Phys. Semicond. 17, 1177 (1983).

THE GROWTH MECHANISMS OF GaSb EPITAXIAL FILM BY MOCVD

H. Y. UENG AND S. M. ChEN
Dept. of Electrical engineering, National
Sun Yat-sen University, Kaohsiung, Taiwain. R. O. C.
Y. K. SU AND F. S. JUANG
Dept. of Electrical engineering, National
Cheng kung University, Tainan, Taiwain. R. O. C.

ABSTRACT

The growth reaction mechanism was experimentally investigated for TEGa and TMSb used for MOCVD GaSb epitaxial growth. The variations of growth rate, substrate temperature , the V/III ratio and the mole-fraction of III and V source gases, were detailly considered and investigated with regard to their effect on electrical and optical properties were measured during the epitaxial process. The experimental results were used to provide an inspection on defects interaction mechanisms and on the type of conductivity. Based on the investigation of growth mechanism, at $_2$ 470°, a high quality GaSb epitaxial film with mobility, 634 cm^2/V*sec and low hole concentration, 1.67×10^{16} cm^{-3} was obtained.

INTRODUCTION

The undoped GaSb grown by a variety of techniques : MBE, MOVPE, LPE, were p-type. Typical free carrier concentrations were in the range 10^{16}-10^{17} cm^{-3} with room temperature mobility of 400-1000 cm^2/V.s. The room temperature mobility for MOVPE grown material is still lower at growth temperature below 550° [4] since the lack of microscopic understanding of the growth mechanisms were used to control the MOCVD process. How to grow high quality GaSb film at lower temperature, a detailed study is necessary to classify the mechanism of GaSb epitaxial process.

In this paper, the GaSb epitaxial film were grown over the range of 470-560 ^{O}K. The growth mechanisms of TMGa and TMSb on GaAs and GaSb will be described. The experimental results were also used to provide an inspection on the intrinsic defects interaction mechanisms and on the type of conductivity.

EXPERIMENTAL

A MOCVD horizontial reactor and the substrates included GaSb (100), GaAs (100) tilt 2^O toward (100) and GaAs (100) were used for the epitaxial growth. Starting materials were triethylgallium (TEGa) and trimethylantimonide (TMSb).

GROWTH MECHANISMS OF GaSb EPITAXIAL FILM

The Growth Temperature

From Fig. 1 , High growth temperature would provide the activation energy for decomposition reaction and surface migration. Region I, below 530°C ,shows that the reaction is surface reaction control process for a fixed TEGa flow rate. It reduces the conversion rate from TEGa to DEGa, the second-order recombination of DEGa with one ethyl radical will form TEGa, therefore, the desorption of TEGa becomes important [4,5]. In region II, it shown that above 530°C, GaSb growth rate was independent on the constant TEGa flow rate. Hence, the reaction process could be the mass transfer control process, and the conversion of the adsorbed TEGa to DEGa was very fast and to approach unity efficiency [4]. For the region III, the fall-off

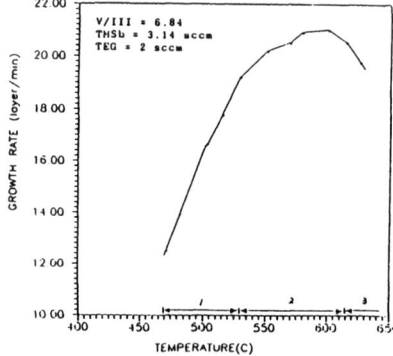

Fig. 1 GaAs growth rate versus growth temperature.

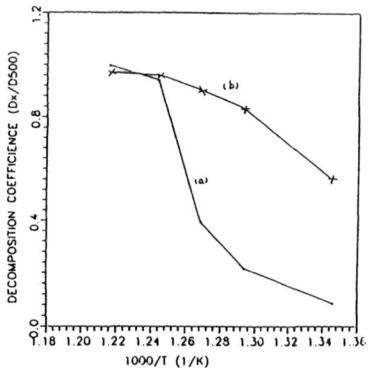

Fig. 2 Decomposition coefficient of TMSb and TEGa versus growth temperature. (a) TMSb (b) TEGa.

in GaSb growth rate with increasing temperature (> 530°C) is then dominated by the DEGa desorption. In this temperature range, a linear relation between growth rate and TEGa flow rate has different slope at different growth temperature.

Mole-Fraction of V and III Sources

Figs.2 and 3 show a linear dependence betwen growth rate and molefraction of TEGa at constant temperature and the growth rate is independent on molefraction of TMSb. Based on the experimental results, the characteristic of TEGa is more important than TMSb in GaSb growth mechanism above the mediate temperature range. Thus, TMSb is still a factor of film quality at low temperature.

The mole fraction of V and III gas source not only affect GaSb growth rate, but also affect the quality of GaSb. In Fig.1, The growth rate of GaSb/GaSb, 16.4 layers / min (GaAs (100), 500°C) is larger than the GaSb/GaAs, 22 layers / min (GaSb (100), 500°C). The hetero-epitaxial GaAs/GaSb layer has larger mismatch of lattice constant(7.74%) and smaller growth rate.

V and III Gas Sources

The dependence of growth rate on temperature could be corresponding to the TEGa decomposition on temperature. The decomposition coefficient couldn't be analyzed since the growth pressure of MOCVD system was about 100 torr. Hence, it was replaced by the normalizayion ratio, D_x/D_{550} and d_x/d_{550}, an alternative method were developed for analysis.

In Fig.4 line-a shows that the conversion of adsorbed TEGa to DEGa is rapid and approaches unity efficiency at temperature above 530°C [4]. However, the desorption of DEGa would reduce the growth rate as the growth temperature decreased from 530°C to 470°C, and the dominant factors of the decreased growth rate could summurized as (1) the conversion from TEGa to DEGa decreased (2) increasing recombination of DEGa, (3) the migration of TEGa related species to proper sites decreased gradually.

In Fig.4 line-b shows that temperature above 530°C the incorporation between TMSb with proper Ga sites varied slowly with growth temperature, Thus, the normalization ratio of TMSb is only a few change. As growth temperature decreased from 530°C to 470°C, the normalization ratio of TMSb decreased abruptly.

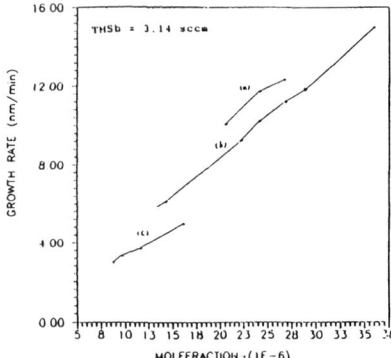

Fig. 3 GaSb growth rate as function TEGa molefraction (flow rate) for different growth temperature and a fixed TMSb flow (3.14 sccm) (a) 470 C (b) 500 C (c) 550 C.

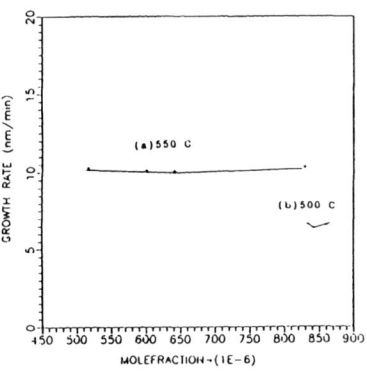

Fig. 4 GaSb growth rate as function TMSb molefraction (flow rate) for different growth temperature and a fixed TEG flow (2 sccm).

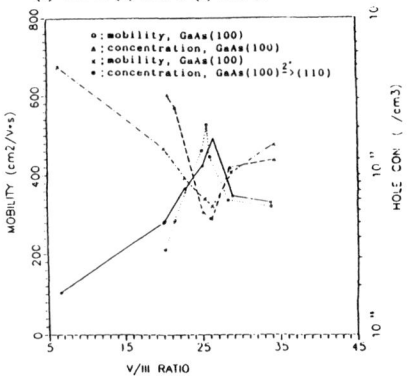

Fig. 5 Hole mobility and hole concentration of GaSb/GaAs and hole concentration as a function of V/III ratio at 550 C.

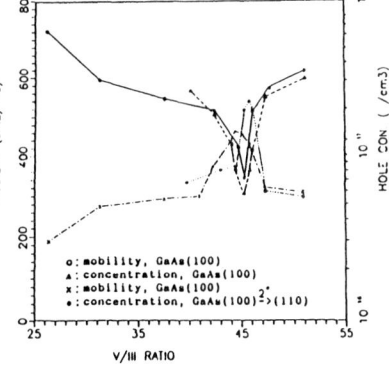

Fig. 6 Hole mobility and hole concentration of GaSb/GaAs and hole concentration as a function of V/III ratio at 470 C.

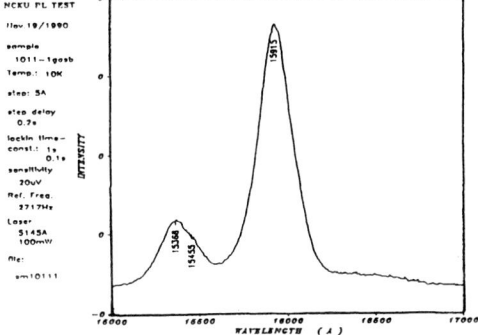

The V/III Ratio

From Fig.5, the highest mobility incorporated with the lowest hole concentration for GaAs (100) substrate and GaAs (100) tilted 2^O toward (110) substrate at growth temperature, $500^O C$. For growth temperature is $470^O C$, in Fig. 6, the same relationship were obtained. It could be concluded that higher mobility and lower hole concentration would be obtained at lower temperature growth. Hence, the carrier concentration and mobility could be obtained a satisifactory values as [h'], $< 1 \times 10^{15}$ cm^{-3} and mobility, > 800 $cm^{-2}/V.s$ for lower temperature $,< 450$ $^O C$ (at optimum V/III ratio) would be possible.

DEFECTS INTERACTION MECHANISM

The as-grown GaSb epitaxial films were p-type and nonstoichiometry. The possible intrinsic defects in GaSb film were basically the Ga interstitial, Sb vacancy and Ga_{Sb} anti-site defects [6]. The dominant defect reactions could be express as

(1) $V_{Sb} + Ga_{Ga} ----> Ga_{Sb} + V_{Ga}$
(2) $Ga_{Sb} + V_{Ga} ----> Ga_{Sb}V_{Ga}$

Hence, the dominant defect could be $Ga_{Sb}V_{Ga}$. From the PL data (Fig. 7), the peak A lie 0.8062 eV could be corresponding to the direct band gap transition, and peak B lie 0.0277 eV above the valence band. The intensity of peak B was high than peak A could be an acceptor level introduced by the complex defect, $Ga_{Sb}V_{Ga}$ during the epitaxial process.

From Fig.6, the mobility and carrier concentration shown an abruptly change near the optimized V/III ratio. It could be revealed that the composition of epitaxial film would approach the stoichiometry at the lower growth temperature. The concentrations of intrinsic defects would reduce to lower values, therefore, the carrier concentration decreased and mobility incseased because of reducing the scattering centers.

REFERENCES

1. A. Saraki, M. Nishiuma and Y. Takeda, Jap. J. Appl. Phys. 19, 1695(1980).
2. A. K. Srivastaua, J. C. Dewinter, C. Caneau, H. A. Pollack and J. L. Zyshind, Appl. Phys. Lett. 48, 903(1986).
3. O. Hildebrand, W. Kuebart, K. W. Benz, M. H. Pilkuhn, IEEE J. Quantum Electron., QE-17(2), 284(1981).
4. S. K. Haywood, N. J. Mason and P. J. Walker, J. of Crystal Growth 93, 56-61 (1988).
5. T.H. Chiu and J.E.Cunnigham,J.of Crystal Growth,95,136 (1989)
6. G. Edelin and D. Mahot, Phil. Mag. B. 42, 95(1980).

UNDOPED GaSb GROWTH by MOCVD

Yan Kuin Su and Fuh Shyang Juang
Department of Electrical Engineering, National Cheng Kung University, Tainan, Taiwan, R.O.C.

ABSTRACT

Undoped GaSb epilayers have been grown on (100) GaSb and S.I. GaAs substrates. The effects of growth temperatures and TMSb/TEGa mole fraction ratios on the epitaxial properties of surface morphology, growth rate, hole concentration and mobility (300K and 77K) have been studied. The lowest concentration 1.8×10^{16} cm^{-3} (77K) and the highest mobility 1447 $cm^2/V.s$ (77K) can be obtained under V/III ratio of 6.64 at 550°C. Photoluminescence intensity was found to be a function of the V/III ratios. When V/III ratios increased or decreased beyond 6-8, the BE peaks disappeared and PL spectra became roughened.
To reduce the effects of large lattice-mismatch in highly strained GaSb/GaAs system (7% mismatch) on the electrical properties, a 10-period $In_{0.3}Ga_{0.7}As/GaAs$ (60A/40A) strain layer superlattice (SLS) has been grown on GaAs substrates as a dislocation filter before the GaSb epitaxial growth. From the comparison of 77K Hall mobility of GaSb/GaAs as a function of growth temperature with that of GaSb/SLS/GaAs, it was clearly observed that the epilayers grown on SLS structures have higher mobility than those grown directly on GaAs substrates. From the TEM analysis, we observed that all dislocations propagated up to the GaSb epilayer surface in the GaSb/GaAs system but some of the dislocations bending before reaching the epilayer surface in the GaSb/SLS/GaAs system.

INTRODUCTION

Gallium-antimonide-based compound semiconductors have received increasing attention recently because the direct bandgap of their alloys corresponds to wavelengths over a wide spectral range from 1.24um (AlGaAsSb)[1]to 4.3um (InGaAsSb)[2-4]. There are two GaSb-based alloys: AlGaAsSb and InGaAsSb. The wavelength of the $Al_xGa_{1-x}As_ySb_{1-y}$ alloys system is between 1.24 and 1.72um which covers the low-loss and low dispersion region for optical fiber communication in silica fibers[5]. So, the AlGaAsSb/GaSb system is a promising alternative candidate for applications in long wavelength lasers and photodetectors besides the InGaAsP/InP system. Extremely low loss fluoride glass fibers may have minimum losses in the 2-4 um wavelength region, 1-2 orders of magnitude lower than that at 1.55 um of the best silica fibers. This can be reached with a suitable choice in the composition of the $In_xGa_{1-x}As_ySb_{1-y}$ alloy system whose emission wavelength would cover the range from 1.7 to 4.3um[3,6,7]. GaSb epitaxial layers have usually been grown by liquid phase epitaxy[8-14] or MBE[15] with only a few reports in the literature of growth by MOCVD[16-23].

EXPERIMENTAL

A self-setup MOCVD system with a horizontal reactor was used for the epitaxial growth. Starting materials were triethylgallium (TEGa, purchased from Alfa Co.) and trimethylantimonide (TMSb, purchased from Yamanaka Co.). The TEGa and TMSb bubblers were maintained at 10 and $-19.5^{\circ}C$, respectively. Pd-diffused hydrogen was used as the carrier gas. The gallium and antimonide alkyls were mixed in a gas manifold (from Thomas Swan Co.). A 10-period $In_{0.3}Ga_{0.7}As/GaAs$ strain layer superlattice was first grown on (100) S.I. GaAs substrates in another MOCVD system. The GaSb epilayers were grown directly on S.I. GaAs or subsequently on the superlattice buffer layer. The TMSb flux was first switched into the reactor at $350^{\circ}C$ to protect the substrates from decomposition before growth[19,20].

The growth was carried out at low pressure 100 torr. The total gas flow rate was 2 SLM. The flow rates of carrier gas through the TEGa and TMSb bubblers were 1-3 and 3.14 sccm, respectively. The V/III ratios were varied between 6 and 7, by altering the alkyl flow, to obtain good quality of epilayers. The growth temperature examined was $550-635^{\circ}C$. Growth time was 1-4 hour with typical growth rate about 1 um/hr.

RESULTS and DISCUSSIONS

Fig.1 shows the dependence of 300 and 77K mobility upon growth temperature between 520 and $635^{\circ}C$ for undoped GaSb grown at 100 torr, V/III ratio of 6.84. Between 520 and $600^{\circ}C$, the mobility is seen to increase with decreasing growth temperature. The highest 77K mobility of the material is 1029 $cm^2/V.s$ for the epilayer grown at $520^{\circ}C$. But when growth temperature is higher than $600^{\circ}C$, the 77K mobility decreased rapidly falling to a very low value at $635^{\circ}C$. From the relationship between electrical properties and growth temperatures, it is indicated that optimum growth temperature is at the low end of the range between 520 and $635^{\circ}C$. The highest mobility (1029 $cm^2/V.s,77K$) and the lowest carrier concentration ($1.4x10^{16}$ cm^{-3}, 77K) for epilayers grown under V/III=6.84 were measured from the sample grown at the lowest substrate temperature $520^{\circ}C$.

The mobilities are also critically dependent on the V/III ratios. The mobility (measured at 300 and 77K) as a function of V/III ratios are presented in Fig.2 for GaSb epilayers grown at $550^{\circ}C$, 100 torr. From Fig.2, it is found that the highest mobility (1447 $cm^2/V.s$, 77K) for $550^{\circ}C$ grown epilayers was obtained under V/III growth ratio of 6.64. As V/III ratio increased above 6.64 or decreased below 6.64, the mobility decreased. The hole concentration and mobility deteriorated more drastically when the V/III ratios decrease below 6.64. This indicates that the optimum value of V/III ratio for 550° grown GaSb epilayers with high electrical quality is about V/III=6.64.

Fig.3 shows the effects of different V/III ratios (3.92, 6.84, 8.06 and 10.25) on the PL spectra for the samples all grown at $600^{\circ}C$ and 100 torr. We observed that the bound-exciton BE peaks only appeared in the spectra measured on the sample grown under V/III=6.84. When V/III ratios increased to 8.06, the BE peaks decayed. When V/III ratios further increased to

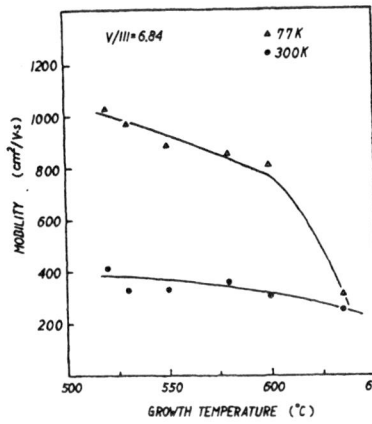

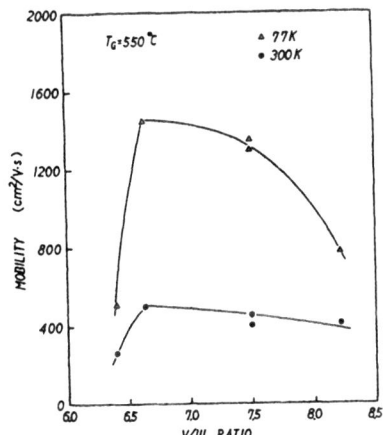

Figure 1. Dependence of 300 and 77K hole mobility of GaSb layers on growth temperature (100 torr, V/III=6.84).

Figure 2. Effects of V/III ratios on hole mobility for GaSb epilayers grown at 550°C, 100 torr.

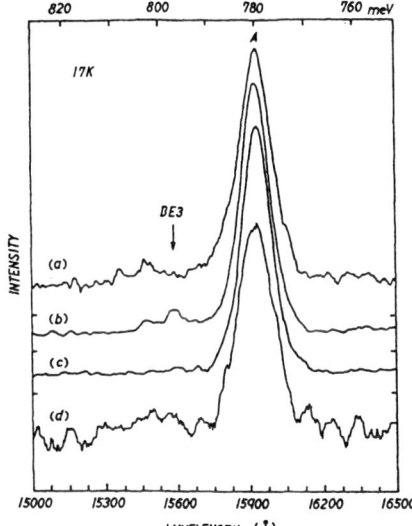

Figure 3. Change of PL spectra of homoepitaxial GaSb layers grown at 600° and 100 torr under various V/III ratios: (a)3.92, (b)6.84, (c)8.06 and (d)10.25 (excitation laser power:100mW).

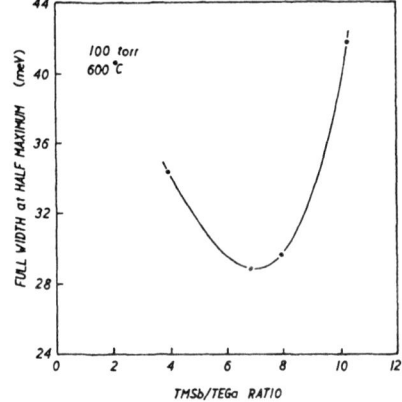

Figure 4. FWHM of the acceptor band (A) in PL spectra as a function of V/III ratios.

10.25 or decreased to 3.92, the BE peaks disappeared and the PL spectra became roughened. Fig.4 shows the dependence of full width at half maximum (FWHM) of acceptor-band peaks on V/III ratios. The FWHM value is smaller for V/III in the range of 6-8 and increased abruptly when V/III>10. This revealed that the good quality epilayers can be obtained at V/III ratios about

6.84, ensuring that obtained from Fig.2, and the quality deteriorated seriously at V/III>10.

The crystallinity of GaSb epilayers grown on GaAs and SLS/GaAs substrates was examined and the lattice mismatch between GaSb epilayers and GaAs substrates was measured by X-ray diffraction. Fig.5 shows the X-ray diffraction patterns of GaSb epilayers grown on GaAs substrates at various growth temperatures (530, 550, 580 and 635°C. At low growth temperature of 530°C, the sharp symmetrical diffraction peaks (Kα_1 and Kα_2) of both GaSb and GaAs are observed. The Bragg angle 2θ of GaSb and GaAs are 60.63 and 65.99°, respectively, for (400) plane. The latice mismatch between GaSb and GaAs was calculated to be 7.74%. As shown in Fig.5, the Kα_1 and Kα_2 peaks of GaSb epilayers combined and deteriorated when the growth temperature exceeded 580°C. This indicates that good quality single crystal

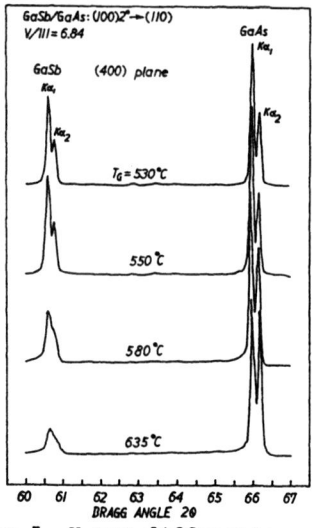

Figure 5. X-ray diffraction patterns for GaSb/GaAs heteroepitaxial samples grown at different temperatures: 530, 550, 580 and 635°C, respectively.

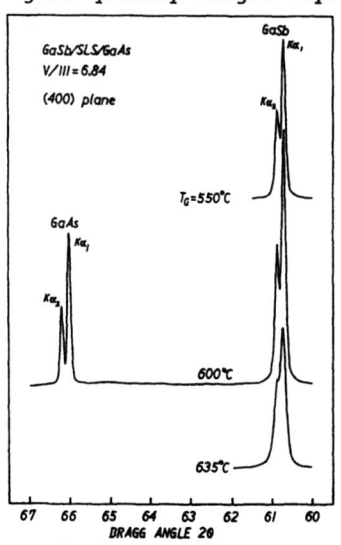

Figure 6. X-ray diffraction patterns for GaSb/SLS/GaAs heteroepitaxial samples grown at different temperatures: 550, 600 and 635°C, respectively, under V/III=6.84.

layers can be obtained at the low end of the range between 530 and 635°C. Fig.6 shows the X-ray diffraction patterns of GaSb epilayers grown on SLS/GaAs substrates at various growth temperatures (550, 600 and 635°C). At growth temperature lower than 600°C, the sharp symmetrical diffraction peaks (Kα_1 and Kα_2) of both GaSb and GaAs are observed. When the growth temperature exceeded 635°C, the Kα_1 and Kα_2 peaks of GaSb epilayers combined and deteriorated, as shown in Fig.6. From Figs.5 and 6, it is observed that the growth temperature used for epitaxial growth on SLS/GaAs substrates can be higher than that grown directly on GaAs substrates.

Fig.7 show the comparisons of 300 and 77K hole mobility of GaSb epilayers grown on SLS/GaAs substrates with those grown

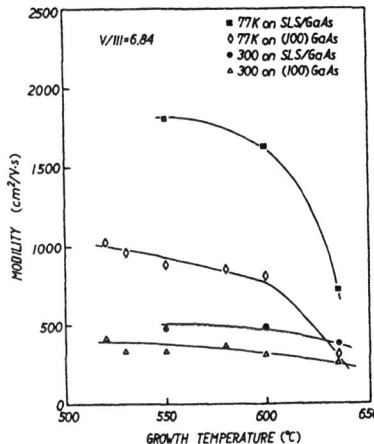

Figure 7. Comparison of 300 and 77K hole mobility of GaSb epilayers grown on SLS/GaAs substrates with those grown dirctly on (100) GaAs substrates at substrate temperature of 550°C.

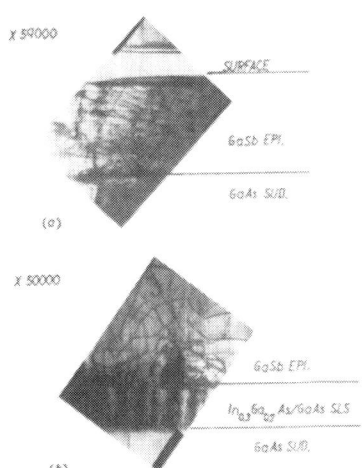

Figure 8. XTEM photographs of dislocation distributions for (a) GaSb/GaAs heteroepitaxial sample and (b) GaSb/SLS/GaAs heteroepitaxial structure.

directly on (100) GaAs substrates. From Fig.7, it is clearly observed that the hole mobilities of GaSb films grown on SLS/GaAs substrates are much higher than those grown directly on (100) GaAs substrates. From the above analyses, it is indicated that the GaSb epilayers grown on SLS/GaAs structure have superior electrical properties than those directly grown on GaAs substrates.

The distributions of dislocations generated in GaSb epilayers grown on (100) GaAs substrates and on SLS/GaAs structures (both grown at 550°C, V/III=6.84) were evaluated by transmission electron microscopy (TEM), as shown in Fig.8(a) and (b), respectively. Fig.8(a) shows the TEM photograph of huge dislocations distributed in GaSb epilayer grown directly on highly lattice-mismatched substrates. It is observed that many dislocations propagate up to the epilayer surface. Fig.8(b) shows the TEM photograph of dislocation distributions in GaSb epilayers grown on SLS/GaAs substrates. From Fig.8(b), it is clearly observed that some dislocations had bent away before reaching the epilayer surface. The superior electrical properties (as shown in Fig.7) of GaSb epilayers grown on SLS/GaAs substrates are related to the bending of dislocations in GaSb epilayers. From above analyses and comparisons, it is found that the strain layer superlattices can be used as buffer layers to block or bend away threading dislocations which form in highly lattice-mismatched system as GaSb/GaAs and the epilayers grown on the SLS/GaAs substrates have better electrical quality.

ACKNOWLEDGEMWNT

The authors wish to express their thanks to Mr. W. Lin for his technical assistance. The project was supported by the National Science Council, Republic of China, under the contract NSC79-0417-E006-04.

REFERENCES

1. A. Saraki, M. Nishiuma and Y. Takeda, Jap. J. Appl. Phys. 19, 1695(1980).
2. A.K. Srivastaua, J.C. DeWinter, C. Caneau, H.A. Pollack and J.L. Zyshind, Appl. Phys. Lett.. 48, 903(1986).
3. T.H. Chiu, J.L. Zyskind and W.T. Tsang, J. Electron. Materials 16, 57(1987).
4. H.J. Cherng, G.B. Stringfellow, D.W. Kister, A.K. Srivastava and J.L. Zyskind, Appl. Phys. Lett. 48, 419(1986).
5. H.D. Law, R. Chin, K. Nakano and R.A. Milano, IEEE J. Quantum Electron. QE-17, 275(1981).
6. C. Caneau, J.L. Jyskind, J.W. Sulhoff, T.E. Glover, J. Centanni, C.A. Burrus, A.G. Dendai and Pollack, Appl. Phys. Lett. 51, 764(1987).
7. A.E. Drakin, P.G. Jeliseev, B.N. Sverdlov, A.E. Bochkarev, L.M. Dolgino and L.V. Druzhinina, IEEEE J. Quantum Electron. QE-23, 1089(1987).
8. H. Miki, K. Segawa and K. Fujibayashi, Jpn. J. Appl. Phys. 13, 203(1974).
9. F. Capasso, M.B. Panish and S. Sumski, IEEE J. Quantum Electron. 17, 273(1981).
10. J.C. DeWinter and M.A. Pollack, J. Appl. Phys. 59, 3593(1986).
11. S.C. Chen, Y.K. Su and F.S. Juang, J. Cryst. Growth 92, 118(1988).
12. S.C. Chen and Y.K. Su, J. Appl. Phys. 66, 350(1989).
13. Y.K. Su, S.C. Chen and F.S. Juang, Solid-state Electron. 32, 733(1989).
14. Y.K. Su and F.S. Juang, J. Material Science 25, 843(1990).
15. M. Lee, D. J. Nicholas, K.E. Singer and B. Hamilton, J. Appl. Phys. 59, 2895(1986).
16. F.S. Juang and Y.K. Su, Prog. Cryst. Growth and Characterization, to be published in 1990.
17. H.M. Manasevit and K.L. Hess, J. Electrochem. Soc.: Solid-State Sci and Techn. 126, 2031(1979).
18. C.B. Cooper III, R.R. Saxena and M.J. Ludowise, J. Electro. Mater. 11,1001(1982).
19. S.K. Haywood, A.B. Henriques, N.J. Mason, R.J. Nicholas and P.J. Walker, Semicon. Sci. Technol. 3, 315(1988).
20. S.K. Haywood, N.J. Mason, P.J. Walker, J. Cryst. Growth 93, 56(1988).
21. E.T.R. Chidley, S.K. Haywood, R.E. Mallard, N.J. Mason, R.J. Nicholas, P.J. Walker and R.J. Warburton, Appl. Phys. Lett. 54, 1241(1989).
22. S.K. Haywood, E.T.R. Chidley, R.E. Mallard, N.J. Mason, R.J. Nicholas, P.J. Walker and R.J. Warburton, Appl. Phys. Lett. 54, 922(1989).
23. T. Kaneko, H. Asahi, Y. Okuno and S. Gonda, J. Cryst. Growth 95, 158(1989).
24. P.D. Dapkus, H.M. Manasevit, K.L. Hess, T.S. Low and G.E. Stillman, J. Cryst. Growth 55, 10(1981).
25. R.M. Biefeld, "Compound Semiconductor Strained Layer Superlattice", p.59(1981).

PART V

III-V-Based Long-Wavelength
Material Processing

DRY ETCHING TECHNIQUES AND CHEMISTRIES FOR III-V SEMICONDUCTORS

S. J. PEARTON
AT&T Bell Laboratories, Murray Hill, NJ 07974

ABSTRACT

Dry etching of III-V materials using both Cl-based (CCl_2F_2, $SiCl_4$, BCl_3, Cl_2) and CH_4/H_2 discharges will be reviewed. The etch rates using chlorine-based mixtures are generally faster than those utilizing CH_4/H_2, but the latter gives smoother surface morphologies for In-containing compounds. The use of microwave (2.45 GHz) electron cyclotron resonance (ECR) discharges minimizes the depth of lattice disorder resulting from dry etching, relative to conventional RF (13.56 MHz) discharges. Recent results on the systematics of ECR plasma etching of both In- and Ga-based III-V semiconductors using CCl_2F_2/O_2 and CH_4/H_2 mixtures will be discussed, including the determination of the maximum self-biases allowable which do not induce near-surface damage to the semiconductor. A further key issue is the prevention of changes in the surface stoichiometry of materials such as InP, where the lattice constituents may have considerably different volatilities in the particular discharge.

INTRODUCTION

The dry etching of GaAs and related compounds is gaining a resurgence of interest, largely due to the need to achieve high resolution, anisotropic etching in device applications [1-4]. There are a variety of requirements of such as fast etch rate for creation of deep (≥ 1 µm) trenches, high selectivity for one material over another (say GaAs over AlGaAs) or conversely equi-rate etching for these materials.

There are two basic classes of gas mixtures used for etching of III-V materials [2,5]. The first is based on chlorine or bromine, particularly the former because gallium and other group III chlorides are volatile at relatively low temperatures, as opposed to the stable gallium fluorides. It is common then to use Cl_2-based etching for III-V materials, in contrast to the F-based etching prevalent for Si. Most of the Cl-containing gases also contain carbon and often problems are encountered with the deposition of polymer films during etching. Some of the gas mixtures used include Cl_2, CCl_4, BCl_3, $SiCl_4/Cl_2$, $SiCl_4/SF_6$, $CHCl_3$, $COCl_2$ and CCl_2F_2 (with O, He or Ar). The advantages of using Freon 12 (CCl_2F_2) are that it is a nontoxic noncorrosive gas which contains both Cl for etching and F to provide an etch stop upon reaching an underlying AlGaAs or AlInAs layer [1]. The etch stop mechanism involves the formation of an involatile AlF_3 layer, allowing selectivities of GaAs-to-AlGaAs and InGaAs-to-AlInAs of several hundred.

The second general class of gas mixture is based on methane or ethane and hydrogen [6]. This has attracted considerable recent attention for etching both Ga- and In-based semiconductors. This nonchlorinated mixture shows controlled, smooth, highly anisotropic etching of all III-V materials. The etch products are thought to be AsH_3 or PH_3 for the group V element and most likely some form of methyl adduct (eq. $(CH_3)_n Ga$) for the group III species.

278

In this paper we will describe some of the characteristics of etching GaAs, InP and related compounds in Cl-Br- and CH_4/H_2-based gas mixtures and detail the use of high density, low damage electon cyclotron resource (ECR) discharges for etching of device structures.

RESULTS

a. Etch Rates

It is common to use the normal boiling points or vapor pressures of the possible etch products as an indication of the suitability of a particular discharge for the dry etching of a III-V semiconductor. Some of these parameters are shown in Table 1 [7]. Strictly speaking, one needs to know the volatility under ion bombardment of these etch products, and in some cases the actual etch product is not known. An example is the group III product for CH_4/H_2 RIE of III-V materials.

Table 1   Normal boiling points and vapor pressures of some of the possible etch products.

Product	Boiling Point (°C)	Vapor Pressure
$AsCl_3$	130	40mm; 50°C
PCl_3	76	1mm; -52°C
PCl_5	162	1mm; 56°C
$InCl$	608	—
$InCl_2$	560	—
$InCl_3$	600	18; 250°C
$GaCl_2$	535	—
$GaCl_3$	201	80mTorr; 25°C
$AlCl_3$	183	1mm; 100°C
$AsBr_3$	221	1mm; 42°C
PBr_3	76	1mm; 8°C
PBr_5	162	—
$InBr_3$	sublimes	—
$GaBr_3$	279	—
$AlBr_3$	263	1mm; 81°C
AsH_3	−55	760mm; −62°C
PH_3	−88	40mm; −129°C

For GaAs, CCl_2F_2 provides convenient, controlled etch rates, as shown in Figure 1 and the selectivity over photoresist can be quite good. The etch rate in small features can be slower than in larger open areas because of difficulty in removing the etch products from a deep, small opening. The temperature dependence of GaAs etch rate in CCl_2F_2/O_2 is shown in Arrhenius form in Figure 2. Above 150°C there appears to be more efficient desorption of $GaCl_3$ relative to $AsCl_3$. There have been a number of reports of non-Arrhenius behavior for CCl_4 and $BCl_3:Cl_2$ RIE of GaAs, where heat sinking of the samples was not performed [8,9]. At low temperatures the main vaporization products are $GaCl_3$ and $AsCl_3$, whereas above ~ 400°C the main products are GaCl and AsCl. The etched surface morphologies go through a series of smooth-to-rough-to-smooth transitions with increasing temperature.

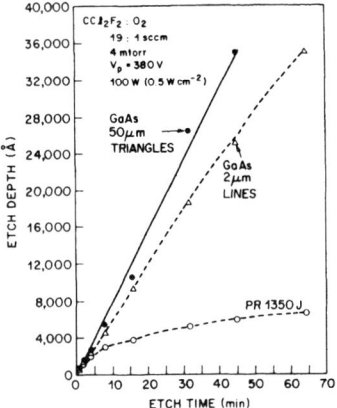

Figure 1. Etch depth of GaAs areas of either triangular shape (50 μm on a side) or lines (20 μm wide) as a function of time in a 19:1 $CCl_2F_2:O_2$ discharge (4 mTorr, 0.56 W cm^{-2}). The amount of the photoresist mask removed is also shown.

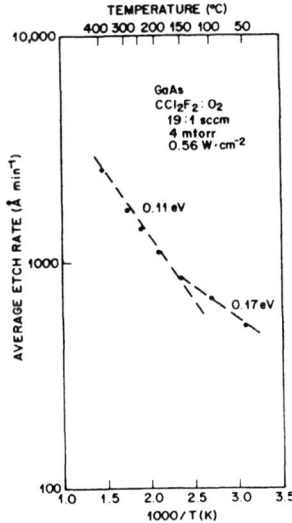

Figure 2. Arrhenius plot of the average etch rate of GaAs under the conditions described in the caption of Fig. 1, as a function of inverse temperature during the RIE.

The Schottky barrier height (ϕ_B) and diode ideality factor (n) are important parameters since they give an indication of the suitability of the etched surface for device fabrications [4]. Figure 3 shows these parameters obtained from n-type GaAs etched in CCl_2F_2/O_2 at temperatures between 50-400°C. Performing the RIE treatment at 50°C led to a deterioration of n, consistent with the introduction of near-surface damage caused by ion bombardment. Increasing the temperature to 150 °C during the RIE produces an n value of 1.00, similar to the situation with elevated temperature ion implantation in GaAs where heating the sample to $\geq$ 150 °C prevents amorphization through dynamic annealing pressures. Above 150 °C, the ideality factor worsens as a result of deviations from stoichiometry and surface roughening. The barrier heights are essentially constant with RIE temperature until 400 °C where polymer deposition and surface roughening leads to a high value of 0.91 eV. The polymer most likely acts as an insulator and the thermally degraded surface is probably a barrier to current flow.

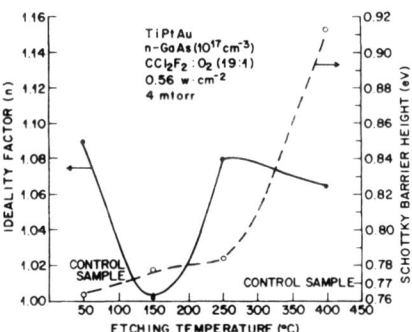

Figure 3.   Schottky barrier heights and ideality factors from TiPtAu Schottky diodes on n-type GaAs etched in a 19:1 $CCl_2F_2:O_2$, 0.56 W cm^{-2} discharge as a function of the temperature of the sample during the RIE treatment.

The Cl-based etching is usually faster for all III-V materials than CH_4/H_2. Comparisons between the etch rates of InP, InGaAs and InAlAs in CCl_2F_2/O_2 and C_2H_6/H_2 are shown in Figure 4, where it is seen that the etch rates with the former mixture are a factor of 3-5 times faster. In many application, notably mesa etching of heterostructures, a slower etch rate is actually an advantage because it gives better control of the etch depth. In other cases such as the etching of via holes right through a thinned-down substrate then obviously a high etch rate is more important. Similar data for RIE of InAs, GaSb and InSb with the same mixtures are shown in Figure 5. From these results it appears that gallium and antimony chlorides are more volatile under RIE conditions than their indium and arsenic counterparts. This result would not be expected from a consideration of the boiling point data in Table 1. Due to the possible restrictions on the production and use of chlorofluorocarbons, we have investigated the etching characteristics of the hydrogenated chlorofluorocabons (HCFCs), notably $CHCl_2F$ (Freon 21) and $CHClF_2$ (Freon 22). Etch depths as a function of time for five In-based materials with these two gases are shown in Figure 6. Based on such data we see that the etch rates with $CHCl_2F/O_2$ are ~ 20% slower for all materials relative to those with CCl_2F_2/O_2 under the same conditions. There was no evidence of an incubation time required before the onset of etching. All of the III-V materials, including GaAs, AlGaAs and GaSb exhibit smooth surface morphologies over a wide range of RIE parameters. Thin (20-30 Å) residue layers containing 3-9 at. % Cl and 1-3 at. % F (24 at. % for AlInAs) are present after dry etching with the HCFC's, although this contamination can be removed by solvent cleaning. The formation of a high concentration of AlF_3 on AlInAs provides a natural etch stop for removal of InGaAs layers in HBTs based on these materials.

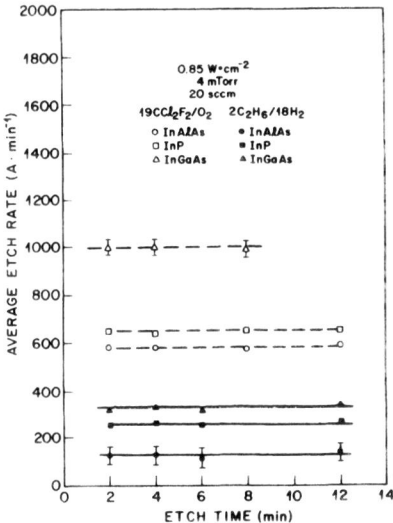

Figure 4.   Average etch rates of InP, InGaAs, and InAlAs as a function of time for either $2C_2H_6:18H_2$ or $19CCl_2F_2:1O_2$ discharges under the conditions of Fig. 1.

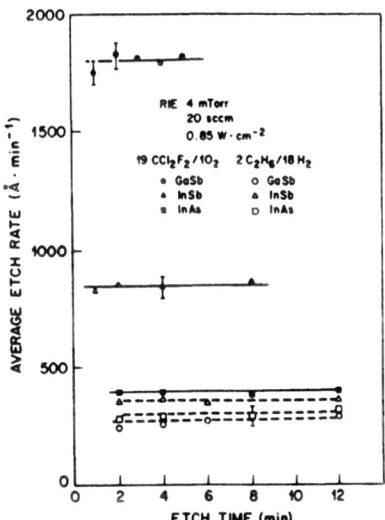

Figure 5.   Average etch rate of GaSb, InSb, and InAs as a function of time for 4 mTorr, 0.85 W cm^{-2} discharges.

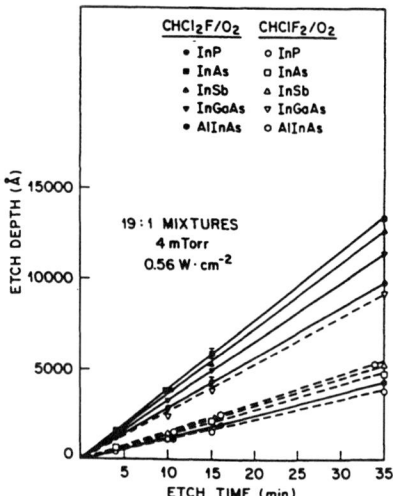

Figure 6. Etched depth in InP, InAs, InSb, InGaAs or AlInAs as a function of exposure time to either 19 CHCl$_2$F:1O$_2$ or 19 CHClF$_2$:1O$_2$ discharges (4 mTorr, 0.56 W cm^{-2}).

Very high rate etching ($\leq$ 10 µm·min^{-1}) of GaAs, AlGaAs and GaSb is feasible in Cl$_2$/Ar and SiCl$_4$/Ar discharges. Highly anisotropic etching is achieved with SiCl$_4$ whereas a greater degree of chemical etching is evident with Cl$_2$. Provided self-bases were $\leq$ 50V, excellent Schottky diode characteristics were exhibited by TiPtAu contacts on SiCl$_4$ or Cl$_2$ etched GaAs surfaces. Photoluminescence decreases of 2-25 times are observed after RIE of GaAs for both types of discharge, depending on the self-bais on the sample. Chlorine residues were typically present to a depth of < 20 Å after etching, with cleaner surfaces obtained with SiCl$_4$ than Cl$_2$. Etch rates for the In-based materials with Cl$_2$ and SiCl$_4$ are considerably slower than for GaAs if the sample temperature is kept low during the plasma exposure. Much faster and smoother etching of InP and related materials are obtained at elevated ($\geq$ 100°C) temperatures, where desorption of the group 111 chlorides becomes easier. Extremely fast etch rates of InP are possible using Cl$_2$ or SiCl$_4$ at these temperatures, although in general this is impractical in many device processing sequences. A list of the comparative etch rates of GaAs, Al$_{0.3}$Ga$_{0.7}$As, InP and InGaAs in the various gas mixtures at low pressure (4 mTorr), and power densities (0.56 W·cm^{-2}, self-biases of ~ 300 V) are shown in Table 2.

Table 2   Typical etch rates of GaAs, InP, $Al_{0.3}Ga_{0.7}As$ and InGaAs in different gas mixtures at 4 mTorr and $0.56W \cdot cm^{-2}$.

	Etch Rate $(\overset{\circ}{A} \cdot min^{-1})$			
Mixture	GaAs	$Al_{0.3}Ga_{0.7}As$	InP	InGaAs
Cl_2/Ar	20,000	5,000	150	150
$SiCl_4/Ar$	5,000	3,000	130	320
CCl_2F_2/O_2	750	500	650	1,000
CH_3Br	600	400	220	300
CH_4/H_2	190	140	220	280
C_2H_6/H_2	220	160	250	320
$CHCl_2F/O_2$	600	300	300	320
$CHClF_2/O_2$	400	250	110	130

b. Damage

One of the major issues with dry etching of GaAs and AlGaAs is the introduction of surface damage by energetic ion bombardment. This has been investigated for a variety of dry etching methods using neutral ion and reactive ion bombardment. The degree of damage has been found to be inversely proportional to the ion mass, and directly proportional to the ion energy [4]. Sidewall damage during dry etching has also been studied [10,11]. The most obvious effect of near-surface damage is a reduction in the carrier concentration between 200-1000 Å from the surface. This is considerably deeper than the projected range of ions crossing the plasma sheath and is ascribed to channeling of these relatively low-energy particles, and recombination-enhanced motion of defects. In at least some cases where H_2 is involved in the etch mixture, there appears also to be passivation of donors and acceptors by hydrogen association. This is a well-known effect from studies of the role of atomic hydrogen in passivating shallow level dopants in semiconductors.

Figure 7 shows the Schottky barrier heights (ϕ_B) and diode ideality factors (n) on CCl_2F_2/O_2 etched GaAs samples. In the case of $CCl_2F_2:O_2$ RIE there is little change in ϕ_B from the unetched control sample ($\phi_B = 0.74$ eV). The ideality factors display more significant changes with post-RIE annealing. After etching the ideality factor rises from 1.045 on the control sample to 1.06. The n value decreases with annealing to 300 °C, where it is actually lower than on the control. This was a reproducible effect, although its origin is not clear. Annealing at ≥ 400 °C led to significant degradation in the n value. This may be related to the beginning of surface deterioration of the GaAs. For the case of ethane-based RIE, the Schottky barrier height was consistently much lower than on the control sample, being in the range 0.57-0.64 eV over the annealing temperatures investigated. The origin of this decrease may possibly lie in the efficient removal of free As and As_2O_3 from the surface by the atomic hydrogen in the discharge through the reactions

$$As + 3H \rightarrow AsH_3 \uparrow ,$$

$$As_2O_3 + 12H \rightarrow 3H_2O \uparrow + 2AsH_3 \uparrow .$$

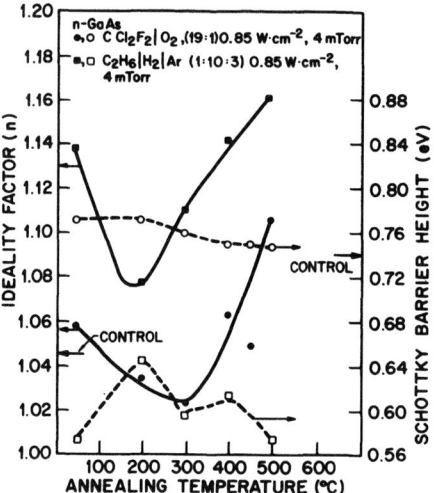

Figure 7.   Schottky barrier heights and ideality factors from TiPtAu Schottky diodes on n-type GaAs etched in either a 19:1 $CCl_2F_2:O_2$ or 1:10:3 $C_2H_6:H_2:Ar$ discharge, as a function of post-RIE annealing temperature (30s anneals).

Free As is thought to produce deep surface states which pin the Fermi level at midgap. Removal of this As should unpin the surface. The ideality factor after ethane-based RIE is also much higher (1.14) than on the control sample, and annealing at 200°C produces a minimum in the n value. Annealing at higher temperatures produces a deterioration in the ideality factor. This may be due to the reactivation of deep levels initially introduced by ion bombardment during RIE, but passivated by the presence of atomic hydrogen. Annealing near 200°C allows redistribution of some of this hydrogen to defect sites, passivating their electrical activity. However, at > 300°C these levels may be reactivated by the irreversible dissociation of the atomic hydrogen from the defect site.

The near-surface carrier profiles in etched GaAs are shown in Fig. 8 as a function of post-RIE annealing temperature. The results for $CCl_2F_2:O_2$ RIE are relatively straightforward, being similar to the data reported by Pang [4]. There is a reduction in the net carrier concentration within 1500 Å of the surface which is recovered by annealing at 300°C. Any form of ion bombardment of GaAs with energies above the threshold for displacement (~ 15 eV) will create midgap deep levels which trap free carriers and are not thermally ionized at room temperature. This effect is widely used in isolating GaAs devices using ion implantation with nondopants ions like O, B, or H. Under our conditions there appears to be annealing of the damage-related levels near 300°C. In the case of $C_2H_6 - H_2 - Ar$ RIE the evolution of the carrier profiles with annealing is more complicated. The reduction in the doping concentration is now due to two effects; damage-related deep levels and passivation of the Si donors in the material by atomic hydrogen. The as-etched sample appears to show an increase in carrier concentration right at the surface, but this is an artifact of the C-V measurement, judging from the phase angle at low reverse biases. A more reliable profile is obtained at depth > 1000 Å. Upon annealing at 200 °C the carrier reduction moves to greater depth, which we ascribe to motion of near-surface atomic hydrogen further into the sample. For annealing above 300 °C the carrier profile recovers towards its unetched shape. Annealing of deep levels again appears to begin at T ≥ 300 °C, whereas reactivation of hydrogen passivated donors occurs at ~ 400 °C. Even after 500 °C annealing the profile is not fully recovered, presumably due to residual near-surface disorder.

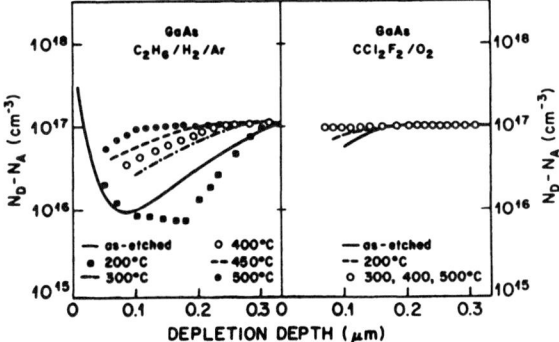

Figure 8.   Carrier profiles in n-type $(n = 10^{17} \ cm^{-3})$ GaAs etched either a 19:1 $CCl_2F_2:O_2$ or 1:10:3 $C_2H_6:H_2:Ar$ discharge (4 mTorr, 0.85 W cm^{-2}), as a function of post-RIE annealing temperature. The profiles before etching were uniform at a level of $10^{17} \ cm^{-3}$.

RIE of GaAs and AlGaAs in two of the most common gas chemistries, $CCl_2F_2:O_2$ and $C_2H_6:H_2:Ar$, provides evidence of both ion-induced damage, and for the latter mixture, hydrogen passivation of Si dopants ions. Annealing at 200 – 300 °C restores the carrier concentration in $CCl_2F_2:O_2$-etched material and slightly higher temperatures are required to produce the most ideal I-V characteristics. In contrast temperatures near 500 °C are required to restore the initial carrier concentrations in $C_2H_6 - H_2 - Ar$ etched material due to the combined effects of hydrogen passivation of donors and more serious damage due to the light H$^+$ ions in the discharge.

Reverse voltage-current characteristics from InP $(n = 6 \times 10^{15} \ cm^{-3})$ etched in C_2H_6/H_2 or CCl_2F_2/O_2 are shown in Fig. 9. Upon RIE, Au contacts no longer showed rectifying behavior for InP etched in C_2H_6/H_2 but were essentially ohmic for either polarity of bias applied to the contact. Once again this is presumably due to the creation of a nonstoichiometric near-surface region. Chemical analysis of the surface showed a deficiency of phosphorus to a depth of ~ 150 Å. By sharp contrast, RIE in a $CCl_2/F_2/O_2$ discharge leads to an approximately fivefold increase in reverse bias current, but the I-V characteristic is still rectifying. Chemical analysis of the near-surface region of InP etched in this type of discharge showed only a slight deficiency of P within 20 Å of the surface. The difference between the two gas mixtures is clearly the high concentration of atomic hydrogen which can preferentially remove phosphorus from InP in the form of PH$_3$. This appears to be a fundamental problem with the C_2H_6/H_2 mixture, since the use of lower hydrogen concentrations leads to increasing polymer deposition.

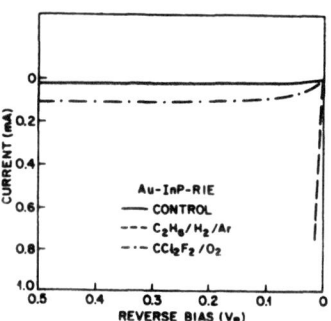

Figure 9. Reverse bias I-V characteristics from Au contacted n-type InP ($n = 6 \times 10^{15}$ cm^{-3}) after etching in $1\,C_2H_6/1\,OH_2$ or $19\,CCl_2F_2/O_2$ discharges prior to Au deposition. The reverse breakdown voltage on the control sample was ~2.5 V.

Lower levels of ion-induced damage are present when microwave Electron Cyclotron Resonance (ECR) discharges are used because of the lower ion energies relative to conventional RF plasmas. An example of the relatively benign nature of ECR etching with regard to InP is given by the forward current-voltage (I-V) measurements shown in Fig. 10. Samples etched under ECR conditions with no additional biasing gave I-V characteristics very close to those of an unetched control sample, with a Schottky barrier height (ϕ_B) of 0.48 eV and diode ideality factor (n) of 1.1, both derived from the forward I-V plots assuming thermionic emission. Once again our past experience with RIE of InP using this type of gas chemistry has been that Au deposition onto the RIE surface results in ohmic behavior, and a rectifying characteristic is not observed until at least 100 Å is removed from the sample by wet etching prior to the Au deposition. With RIE we also observe substantial In enrichment of the near-surface region, but to much greater depths than with ECR. Even with the addition of 100 V substrate bias during the ECR etching we observe only a relatively small reduction of the Schottky barrier height to 0.44 eV, while the ideality factor shows a greater degradation, to a value of 1.6. This is a convincing demonstration of the much lower degree of disruption to the semiconductor surface using ECR discharges compared to conventional RIE.

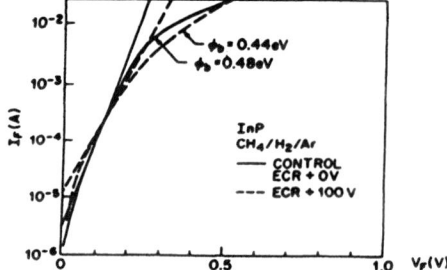

Figure 10. Forward current-voltage characteristics from Au-InP Schottky diodes etched in ECR 5 CH$_4$/15 H$_2$/7 Ar discharges (0 or 100V substrate self-bias) prior to deposition of the Au contacts. The straight lines in each case are used to give the intercept and slope of the characteristic. The ECR + 0 V sample had forward I-V curves very close to those of an unetched control sample.

c. Surface Morphologies

In general we observe smoother, more anisotropic etching of GaAs, AlGaAs and GaSb with SiCl$_4$/Ar than with Cl$_2$/Ar. At the top of Fig. 11 we show two SEM micrographs of GaAs etched for 4 min. in a 10 SiCl$_4$/5 Ar, 20 mTorr discharge with a dc bias on the cathode of 100 V. The etching is anisotropic with some roughness evident for the morphology of the etched surface. This is typical of all of the GaAs etched in SiCl$_4$/Ar discharges. There also is clear "ribbing" on the sidewalls, which is often observed both with Cl- and CH$_4$-based RIE of III-V materials, and which may be due to the etching process replicating features from ragged mask edges. The use of Cl$_2$/Ar discharges leads to rougher surface morphologies and more undercutting of the sidewalls due to the greater degree of chemical etching occurring relative to SiCl$_4$/Ar RIE. The lower two pictures of Fig. 11 show features etched in GaAs using a 10 Cl$_2$/5 Ar, 100 V discharge either at 20 mTorr (at left) or at 100 mTorr (at right). The increase in undercutting in the latter case is apparent. Crystallographic etching occurs under high pressure (100 mTorr) and low-bias conditions with Cl$_2$/Ar.

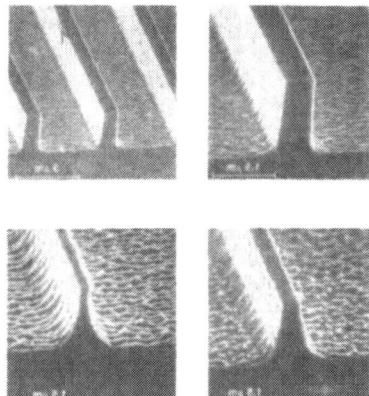

Figure 11.   SEM micrographs of GaAs etched in a 10 SiCl$_4$/5 Ar, 20 mTorr discharge with 100 V dc bias (top left and right), or in 10 Cl$_2$/5 Ar, 20 mTorr (at bottom left) or 100 mTorr (at bottom right).

Smooth etching of all of the III-V materials is obtained with CH$_4$/H$_2$ or C$_2$H$_6$/H$_2$ mixtures provided that the CH$_4$-to-H$_2$ ratio is kept between approximately 1:1 to 1:3. For high CH$_4$ concentrations, polymer deposition on the sample leads to micromasking and rough surfaces, whereas at high H$_2$ concentrations there is a preferential loss of phosphorus or arsenic relative to indium or gallium and this also leads to rough surface morphologies. An initial problem with ECR etching of InP with CH$_4$/H$_2$ mixtures was the roughening of the surface because of an imbalance between the active hydrogen and methyl species in the discharges at high microwave powers. We have overcome this problem by limiting the microwave power to 150W — under these conditions the etching of InP in 5 CH$_4$/17 H$_2$/8 Ar, 1 mTorr, 100V discharges is just as smooth under RIE conditions (zero microwave power) as under ECR conditions, as shown in Figure 12.

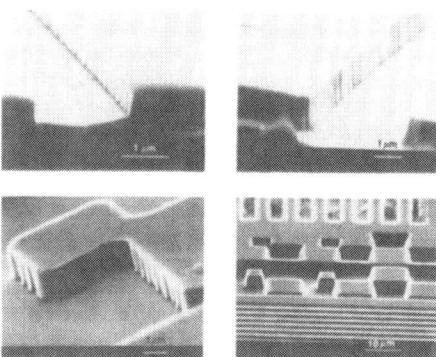

Figure 12. SEM micrographs from InP etched in $CH_4/H_2/Ar$ discharges under RIE conditions (top and bottom left) or under ECR conditions (top and bottom right).

The etching of all of the III-V materials in the Freon-based mixtures is relatively smooth under RIE or ECR conditions when Ar, He or O is used as the dilutent. As mentioned earlier these mixtures usually leave a surface residue 20-50 Å thick, although solvent cleaning removes essentially all of this film. Table 3 lists the depth of the various components of the near-surface residue found on $Al_{0.3}Ga_{0.7}As$ after RIE in CCl_2F_2-based mixtures.

Table 3   Depth of various components of the near-surface residue found after RIE of AlGaAs with a CCl_2F_2 discharge for 4 min at 4 mTorr (in Å).

Component	Control	A	B	C	D
Fluorocarbons	none	<10	<10	<10	<10
As_2O_3	<10	<10	<10	<10	<10
Ga_2O_3	10-20	30-40	30-40	40-50	20-30
GaF_3	none	none	10-20	20-30	20-30
Al_2O_3	60-70	60-70	110-120	100-110	90-100
AlF	none	40-50	80-90	60-70	50-60

A: CCl_2F_2/O_2, 10:10 sccm, 0.85 W cm^{-2}

B: CCl_2F_2, 20 sccm, 0.85 W cm^{-2}

C: CCl_2F_2/O_2, 19:1 sccm, 0.85 W cm^{-2}

D: CCl_2F_2/O_2, 19:1 sccm, 1.3 W cm^{-2}

d. ECR Etching

Various methods have been developed for reducing ion energies in the discharge while trying to maintain anisotropic etching. These include the so-called triode reactor in which a second plasma-generating electrode is included within the process chamber, or the addition of magnetic fields configured to reduce electron loss from the discharge and thus reduce the potential between it and the sample. This form of magnetically enhanced etching is generally divided into two types — magnetron RIE or electron cyclotron resonance (ECR) plasma etching. In ECR discharges, free electrons in the plasma are forced to orbit about magnetic field lines while absorbing microwave energy. At the cyclotron resonance condition outer shell electrons from gas molecules in the discharge may also be liberated, leading to a very high degree of ionization in the plasma [12]. Since the motion of the electrons is constrained by the external magnetic field, fewer are lost by collisions with the reactor walls than in a conventional radio frequency (rf) plasma and therefore the plasma potential relative to ground is much lower. The resultant energies of ions reaching the sample to be etched are typically ≤ 15 eV. Since this is less than the displacement threshold for damage in most semiconductors, ECR etching should lead to much lower levels of damage than conventional RIE processes. A schematic of the ECR reactor we use is shown in Fig. 13 (Plasma-Therm SL720). The sample is manually loaded into the load lock, and then transferred into the etch chamber on a robotic arm. The system utilizes a 2.45 GHz microwave excitation source with additional rf bias superimposed at the wafer position (13.56 MHz). Pumping is accomplished through a very high conductance pump manifold linking the process chamber to a 1000 ℓs^{-1} turbomolecular pump.

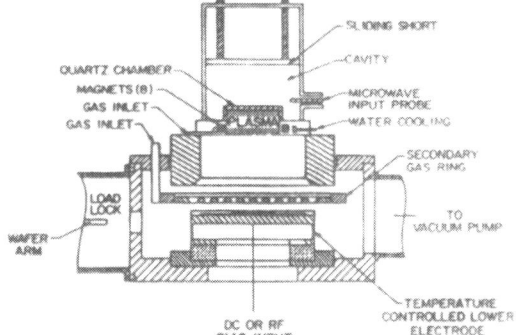

Figure 13.   Schematic diagram of multipolar ECR plasma-etching system.

Electron cyclotron resonance is provided by a plasma source of the multipolar, tuned cavity design. The microwave cavity plasma source has been described extensively elsewhere, [12] and only a brief description is given here. The resonant cavity is 17.8 cm i.d. brass cylinder terminated at the top by an adjustable short (Fig. 13). A variable length launching probe enters the resonant cavity at the side, impressing microwave energy from the magnetron/waveguide assembly to an evacuated 100 mm diameter Quartz "cup" where the plasma resides. The brass resonant cavity is at atmosphere. Within the baseplate are eight high strength rare-earth magnets which produce the B field level of 0.0875 T necessary for resonance.

Cyclotron resonance occurs on an ECR surface within the quartz cup. In as much as large solenoidal magnets are not used, there is no degrading B field to Coulombically accelerate ions to the wafer position. Therefore, O V rf bias at the wafer will indeed correspond to zero additional acceleration energy on the etching species; arrival mechanisms for 0 V rf are dominated by ambipolor and free fall diffusion. The ions will still however be accelerated through a sheath potential even with zero rf biasing. We have demonstrated low-damage, dry etching of III-V materials using $CH_4/H_2/Ar$ and CCl_2F_2/O_2 ECR discharges. In essence, the provision of microwave power provides faster etch rates at the same bias relative to conventional RIE, or equivalently with ECR discharges one can use lower self-biases and still obtain useable etch rates.

ACKNOWLEDGMENTS

The author acknowledges the valued collaboration of W. S. Hobson, C. R. Abernathy, U. K. Chakrabaiti, K. S. Jones, C. Constantine, F. Ren and T. R. Fullowan.

REFERENCES

[1]   E. L. Hu and R. E. Howard, Appl. Phys. Lett. *37*, 1022 (1980).

[2]   R. H. Burton, R. A. Gottscho and G. Smolinsky, Dry Etching for Microelectronics, ed. R. A. Powell (Elselvier, NY 1984).

[3]   R. A. Gottscho, G. Smolinsky and R. H. Burton, J. Appl. Phys. *54*, 5908 (1982).

[4]   S. W. Pang, J. Electrochem. Soc. *133*, 784 (1986).

[5]   D. L. Flamm, Plasma Etching — an introduction, ed. D. M. Manos and D. L. Flamm (academic Press NY, 1989).

[6]   U. Niggebugge, M. Klug and G. Garus, Inst. Phys. Conf. Ser. *78*, 367 (1985).

[7]   Handbook of Physics and Chemistry, 70th Edition (CRC Press, Florida 1990).

[8]   R. J. Contolini, J. Electrochem. Soc. *135*, 929 (1988).

[9]   N. Furutata, H. Miyamoto, A. Okamoto and K. Ohata, J. Appl. Phys. *65*, 168 (1989).

[10] C. M. Knoedler, L. Osterling and H. Shkitman, J. Vac. Sci. Technol. *B6*, 1573 (1988).

[11] A. Scherer, H. G. Craighead and E. D. Beebe, J. Vac. Sci. Technol. *B5*, 1599 (1987).

[12] J. Asmussen, J. Vac. Sci. Technol, *A7*, 883 (1989).

MeV ION BEAM APPLICATIONS IN III-V SEMICONDUCTORS

R.G. ELLIMAN*, M.C. RIDGWAY, S.T. JOHNSON* and J.S. WILLIAMS*
Department of Electronic Materials Engineering, Research School of Physical Sciences,
Australian National University, Canberra, Australia
* Also Microelectronics and Materials Technology Centre, Royal Melbourne Institute of
Technology, Melbourne, Australia

ABSTRACT

This paper reviews some key areas where MeV ion beams can be applied to III-V semiconductor materials. In particular, ion damage is assessed for various III-V materials in terms of implantation parameters, especially substrate temperature and dose rate. Implant isolation, involving the introduction of damage to remove carriers and achieve highly-resistive layers, is assessed for MeV irradiation. It is concluded that MeV ions can provide deep, uniform damage with a single-energy implant. Finally, improved epitaxy of amorphous InP with MeV ions is demonstrated.

BACKGROUND

Energetic ion beams are now widely used for selectively modifying the resistivity of III-V compound semiconductors. There are two major applications: i) the formation of conductive layers or junctions by introducing shallow n-type or p-type impurities; and ii) the generation of highly resistive regions for device isolation, mainly achieved by producing damage-related or impurity-related deep levels within the semiconductor band gap.

In the former application, ion implantation of dopants produces damage in the semiconductor and it is the removal of this damage and the associated optimum activation of the dopant which is a key issue in III-V semiconductor technology [1-3]. It has been shown [2,3] that it is often better to avoid the formation of amorphous layers during implantation of GaAs and InP if maximum electrical activity is sought. This usually requires elevated temperature implants unless light dopants such as Be (p-type) or Si (n-type) are employed [4]. Nevertheless, annealing temperatures of greater than 700°C are required to achieve satisfactory dopant activation in III-V semiconductors. Despite the fact that damage removal processes in GaAs and, to a lesser extent, InP have been extensively studied (see, for example, [1-3]), the relationships between residual defects and dopant atoms, which affect activation, are not well understood [4]. In other III-V materials, such as AlGaAs [5], GaP [2] and more complex In-based compounds, the damage-related processes involved in implantation doping have not yet been sufficiently studied let alone understood. However, it is clear that the nature and quantity of disorder introduced during implantation play major roles in determining ultimate electrical properties of III-V structures. More systematic studies of implant disorder in various III-V materials, and dependence of disorder on implant temperature and other implantation parameters, are required. Here, the more controlled damage profiles produced by MeV implantation can be used to advantage in such studies, as is discussed below. The use of MeV ion beams for direct (deep) doping has had limited application to date [6] but could have increased interest for producing deep conducting layers in III-V multilayers.

In the production of highly resistive layers by ion bombardment, it is most often the damage introduced by the ion beam which achieves the desired property change. Direct implant damage usually traps carriers and electrical transport is via hopping conduction between damage centres [4]. An annealing step is required to achieve maximum resistivity by removing gross damage and reducing hopping conduction. Ideally, carriers are trapped at deep, damage-related levels, thus pinning the Fermi level at mid-gap. This situation can be achieved with GaAs but not as effectively in other III-V materials [2,4], necessitating other isolation methods, some of which are mentioned later. As with implantation doping, further systematic studies of ion damage processes in III-V's are required to optimise implant isolation conditions and to begin to understand the various defect trapping processes relevant to specific III-V materials. Again, as we illustrate below, MeV ion beams have significant advantages, not only for aiding in the optimisation and understanding of implant isolation, but also for simplifying the implantation schedule for efficient isolation.

During ion bombardment, atomic displacements which are created by collision processes can be stable and accumulate, ultimately leading at low temperatures to total lattice disruption (amorphization). As the substrate temperature is raised, defect migration during implantation becomes more probable and this can lead to dynamic annealing. If the temperature is sufficiently high, amorphous layers cannot be created. It might be expected that different III-V materials have different temperature-dependent ion disordering behaviour, as discussed below. Under some situations [7,8], energetic ion beams can remove pre-existing amorphous layers in III-V semiconductors by providing mobile defects. MeV ion beams are particularly attractive for studying such damaging and dynamic annealing processes in III-V materials.

In this paper, we review some of the potential uses of MeV ion beams in III-V semiconductors, concentrating on understanding damage production; and its subsequent removal, and in applications for implant isolation.

RESIDUAL DAMAGE STUDIES

Since there have been few studies of residual damage to III-V semiconductors following MeV ion implantation, we first briefly review behaviour under keV ion bombardment. Recently, Jones and Santana [9] have studied the amorphization of a range of III-V semiconductors under 20keV Si^+ bombardment at 77K. A selection of their data is reproduced in Fig.1. Here, the minimum damage density necessary to amorphize the material is shown for various semiconductors. Apart from the II-VI material ZnSe, AlAs is the most difficult material to amorphize, followed by GaAs. The In-based materials (InP and InAs) are amorphized at the lowest deposited damage density. Care was taken to minimise or take account of dynamic annealing effects (by low temperature implantation and careful observation of crystalline defect in TEM). In general, substrates with higher average mass number are more easily amorphized (as collisional considerations would suggest [10]) but the reverse is not universally true. No current bond energy or other single models were found to adequately explain the trends. However, the authors did observe evidence for dynamic annealing in AlAs (even at 77K), in agreement with the difficulty in amorphising AlAs compared with GaAs that Poate and co-workers [11] observed for below room temperature implantation.

It is important to note that the data in Fig.1 were taken under constant dose ($10^{15}cm^{-2}$), dose rate and implant temperature conditions. The authors noted that lower and higher doses produced incomplete amorphization or increased dynamic annealing, respectively. Presumably, changing the implant temperature and dose rate would also have altered the results. For example, increasing the implant temperature would have enhanced dynamic annealing, leading, for example, to an inability to amorphize AlAs [11]. Dose-rate-dependent damage has been observed previously in GaAs [2]. We illustrate this situation in Fig.2 for 80keV Ar^+ bombardment of GaAs at 75°C. Optical reflectivity measurements during implantation [12] indicate that, the higher the dose rate, the more easily GaAs is rendered amorphous with dose (disorder level of 1.0). At the lowest dose rate ($0.04\mu A\ cm^{-2}$), only disordered crystal is produced up to a dose of $2.3 \times 10^{15}cm^{-2}$: higher doses exceeding $10^{16}cm^{-2}$ (not shown) still do not render the GaAs amorphous. These data indicate that the rate of dynamic annealing dominates damage production at low dose rates and vice versa at high dose rates for 75°C bombardment.

Fig.3 schematically illustrates the temperature-dependent ion disordering behaviour for semiconductors. For a given ion species, energy, dose and dose rate, a critical temperature T_c can be obtained above which amorphization does not occur. T_c indicates the temperature at which the rate of dynamic annealing just balances disorder production. Under the same irradiation conditions, T_c will vary for different III-V semiconductors since defect migration energies (and defect annihilation efficiencies) would be different. Thus, under

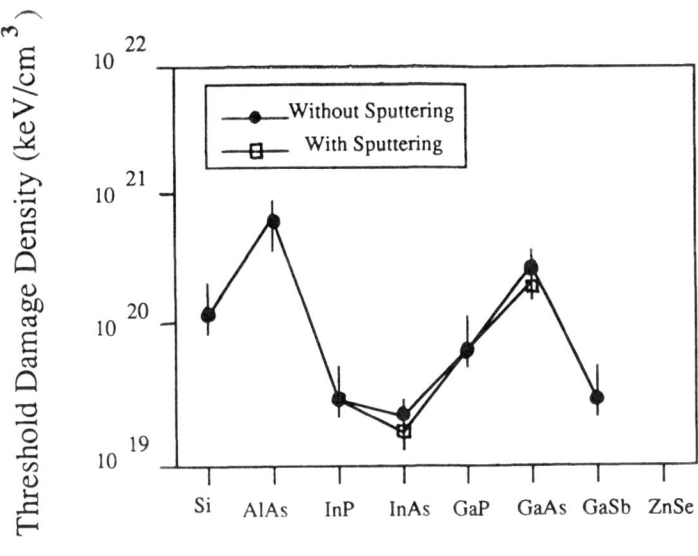

Fig.1   Threshold damage density for amorphization for various semiconductors implanted with 20keV Si+ ions at 77K. After Ref. [9]

conditions when dynamic annealing is important, the ease of amorphizing the various compound semiconductors may not follow the trends given in Fig.1. Indeed, Wie et.al. [13] carried out 15MeV Cl+ irradiation of both GaAs and GaP at room temperature and found damage saturation (more extensive disorder) in GaP compared with GaAs. These authors concluded that irradiation-induced defects were essentially immobile in GaP at room temperature but mobile in GaAs. They based their conclusion on the work of Lang et.al. [14] who studied the removal of discrete defects in various III-V materials as a function of temperature. As the temperature was increased from 80K, Lang et.al [14] found that defects became mobile (and annihilated) first in InSb followed by InAs, GaP, GaAs and then GaP. Based on defect mobility (the ease of dynamic annealing) this suggests that InSb should be more difficult to amorphize under bombardment at 300K, compared to GaAs and GaP. This contradicts the order observed at 77K by Jones and Santana [9] and highlights the importance of temperature in determining the relative ease of amorphization in various III-V materials.

Fig.3 also indicates the temperature range over which dose rate effects are important. When the implant temperature is much lower or higher than T_c, dose rate effects are usually not observed since either damage production or dynamic annealing, respectively, dominate the final damage structure. Close to T_c, the balance between damage production and dynamic annealing is critically dependent on the dose rate. Higher dose rates raise T_c and induce amorphization, lower dose rates favour dynamic annealing (effectively lowering T_c) resulting in defective crystal as illustrated in Fig.2. Evidence for such behaviour can be readily found from the literature for room temperature irradiated GaAs. For example, Kular et.al. [15] observed the formation of amorphous layers for Si+ implanted GaAs but Fletcher et.al. [16] did not, even though all implant conditions except dose rate were similar. Such discrepancies in observed residual damage are not uncommon [2,17], and highlight the importance of dose rate at temperatures close to Tc. Such behaviour has been well characterized in MeV irradiated Si [18] but no systematic data are available for III-V's.

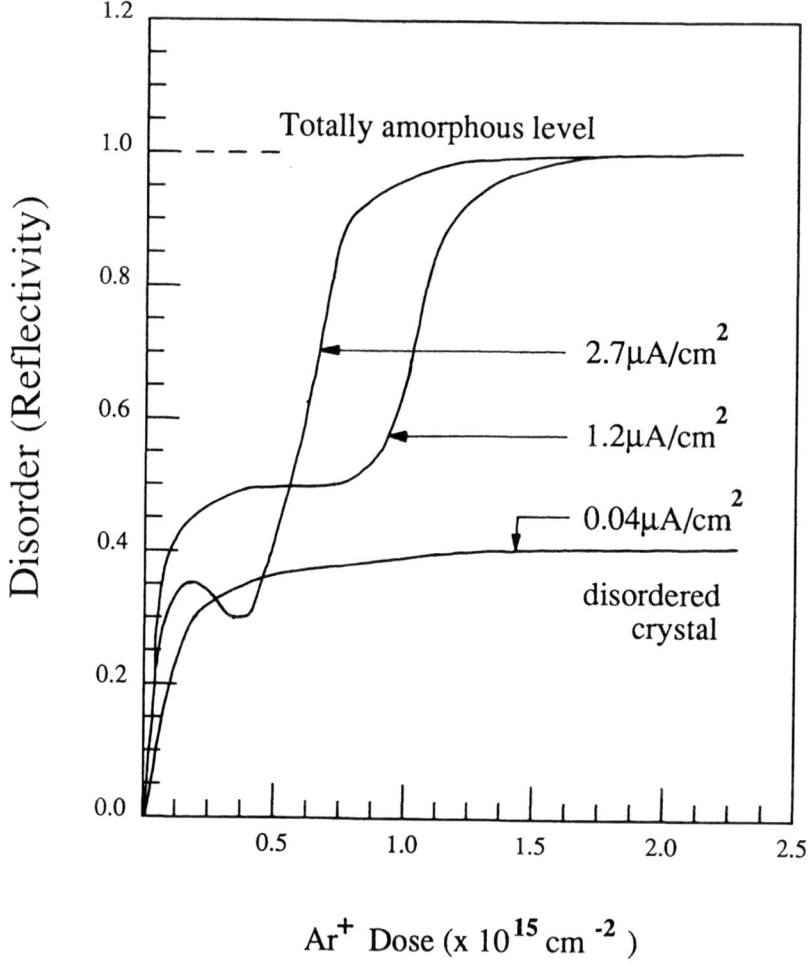

Fig.2   Relative lattice disorder as measured using reflectivity of He-Ne laser light as a function of Ar dose during 100keV Ar+ irradiation of GaAs at 75°C and for 3 different dose rates. After Ref. [12].

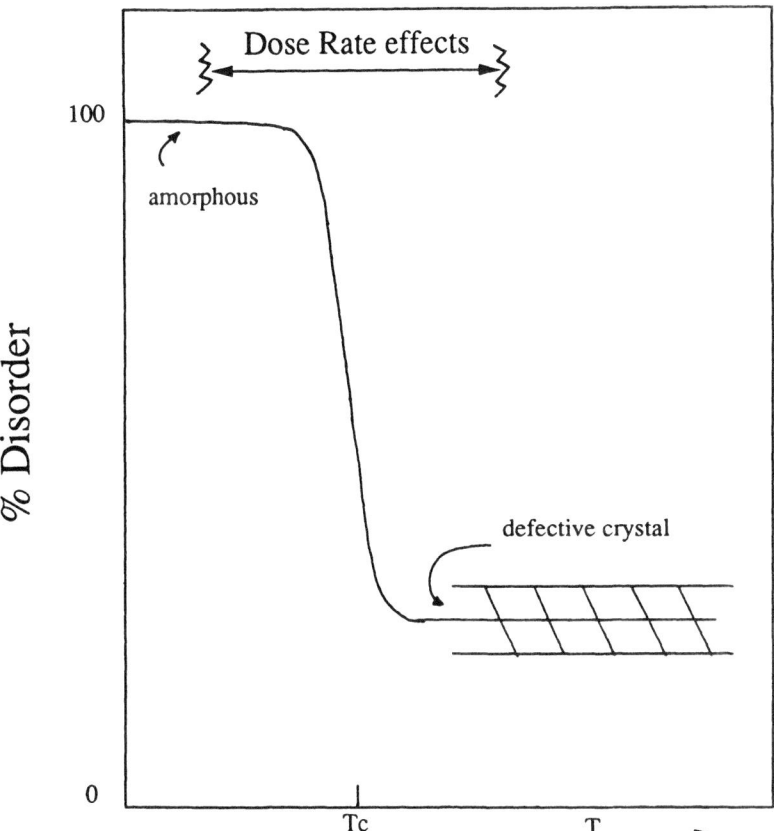

Implantation temperature

Fig.3   Schematic illustration of residual disorder in semiconductors vs. implant temperature (at constant dose, dose rate and species), showing the transition temperature Tc and the region over which dose rate effects can be observed.

DAMAGE ISSUES IN IMPLANT ISOLATION

The production of high resistivity layers in III-V's by introducing implant damage is most effective in large band gap semiconductors and when Fermi level pinning to mid-gap is achieved by introduction of appropriate defect levels [2,4]. The introduction of deep level impurities (such as Cr or O in GaAs or Fe in InP) can also result in Fermi level pinning to mid-gap and produce high resistance layers. This latter (chemical) doping process is more effective than damage isolation in InP where shallow damage levels [2,4] can limit the ultimate resistivity. However, damage isolation is entirely adequate for GaAs/AlGaAs structures where resistivities exceeding $10^8 \Omega/\square$ are achievable [4]. Little data is available for other III-V's although it is unlikely that implant isolation will be a viable technology for small band gap materials such as InGaAs. However, for III-V superlattices and optoelectronic structures, it is often sufficient to destroy the layered composition by inducing interdiffusion under ion irradiation [19,20]. In such structures, ion bombardment may be adequate for locally destroying electrical or optical properties.

The central issues in damage isolation are depth of penetration, isolation efficiency and isolation uniformity. Since most structures to be isolated are of the order of 0.5 to 4μm in thickness, light ions such as protons [2,4] have been employed using conventional < 200kV implanters, even though the isolation efficiency of protons is low, requiring doses of the order of 10^{15}cm^{-2}. Producing deep, uniformly damaged layers can be tedious: up to 9 individual implant steps have been used [4] to achieve optimum isolation in GaAs/AlGaAs HBT's. We suggest that MeV heavy ion irradiation has a potentially important role to play in reducing the complexity of implant isolation processing. For example, MeV ions exhibit a rather uniform rate of energy deposition into damage processes over the first 80% or so of their ion range. This can potentially result in near-uniform damage over the first few microns of a multilayer substrate.

In Fig.4 we show ion channeling spectra [12,21] to illustrate the damage build up for 1.5MeV Ne$^+$ irradiation of GaAs at -40°C (Fig.4a) and 150°C (Fig.4b). In both cases observable damage builds up at a depth of about 1.4μm and extends back to the surface with increasing dose. Considerable dynamic annealing occurs at the higher temperature resulting in significantly less damage than at -40°C. The nature of damage is also different, being essentially amorphous at high doses for the -40C°C case but remaining defective crystal at 150°C. Particularly at the lower doses, the energy deposited into damage production is uniform within 15% over the first 0.5μm of ion penetration, as indicated by TRIM calculations [22]. This situation is ideal for implant isolation.

Fig.5 shows the resistivity behaviour of a 0.2μm n$^+$:0.6μm n: semi-insulating GaAs structure following room temperature bombardment with 1.5MeV Ne$^+$ ions to a dose of 10^{15}cm^{-2} [12]. Although the implant dose and substrate temperature have not been optimised, the isolation properties for a single MeV implant following annealing at 500°C (> $10^8 \Omega/\square$) are entirely satisfactory. Indeed, the sheet resistance following irradiation (~ $10^4 \Omega/\square$) is consistent with gross damage and hopping conduction whereas the higher resistance achieved at 450-600°C is indicative of low damage and defect-related carrier trapping [4,12]. Previous studies [4] have also indicated acceptable isolation properties for MeV O$^+$ implantation of HBT structures. This success warrants further studies of optimum implant isolation conditions for MeV irradiation. The more uniform energy deposition of MeV ions can also aid in understanding of defect-carrier interactions.

ION BEAM INDUCED EPITAXIAL CRYSTALLIZATION

As indicated earlier, ion implantation damage to III-V materials is difficult to remove during subsequent annealing. There are several review papers [1-4, 17,23,24] which discuss the relationships between the nature of implant damage and its influence on subsequent removal and dopant activation. In comparison with Si, amorphous layers are particularly difficult to recrystallize epitaxially in III-V materials [2,17]. In general, recrystallization of both (100) GaAs and InP leads to a high density of microtwins and other defects which adversely affect electrical properties [2,17,23]. Recently, we have shown that

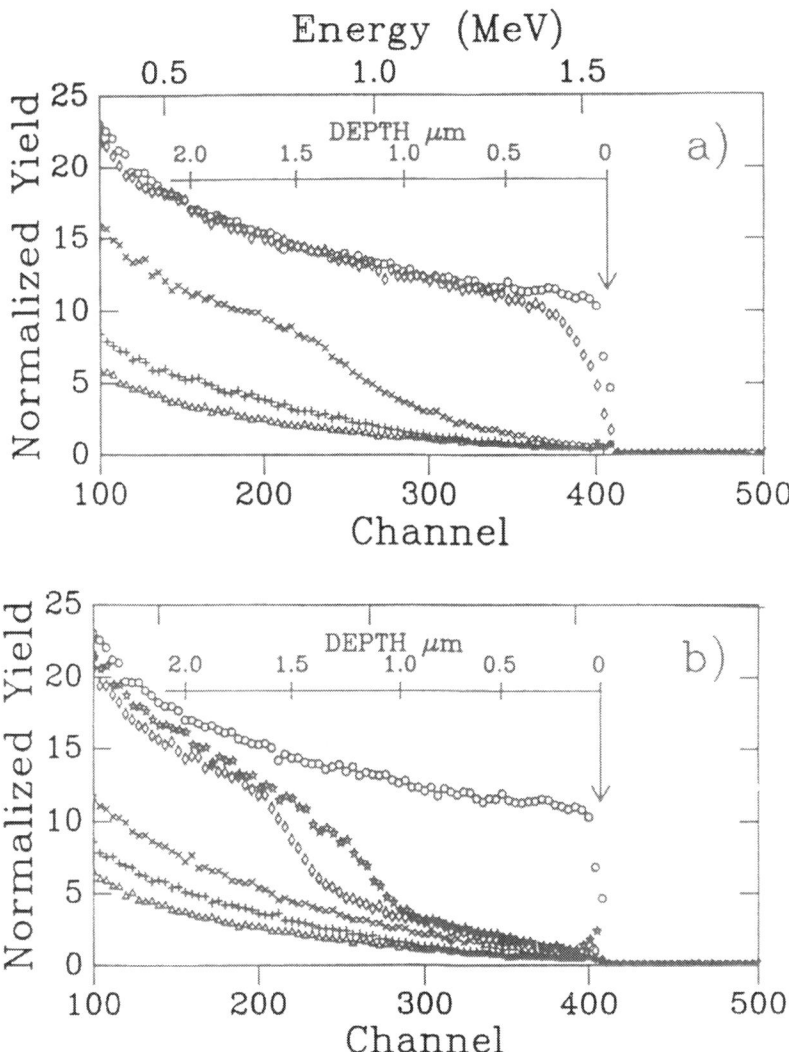

Fig.4 2MeV He ion channeling spectra of GaAs following irradiation with 1.5MeV Ne+ at -40°C (a) and 150°C (b). The open circles and triangles refer, respectively, to random and unirradiated (100) aligned spectra. The aligned spectra in a) refer to Ne+ doses of $10^{13}cm^{-2}$ (+), $10^{14}cm^{-2}$ (x) and $10^{15}cm^{-2}$ ( ); and in b) to $10^{13}cm^{-2}$(+), $10^{16}cm^{-2}$ (x), 3 x $10^{16}cm^{-2}$ ( ) and $10^{17}cm^{-2}$ ( ). After Ref. [12].

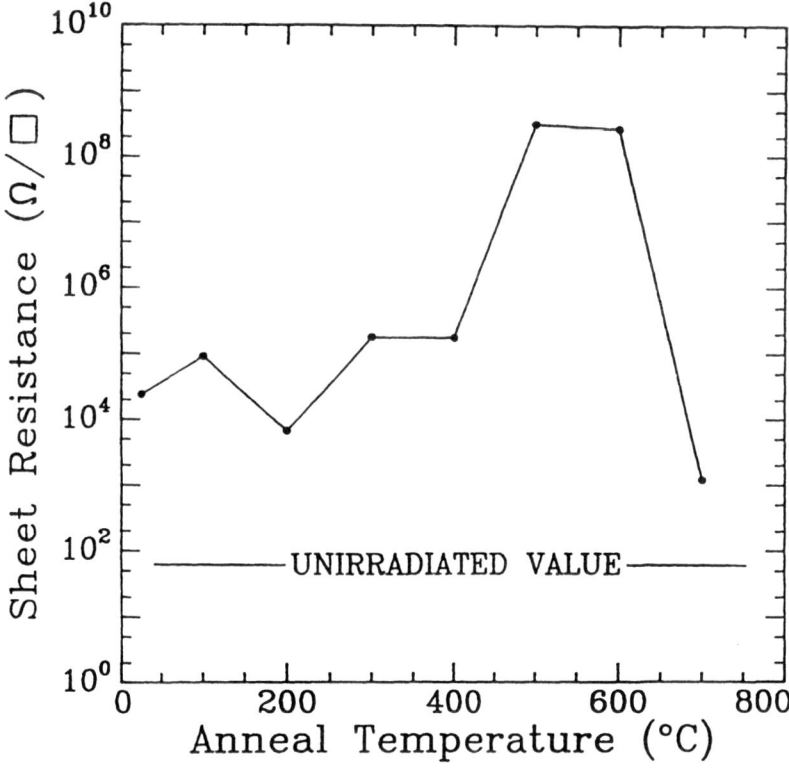

Fig.5 Measured resistivity vs. anneal temperature after irradiation of n+/n/semi-insulating GaAs with 1.5 MeV Ne+ at room temperature to a dose of 1 x 10^{15}cm^{-2}.

MeV ion irradiation of amorphous layers in both GaAs [8] and InP [25] at elevated temperatures can lead to improved epitaxy.

Figs.6 and 7 are ion channeling spectra for InP and GaAs, respectively, which compare MeV-ion-beam-induced crystallization with thermally-induced crystallization. The original amorphous layers were ~ 650Å for GaAs (formed by 15-25keV Si+ implantation at 77K) and ~ 1000Å for InP (formed by 50keV P+ implantation at 77K). Fig.6 clearly shows much improved epitaxial quality following 1.5MeV Ar+ ion beam annealing at 145°C compared with furnace annealing at 260°C [25,26]. Fig.7 shows similar behaviour for GaAs annealed using 1.5MeV Ne+ at 75°C compared with furnace annealing at 250°C. For GaAs, further Ne+ irradiation causes a breakdown in epitaxy and extensive near-surface residual disorder.

Fig.8 more directly compares ion beam annealing of GaAs and InP under similar 1.5MeV Si irradiation at 160°C [27]. The original amorphous layers (~ 1000Å) were both formed by 50keV Si implantation at 77K. Clearly, Fig.8a) shows that epitaxy proceeds to completion with InP, the growth front progressing linearly with increasing Ar dose. There is, however, some residual disorder remaining and Fig.9b indicates that this can be attributed to some tendency for microtwinning, although the level of such disorder is far lower than for furnace annealing. For GaAs (Fig.8b) epitaxy breaks down after only about 300Å of growth. TEM (Fig.9a) shows that polycrystalline material remains in the near-surface in the case of GaAs.

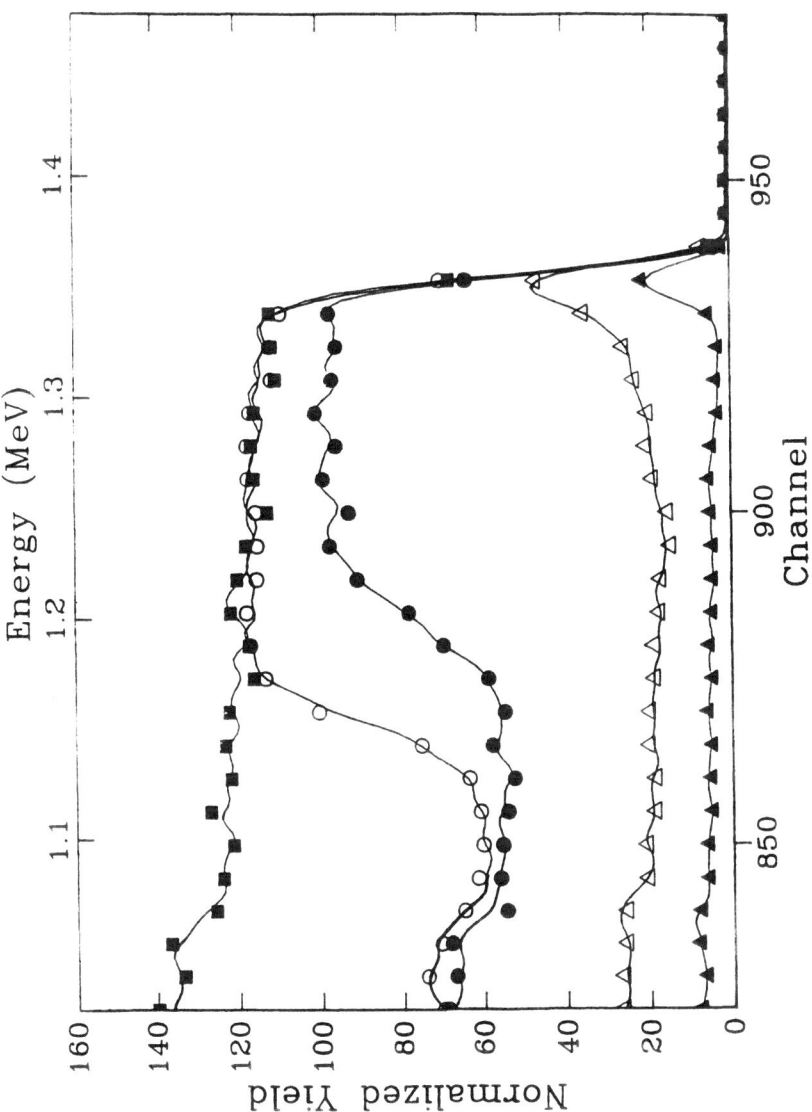

Fig.6   Ion channeling spectra illustrating the quality of crystallization of an ~ 1000Å amorphous layer on (100) InP (open circles) during furnace annealing at 260°C (solid circles) compared with 1.5MeV Ar+ ion beam annealing at 145°C (open triangles). A virgin InP spectrum is also shown (solid triangles). After Ref. [26].

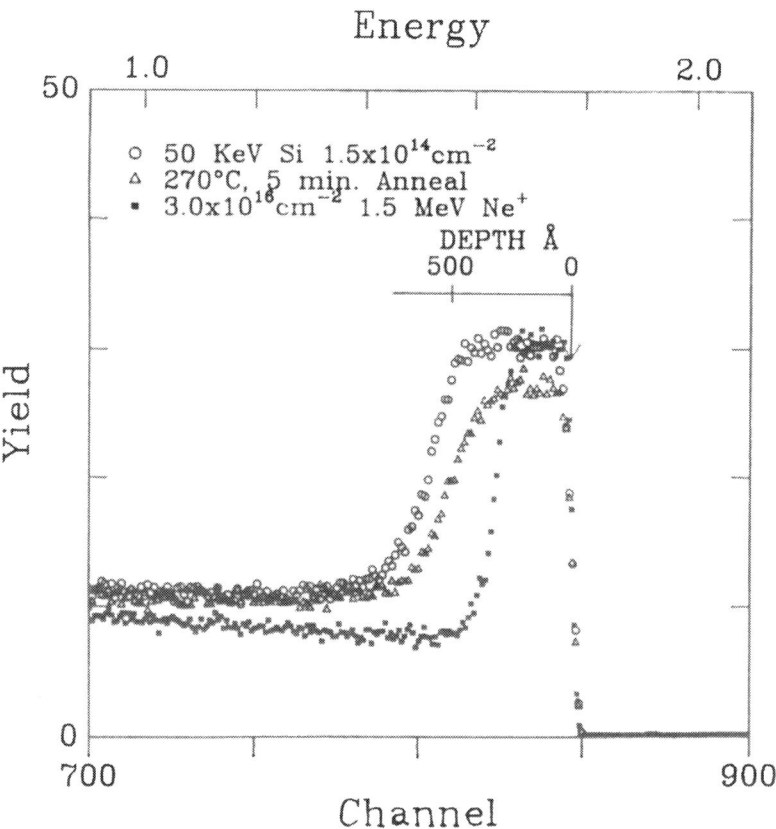

Fig.7   Ion channeling spectra showing the extent of epitaxial growth in GaAs initially
containing a 650Å amorphous layer (open circles): after thermal annealing at 270°C
(triangles) and after partial epitaxy using 1.5MeV Ne+ ions at 75°C (solid squares).
After Ref. [26].

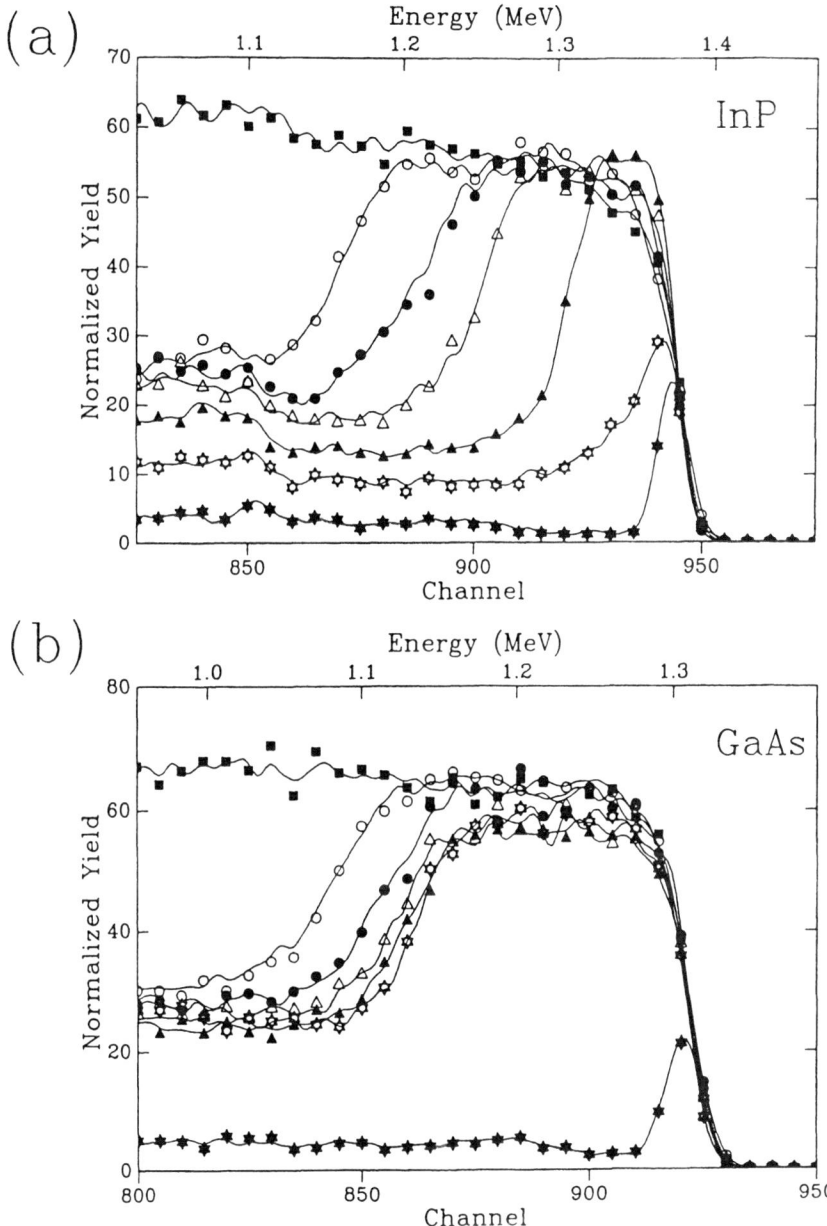

Fig.8   Ion channeling spectra for InP (a) and GaAs (b) irradiated with 1.5MeV Si ions at
160°C to doses of a) 0 (O), 1.7 (●), 5.2 ( ), 8.5 (▲) and 16.9 ( ) x 10^{15}cm^{-2};
and b) 0 (O), 1.2 (●), 1.6 ( ), 2.7 (▲) and 4.9 ( ) x 10^{15}cm^{-2}. Also included
are spectra for random (■) and unimplanted ( ) samples.

a)

b)

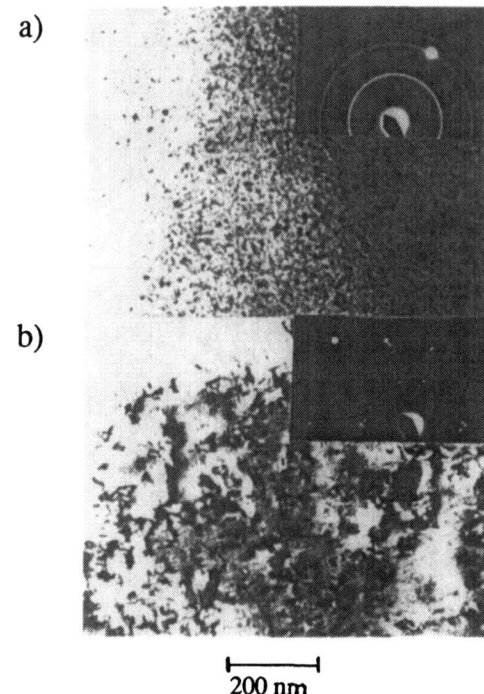

|————————————|
200 nm

Fig.9   Plan view TEM image and diffraction pattern inset (upper right) of amorphized
        GaAs (a) and InP (b) recrystallized at 160°C with 1.5 MeV Si ions to doses of 4.9 x
        $10^{15}cm^{-2}$ and 16.9 x $10^{15}cm^{-2}$, respectively.

In conclusion, MeV ion irradiation can result in substantial improvement in the
quality of epitaxy for InP but results in poor recrystallization for GaAs, where the mode of
crystallization (polycrystalline) appears to differ from the thermal case (heavily twinned).
Similar studies in other III-V's may be warranted to provide a better understanding of beam-
annealing in compound semiconductors. At present, no satisfactory explanation for the
improved InP epitaxy or the different modes of GaAs crystallization are available.

CONCLUSIONS

In this paper we have highlighted a number of areas where MeV ion beams have
application in III-V semiconductors.

(1) Little data exists to assess the nature and quantity of ion damage in various III-V
materials. The ion dose rate is an important parameter which can control whether
amorphization is achieved, particularly when irradiating at temperatures where irradiation
induced defects are mobile. MeV ions can enable controlled experiments on damage
assessment to be undertaken.

(2) MeV ion beams can deposit energy into collisional processes uniformly over the
first few microns of penetration, making them attractive for damage isolation applications.

(3) Irradiation with MeV ions can induce epitaxial crystallization of amorphous III-V layers. For InP, ion beam annealing offers substantially improved epitaxy over conventional furnace annealing. For GaAs, epitaxy is poor for both furnace and ion beam cases but the mode of crystallization appears to be different in the two cases.

Finally, MeV ion beams have much to offer in III-V semiconductors, from aiding the understanding of ion beam damaging processes, through improved epitaxy of amorphous layers to potentially important applications in single step implant isolation.

References

1. K.G. Stephens, Nucl.Instr.Meth. 209/210, 589 (1983).

2. J.P. Donnelly, in "III-V Semiconductor Materials and Devices", ed. R.J. Malik, Elsevier Science Publishers, Amsterdam (1989) p.331.

3. S.J. Pearton, J.S. Williams, K.T. Short, S.T. Johnson, D.C. Jacobson, J.M. Poate, J.M. Gibson and D.O. Boerma, J.Appl.Phys. 65, 1089 (1989).

4. S.J. Pearton, Proceedings of IBMM '90. To be published in NIM B (1991).

5. S.J. Pearton, W.S. Hobson, A.E. Von Neida, N.M. Haegel, K.S. Jones, N. Morris and B.J. Sealy, Mat.Res.Soc.Symp.Proc. 157, 665 (1990).

6. D.E. Davies, J.Cryst.Growth 54, 150 (1981); D.E. Davies, Nucl.Instr.Meth. B7/8, 387 (1985).

7. D.K. Sadana, H. Choksi, J. Washburn, P.F. Byrne and N.W. Cheung, Appl.Phys. Lett. 44, 301 (1984).

8. S.T. Johnson, J.S. Williams, E. Nygren and R.G. Elliman, J.Appl.Phys. 64, 6567 (1988).

9. K.S. Jones and C.J. Santana, J.Mat.Res. (submitted).

10. D.K. Brice "Ion Implantation Range and Energy Deposition Distributions", (Plenum Press, N.Y., 1975) p.1.

11. A.G. Cullis, N.G. Chew, C.R. Whitehouse, D.C. Jacobson, J.M. Poate and S.J. Pearton, Appl.Phys.Lett. 55, 1211 (1989).

12. S.T. Johnson, Ph.D. thesis, University of Melbourne (1989).

13. C.R. Wie, T. Vreeland and T.A. Tombrello, Nucl.Inst.Meth. B16, 44 (1986).

14. D.V. Lang, R.A. Logan and L.C. Kimerling, Phys.Rev. B15, 4874 (1977).

15. S.S. Kular, B.J. Sealy, K.G. Stephens, D. Sadana and G.R. Booker, Solid State Elect. 23, 831 (1980).

16. J. Fletcher, J. Narayan and D.H. Lowndes, Inst.Phys.Conf.Ser. 60, 121 (1981).

17. J.M. Poate and J.S. Williams, in "Ion Implantation and Beam Processing", ed. J.S. Williams and J.M. Poate (Academic Press, 1984) p.13.

18. R.G. Elliman, J. Linnros and W.L. Brown, Mat.Res.Soc.Symp.Proc. 100, 363 (1988).

19. D.G. Deppe and N. Holonyak, J.Appl.Phys. 64, R93 (1988).

20. P. Mei, T.N.C. Venkatesan, S.A. Schwarz, N.G. Stoffel, J.P. Harbison, D.L. Hart and L.A. Florez, Appl.Phys.Lett. 52, 1487 (1988).

21. J.S. Williams, R.G. Elliman, S.T. Johnson, D.K. Sengupta and J.M. Zemanski, Mat. Res.Soc.Symp.Proc. 144, 355 (1989).

22. TRIM (1988). J.F. Ziegler, J.P. Biersack and U. Littmark, "The Stopping and Range of Ions in Solids", Vol.1, Pergamon Press, New York, (1986) p.141.

23. D.K. Sadana, Nucl.Instr.Meth. B7/8, 375, (1985).

24. J.S. Williams and S.J. Pearton, Mat.Res.Soc.Symp.Proc. 35, 427 (1985).

25. M.C. Ridgway, G.R. Palmer, R.G. Elliman, J.A. Davies and J.S. Williams, Appl. Phys.Lett. (in press).

26. J.S. Williams, M.C. Ridgway, R.G. Elliman, J.A. Davies, S.T. Johnson and G.R. Palmer, Nucl.Instr.Meth. (in press, 1990).

PROPERTIES OF PT/TI CONTACTS TO III-V MATERIALS

A. Katz*, S. Nakahara*, S. N. G. Chu*, B. E. Weir*, C. R. Abernathy*, W. S. Hobson*,
S. J. Pearton*, W. Savin**
*AT&T Bell Laboratories, Murray Hill, NJ 07974
**New Jersey Institute

ABSTRACT

Pt/Ti contact to variety of binary III-V and related ternary semiconductor materials were established. These contacts were formed by electron beam evaporation and subsequent rapid thermal processing in order to sinter the metal-semiconductor systems. The contacts to p-type InAs, GaAs, $In_{0.53}Ga_{0.43}As$, $In_{0.52}Al_{0.48}As$ and $Ga_{0.7}Al_{0.3}As$ were ohmic, as a result of heating at temperatures of 450°C or higher. The Pt/Ti contacts to InP and GaP displayed Schottky behavior as-deposited and preserved the rectifying nature through heat treatments, regardless of the processing conditions. The electrical properties and the microstructure evolution in these 7 systems is discussed in this paper.

INTRODUCTION

The issues involved in the thermal processing of high quality ohmic contacts to III-V and related materials have been widely reported and are of great interest in conjunction with device manufacturing technology. The metallization system of choice for the ohmic contacts has to provide the correct electrical link between the active region of the semiconductor and the external circuit while enabling a low energy carrier transport mechanism through the thin interface region and ensuring a negligible series resistance in it under the device operation conditions. Often the contact has to be sintered under relatively high temperature conditions in order to drive the required metal-semiconductor interfacial reaction, which accounts for the decomposition of the interfacial oxides and contaminations and for the formation of some narrow band-gap interfacial compounds [1,2]. These reactions, however, have to be limited and controlled in order to eliminate the formation of spiky interface, which leads to nonuniform current flow and degraded microstructure, and therefore short heating durations by means of RIP are attractive. Moreover, since the contact sintering process takes place at the final stage of the device manufacturing sequence, a moderate heat treatment such as rapid thermal processing (RTP) is essential to minimize the spillover of the dopants into the adjoining layers of the heterostructure devices and the occurrence of different interfacial reactions and diffusion processes.

In addition, another common approach for improving the contact planarity is to use metallization systems which have relatively high eutectic melting points with the III-V semiconductors, and thus provide a thermodynamically stable system and inert metal-semiconductor interfaces.

In this work we provide a detailed summary of the properties of the Pt/Ti bilayer contact to variety of binary and ternary compound semiconductors.

EXPERIMENTAL

<100> Semi-insulating InP and GaAs wafers were used as substrates for the epitaxial growth of heavily doped ($Zn \sim 5 \times 10^{18}$ cm^{-3}) binary and ternary layers, all of which had a thickness of about 0.5 µm. The InP film was grown by liquid phase epitaxy (LPE) [3], the $In_{0.53}Ga_{0.47}As$ (lattice matched to InP) was grown by hydride vapor phase epitaxy (VPE) [2]. The GaAs, GaP, InAs and $In_{0.52}Al_{0.48}As$ (lattice matched to InP) films were grown by metalorganic chemical vapor deposition (MOCVD) [4,5], while $Al_{0.3}Ga_{0.7}As$ (lattice matched to GaAs) film was grown by metalorganic molecular beam epitaxy (MOMBE) [6]. Ti(50 nm)/Pt(60 nm) bilayer metallizations were evaporated by electron-beam with a background vacuum better than 1×10^{-7} Torr and deposition rates of 5 and 10 Å/sec, respectively.

Mat. Res. Soc. Symp. Proc. Vol. 216. ©1991 Materials Research Society

The samples were rapid thermally processed using an A. G. Associates 410T Heatpulse™ annealer under controlled forming gas ambient (15% H_2) at temperatures between 300°C and 600°C. The proximity heating approach [7] was used to prevent substrate decomposition by placing the samples with the metallized side up on a silicon wafer contained within the furnace chamber.

For the electrical measurements, using the transmission line method (TLM), the metal films were deposited onto square openings (200 × 200 μm^2) linearly spaced (with intervals of 10-50 μm), that were wet etched in a plasma-deposited SiO_2 layer (300 nm thick). Subsequently, semiconductor mesas were etched to give the required one dimensional current flow. In addition, broad-area monitor semiconductor layers were used for metallurgical studies.

The analytical examinations involved a variety of techniques such as high resolution field emission scanning electron microscopy (SEM), transmission electron microscopy (TEM) both in cross sectional (XTEM) and top-view (flat-on) modes, Auger electron spectroscopy (AES) with sputter depth profiling, 1.8 and 1.5 MeV $^4He^+$ Rutherford backscattering spectrometry (RBS) and secondary-ion mass spectrometry (SIMS) with Cs^+ primary beam and detecting positive secondary ions.

The electrical characterization involved measurements of current-voltage-temperature (I-V-T) and specific contact resistance by means of TLM, which have been described in detail elsewhere [8].

RESULTS AND DISCUSSION

A. Interfacial microstructure evolution sequence

1. Binary systems

The complicated nature of the phases formation and reactions which take place in metal/III-V semiconductor interfaces, has been widely studied and reported [9]. These reactions in the Pt/Ti metallization scheme to the binary semiconductors, such as InP, InAs, GaP and GaAs were also evaluated [3-6,10], and the scope of this paper is to present them in a summarizing manner.

Figure 1 shows AES depth profiles of the Pt(60 nm)/Ti(50 nm) contacts to InP, InAs, GaP and GaAs substrates, for the as-deposited samples and after sintering by RTP at various temperatures. Some similar trends in the interdiffusion behavior of the elements in the four systems may be observed. In all the systems a Pt-Ti intermixing layer is formed as a result of RTP at the lowest temperature of 300°C. The reaction between the Ti layer adjacent to the substrates and the semiconductor elements is first observed as a result of RTP at the temperature range of 400-450°C.

For the GaAs and GaP, moderate reactions and outdiffusion of Ga at the Ti-substrate interface take place at 400°C. Those reactions are even more pronounced after RTP at 450°C, leading to a binary reaction between the Ti and Ga. Heating the samples to 500°C or higher temperatures enhances the reaction between the Ti and the group V element at the original Ti-semiconductor interface, and causes a migration of Pt and Ga into the intermediate Ti region. This process causes Ga to concentrate at the Pt/Ti original interface and enhances its reaction with the Pt. Heat treatment at 550°C led to the outdiffusion of the excess Ga and Ti to the top metallization surface.

For the InAs and InP there are indications for a considerable reaction at the Ti-substrate as a result of RTP already at 350°C. These reactions are further pronounced at higher sintering temperatures, in which both elements of the substrate undergo an outdiffusion. In is found to accumulate at the original Ti-semiconductor interface, suggesting an extensive reaction with the Ti, and also accumulates at the same interface. The volatile group V element is detected in higher concentration at the interface as well, but RTP at temperatures of 500°C or higher lead to its outdiffusion and spreading toward the surface of the metal contact.

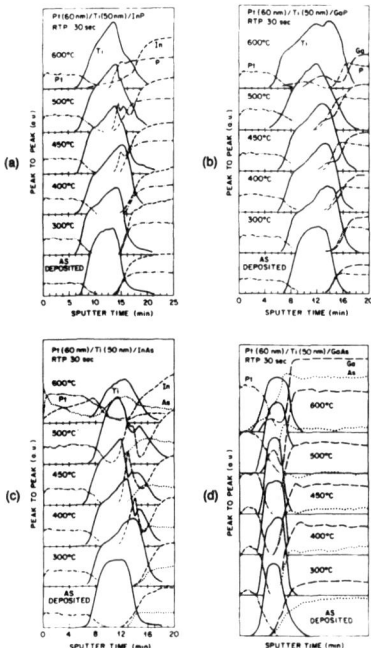

Figure 1   AES depth profiles of Pt(60 nm)/Ti(50 nm) contacts to (a) InP, (b) GaP, (c) InAs, and (d) GaAs, for as-deposited samples and samples that were sintered by RTP at elevated temperatures for durations of 30s.

RBS profiles of the Pt/Ti contacts to the 4 binary semiconductors were taken. These results provide further verification of the above mentioned interfacial reactions. The RBS spectra at 170° were taken at 1.8 MeV on the as-deposited samples and after RTP at 300, 350, 400, 450 and 500°C. All spectra were collected at low counting rates and were reproducible. The RBS utilities and manipulation package (RUMP) simulation was done for each of the spectra and were used to determine the systematic trend for layer intermixing. Figure 2 summarizes in a schematic manner the RUMP analysis. As a generality for all the systems, the Ti and Pt layers were shown to start mixing due to RTP at 300°C, resulting in the formation of a very thin (less than 5 nm) interfacial layer. The thickness of the intermixed layer expanded as the sintering temperature was elevated, but the Pt-Ti layer never exceeded a thickness of about 20 nm. The As from both InAs and GaAs substrates underwent an outdiffusion and penetrated into the Ti layer as a result of RTP at 350°C. Both substrate elements in the two semiconductor systems were detected in a considerable amount in the ternary interfacial layer (10 nm thick), formed in between the Ti layer and the InAs or GaAs. RTP at 450°C led to the thickening of this layer to about 25 nm, as well as significant outdiffusion of the As in the InAs system, and Ga in the GaAs system, through the Ti layer to accumulate at the interfacial layer between the Pt and Ti. This type of outdiffusion of one of the substrate elements through the Ti layer was not observed either in the InP or in the GaP systems. RTP at 500°C and higher temperatures led to a significant expansion of the overall ternary mixed layers (about 60-80 nm thick), which exist between the substrate and the metals top layers. A complete collapse of the layered structure was observed in all the systems after RTP at 550°C, which resulted in an extensive outdiffusion of both substrate elements to the contact surface, as well as a relatively deep penetration of Ti into the substrate and the formation of a wide metal-semiconductor reacted layer.

308

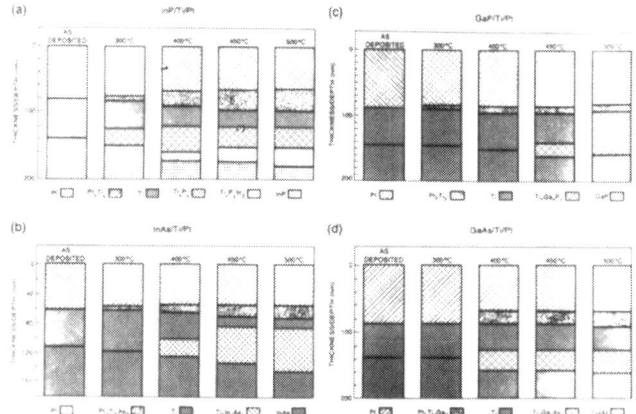

Figure 2   Schematic representation of the layer composition and thickness of
Pt(60 nm)/Ti(50 nm) contacts to (a) InP, (b) InAs, (c) GaP, and (d) GaAs, as-
deposited and after RTP at 300, 400, 450 and 500°C for durations of 30s.

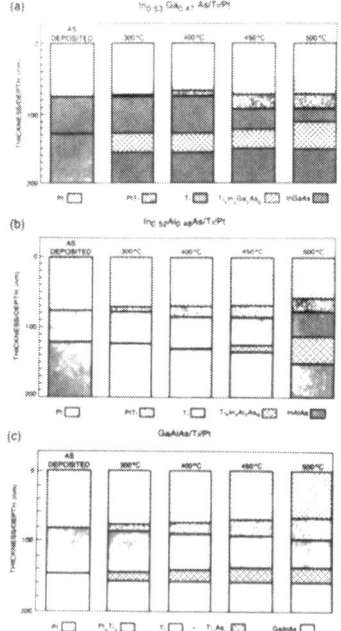

Figure 3   Schematic representation of the layer composition and thickness of the
Pt(60 nm)/Ti(50 nm) contacts to (a) $In_{0.53}Ga_{2.47}As$, (b) $In_{0.52}Al_{0.48}As$, and
(c) $Ga_{0.7}Al_{0.3}As$, as-deposited and after RTP at 300, 400, 450 and 500°C for
durations of 30s.

From the AES and RBS observations, one can conclude that the main phase evolution in all the studied systems took place as a result of RTP at temperatures of 400°C or higher. TEM analysis of specially prepared Pt(10 nm)/Ti(10 nm)/III-V-substrate samples provided further information regarding the identification of the formed phases and were reported elsewhere [4,5,11].

2. Ternary systems

The same analytical tools were applied to study the Pt/Ti contacts to 3 ternary compound semiconductors substrates, $In_{0.53}Ga_{0.47}As$, $In_{0.52}Al_{0.48}As$ and $Ga_{0.7}Al_{0.3}As$. A detailed discussion of the results of these studies are provided elsewhere [2,6,12,13]. Figure 3 schematically summarizes the RUMP analysis, which were carried out on each of the samples in these 3 systems. The results were verified by the AES analysis as well as by TEM studies. The Pt and Ti layers were shown to start mixing after RTP at 300°C. The thickness of the Pt-Ti intermixing layer in all systems increased as the sintering temperature was elevated, while the most limited reactions took place at the Pt/Ti/$In_{0.53}Ga_{0.47}As$ system, and gradually expanded through the Pt/Ti/$In_{0.52}Al_{0.48}As$ to the Pt/Ti/$Ga_{0.7}Al_{0.3}As$ system. In the former system a quaternary intermixing layer, about 20 nm thick, containing Ti, In, Ga and As was observed at the original Ti/$In_{0.53}Ga_{0.47}As$ interface after RTP at 300°C. After RTP at 500°C the thickness of this layer was about 50 nm, almost adjacent to the Pt-Ti intermixing layer, which was formed at the Pt/Ti original interface. An interfacial quaternary intermixing layer was detected at the Pt/Ti/$In_{0.52}Al_{0.48}As$ system, as well, containing Ti, In, Al and As, but began to develop only as a result of RTP at 450°C. The Pt/Ti/$Ga_{0.7}Al_{0.3}As$ system exhibited a different behavior through the thermally driven reaction. In this system only the As element reacted with the subsequent Ti to form a binary interfacial mixing layer, widening from about 15 nm as a result of RTP at 300°C to about 30 nm after sintering at 500°C.

B. Electrical properties of the Pt/Ti contacts

V-I measurements of the e-gun evaporated Pt/Ti contacts to the 5×10^{18} cm^{-3} Zn-doped binary and ternary substrates were carried out. The only binary systems that showed a linear V-I curve, already as-deposited, was the Pt/Ti/InAs, which suggest that Pt/Ti formed an ohmic contact to this substrate without the need of any interfacial mixing due to sintering process. Some reduction of the contact resistance, reflected from the lesser steep slope of the V-I curves, was observed, at the sampler after RTP, with a minimum value after sintering at 450°C. The Pt/Ti contacts to InP and GaP substrates were Schottky as-deposited and preserved this rectifying nature through all the RTP treatments, regardless of the sintering temperature. The Pt/Ti contacts to the GaAs substrate underwent transition from a rectifying contact, as-deposited, to an ohmic contact, as a result of RTP at temperatures higher than 450°C.

The Pt/Ti contacts to all the ternary systems exhibited a linear V-I behavior after RTP at temperatures higher than 300°C. The contacts to the substrates which contain Al were rectifying as-deposited, but underwent the transition to ohmic contacts when sintered at this low temperature for a short duration (30 sec). For the Pt/Ti contacts to the $Ga_{0.7}Al_{0.3}As$ the transient may have been attributed to the binary Ti-As interfacial reaction which took place at 300°C (see Fig. 3). For the contact to $In_{0.52}Al_{0.48}As$ the explanation may be more complicated, since the metal-semiconductor interface showed an inert behavior through heating up to 450°C. In both cases the decomposition of the metal-semiconductor interfacial native oxides as a result of RTP may have been the major contributor to the decrease in the contact resistance and thus for the transition from Schottky to ohmic contacts.

Figure 4 shows the dependence of the specific resistance of the Pt/Ti contacts to the 5×10^{18} cm^{-3} Zn-doped InAs, InP, $In_{0.53}Ga_{0.47}As$, $Ga_{0.7}As_{0.3}As$ and $In_{0.52}Al_{0.48}As$, on the RTP temperature. All the contacts, regardless of the semiconductor type, had the lowest specific resistance after sintering at 450°C. The lowest value of about 1.1×10^{-6} $\Omega \cdot$ cm^{-2} was measured for the contact to InAs, and increased gradually through the contacts to $In_{0.53}Ga_{0.47}As$, GaAs, $In_{0.52}Al_{0.48}As$ and $Ga_{0.7}Al_{0.3}As$, having a highest value of 3×10^{-5} $\Omega \cdot$ cm^{-2}.

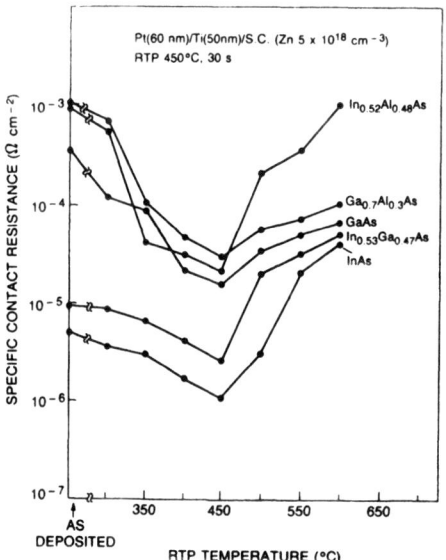

Figure 4   Specific resistance as a function of RTP temperature (t = 30s) of the Pt/Ti contacts to 5×10^{18} cm^{-3} Zn doped InAs, In$_{0.53}$Ga$_{0.47}$As, GaAs, Ga$_{0.7}$Al$_{2.3}$As, and In$_{0.52}$Al$_{0.48}$As, measured at 25°C.

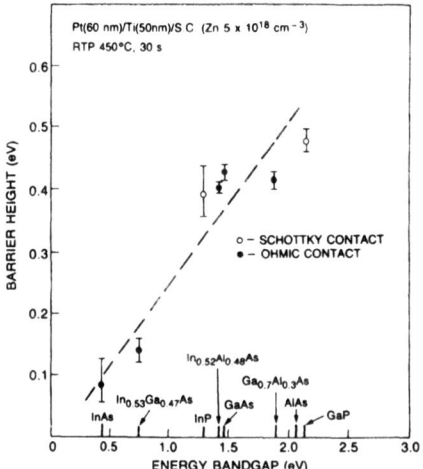

Figure 5   Apparent barrier heights of the Pt/Ti contacts to 5×10^{18} cm^{-3} Zn doped III-V binary and related ternary semiconductors, as a function of the semiconductor bandgap energy.

Evaluation of the exponential dependence of the contact resistance as a function of either the reciprocal of the measured temperature or the doping concentration provides a general indication with regard to the primary carrier transport mechanism mode across the interfacial barrier. A better approach for modeling the transport across the highly activate interface of metal contacts to III-V semiconductors is to consider that simultaneous transport mechanisms occur, and each of these processes partially contribute to the overall transport. We have demonstrated the viability of this concept for different metal-semiconductor systems [2,5]. This model accounts for the intensive interfacial reaction between the metal and the group III low melting point and group V volatile elements, which produce a complicated boundary layer. Moreover, modeling and fittings of the apparent measured barrier heights of the various studied contacts to the calculated one were done and are presented in Figure 5, as a function of the semiconductor bandgap width. A clear dependence of the Pt/Ti/semiconductor as-deposited contact barrier height on the semiconductor energy bandgap is observed. Fitting the measured results to a linear line provide a slop of about 0.25, which is by itself an interesting behavior trend of the given Pt/Ti bilayer metallization scheme to the various III-V binary and related ternary compound semiconductors.

SUMMARY AND CONCLUSIONS

The interfacial reaction of the Pt/Ti contacts to III-V binary and related ternary compound semiconductors have been studied by means of TEM, AES and RBS. Phase formation, film microstructure, interfacial morphology, and their effect on the electrical properties were considered. It was found that the Pt/Ti contacts to InAs and $In_{0.53}Ga_{0.47}As$ were ohmic as-deposited while the same metallization scheme to GaAs, GaP, InP, $In_{0.52}Al_{0.48}As$ and $Ga_{0.7}Al_{0.3}As$ provided a rectifying contact as-deposited. Applying RTP at temperature of 300°C or higher for the ternaries, and the 450°C or higher for GaAs, led to a transition of these contacts to ohmic behavior. The Pt/Ti contacts to InP and GaP remained Schottky regardless of the applied sintering cycle. The apparent barrier height energies of these contacts were correlated to the energy bandgap of the various semiconductors and a fitting line with a slope of about 0.25 was obtained.

REFERENCES

[1] R. Singh, R. P. S. Thakur, A. Katz, A. J. Nelson, S. C. Gebhard and A. B. Swartzlander, in Appl. Phys. Lett., 57, 1239 (1990).

[2] S. N. G. Chu, A. Katz, T. Boone, P. M. Thomas, V. G. Riggs, W. C. Dautremont-Smith and W. D. Johnston, Jr., J. Appl. Phys., 67, 3754 (1990).

[3] A. Katz, B. E. Weir, S. N. G. Chu, P. M. Thomas, M. Soler, T. Boone and W. C. Dautremont-Smith, J. Appl. Phys., 67, 3872 (1990).

[4] A. Katz, S. Nakahara, W. Savin and B. E. Weir, J. Appl. Phys., 68, 4133 (1990).

[5] A. Katz, S. N. G. Chu, B. E. Weir, W. C. Dautremont-Smith, R. A. Logan and T. Tanbun-EK, J. Appl. Phys., to be published.

[6] A. Katz, C. R. Abernathy, S. J. Pearton and B. E. Weir, J. Appl. Phys., to be published.

[7] S. J. Pearton and R. Caruso, J. Appl. Phys., 67 (1989).

[8]   A. Katz, P. M. Thomas, S. N. G. Chu, W. C. Dautremont-Smith, R. G. Sobers and S. G. Napholtz, J. Appl. Phys., **67**, 884 (1990).

[9]   See for example J. Woodal, N. Braslan and J. L. Freeouf, in "Physics of Thin Films", Ed. by M. H. Francombe and J. L. Vossen (Academic Press Inc., New York, 1987), 199-225.

[10] A. Katz and W. C. Dautremont-Smith, J. Appl. Phys., **67**, 6237 (1990).

[11] A. Katz and S. Nakahara, unpublished results.

[12] A. Katz, W. C. Dautremont-Smith, S. N. G. Chu, P. M. Thomas, L. A. Koszi, J. W. Lee, V. G. Riggs, R. L. Brown, S. G. Napholtz, and J. L. Zilko, Appl. Phys. Lett., **54**, 2306 (1989).

[13] A. Katz, B. E. Weir and W. C. Dautremont-Smith, J. Appl. Phys., **68**, 1123 (1990).

MULTIWAFER PRODUCTION OF LONG WAVELENGTH EPITAXIAL MATERIAL

M. HEYEN, D. SCHMITZ, G. STRAUCH AND H. JÜRGENSEN
AIXTRON GmbH, Kackertstraße 15 – 17, D – 5100 Aachen, FRG

INTRODUCTION

Low pressure MOVPE in a horizontal reactor has proven to be capable of yielding uniform InP, GaInAs and GaInAsP layers [1, 2]. However, the complexity of some devices as MQW lasers and HEMTs require even further improvement in thickness and compositional uniformity. This can be achieved by using the technique of sub – strate rotation to overcome gas phase depletion problems and geometry related non uniformities.

Techniques for the pratical realization of such systems using a mechanical drive have been described earlier [3]. However, the technical solution for substrate rotation, using mechanical feed throughs was only applied to small, laboratory scale systems.

In this paper we will discuss the use of the new "gas foil rotation" technology for growth in a low pressure reactor. In this technique the wafer holder is floating on a foil of high purity hydrogen. Since the hydrogen is forced to flow with a circular compo – nent, the plate with the substrate is put into rotational motion [4, 5].

InP, GaInAs and GaInAsP layers grown in a single wafer reactor on 2" substrates applying this substrate rotation show film thickness variations less than 2% over the entire wafer. Resistivity measurements on doped binary layers showed variations of less than 2%. The lattice mismatch variations of ternary and quaternary layers was below 5×10^{-4}. The electrical properties of GaInAs and InP as residual carrier con – centration and electron mobility in undoped layers were identical to those measured on reference layers grown in the same reactor on a static susceptor. Growth para – meters had not to be modified when using this technique. The gas foil rotation as – sembly does not require mechanical drive parts, feed through etc., and thus does not generate particles; it can easily replace the standard static susceptor with only minor changes in the gas supply system.

The method has also been applied in multiwafer reactors, where the wafers rotate in "planetary" motion. Similar results have been obtained in a horizontal reactor with 5 wafers rotating in planetary motion.

The growth of GaAs/AlGaAs heterostructures and InGaAs with outstanding film pro – perties and high requirements on film homogeneities such as HEMTs [1, 2] has been demonstrated in an atmospheric pressure reactor designed for 7 wafers rotating in planetary motion in a radialsymmetric horizontal flow system.

EXPERIMENTAL

The principle of gas foil rotation has been described in fig. 1. The susceptor consists of the lower graphite body and a cylindrically shaped graphite wafer carrier. A system of circularly shaped grooves at the bottom of the depression allows introduction of gas in such a way that the plate is floating on a gas foil. Since the gas is forced to flow with a circular component, the plate with the substrate is put into rotational mo-tion. A circular groove at the outer rim and a draining channel collect the gas and lead it to the exhaust. By doing so, it is avoided, that the driving gas is released into the growth atmosphere above the substrate and changes its composition.

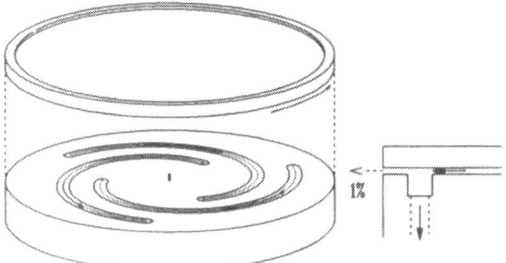

Fig. 1   Schematic of gas foil rotation susceptor

The driving gas is purified H_2 which is also used as carrier gas. The amount of H_2 to achieve a rotating speed of $2\ s^{-1}$ (1 to 2 revolutions per monolayer; deposition tem-perature 640 °C; total pressure 20 mbar) is around 10 ml/min. This flow is much smaller than the total gas flow in the reactor (6 slm).

The susceptor described above has been used in a commercial single wafer reactor [1, 2] by simply replacing the static susceptor. Optimum quality of epitaxial layers could be achieved without changing any growth parameters.

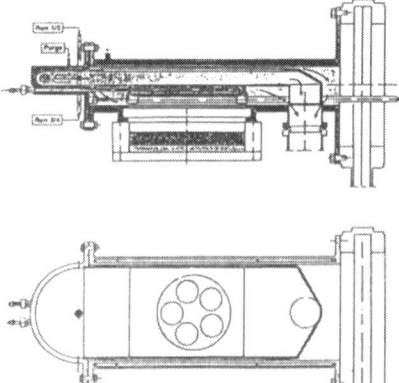

Fig. 2   Schematic of multiwafer reactor for 5x2", 3x3", or one wafer up to 6" diameter

The schematic diagram of a multiwafer reactor has been shown in fig. 2. The sche – matic shows a wafer holder for 5x2" wafers. In this case two different gas flows drive the large disc and the small discs independently. The large disc can be easily replaced by different inserts for growth on 3x3" wafers or one wafer up to 6".

The third reactor concept is described in fig. 3. In this case a central gas inlet is used. Again a main rotating plate is used, with individual rotating wafer holders for 7x2" or 5x3" wafers rotating in planetary motion [5].

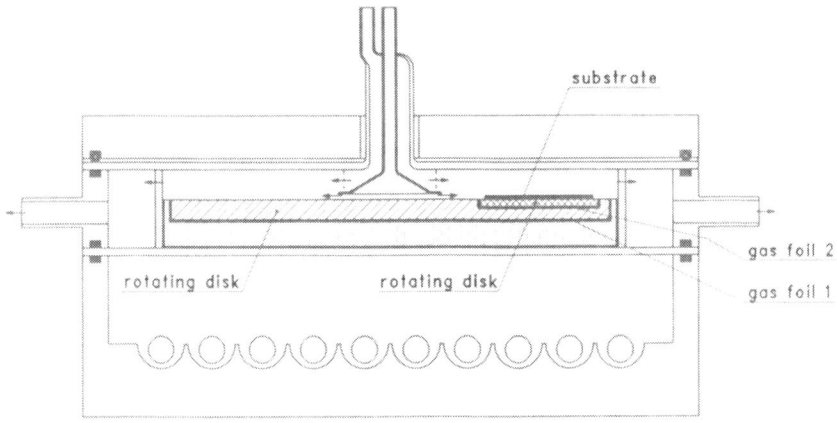

Fig. 3   Schematic of "planetary" reactor for 7x2" or 5x3" wafers

RESULTS AND DISCUSSION

As expected, the results on uniformity of film thickness, carrier concentration and composition could be further improved by applying the gas foil rotation technique to the single wafer reactor. The variations in InP, GaInAs and GaInAsP film thickness could be limited in the order of 1% to 2%. Similar data were obtained in the GaAs/AlGaAs system.

The lattice mismatch Δ a/a of GaInAs varies less than $\pm$ 2x10^{-4}. Wide band gap ma – terials in the Ga – In – As – P alloy system (λ = 1.05 µm) showed variation of less than 2 nm over the entire area of a 2" wafer.

The gas foil rotation principle has also been applied in the 5x2" wafer growth chamber with planetary motion of the wafers. Compound semiconductors in the Al – Ga – As and Ga – In – As – P alloy systems were grown.

Figs. 4a and 4b shows histograms of the material properties over the full load of wa – fers possible in one run depositing AlGaAs layers on GaAs substrates. The determi – nation of the AlGaAs film thickness was performed by using a selective etching tech –

nique and DEKTAK measurement. The alloy composition was measured by DCXD.

The process parameters for this growth run were chosen to meet the minimum con – dition of rotation speed (more than one revolution per deposition time of one lattice constant of the appropriate material). The film thickness variation of the average values over the full susceptor load is less than ± 1% as shown in fig. 4a. This also represents the uniformity of GaAs layers grown in the same reactor in a similar way. The histogram in the figure shows the distribution of single measured values over all five wafers.

Fig. 4b shows the variation of alloy composition in the same run. As can be seen from the histogram the variation over all five wafers is less than 2% in all measured points. Taking into account the resolution of the measurement method this can be conside – red an excellent result.

Similar experiments were carried out for the more critical Ga – In – As – P alloy system on compositions for various emission wavelengths. As an example the properties of the λ = 1.3 µm emitting GaInAsP grown in the described reactor are presented. From the results that have been obtained from the GaAs based materials the experimental conditions for the rotating speed in relation to the growth rate were considered to follow similar rules in order to obtain optimum results.

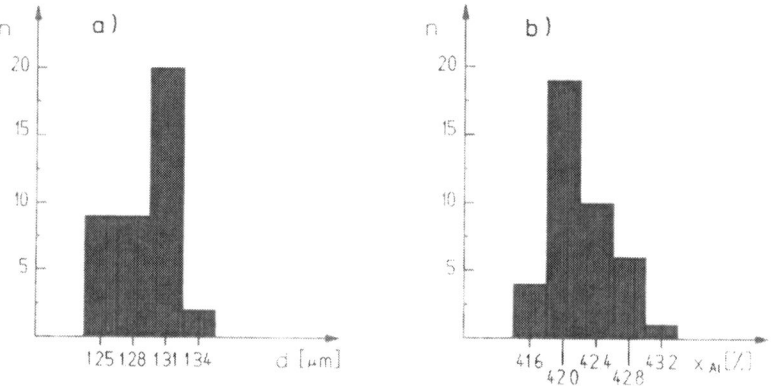

Fig. 4 a) Variation of AlGaAs film thickness over 5x2" wafers
       b) Variation of AlGaAs film composition over 5x2" wafers

In fig. 5a the film thickness distribution over a full load of five 2" wafers in this reaction cell is plotted. The distribution of the measured values, as shown in the histogram, shows thickness variations of less than 2% over the entire load and less than 1% in the average between different wafers.

The alloy composition in the layers described before is represented by the plot of the emission wavelengt distribution in fig. 5b. Variations in λ of less than 2 nm around the average are obtained in the described run. The histogram shows that not only wafer to wafer reproducibility is excellent but also the uniformity on one single wafer is meeting the requirements for device fabrication. This is a basic progress once the requirements for optical devices have become more stringent by the development of more complicated device structures.

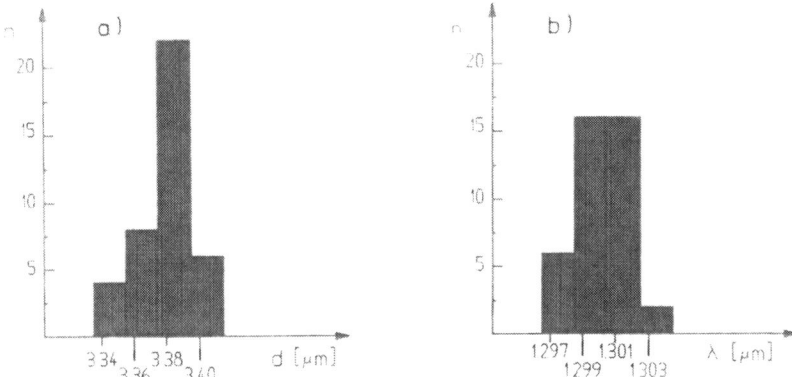

Fig. 5 a)   Variation of GaInAsP film thickness over 5x2" wafers
     b)   Variation of GaInAsP emission wavelength over 5x2" wafers

The planetary reactor for seven 2" wafers has a design as described earlier by P. Frijlink [5]. In difference to the reactor types discussed above it uses a radialsymme -tric gas inlet. Also in this case the gases pass horizontally over the wafers as in the other reactors. The main advantage of this design is, that further upscaling of the al -ready very high capability of seven 2" or five 3" wafers is easily possible.

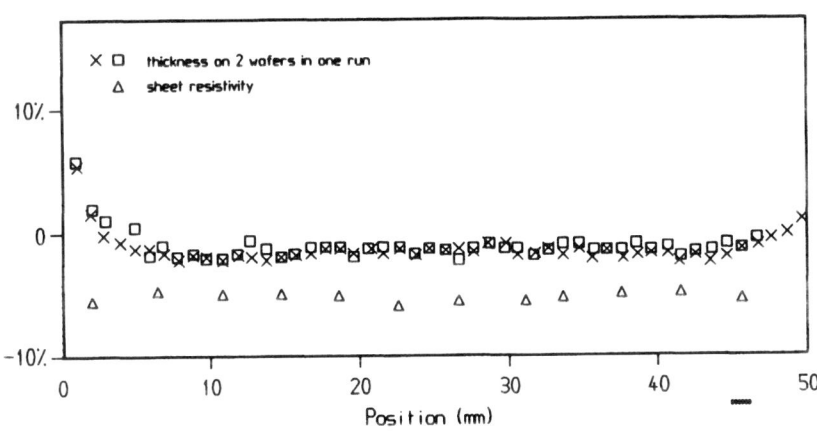

Fig. 6   Distribution of film thickness and sheet resistivity over the area of 2x2" wafers grown in the 7x2" planetary reactor [5]

The uniformity results in film thickness and sheet resistivity distribution over samples of Si doped GaAs wafers are better than 2% over the main part of the wafer area. Fig. 6 shows the distribution of the layer properties measured over a diameter of two sample wafers prepared in one run. The residual rim effects occurring here in a re- latively close distance to the wafer edges are more likely due to gas flow dynamic problems close to the wafer edges than to depletion effects.

Microwave devices such as HEMT structures have been prepared in the planetary reactor. The typical mobility versus temperature dependence of such a HEMT struc- ture is shown in fig. 7. An electron mobility of more than 650.000 cm^2/Vs could be measured at LHe temperature in the 2DEG.

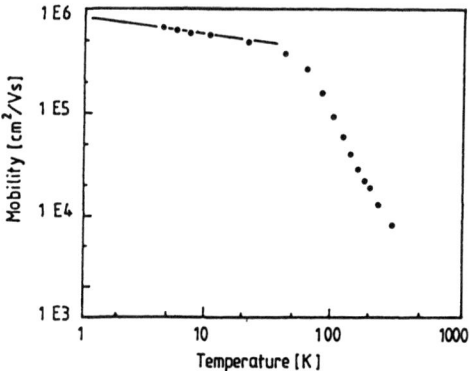

Fig. 7   Dependence of electron mobility on temperature in GaAs/AlGaAs 2DEG structure grown in the planetary reactor [6]

The verification of real HEMT devices on entire 2" wafers is strongly limited by the device preparation process. So the yield and device performance uniformity of a processed epistructure gives a figure that is mainly contributed by the processing. Fig. 8 shows the current gain cutoff frequency distribution of HEMTs processed on a 2" wafer, that was grown in the planetary reactor.

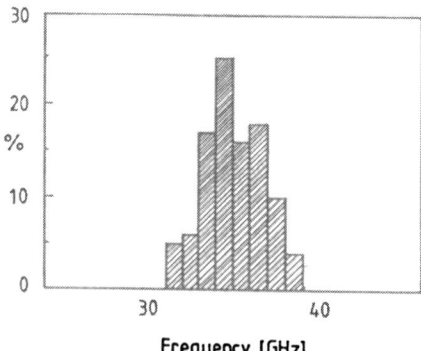

Fig. 8   Histogram of HEMT current gain cutoff frequency in devices prepared on a 2" wafer from the planetary reactor [6]

CONCLUSION

The principle of wafer rotation in a laminar horizontal gas flow reactor has been proven to improve the uniformity of the grown layers to a number that satisfies most of the requirements for advanced optical and microwave device production. The principle of gas foil rotation introduced by P.M.Frijlink is a simple method that can be easily retrofitted to a variety of commercial reaction chambers using laminar horizontal gas flow.

The results obtained with such reactors were uniformities of better than 2% in the important electronical, optical and dimensional material properties in epitaxial struc-tures over almost entire 2" wafer areas. Very high quality microwave device results qualify the reactors for mass production of device structures on GaAs and InP.

The relatively simple realization possible for such a variety of different sizes emphasize the efforts to create even larger reaction cells capable for large area epitaxial growth of different material systems. The use of silicon wafers as substrates for advanced device structures is no longer limited to the 2" to 3" sizes.

REFERENCES:

[1]   M. Heyen, M. Heuken, G. Strauch, D. Schmitz, H. Jürgensen and K. Heime: Proc. of MRS Spring Meeting, San Diego CA (1989) 245.

[2]   D. Schmitz, G. Strauch, H. Jürgensen, M. Heyen and P. Harde: Proc. "1st Int'l. Conf. on InP and Rel. Comp.", Norman OK (1989).

[3]   A. Mircea, R. Mellet, B. Rose, P. Dasté and G. Schiavini: J. Cryst. Growth 77 (1986) 340.

[4]   E. Woelk and H. Beneking: J. Cryst. Growth 93 (1988) 216.

[5]   P.M. Frijlink: J. Cryst. Growth 93 (1988) 207.

[6]   P.M. Frijlink: Accepted for publication in J. Cryst. Growth "Conf. Proc. IC MOVPE 5", Aachen FRG (1990)

MATERIALS ISSUES AND DEVICE-DEGRADATION IN THE InGaAs(P)/InP SYSTEM

O. UEDA
Fujitsu Laboratories Ltd., 10-1 Morinosato-Wakamiya, Atsugi 243-01, Japan

ABSTRACT

This paper reviews the current status of material issues in the InGaAs(P)/InP system and our understanding of degradation in InGaAsP/InP double-heterostructure lasers and LED's.

Among the materials issues for this system are the generation of defects and the thermal stability of the material. Crystal growth-induced defects can be classified into interface and bulk defects.; the former group includes misfit dislocations and inclusions and the latter structural and non-structural precipitates. Thermal stability can lead to structural imperfections such as quasi-periodic modulated structure due to spinodal decomposition of the crystal either at the liquid/solid interface or growth surface, and atomic ordering which also occurs on the growth surface through migration and reconstruction of deposited atoms. Diffusion processes in InP often lead to the generation of microdefects in this material. Finally, in metallization procedures, alloy reactions between the semiconductor and electrode occasionally take place forming non-planar/non-uniform interfaces.

Three major degradation modes, rapid degradation, gradual degradation, and catastrophic failure in InGaAsP/InP double-heterostructure lasers and LED's, are discussed. For rapid degradation, recombination-enhanced dislocation climb and glide, which are responsible for degradation, do not occur easily InGaAsP/InP system . Differences in the ease with which these phenomena occur in InGaAsP/InP and other systems are presented. Based on the results, dominant parameters involved in the phenomena are discussed. Gradual degradation takes place presumably due to recombination enhanced point defect reaction in GaAlAs/GaAs-based optical devices. However, we do not observe this mode in InGaAsP/InP-based optical devices. Catastrophic failure due to catastrophic optical damage at a mirror or at a defect in GaAlAs/GaAs DH lasers, is not found in InGaAsP/InP DH lasers. Degradation of InGaAsP/InP double-heterostructure LED's with large current application is also presented. These results indicate that InGaAsP/InP optical devices ensure very high reliability in long-wavelength communication systems.

INTRODUCTION

Since the discovery of semiconductor lasers in the early 1970's [1], a variety of optical devices have been developed from III-V alloy semiconductors. Among the many fields of application are fiber optical communication systems, digital audio systems, and optical printers. In the development of these devices, high reliability has been one of the most important goals. Degradation is a major problem in this regard. It is now understood that defects in the crystal often play a large role in device-degradation.

Regarding InGaAsP/InP double-heterostructure (DH) lasers and LED's, they have been extensively developed since the wavelength range where silica transmission fibers exhibit low loss and chromatic dispersion moved to the region longer than 1.0 μm. In addition, it has been found that they do not rapidly degrade as compared with the GaAlAs/GaAs optical devices, which is favorable for practical use in optical fiber communication systems.

In this paper, we describe materials issues in InGaAs(P)/InP system, i.e., generation of defects during growth and device-processes and phenomena induced by thermal instability of the material, and our current understanding of degradation in InGaAsP/InP DH lasers and LED's in terms of three major degradation modes, rapid degradation, gradual degradation, and catastrophic failure.

Mat. Res. Soc. Symp. Proc. Vol. 216. ©1991 Materials Research Society

MATERIALS ISSUES IN InGaAs(P)/InP SYSTEM

Materials issues in the growth and processing of the InGaAs(P)/InP system are focused on. We describe the generation of various kinds of defects during growth, generation of modulated structures due to spinodal decomposition, and atomic ordering associated with growth kinetics on the surface. Defects induced in diffusion and metallization are also presented.

Defect generation during crystal growth

Defects introduced during crystal growth are classified into two types; interface defects and bulk defects. Defects belonging to the former type are inclusions and misfit dislocations. The latter type of defects are generated by local segregation of dopant atoms or native point defects. They are non-structural precipitates in heavily doped liquid phase epitaxial (LPE) InGaAsP crystals and structural precipitates in heavily doped InP (Fe) crystals by metalorganic vapor phase epitaxy (MOVPE).

Inclusions—In InGaAsP/InP DH materials grown by LPE, inclusions are observed in the InGaAsP layer. They are spherical shaped and 0.5-1.0 µm in diameter, originating from the interface between the InP cladding layer and the InGaAsP active layer. By EDX analysis, it is found that they are In and/or P rich in composition. Thus, they are presumably caused by local segregation of solute atoms at the liquid-solid interface. The presence of these inclusions gives rise to rapid degradation of LED's at high temperature [2] and catastrophic failure of lasers [3].

Misfit dislocations—If there is difference in lattice constant between the epi-layer and the substrate, misfit dislocations are generated when the film thickness exceeds the critical thickness t_c, which depends on the structure and elastic properties of the materials, and the type of dislocations. Possible mechanisms for misfit dislocation generation are; a) gliding out or bending out of threading dislocations from the substrate to the wafer edge; b) termination of two dislocations with the same Burgers vectors in the substrate into one misfit dislocation at the hetero-interface, forming a half-loop. If the dislocation density in the substrate is low, another mechanism is expected: i) nucleation of micro-half-loops at the surface; ii) expansion of loops down to the interface; iii) gliding out of the threading segments to the edge [4]. In InGaAsP/InP DH matarial, when the InGaAsP layer becomes thick (>3 µm), misfit dislocations are occasionally generated in the upper InP layer close to an interface between the InP layer and the InGaAsP active layer [5]. In most cases, the misfit dislocations are of the 60° type and pure edge or screw dislocations are observed very rarely. When the dislocation density increases, reaction between dislocations takes place. Figure 1(a) shows a TEM image of typical misfit dislocations caused by annihilation of dislocations at the intersecting point. Contrast experiments show that both of the dislocations lying in the <110> and the <$\bar{1}$10> directions are 60° dislocations with the same Burgers vectors. The separated segment is L-shaped. These dislocations may be formed by glide motion at the edge of the L-shaped segment on the two different <110>/{111} glide systems after the annihilation reaction [6]. If the Burgers vectors of the original dislocations are different to each other, a H-shaped dislocation network is formed after the reaction as shown in Fig. 1(b). At each node in the network, the total Burgers vector is conserved.

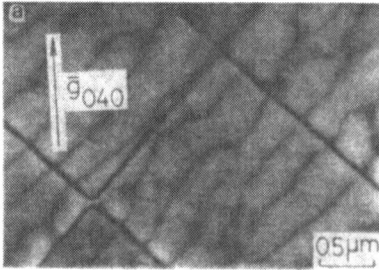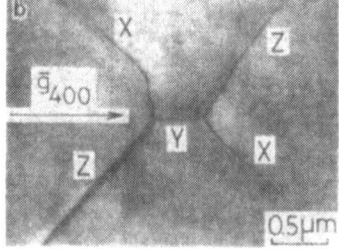

Fig. 1 Reaction of misfit dislocations. (a) Annihilation reaction; (b) network formation.

Non-structural precipitates in heavily doped LPE-InGaAsP crystals—When supersaturated dopants are present in the crystal, they tend to segregate locally during growth or the cooling process after growth, forming precipitates and/or dislocation loops [7,8]. In heavily doped InGaAsP crystals (λ=1.3 µm), non-structural precipitates are occasionally observed although dislocation loops are not observed [7]. In InGaAsP (Zn) and InGaAsP (Sn), a very high density (1×10^9 cm^{-3}) of lattice-mismatched precipitates accompanied by some dislocation tangles and/or strain field are found to be distributed uniformly (see Fig. 2). However, in InGaAsP (Cd), only small particle-like defects are observed. In InGaAsP (Te), precipitates are few in number, although extremely heavy doping of Te into InGaAsP causes cluster-like defects with strong absorption contrast.

Fig. 2 TEM images of a non-structural precipitate in heavily doped InGaAsP (Zn).

FeP-precipitates in Fe-doped MOVPE-InP—Semi-insulating epitaxial InP doped with Fe is one of the important materials for opto-electronic devices. However, heavy doping of Fe into InP often induces iron-phosphorous precipitates in the crystal [9]. Figure 3(a) illustrates a plan-view bright field TEM image of precipitates in heavily doped InP with Fe grown by MOVPE [9]. The precipitates are spherical FeP particles of 4-20 in diameter and are uniformly distributed in the crystal. As shown in the high resolution TEM image (see Fig. 3(b)), they are coherent precipitates with certain orientation relationship to the InP. In crystals grown with Fe-doping gas flow rates of less than 5 ml/min, precipitates are not found. Crystals grown with gas flow rates of 5 ml/min contain an Fe-concentration of 10^{17} cm^{-3} which is the solubility limit of Fe in InP at the growth temperature of 650°C. From these results, it is suggested that when the doped Fe concentration exceeds the solubility limit, excess Fe tend to condense in the matrix forming FeP precipitates.

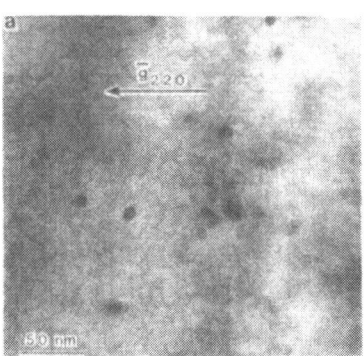 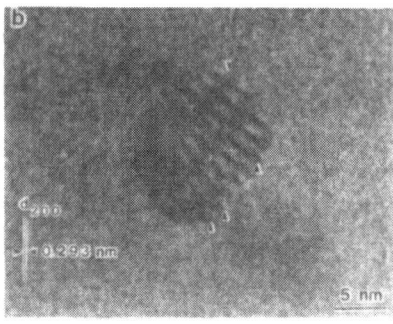

Fig. 3 FeP-precipitates in MOVPE-grown InP heavily doped with Fe.
(a) a plan-view bright field TEM image; (b) a high resolution TEM image.

Structural imperfections due to thermal instability of the crystals

Modulated structures — These structures are generated in alloy semiconductors such as InGaAs, InGaP, InGaAsP, and GaAsSb whose compositions are inside the "spinodal region"[10]. They are observed as quasi-periodic diffraction contrast in bright or dark field images under two-beam or multi-beam conditions. Figure 4 shows a plan-view and a (110) cross-section dark-field TEM images of modulated structures in MBE-grown InGaAs on (001) InP [11]. They develop in the two equivalent directions of <100> and <010> (Fig. 4(a)), and they are columnar shaped in the <001> growth direction (Fig. 4(b)). In spinodally decomposed alloys which have asymmetry in elastic coefficients, modulated waves extend so as to minimize the strain contribution to free energy in the solid solution, i.e. the modulated structures are induced in the direction of the [hkl], which minimize the elastic coefficient Y_{hkl}. From calculations of the values Y_{100}, Y_{110}, and Y_{111} for GaAs, GaP, InAs, and InP, it has been found that in all cases, $Y_{100} < Y_{110} < Y_{111}$ [12]. This can well explain the preferential direction of elongation of the structures. It is also assumed that they are very stable once they are formed, propagating in the <001> growth direction. Furthermore, it is clarified that both structures with long (50-200 nm) and short (5-20 nm) periodicities are formed when the composition of the crystal is inside the spinodal region, whereas only the latter ones are formed when the composition is outside the spinodal region. Thus, one can expect that the former structures are formed during growth by spinodal decomposition and that the latter ones during the cooling process after growth. Moreover, the compositional fluctuations along the structure measured by EDX are in the range 2-3% in content, which is far smaller than expected. This may be due to the fact that in semiconductors, atomic diffusion in the bulk or on the surface is very slow compared with the case of metal alloys. Their generation does not depend on the growth method. The decomposition process may be enhanced by atomic diffusion at the liquid-solid interface for LPE growth and at the growth surface for MOVPE, VPE, and MBE. From the results described above, it is suggested that modulated structures can be eliminated by the growth of the crystal at temperatures where the composition of the crystal is outside the spinodal region and subsequent rapid cooling or quenching of the crystal.

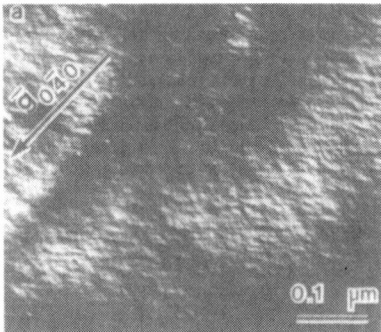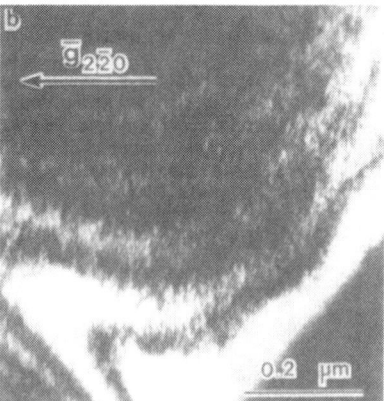

Fig. 4 TEM images of modulated structures in MBE-grown InGaAs grown on (001) InP substrate.
(a) (001) plan-view (b) (110) cross-section.

Ordered structures — Recently, ordered structures (or natural superlattices) in which two kinds of element atoms in column III or V lattice periodically arrange, have been observed in various III-V alloy semiconductors [13]. In most cases, the ordered structures are generated locally in the crystal, and the degree of ordering is strongly affected by the growth method and growth conditions. It is known that there are three types of substitutionally ordered structures in the fcc lattice of AB type binary alloy systems. They are the types of CuAu-I (Ll_0), CuPt (Ll_1), and Chalcopyrite (El_1), whose atomic arrangements and corresponding reciprocal lattices are schematically shown in Fig. 5. Typical features of the ordered structures are: a) the most common structure in crystals grown on a (001) substrate is CuPt-type (see Fig. 5), and other structures are only occasionally observed; b) they are generated very often in crystals grown by MOVPE, VPE, and MBE, whereas no ordering takes place in LPE-grown crystals (except for one report for ordered structures in InGaAs by Nakayama et al.[13], which is not confirmed by others); c) regarding CuPt-type structures, one can observe only two variants in the (110) cross-section (atomic steps on the growth surface are believed to play an important role in the generation of the ordered structures since one of the two variants is preferentially enhanced when substrates tilted toward the <110> direction are used); d) the ordering is not perfect and the ordered regions are plate-like microdomains lying on planes nearly parallel to the growth surface; e) defects such as antiphase boundaries are often generated in the ordered regions; f) the degree of ordering depends on the growth temperature, V/III partial pressure ratio (in the case of MOVPE and MBE), and rotation velocity of the substrate; g) only strongly ordered InGaP crystals grown by MOVPE exhibit abnormal band gap energies (up to 50 meV lower than the normal value). From these results, it is now understood that the ordered structures are generated by the migration and reconstruction of the deposited atoms on the growth surface, and not formed under thermal equilibrium conditions. Thus, one can control the introduction of these structures, i.e., fabrication of nearly perfectly ordered structures or complete elimination of them, by choosing appropriate growth conditions. Recently, we have studied ordering in MBE-grown InGaAs on (110) InP substrate. We have found that CuAu-I type ordered structure is formed in the crystal by electron diffraction analysis and high resolution TEM observation [14] (see Fig. 6). Dark field imaging with one of the superstructure spots indicates that the ordered regions are plate-like microdomains lying on planes slightly tilted from the (110) plane. Strong ordering can be achieved by control of growth conditions and substrate misorientation, which enhances the electron mobility in the crystal [14,15].

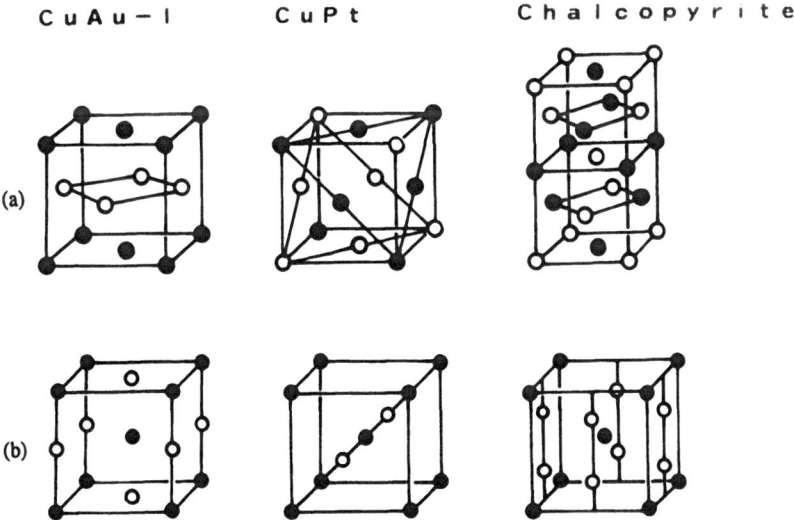

Fig. 5 Three types of ordered structures in fcc lattice. (a) unit cells; (b) reciprocal lattice.

326

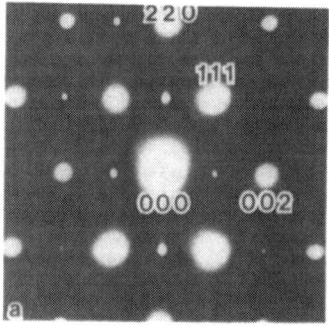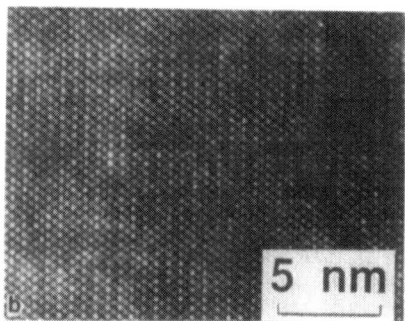

Fig. 6 Atomic oredering of InGaAs grown on (110)InP by MBE.
(a) A TED pattern from the (1̄10) cross-section; (b) a high resolution TEM image of CuAu-type ordered structure.

Defect generation in diffusion processes

Diffusion of Cd and Zn into n-InP substrates or epitaxial layers for forming p-n junctions is one of the important techniques in fabrication of opto-electronics devices. Several problems exist in this process: i) generation of stress due to different thermal expansion coefficients of the diffusion mask and the epitaxial crystal; ii) nucleation of defects due to segregation of dopant atoms at high concentrations; iii) generation of point defects or defect complexes in the crystal.

Figure 7 is a typical bright-field TEM image of a Cd-diffused InP substrate [16]. The micrograph shows the presence of a high-density of microdefects (1×10^{11} cm^{-3}). They are 10-100 nm in diameter and most of them are non-structural, some being rectangular or circular. These microdefects are found to be distributed uniformly over the whole diffused region. The contrast of the defects is opposite to the contrast of the background extinction contour. This is perhaps caused by the changes in the intensity of the transmitted electrons due to the shortening of the path of the incident electrons at the defect region. Therefore, these microdefects are expected to be clusters of excess point defects, i. e., vacancies or interstitials. These microdefects may act as non-radiative recombination centers for minority carriers, and it is probable that the minority-carrier diffusion length in the Cd-diffused region is shorter than that in as-grown p-type InP epitaxial layers. Another two types of defects are occasionally observed in the surface area of the Cd-diffused region. They are precipitate-like defects and dislocation networks. The former defects are one to few μm in diameter and are accompanied by secondarily-generated dislocation clusters. These defects are presumably caused by the anomalous precipitation of Cd or Cd-alloy. The dislocation networks are generated over the whole of a wafer in the form a net. These networks are expected to be generated by relaxation of stress which is induced by the formation of a non-uniform oxide film on the substrate prior to diffusion. Thus, sufficient chemical polishing before diffusion may solve this problem.

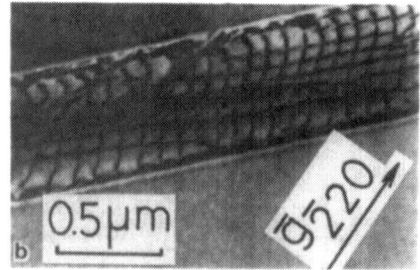

Fig. 7 TEM images of defects in Cd-diffused InP. (a) Microdefects; (b) dislocation network.

Materials issues in metallization

Metallization techniques are very important in establishing fabrication technologies for devices with sufficiently high performance and reliability. The requirements for opto-electronic devices are; i) acceptable electrical characteristics, i.e., low contact resistance for ohmic contacts and large barrier height for Schottky contacts; ii) high-uniformity over the entire wafer area; iii) high reproducibility in processes such as patterning and alloying; iv) high-stability under thermal and electrical stresses leading to overall high reliability. Due to the larger Schottky barrier height for p-type InP-based materials compared with n-type, the formation of p-type ohmic contacts is more important. Also since more severe electric field and current stresses are expected at p-contacts in many opto-electronic devices, the reliability of such contacts is of more practical significance. Au-based contacts have widely been used for both n- and p-type materials. We concentrate on the strong reaction between Au and InP, and the reliability of Au/Zn/Au p-contacts for InGaAsP/InP DH LED's [17].

Thermal reaction of InP with Au—We have studied the thermal reaction of InP with an evaporated Au layer within a variety of temperarue and duration ranges. The Au-InP system was annealed at temperatures in a range 345-500°C under a flow of purified nitrogen for a duration of 5-55 min. Below 345°C, no appreciable reaction was detected. By increasing the temperature, a penetration of reacted alloy into the InP bulk became observable and the whole region below the Au film reacted at 380°C. Further increase of temperature resulted in an extremely rapid reaction which often formed voids in the InP near the periphery of Au stripe, presumably indicating the formation of volatile material. Annealing above a temperature of 450-460°C causes a drastic change in the structure. The Au layer completely reacts with InP to generate a nearly planar surface. An X-ray diffraction measurement indicates that at temperatures of 355-420°C, Au reacts with both In and P to form binary compounds, and the Au-In compound changes phase to form $AuIn_2$ at temperatures above 450°C.

Au/Zn/Au p-type contact—The Au penetration is suspected to occur in practical InGaAsP/InP DH lasers and LED's with Au-based alloy p-contacts, to accelerate the diode degradation. Previous investigation of accelerated aging of InGaAsP/InP DH LED's, with Au/Zn/Au p-contact at a current density of approximately 8 kA/cm^2 revealed a thermally activated operating lifetime with an activation energy of 1.0 eV in the emission wavelength range of 1.15-1.5 μm. This activation energy is completely different from the value of 2.31 eV determined for the Au-InP reaction and the LED life at moderate operating conditions is not necessary related to the Au penetration into the InGaAsP and InP layers. However, when the LED is stressed by a large operating current, Au penetration through the p-InGaAsP contact layer into the p-InP cladding layer and even into the InGaAsP active layer is actually detected (to be describe later). By considering these results, we have developed a new technique in which an ultra-thin Au/Zn/Au layer is used as a p-contact. In a conventional contact, this layer is about 300 nm thick and is capped by a reaction barrier. The entire structure is covered with plated Au. In the ultra-thin process, the alloy layer thickness is typically 50 nm. As in the conventional contact, a reaction barrier is deposited through and opening in the SiO$_2$ protective coating. The advantage of the thin layer is that the Au is completely consumed during alloying, leaving no fresh Au available to penetrate the contact layer during operation. A substantial improvement in the threshold current can be achieved by using an ultra-thin Au/Zn/Au layer over a thick layer.

DEVICE DEGRADATION

In this section, degradation phenomena in InGaAsP/InP DH lasers and LED's are compared with those in GaAlAs DH lasers and LED's, in terms of three major modes; rapid degradation, gradual degradation, and catastrophic failure. On the basis of these results, differences in the degradation behavior in these two major systems are discussed and methods for the elimination of device-degradation are also considered ([18] and references therein).

Rapid degradation

Characteristic phenomena in rapid degradation are sudden decrease in output power and formation of non-radiative regions, i.e., "dark-line defects" (DLD's) or "dark-spot defects" (DSD's) in the active region. The half-life o such devices is less than 100h a room temperature. This degradation is explained by recombination-enhanced dislocation climb (REDC) or recombination-enhanced dislocation glide (REDG). In this paper, we only focus on REDC.

In a rapidly degraded GaAlAs/GaAs DH laser, dislocation dipoles are observed corresponding to <100>DLD's. The dipoles are of interstitial type with Burgers vectors of the type (a/2)<011> 45° inclined to the junction plane. They develop from threading dislocations which propagate from the substrate during growth. Similar results are obtained in rapidly degraded GaAlAs/GaAs DH LED's. However, in InGaAsP/InP DH lasers, rapid degradation due to REDC is not observed even at high temperature. Moreover, similar rapid degradation is not observed in InGaAsP/InP DH LED's. Although dark defects were observed in the light emitting region after 100h of operation at room temperature, they were all shown to have been induced before operation, i.e. during growth or fabrication (see Table 1).

As described above, the ease with which rapid degradation by REDC occurs depends on the material. To explain this effect, two models are most widely accepted although many have been proposed. The first model is the "extrinsic defect model" shown in Fig. 8(a) [19]. Here, only one type of interstitial atom, e.g. Ga, is required for climb motion. In the second model, the "intrinsic defect model" (see Fig. 8(b)) [20], emission of two types of vacancies, e.g. Ga-and As-vacancies in the case of GaAs, is needed. Based on these models, one can suggest several candidates responsible for the REDC. First of all, since the non-radiative recombination event is thought to be caused at a defect by an energy transformation from a localized multiple phonon mode excited by the absorbed light to lattice vibrational mode, the REDC may be dominated by band gap energy. This can well explain the REDC in GaAs- and GaP-related materials and difficulty in REDC in the InGaAsP system, but not for InP. The second candidate is a deep energy level and a non-radiative recombination rate associated with defects such as dangling bonds or native point defects. This is supported by the fact that deep levels are generated in Ga(Al)As and InGaAsP on GaAs where REDC occurs easily and that they are not generated in InGaAsP on InP where REDC does not occur easily. Moreover, since point defects must be absorbed to dislocations or emitted from the dislocation core in REDC, the energies required for their generation and migration are also the key parameters for REDC.

Table 1 Dark defect and corresponding defect structures in InGaAsP/InP DH LED's.

Dark defects	Defect structure	origin or cause
Cross-hatched <110>DLD's	Misfit dislocations	Thermal stress
Regular tetragonal <100>DLD's	Stacking faults	Contamination
Lattice-point-shaped DSD's	Spherical precipitates	Unknown
Dark-band defects (DBD's)	Mechanical damage	Scratches
DSD's with bright circumferences	Alloyed regions	Reaction at the contact

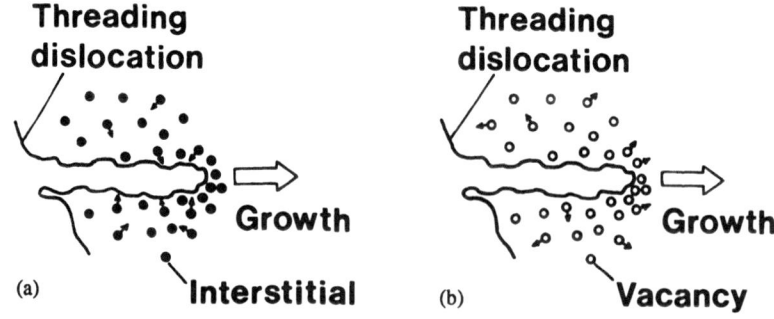

Fig. 8 Two proposed models for dislocation climb in degraded optical devices.
(a) An extrinsic defect model; (b) an intrinsic defect model.

Gradual degradation

Typical characteristics of gradual degradation are slow decrease of output power, uniform darkening of the active region or formation of DSDs, and an increase in deep levels in GaAlAs/GaAs optical devices. TEM observations reveal that numerous Frank-type interstitial microloops and point-like defects (possibly point defect clusters) are generated throughout the active region. Two types of hole traps A and B are present in the active region before operation. During operation, both deep levels undergo a two-step increase. Thus, defect complexes associated with the deep levels could be the original source of the point defect reaction. Gradual degradation might then proceed as follows: i) non-radiative recombination occurs at some defects, which causes a point defect reaction and fresh point defect generation; ii) the new defects can also act as non-radiative recombination centers, thus we have positive feedback for these two processes; iii) the generated point defects migrate and condense at some nucleation centers; iv) defect clusters and/or microloops are formed as byproducts.

In InGaAsP/InP optical devices, the phenomena described above were not observed during accelerating aging at high temperature. DSD's are occasionally observed in the light emitting region but no change in light output is observed at all. Figure 9(a) shows an EL image of a light emitting region of a diode operated at 200°C for 100h. Many DSD's are observed in the light emitting region A in Fig. 9(a). A TEM image of precipitates corresponding to the DSD's in region A is shown in Fig. 9(b). It is also found that one or more precipitates correspond to one DSD. The precipitates are bar-shaped and are lying along the <100>-(precipitate denoted by a) or <110>-direction (precipitate denoted by b). No elements other than the matrix elements are detected by EDX measurement from the precipitate region. However, Chin et al. have claimed that the precipitates are associated with migration of Au atoms from the electrode by similar analysis [21]. At any rate, in this material, generation of dislocation loops by recombination-enhanced point defects generation and migration during operation, can not occur easily even at high temperature.

From these results, one can conclude that the difference in gradual degradation between GaAlAs/GaAs and InGaAsP/InP optical devices is very similar to the case for rapid degradation, i. e., it reflects the difference in the recombination-enhanced point defect reaction in the crystals. Consequently, one can conclude that to eliminate the gradual degradation effect, tight control of stoichiometry (to reduce the deep level population), careful lattice matching in the heterostructure and avoidance of stress introduction during fabrication, are required.

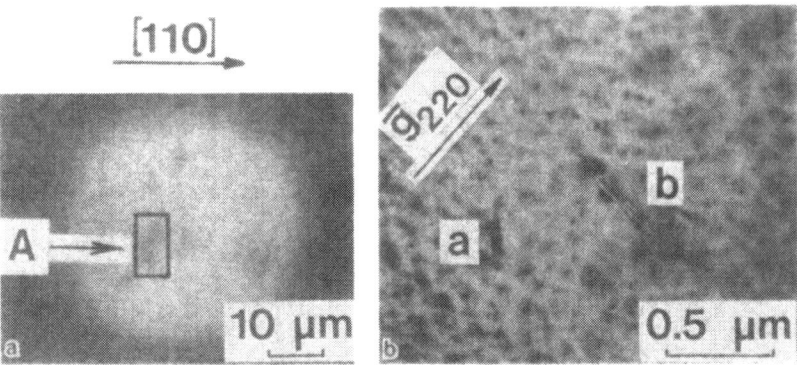

Fig. 9 (a) An EL image of DSD's appeared in an InGaAsP/InP DH LED operated at high temperature. (b) a TEM image of defects corresponding to the DSD's.

330

Catastrophic failure

Lasers—This phenomenon takes place accidentally by current surge, i.e. at high power density in lasers. The degradation occurs predominantly at the mirror surface by "catastrophic optical damage", COD. The COD can also be caused by strong optical excitation. In PL images of catastrophically degraded GaAlAs/GaAs DH lasers, one or more <110>DLD's generated at the mirror surface are observed. From TEM studies, these are seen to correspond to arrays of dislocation networks or multiple dislocation loops connected by dark knots. The proposed mechanism for COD is as follows: i) when the output power density reaches a critical value, strong optical absorption occurs at the mirror surface, leading to local heating of the crystal; ii) the band gap energy shrinks at this region, causing further optical absorption; iii) these processes give rise to very rapid thermal runaway and finally the melting of the crystal; iv) subsequent rapid cooling generates residual defects in the degraded area. In the case of pulse operation, these processes repeat and the molten region propagates from the mirror to the inner crystal. In InGaAsP/InGaP DH lasers lattice matched to GaAs, emitting in the short wavelength range, we have also found COD during operation at high output power density.

In InGaAsP/InP DH lasers, however, there have been no reports on COD events for any operating conditions. In order to confirm the absence of COD, pulsed high currents were applied to InGaAsP/InP v-grooved substrate buried heterostructure (VSB) lasers. After the degradation, the diodes stops lasing and becomes ohmic. Figure 10(a) is a typical SEM cross-sectional image of the diode after degradation with application of a pulsed large current (2.2 A, forward, 0.12 ms of pulse width). A heavily damaged area is clearly observed from the contact to the n-InP buffer layer. Figure 10(b) shows an EDX image of the same area obtained by the P K_α line. A considerable decrease in phosphorous content is observed in the degraded region. This is presumably caused by the abrupt passage of a large current along the mirror surface. Penetration of electrode metals into the epitaxial layer is also found to be associated with the degradation. The <110>DLD's, which are peculiar to catastrophic degradation in GaAlAs/GaAs DH lasers, are not observed in any part of the active stripe region in degraded VSB lasers, indicating the absence of catastrophic optical damage in these lasers. Thus, one can assume that the COD level, i.e. critical output power density for causing COD in this laser is very high. Since COD occurs initially due to optical absorption, the difference in COD level in different materials is explained by different surface recombination velocities. These results lead us to the conclusion that protection of mirrors with dielectric films (SiO$_2$ etc.) and reduction of inclusions and/or precipitates during growth is essential for eliminating catastrophic failure in lasers.

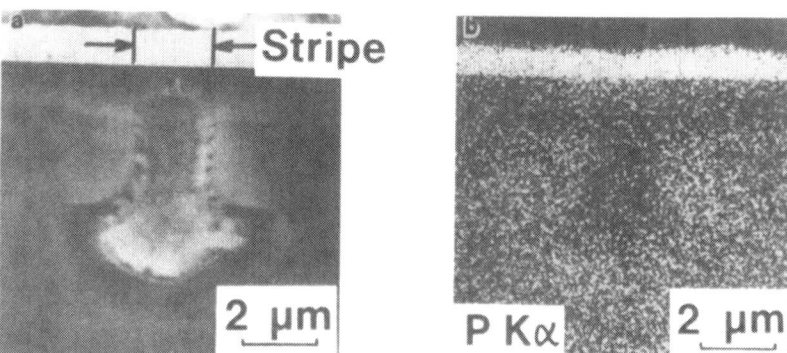

Fig. 10 (a) An SEM image of a mirror of a degraded InGaAsP/InP VSB laser by application of large current pulses. (b) An EDX image of the degraded region in Fig. 10(a).

LED's—In InGaAsP/InP DH LED's, it is assumed that there is no COD-related degradation since there is no mirror. However, a different type of catastrophic degradation due to application of pulsed large current have been observed in these LED's. Figure 11 shows results of EL and TEM analyses of degraded InGaAsP/InP under heavy current application. The diode was stressed at a current of 0.75 A (60 kA/cm^2) with a current pulse width of 300 µs. During current application, a number of DSD's appeared in the EL image and these DSD's grew very rapidly to form a widespread dark region as shown in Fig. 11 (left). Structural analysis of the dark region by TEM was also carried out, and the results are illustrated in Fig. 11 (right). From these data, it has been established that the DSD's or the dark regions are associated with a) an amorphous area of the matrix, b) micrograins of the matrix crystal, and c) regions where the matrix crystal was alloyed with the metals of the electrode. Furthermore, EDX analysis indicated that the DSD's or the dark regions contained a large amount of Au. These defects were considered to be generated by the alloy reaction between the matrix crystal and the metals of the electrode, especially Au. Based on these results, Au penetration under large current application is interpreted as follows. The non-planar and non-uniform interface of the alloyed contact results in an inhomogeneous contact resistance. Assuming the underlying semiconductor bulk to be highly conductive, Joule heating is enhanced locally at the low-contact resistance parts of the interface. This causes strong reaction between Au and the matrix crystal, leading to Au penetration there. This process repeats until the reacted area expands to the entire emitting region forming a large dark region. Therefore, in order to eliminate the alloy reaction, i.e., to achieve higher degradation lebel, the following methods are essential: i) using ultra thin Au/Zn/Au p-contacts as described previously; ii) using non-alloy contacts with low resistivity; iii) using a thick contact layer.

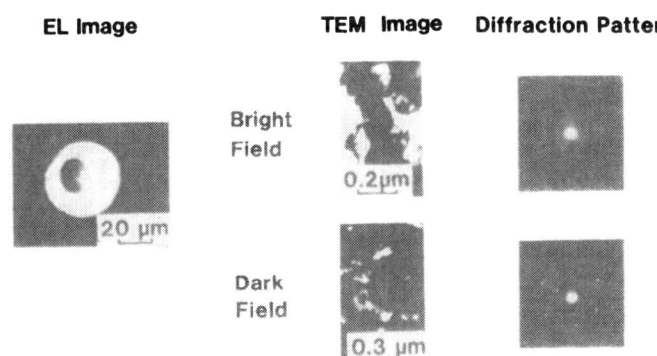

Fig. 11 EL and TEM analyses of a degraded InGaAsP/InP DH LED by large current application.

Influence of modulated and ordered structures on optical devices

As described previously, modulated structure is associated with quasi-periodic compositional variation presumably due to spinodal decomposition of the system. However, the compositional fluctuation measured by EDX is 2-3%, and the domains are columnar shaped and quite stable. Thus, they do not strongly affect optical and electrical device properties. Furthermore, in optical devices, the periodicity and amplitude of modulation do not change following degradation by any of the modes discussed earlier. However, dislocation loops formed as byproducts of modulated structure generation can cause rapid degradation, and point defects are associated with composition modulation in some materials.

Ordered structures, on the other hand, only affect the PL peak energies in strongly ordered InGa(Al)P crystals grown by MOVPE. In other alloy semiconductors such as InGaAs, InAlAs, GaAsSb, the degree of ordering is lower and PL peak energies have normal values. The ordered structures themselves are monolayer superlattices, for example, of (InP/GaP) on the (111) plane in CuPt structure. Thus, provided that the dislocations themselves do not act as non-radiative recombination centers, one can expect that ordered structures do protect dislocations from REDC, supressing rapid degradation. However, such structures may involve lattice strain and it is also possible that stress is associated with the boundary between ordered and non-ordered regions and with antiphase boundaries (APD's) which are often observed in the ordered region. These strains influence gradual degradation, i.e., by generation of point defects through "recombination-enhanced local disordering".

Table 2 Influence of modulated and ordered structures on optical devices.

	Device characteristics	Degradation		
		Rapid	Gradual	Catastrophic
Modulated structure	No effect	None*	None	None
Ordered structure	Eg: smaller μ: higher I_{th}, η: Not clear	Supress?	Disordering?	Not clear

*Formation of loops as byproducts can lead to degradation.

CONCLUSIONS

The current status of evaluation of defects and thermal instability in InGaAs(P)/InP by TEM and understanding of degradation modes in InGaAsP/InP optical devices have been reviewed. The detailed nature of growth- and process-induced defects and modulated and ordered structures are described. The roles of these defects as well as point defects and their complexes in the degradation of optical devices are presented in terms of three major degradation modes. Based on these results, techniques for eliminating the degradation problem are discussed.

REFERENCES

1. I. Hayashi, M. B. Panish, P. W. Foy, and S . Sumski, 17, 109 (1970).
2. O. Ueda, S. Komiya, S. Yamakoshi, and T. Kotani, Japan. J. Appl. Phys. 20, 1201 (1981).
3. O. Ueda, K. Wakao, S. Komiya, A. Yamaguchi, S. Isozumi, and I. Umebu, J. Appl. Phys. 58, 3996 (1985).
4. P. M. J. Maree, J. C. Barbour, J. F. van der Veen, K. L. Kavanagh, C. W. T. Bulle-Lieuwma, and M. P. A. Viegers, J. Appl. Phys. 62, 4413 (1987).
5. O. Ueda, S. Komiya, S. Yamazaki, Y. Kishi, I. Umebu, and T. Kotani, Japan. J. Appl. Phys. 23, 836 (1984).
6. W. Hagen, Appl. Phys. 17, 85 (1978).
7. O. Ueda, I. Umebu, and T. Kotani, J. Crystal Growth 62, 329 (1983).
8. T. Kotani, O. Ueda, K. Akita, Y. Nishitani, T. Kusunoki, and O. Ryuuzan, J. Crystal Growth 38, 85 (1977).
9. O. Ueda, K. Nakai, S. Yamakoshi, and I. Umebu, Mat. Res. Soc. Symp. Proc. 138, 509 (1989)
10. J. W. Cahn, Acta Met. 9, 795 (1961).
11. O. Ueda, Y. Nakata, and T. Fujii (unpublished).
12. O. Ueda, S. Komiya, and S. Isozumi, Japan. J. Appl. Phys. 23, L241 (1984).
13. See for example, T. S. Kuan, T. F. Kuech, W. I. Wang, and E. L. Wilkie, Phys. Rev. Lett. 54, 208 (1985); H. Nakayama and H. Fujita, Inst. Phys. Conf. Ser. 79, 289 (1986); O. Ueda, M. Takikawa, J. Komeno, and I. Umebu, Japan. J. Appl. Phys. 26, L1824 (1987).
14. O. Ueda, Y. Nakata, T. Nakamura, O. Ueda, and T. Fujii, this Meeting, paper K7.4.
15. Y. Nakata, O. Ueda, and T. Fujii (unpublished).
16. O. Ueda, H. Ishikawa, and I. Umebu, Japan. J. Appl. Phys. 23, 1551 (1984).
17. O. Wada and O. Ueda, Mat. Res. Soc. Symp. Proc. 181 (1990).
18. O. Ueda, J. Electrochem. Soc. 135, 11C (1988).
19. P. M. Petroff and L. C. Kimerling, J. Appl. Phys. 29, 461 (1976).
20. S. O'Hara, P. W. Hutschinson, and P. S. Dobson, Appl. Phys. Lett. 30, 368 (1977).
21. A. K. Chin, C. L. Ziepfel, S. Mahajan, F. Ermanis, and M. A. DiGiuseppe, Appl. Phys. Lett. 41, 555 (1982).

PART VI

III-V-Based Long Wavelength Material Growth and Characterization

PHOTOREFLECTANCE AND DOUBLE CRYSTAL X-RAY STUDY
OF STRAINED InGaAsP LAYERS ON InP SUBSTRATES

J. R. FLEMISH*, H. SHEN**, M. DUTTA*, K.A. JONES*, and V.S. BAN***
*   U.S. Army Electronics Technology & Devices Laboratory, Fort Monmouth, NJ. 07703
**   Geo-Centers, Incorporated, Hopatcong, NJ 07849
***   Epitaxx, Inc., Princeton, NJ 08540

ABSTRACT

Determining the composition of quaternary epitaxial films requires accurate measurements of both the lattice parameter and the bandgap energy. Complications arise in lattice-mismatched material, because the mismatch produces tetragonal distortion of the epi-layer and splitting of the valence band energies in a manner which depends on the film composition. We present studies on strained InGaAsP grown on (100) InP. Using room temperature photoreflectance (PR) we observe shifting of the band gap and splitting of the valence band energies, and using the (115) and (004) reflections from double crystal x-ray diffraction (DXRD) we determine the values of the parallel and perpendicular lattice constants. By combining the lattice parameter measurements with band splitting data, we accurately determine the quaternary composition from a self-consistent model using an iterative procedure. By linear interpolation of the elastic-stiffness constants, C_{11} and C_{12}, as well as the shear and hydrostatic deformation potentials for the four binary compounds in the InGaAsP system, we relate the state of biaxial stress to the induced shifts in the valence band energies.

INTRODUCTION

The composition of $In_{1-x}Ga_xAs_{1-y}P_y$ alloys can be determined nondestructively by measuring both the lattice parameter, a_o, and the bandgap energy, E_g, if the compositional dependence of each property is known. Values of a_o for the binary compounds, InP, GaAs, GaP, and InAs have been well documented, and values of a_o for the ternary and quaternary alloys are generally assumed to be linearly dependent on composition (Vegard's Law). Several investigators have reported on the bandgap energies of ternary alloys and have introduced an empirical bowing parameter in order to relate E_g as a function of either x or y [1-4]. Moon et al. [5] have used the results from the ternary systems as boundary conditions to develop an equation for $E_g(x,y)$ for quaternary alloys. Others have since presented equations for $E_g(x,y)$, and although discrepancies exist, all seem to agree that equations derived by interpolating data from ternary systems exhibit excessive bowing for quaternary alloys [6-8] For example, the E_g value obtained by this interpolation can differ from the measured value by more than 50 meV when y = 0.5. Independent investigations using electron microprobe analysis (EMPA) have determined quite similar empirical equations for $E_g(y)$ of $In_{1-x}Ga_xAs_{1-y}P_y$ lattice matched to InP [6,7]. For a more general case (i.e., not necessarily lattice matched) we have modified Moon's equations to

$$
\begin{aligned}
E_g \text{ (eV)} = \quad & 0.35 + 1.09x + y + 0.33xy - (0.45+0.28y)x(1-x) \\
& - (0.101+0.109x)y(1-y) + 0.05 \sqrt{xy(1-x)(1-y)}
\end{aligned}
\tag{1}
$$

We introduce the square root term empirically so this equation is consistent with the EMPA results.

Determining the composition of mismatched InGaAsP films via measurements of E_g and a_o is more complicated. As shown in Fig. 1 for an epitaxial layer in tension, lattice mismatch results in biaxial stress which affects the valence band structure, as well as the lattice parameters in directions perpendicular and parallel to the interface, denoted $a_\perp$ and $a_{\parallel}$. Biaxial stress shifts the valence bands by an amount, ΔE_H, proportional to the hydrostatic deformation potential, a, and splits the degeneracy of the valence bands by an amount, $2\Delta E_S$, proportional to the shear deformation potential, b [9,10]. We denote the energies of these valence bands by E_1 and E_2 referring to the levels (J = 3/2, m_j = ±1/2) and (J = 3/2, m_j = ±3/2), respectively. The resulting energy gaps between the conduction band and the valence bands are

Mat. Res. Soc. Symp. Proc. Vol. 216. ©1991 Materials Research Society

336

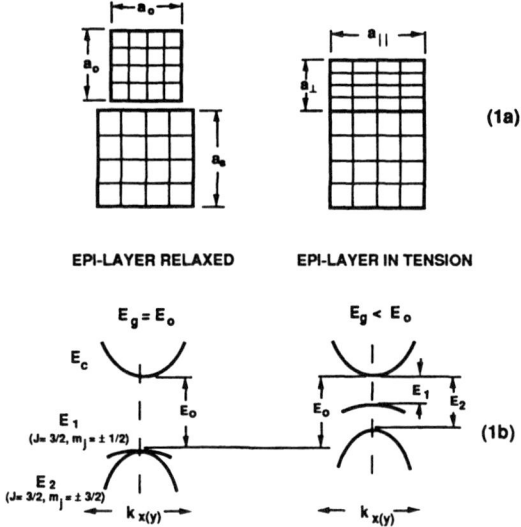

Figure 1. Representation of lattice dimensions (a) and energy band diagram (b) for an
epitaxial III-V layer in a relaxed state and in a strained state of tension.

$$E_1 = E_o + \Delta E_H + \Delta E_S \qquad (2)$$

$$E_2 = E_o + \Delta E_H - \Delta E_S \qquad (3)$$

where E_o is the unstrained direct band gap of the epi-layer and the values of ΔE_H and ΔE_S are given in terms of the perpendicular strain, $\varepsilon_z = (a_\perp - a_o)/a_o$, and the elastic-stiffness constants, C_{ij}, by

$$\Delta E_H = -2a[(C_{11}-C_{12})/C_{11}]\varepsilon_z \qquad (4)$$

$$\Delta E_S = -b[(C_{11}+2C_{12})/C_{11}]\varepsilon_z \qquad (5)$$

For the case of strained InGaAsP layers one can convert the measured values of the lattice parameters and E_g to the corresponding unstrained values, if one knows accurate values of the stiffness constants and the deformation potentials. Based on photoluminescence experiments the effects of strain on the measured bandgap energies of ternary alloys have been reported previously [11,12].

This paper presents studies on strained InGaAsP grown on (100) InP. Using room temperature photoreflectance (PR) we observe strain-induced spitting of the valence band energies as well as a shift in the measured E_g. We determine the composition of strained quaternary layers from measured values of $a_\perp$, $a_{||}$, E_1, and E_2 using values of C_{11}, C_{12}, a, and b obtained by interpolation of the corresponding constants for the binary constituents.

EXPERIMENTAL

Epitaxial layers of InGaAsP were grown on InP substrates at 700°C and atmospheric pressure by hydride vapor phase epitaxy [13]. The substrates were oriented 2° off the (001) toward a nearest (011). The reactant flows were set to produce nearly lattice matched compositions. Strained compositions were produced by injecting various amounts of HCl into the deposition zone of the reactor to shift the film stoichiometry toward compositions more rich in Ga and As [14].

Lattice mismatch was determined using a Blake double crystal x-ray diffractometer. X-ray rocking curves of the (004) reflection and the four nearest {115} reflections were obtained using $CuK\alpha_1$ radiation separated from the $K\alpha_2$ component by a (001) Ge first crystal. A slit width of 0.5 mm was positioned to block the separated $K\alpha_2$ radiation.

The photoreflectance (PR) apparatus has been described in the literature [15]. The pump beam was the 514.5 nm line of an air cooled Ar^+ laser. The probe beam is from a lamp filtered by a one-quarter-meter monochromator. The PR signal ($\Delta R/R$) was recorded by a liquid N_2 cooled Ge detector and lock-in combination and was normalized by division. The room temperature photoluminescence (PL) peak energy of each sample was also measured.

The compositions of the strained layers were also evaluated by x-ray energy dispersive spectroscopy (EDS) using the lattice matched layers with $E_g = 0.855$ and 0.953 eV as standards.

DATA ANALYSIS AND RESULTS

From the (004) reflection and the four nearest {115} reflections, values of perpendicular and parallel mismatch, $(\Delta a_\perp /a_s)$ and $(\Delta a_{||}/a_s)$, were determined in a manner described by Matsui et al. [16]. The relaxed lattice parameter, a_0 was found from

$$\left(\frac{a_0 - a_s}{a_s}\right) = \left(\frac{1-v}{1+v}\right)\left(\frac{\Delta a_\perp}{a_s}\right) + \left(\frac{2v}{1+v}\right)\left(\frac{\Delta a_{||}}{a_s}\right) \tag{6}$$

as described by Chu et al. [17], where a_s is the substrate lattice parameter. Poisson's ratio is calculated from $v = C_{12}/(C_{11}+C_{12})$.

A typical room temperature PR spectrum of a InGaAsP layer in tension is shown by the dotted line in Fig. 2a. The energies E_1 and E_2 are found from a least squares fit (solid line) to the third derivative function (TDFF) of a three-dimensional critical point [18].

$$\frac{\Delta R}{R} = \sum_{j=1}^{n} Re \left(A_j \exp(i\theta_j) (E - E_j + i\Gamma_j)^{-5/2} \right) \tag{7}$$

Here n equals 2, the number of features involved, A_j is the amplitude, θ_j is the phase factor, E is the photon energy, E_j is the energy gap, and Γ_j is the broadening parameter. The obtained values of E_j (j=1,2) correspond to the energy gaps E_1 and E_2 and are indicated by arrows in the top of Fig. 2.

The composition of each sample was calculated from the DCXRD and PR data as follows. In an iterative procedure, the values of the stiffness constants and the deformation potentials were obtained by estimating the composition and interpolating from the values corresponding to the binary compounds, listed in Table I. The deformation potentials are the same as those used by Asai [11] et al. and by Pan et al. [19]. The elastic stiffness constants are those given by Hornstra and Bartels [20] with a corrected value of C_{12} for GaP. The values of a_0, ϵ_z, ΔE_H, and ΔE_S are then calculated using the interpolated parameters. The strain-free band gap, E_0 is then estimated from Eqs. 2 and 3, so that Eq.1 can be solved simultaneously with Vegard's law producing a new estimate of the composition variables, x and y. This procedure converges rapidly to yield the final composition.

The DCXRD and PR results are summarized in Table II along with the calculated InGaAsP compositions. All of the epitaxial layers are in tension and are relatively coherent with the substrate, since $|\Delta a_{||}| << |\Delta a_\perp|$. Strain-induced splitting of the valence band is observed for $|\Delta a_\perp/a_s|$ greater than approximately 0.2%. The experimentally observed splits in the valence bands ΔE_S^m are slightly greater than those predicted using the interpolated parameters and measured strains, ΔE_S^c. This discrepancy may be due, in part, to these parameters not being linear functions of the composition. In particular, it is possible that the deformation potentials for the quaternary alloys are greater in magnitude than is predicted by interpolation. For sample G the InP substrate was etched preferentially using a 1:1 solution of $HCl:H_3PO_4$, and a second PR spectrum was taken (Fig. 2b). The resulting spectrum showed only a single feature at 0.892 eV. This value is essentially the same as the relaxed value $E_0 = 0.893$ eV obtained from the calculations. Unfortunately, removal of the substrate was not successful for all specimens due to cracking of the epi-layers.

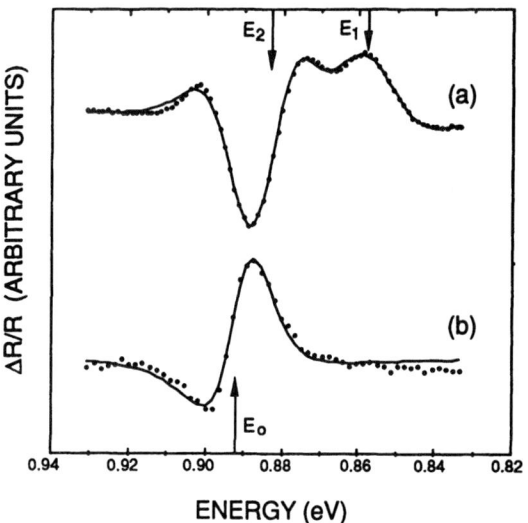

Figure 2.   (a) PR spectrum with least squares fit for strained layer G, and (b) PR spectrum with least squares fit for same layer with substrate removed.

Table I   Values of lattice parameter (a_0), elastic-stiffness (C_{ij}), hydrostatic deformation potential (a), and shear deformation potential (b) for binary III-V compounds.

compound	a_0 (Å)	C_{11} (dyn/cm^2)	C_{12} (dyn/cm^2)	v	a (eV)	b (eV)
GaAs	5.6533	1.188×10^{12}	0.532×10^{12}	0.3093	-9.8	-1.76
GaP	5.4512	1.412×10^{12}	0.625×10^{12}	0.3068	-9.5	-1.3
InP	5.8686	1.022×10^{12}	0.576×10^{12}	0.3605	-8.0	-1.55
InAs	6.0584	0.833×10^{12}	0.453×10^{12}	0.3523	-5.8	-1.8

Table II   Results of DCXRD and PR analysis of strained $In_{1-x}Ga_xAs_{1-y}P_y$ layers on InP.

sample	$(\Delta a_\perp/a_s)$	$(\Delta a_\parallel/a_s)$	E_1	E_2	ΔE_S^m	ΔE_S^c	E_0	x	y
A	7.6×10^{-4}	5.0×10^{-4}	0.953*		-	0.0026	0.953	0.268	0.406
B	-2.49×10^{-3}	-1.0×10^{-5}	0.948	0.958	0.010	0.0085	0.962	0.302	0.383
C	-5.78×10^{-3}	-7.3×10^{-4}	0.942	0.964	0.022	0.0171	0.972	0.347	0.348
D	-1.03×10^{-2}	-1.03×10^{-3}	0.955	0.993	0.038	0.0313	1.010	0.380	0.352
E	5.43×10^{-4}	3.0×10^{-5}	0.855*		-	0.0018	0.855	0.356	0.223
F	-3.82×10^{-3}	-4.0×10^{-4}	0.858	0.869	0.011	0.0118	0.877	0.403	0.196
G	-5.89×10^{-3}	-2.0×10^{-5}	0.858	0.883	0.025	0.0201	0.893	0.412	0.203
H	-7.35×10^{-3}	-3.0×10^{-4}	0.856	0.884	0.027	0.0241	0.898	0.434	0.185
I	-1.17×10^{-2}	-3.2×10^{-3}	0.888	0.938	0.050	0.035	0.954	0.455	0.218

* only one feature was observed in PR spectrum. E_0 taken as measured E_g. All energy values in eV.

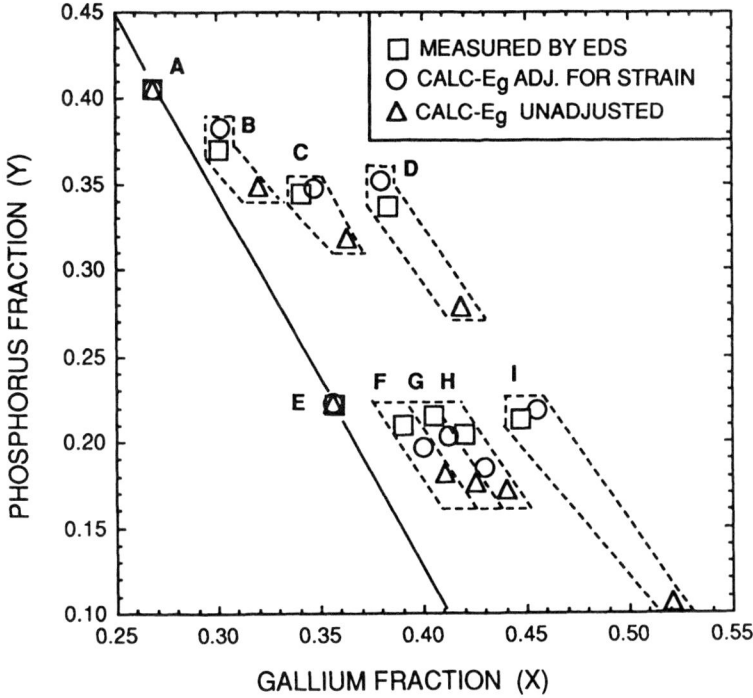

Figure 3. Composition of InGaAsP samples as caluculated using E_g values as measured compared to using E_g values adjusted for strain. Also shown are compositions determined by EDS using lattice matched layers as standards. Solid line represents compositions lattice matched to InP.

The compositions of the epitaxial layers listed in Table II are plotted in Fig. 3. Superimposed on the plot are calculated compositions which result when the bandgap is taken as the PL peak energy and is not corrected for the effects of strain. Also shown are the layer compositions obtained by EDS measurements using the two lattice-matched compositions as standards. Using either standard to evaluate the other by EDS, gave results which agreed to within 1 atom-%. For the case where the bandgap is not corrected for strain, the calculated values of x and y can each differ from the values determined from EDS by up to 10 atom-% when $|\varepsilon_z| > 0.4\%$. For all of samples in this study, calculations using an uncorrected E_o, the energy of the photoluminescence peak, will result in a composition which is too high in Ga and As. When the bandgap energy is adjusted to account for the effects of strain, the calculated compositions of the layers agree more closely with the EDS results. This agreement is within 1 atom-% for samples A through D and within 2 atom-% for samples E through I. On this basis, it is clear that for determination of the composition of strained InGaAsP layers from measured values of E_o and a_o, correction of the measured bandgap for the effects of strain significantly improves the accuracy of the results.

ACKNOWLEDGEMENTS

JRF gratefully acknowledges the support of the National Research Council. VSB wishes to acknowledge the SBIR support of the Army under contract DAAL01-89-C-0938.

REFERENCES

1. G.A. Antypas and T.O. Yep, J. Appl. Phys. 42, 3201 (1971).

2. T.Y. Wu and G.L. Pearson, J. Phys. Chem. Solids 33, 409 (1972).

3. R. J. Nelson and N. Holonyak, J. Phys. Chem. Solids 37, 629 (1976).

4. A.G. Thompson, M. Cardona, K.L. Shaklee, and J.C. Woolley, Phys. Rev. 146, 601 (1966).

5. R.L. Moon, G.A. Antypass, and L.W. James, J. Electron. Mater. 3, 635 (1974).

6. R.E. Nahory, M.A. Pollack, and W.D. Johnston, Jr., Appl. Phys. Lett. 33, 659 (1978).

7. Y. Yamazoe, T. Nishino, and Y. Hamakawa, IEEE J. Quant.Elect. QE-17, 139 (1981).

8. K. Nakajima, A. Yamaguchi, K. Akita, and T. Kotani, J. Appl. Phys. 49, 5944 (1979).

9. F.H. Pollak and M. Cardona, Phys. Rev. 172, 816 (1968).

10. F.H. Pollak, Surf. Sci. 37, 863, (1973).

11. H. Asai and K. Oe, J. Appl. Phys. 54, 2052 (1983).

12. C.P. Kuo, S.K. Vong, R.M. Cohen, and G.B. Stringfellow, J. Appl. Phys. 57, 5428 (1985).

13. G.H. Olsen, IEEE J. Quant. Elect. QE-17, 128 (1981).

14. J.R. Flemish, K.A. Jones, A. Tripathi, V. S. Ban and C. H. Park, submitted to J. Electrochem. Soc.

15. H. Shen, P. Parayanthal, Y.F. Liu, and F.H. Pollak, Rev. Sci. Instrum. 58, 1429 (1987).

16. J. Matsui, K. Onabe, T. Kamejima, and I. Hayachi, J. Electrochem. Soc. 126, 664 (1979).

17. S.N.G. Chu, A.T. Macrander, K.E. Strege, and W.D. Johnston, Jr., J. Appl. Phys. 57, 249 (1985).

18. M. Cardona, in Modulation Spectroscopy , edited by F. Seitz, D. Turnbull, and H. Ehrenreich (Academic, New York, 1969).

19. S.H. Pan, H. Shen, Z. Hang, F.H. Pollak, W. Zhuang, Q. Xu, A.P. Roth, R.A. Masut, C. Lacelle, and D. Morris, Phys. Rev. B 38, 3375 (1988).

20. J. Hornstra and W.J. Bartels, J. Cryst. Growth 44, 513 (1978).

TEMPERATURE DEPENDENCE OF THE LINEWIDTH AND PEAK POSITION OF THE INTERSUBBAND INFRARED ABSORPTION IN GaAs/Al$_{0.3}$Ga$_{0.7}$As QUANTUM WELLS

M. O. Manasreh*, C. E. Stutz*, K. R. Evans*, F. Szmulowicz**, and D. W. Fischer***
*Electronic Technology Laboratory (WRDC/ELRA), WPAFB, OH 45433.
**University of Dayton Research Institute, 300 College Park Avenue, Dayton , OH 45469
***Materials Laboratory (WRDC/MLPO), WPAFB, OH 45433.

ABSTRACT

The linewidth and peak position (v_0) of the intersubband transition (IT) in GaAs/Al$_{0.3}$Ga$_{0.7}$As multiple quantum wells are studied as a function of temperature using the infrared absorption technique. We find that electrons in the GaAs well are weakly coupled to the GaAs normal optical phonon mode. The total integrated area of IT absorption is found to be approximately constant in the samples that were doped in the well but temperature dependent in the samples that were doped in the barrier. We also find that v_0 increases as the temperature decreases. This blue shift is found to increase as the dopant concentration is increased. We calculated the absorption spectrum in a nonparabolic-anisotropic envelope function approximation including temperature dependent effective masses, nonparbolicity, conduction band offsets, the Fermi level, and lineshape broadening. Our results indicate that a large many-body correction, in particular an exchange interaction (E$_{exch}$) for the ground state, is necessary to account for the observed blue shift as the dopant concentration increases.

INTRODUCTION

Infrared absorption due to the electronic transition between the confined ground and the first excited states in a GaAs/AlGaAs quantum well was first reported by West and Eglash[1] and then by many others (see for example Refs. 2 and 3 and references therein). Infrared detectors[4] and optical modulators[5,6] based on this intersubband transition (IT) have been fabricated. We have found that the full width at half maximum (FWHM), total integrated area, and the peak energy position (v_0) are temperature dependent. The configuration coordinate model and the temperature dependence of the FWHM are used to estimate the electron-phonon coupling strength and the phonon energy involved in the intersubband transition. The temperature-dependence of the total integrated area of IT is used to obtain an expression for the two-dimensional electron gas (2DEG) density (σ). The Fermi energy (ε_F) is also found to be temperature dependent. The blue shift of v_0 cannot be explained by the intersubband transition energy calculated with the envelope function approximation (EFA) method[7] which includes the temperature-dependence of the electron effective mass, the conduction band nonparabolicity, the Fermi level, and the temperature dependent conduction band offsets. The electron-electron exchange interaction[8] is not negligible and it may give rise to a significant blue shift as the temperature is decreased.

EXPERIMENTAL TECHNIQUE

The samples were grown on a semi-insulating GaAs substrate by molecular beam epitaxy and consisted of a 1000 Å undoped GaAs buffer layer followed by 75 Å undoped GaAs quantum wells and 100 Å barriers of Al$_{0.3}$Ga$_{0.7}$As. This was followed by a 75 Å undoped GaAs cap layer. The characteristics of the three samples used in the present investigation are shown in table I. The infrared absorption was recorded at a 60° or 73° angle from the normal using a BOMEM DA3 Fourier-transform interferometer. The sample was cooled to 5 K using a continuous flow cryostat and the temperature was controlled between 300 and 5 K within ±1 K.

Table I. Characteristics of the samples used in the present study. Samples no. 1 and 2 were doped in the well while sample no. 3 was doped in the barrier. S is Huang-Rhys factor and $\hbar\omega$ is the phonon energy. Period indicates the number of quantum wells.

Sample No.	[Si] ($\times 10^{18}$cm^{-3}).	S (10^{-2})	$\hbar\omega$ (meV)	Period
1	5	1.66	37.9	100
2	1	1.45	36.5	100
3	1	0.66	37.0	70

RESULTS AND DISCUSSIONS

Typical infrared absorption spectra of the IT taken at 296 and 5 K are shown in Fig. 1. The spectra (solid lines) were fitted with a Lorentzian lineshape (dashed lines) according to the form $\alpha(\nu) \propto \Gamma_0/(\Gamma_0^2 + (\nu - \nu_0)^2)$ where $\alpha(\nu)$ is the absorption coefficient and is related to the absorbance, $2\Gamma_0$ is the FWHM, ν is the wave number (cm^{-1}), and ν_0 is the IT peak position. Γ_0 and ν_0 were used as fitting parameters. The FWHM ($2\Gamma_0$) was measured for the three samples and the results are shown in Fig. 2.

The configuration coordinate model in the Gaussian approximation is used to fit the FWHM according to the following equation[9]

$$2\Gamma_0 = \sqrt{8\ln 2}\, \sqrt{S}\, \hbar\omega\, \sqrt{\coth(\hbar\omega/2kT)}\,, \tag{1}$$

where S is the Huang-Rhys factor which is a measure of strong (S>>1) or weak (S<<1) electron-phonon coupling, $\hbar\omega$ is the energy of the phonon that is coupled to the electrons, k is Boltzmann's constant, and T is the absolute temperature. The experimental data in Fig. 2 are

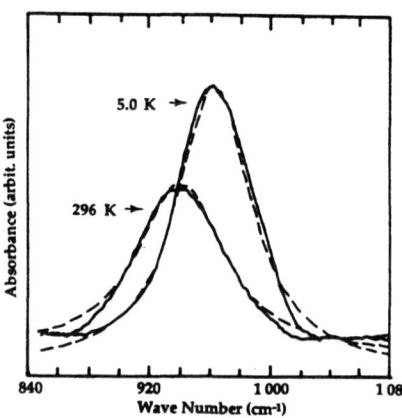

Fig.1. Infrared absorption spectra (solid lines) due to the intersubband transition in a GaAs/Al$_{0.3}$Ga$_{0.7}$As quantum well obtained at 296 and 5 K. The dashed lines are the result of a Lorentzian lineshape fit.

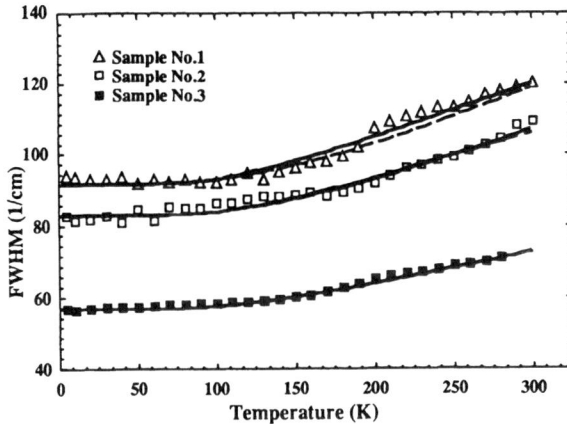

Fig. 2. Full width at half maximum (FWHM or 2Γ_o) of the intersubband transition as a function of temperature. The solid line is due to the fitting of the experimental data by using Eq. (1). The dashed line represents the fitting of the data using Eq. (2).

fitted by Eq. (2) with S and $\hbar\omega$ as fitting parameters and the result is shown as the solid line. The S and $\hbar\omega$ values are shown in table I. The small value of S indicates that the electon-phonon coupling is very weak in the present quantum well structure. In addition, the value of $\hbar\omega$ is in excellent agreement with the LO phonon mode[10] of GaAs. This suggests that the broadening of Γ_o as a function of temperature is due to a weak interaction between electrons which undergo the IT and the LO phonon mode of the well material. A similar result for $\hbar\omega$ is obtained when the data in Fig. 2 are fitted by a different expression[11] (which was used to fit the FWHM of a quantum well exciton transition[12]) defined as

$$\Gamma_o(T) = \gamma_1 + \gamma_2/[\exp(\hbar\omega/kT) - 1] \qquad (2)$$

where γ_1 and γ_2 are constants. Using Eq. (2) with $\hbar\omega$ as a fitting parameter, we obtained $\hbar\omega \approx 34\pm1$ meV for the three samples and the result is shown in Fig. 2 as the dashed line. It is evident that there is an excellent agreement between the data and the above two approaches.

The total integrated area of the IT was also measured as a function of temperature. The results are shown in Fig. 3. It is observed that the integrated area of samples no. 1 and 2 remains constant as a function of temperature. On the other hand, the integrated area of sample no. 3 which is doped in the barrier decreases as the temperatures is increased. This behavior indicates that the Fermi level is shifted in sample no. 3 as the temperature is varied which may correspond to the electron being excited out of the well.

The IT peak energy position, v_o, is observed to shift to a higher energy (blue shift) as the temperature is reduced. This shift was reported earlier[1-3] for GaAs/AlGaAs as well as for InGaAs/AlGaAs quantum wells.[13] In Fig. 4 we plot the peak position as a function of temperature for the three samples. The one-electron energy band structure for the well as a function of wave vector is calculated using Ekenberg's nonparabolic-anisotropic envelope function formalism.[7] We used temperature dependent electron effective masses for both the well and the barrier, which result from their dependence on the band gap[14] [see Eq. (63) of Ref. 14]; the band gap was calculated using the Varshni parameterization.[15] The nonparabolicity parameter was taken as temperature dependent [see Eqs. (64) and (65) of Ref. 14] with the low temperature value of $\alpha' = 0.600$ eV^{-1}; the anisotropy parameter $\beta' = 0.702$ eV^{-1} was kept constant (see Ref. 12 for the definition of α' and β'). The conduction band offset was taken as temperature dependent[16] with the temperature gradient of

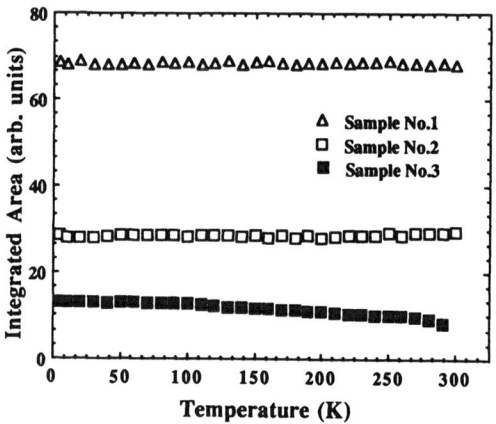

Fig. 3. The total integrated area of the intersubband transition. The integrated area of the samples doped in the well is found to be temperature-independent, while it is temperature-dependent in the case of the sample doped in the barrier.

$(-0.57) \cdot (1.15 \times 10^{-4}) \cdot (x)$ where 0.57 reflects the 57/43 conduction to valence band offset ratio and x is the alloy composition. Our calculations show that the excited state has a lower curvature than the ground state, which contributes to the width of the calculated absorption spectrum.

The linear absorption coefficient $\alpha(v)$ is calculated from the following expression

$$\alpha(v) \sim (1/v)\int d^2k \; \delta(E_2-E_1-hv) \; f(E_1) \; [1-f(E_2)] \tag{3}$$

where the integral is over the Brillouin zone, E_1 and E_2 are the ground and excited state energies as a function of wave vector in the plane of the well, and $f(E_i)$ are the Fermi-Dirac distribution functions. Subsequently, we broadened the calculated $\alpha(v)$ using the Lorentzian lineshape with Eq. (2) as the broadening function.

In Fig. 4 we show the experimental peak position together with our theoretical results for sample no. 3 which include the calculated peak position.[17] We first note that the nonparabolicity lowers the IT energy from data represented by the solid line in Fig. 4 to the data represented by the dashed line. Since broadened theoretical curves, unlike the experimental spectra in Fig. 1, are slightly asymmetric, a further study of broadening mechanisms is necessary. On the basis of the present calculations, we conclude that a one-electron EFA calculation gives IT energies lower than that observed experimentally and predicts no significant shift with temperature.

The work of Bandara et al.[8] shows that the ground state energy of a well filled with electrons is lowered by the exchange interaction which depends on the density of electrons, σ, in the well. Their expression for the exchange interaction energy which lowers the ground state energy does seem to provide a qualitative explanation for the blue shift in v_0. Using the values of σ obtained from sample no. 3 and the expression for the exchange energy (E_{exch}) from Bandara et al. we estimated E_{exch} (k=0) to be -23.10 and -20.84 meV at 5 and 300 K, respectively. Also, adding the exchange correction to the calculated one-electron IT peaks would bring the theory into good agreement with data of sample no.3. From the Bandara et al. results, one would expect a smaller Coulomb correction to the present calculated results. The blue shift of v_0 as a function of temperature in the sample doped in the barrier can be easily explained by exchange interaction energy mainly because the integrated area was found to increase as the temperature is decreased. However, the fact that the total integrated area was found to be temperature-independent in samples doped in the well is in disagreement with the

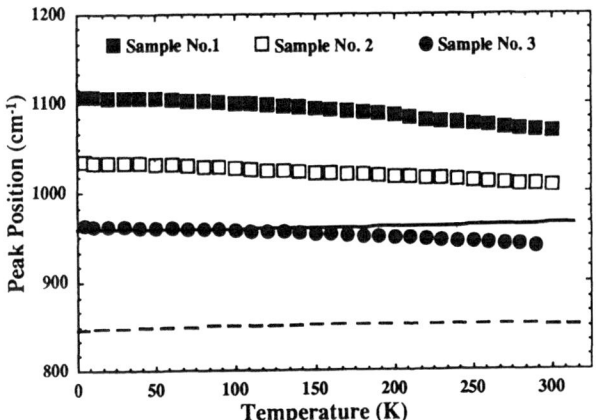

Fig. 4. The peak position, v_o, of the intersubband absorption as a function of temperature for the three samples. The solid line represents the theoretical calculations using the EFA method considering only the electron effective mass as a function of temperature. The dashed line is the same as the dashed line but nonparabolicity, anisotropy, conduction band offset, Fermi level, and lineshape broadening effects as a function of temperature are included.

above picture. This discrepancy may be easily solved if one consider that the electron-electron exchange interaction also exists in the excited state and the occupancy of both the ground and excited states is temperature dependent.

In order to test the validity of the many-body correction hypothesis, we studied samples with different dopant concentrations as shown in Fig. 4. In this figure, it is clear that the IT energy of sample no. 1 is higher than that of sample no. 2 and the IT energy of sample no. 2 is higher than that of sample no. 3. It should be mentioned that even though the dopant concentration is the same in both samples no. 2 and 3, the density of 2DEG in the quantum well should be different because sample no. 2 was doped in the well while sample no. 3 was doped in the barrier.

CONCLUSIONS

In conclusion, the infrared intersubband transition in a GaAs/AlGaAs quantum well is studied as a function of temperature. It is predicted from the configuration coordinate model and the temperature-dependence of the linewidth that electrons in the well are weakly coupled to the LO normal phonon mode of GaAs. The total integrated absorption was found to be temperature dependent for a sample doped in the barrier and temperature independent for samples doped in the well. The intersubband transition energies and the absorption spectra were calculated as a function of temperature using the EFA method. The nonparabolicity effect, which is included in our calculations, cannot explain the blue shift observed in v_o as the temperature is reduced. The electron-electron exchange interaction energy predicted by Bandara et al.[8], on the other hand, does seem to account qualitatively for the blue shift of v_o observed in the sample doped in the barrier mainly because the total integrated area is temperature-dependent. However, further analysis is needed to explain the same shift observed in samples doped in the well where the integrated area of the intersubband transition is temperature-independent.

ACKNOWLEDGMENTS -- This work was partially supported by the Air Force Office of Scientific Research. F.S. was supported under USAF contract F33615-88-C-5423. We would like to thank E. Taylor and J. Ehret for their technical support and R. E. Perrin and

T. Vaughan for their help in the computer analyses. We also would like to thank G. Bambakidis, H. Taylor, and G. Hasnain for useful discussions.

References

1. L. C. West and S. J. Eglash, Appl. Phys. Lett. 46, 1156 (1985).
2. B. F. Levine, C. G. Bethea, K. K. Choi, J. Walker, and R. J. Malik, Appl. Phys. Lett. 53, 231 (1988).
3. B. C. Covington, C. C. Lee, B. H. Hu, H. F. Taylor, and D. C. Streit, Appl. Phys. Lett. 54, 2145 (1989).
4. B. F. Levine, C. G. Bethea, G. Hasnain, J. Walker, and R. J. Malik, Appl. Phys. Lett. 53, 296 (1988); B. F. Levine, K. K. Choi, C. G. Bethea, J. Walker, and R. J. Malik, Appl. Phys. Lett. 50, 1092 (1987).
5. A. Harwit and J. S. Harris, Jr., Appl. Phys. Lett. 50, 685 (1987).
6. N. F. Johnson, H. Ehrenreich, and R. V. Jones, Appl. Phys. Lett. 53, 180 (1988).
7. U. Ekenberg, Phys. Rev. B 36, 6152 (1987).
8. K. M. S. V. Bandara, D. D. Coon, O. Byungsung, Y. F. Lin, and M. H. Francombe, Appl. Phys. Lett. 53, 1931 (1988).
9. R. K. Watts, Point Defects in Crystals (Wiley, New York, 1977), chap. 3, p. 75.
10. J. L. T. Waugh and G. Dolling, Phys. Rev. 132, 2410 (1963).
11. D. S. Chemla, D. A. B. Miller, P. W. Smith, A. C. Gossard, and W. Wiegmann, IEEE J. Quantum Electron. QE-20, 265 (1984).
12. M. Wegener, I. Bar-Joseph, G. Sucha, M. N. Islam, N. Sauer, T. Y. Chang, and D. S. Chemla, Phys. Rev. B 39, 12794 (1989).
13. X. Zhou, P. K. Bhattacharya, G. Hugo, S. C. Hong, and E. Gulari, Appl. Phys. Lett. 54, 855 (1989).
14. J. S. Blakemore, J. Appl. Phys. 53, R123 (1982).
15. Y. P. Varshni, Physica 34, 149 (1967).
16. S. Adachi, J. Appl. Phys. 58, R1 (1985).
17. The IT peak position is obtained from the broadened $\alpha(\nu)$, $[\overline{\alpha}(\nu)]$, and is given by:

$$\overline{\alpha}(\nu) = \int (1/\pi) \frac{\Gamma_0}{\Gamma_0{}^2 + (\nu - \nu')^2} \alpha(\nu') d\nu',$$ where $\alpha(\nu')$ is the same as Eq. (3).

LOW TEMPERATURE INFRARED MEASUREMENTS AND PHOTO-INDUCED
PERSISTENT CHANGES OF INTERSUBBAND TRANSITIONS IN
GaAs/AlGaAs MULTIPLE QUANTUM WELLS.

B. Dischler, J.D. Ralston, P. Koidl, P. Hiesinger, M. Ramsteiner, and M. Maier

Fraunhofer-Institut für Angewandte Festkörperphysik,
Tullastrasse 72, D-7800 Freiburg, West Germany.

ABSTRACT

Utilizing infrared absorption and Hall measurements, a detailed study is presented of the temperature dependence (10 K < T < 300 K) of the intersubband transition and related doping behaviour in GaAs/Al$_{0.32}$Ga$_{0.68}$As multiple quantum well structures with Si doping concentrations between 1 x 10^{18} and 8 x 10^{18} cm^{-3} in the quantum wells. The intersubband transition frequency increases with decreasing temperature, which can be adequately modelled by considering the different effective masses and the thermal populations of the two electron subbands. This model also accounts for the decrease in linewidth of the infrared absorption resonance upon cooling. In Hall measurements a persistent photo-induced decrease in the free electron concentration within the GaAs quantum wells is observed for Si doping levels exceeding 5 x 10^{18} cm^{-3}. Corresponding decreases are observed both in the integrated absorption and the absolute frequency of the intersubband transition , the latter arising from a reduction in the depolarization shift. The spectral dependence of the photo-induced changes over the energy range 1.2 eV < hν < 2.5 eV, along with thermal regeneration over the temperature range 70 K < T < 150 K, have been studied.

INTRODUCTION

Intersubband absorption in GaAs/AlGaAs quantum wells (QW) has become a field of increasing interest [1-6]. For photoconductive detection of 8-12 μm infrared (IR) radiation, typical device operating temperatures are 77 K or below to reduce the dark current [2]. To date, however, very few measurements have been published on the temperature dependence of the intersubband transition [1,5,6].

EXPERIMENTAL

Samples utilized in the present study were grown on undoped, semi-insulating (100) GaAs substrates in both Riber 2300 and Varian GEN II molecular beam epitaxy (MBE) machines. A 20 period undoped superlattice buffer was first grown at a substrate temperature of 690 °C. The active region, consisting of a stack of ten 7-nm GaAs QW's, separated by 14-nm Al$_{0.32}$Ga$_{0.68}$As barriers, was next grown at a substrate temperature of 610 °C. The samples were capped by 5 nm of undoped GaAs. An initial series of samples was grown with the center 5 nm of the GaAs QW doped with nominal Si concentrations ranging from 1 x 10^{18} to 8 x 10^{18} cm^{-3}. A second series was grown with the center 5 nm of the AlGaAs barrier layer doped with Si at concentrations ranging from 4 x 10^{18} to 8 x 10^{18} cm^{-3}. Additional QW- or barrier-doped samples were grown with 1 μm of undoped GaAs between the superlattice buffer and the active region, in order to exclude any possible contribution of back-gating to the observed temperature or illumination dependence. For further comparison 300 nm thick GaAs and $Al_{0.32}Ga_{0.68}As$ samples were grown, doped uniformly with Si at 8×10^{18} cm^{-3}. Infrared absorption was measured with a dual beam grating spectrometer. The samples were mounted in a helium cryostat with temperature stabilization between 10 K and 300 K and were oriented at Brewster's angle (73°) relative to the polarized probing light to allow for intersubband transitions [1]. Short wavelength spectrometer light was blocked using germanium filters. For the illumination experiments a tungsten lamp and a monochromator with 0.5 mm slits were used. DC Hall measurements were made using van der Pauw samples with photolithographically etched 3 mm cross patterns. Secondary ion mass spectroscopy (SIMS) was performed with an Atomica spectrometer using 3 keV Cs primary ions. SIMS depth profiles confirmed that the Si doping was at the intended position and had the intended doping level.

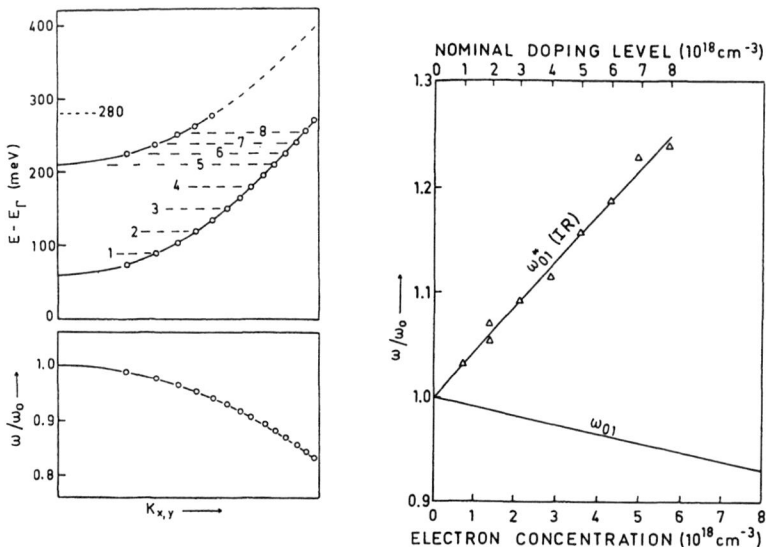

Fig. 1. Dispersion with respect to the in-plan wave vector energy (above) and of the intersubband transition frequency (below) for a 7 nm wide GaAs QW between $Al_{0.32}Ga_{0.68}As$ barriers. The conduction band offset is taken as 280 meV. Numbers indicate the band filling for electron concentrations in units of 10^{18} cm^{-3}.

Fig. 2. Change of intersubband transition frequency with electron concentration for the QW system shown in Fig. 1. The curve for the single particle mode ω_{01} is calculated. The curve for the collective plasmon-phonon mode ω^{*}_{01} is obtained by adding the experimental depolarization shift [4] to the single particle mode ω_{01}. The upper scale shows the nominal Si concentration in the doped (5 nm) layer, which has been converted to electron concentration in the 7 nm QW, assuming full activation.

RESULTS AND DISCUSSION

The influence of the Si doping level on the intersubband transition energy has been previously studied by Raman scattering and IR absorption [4]. Fig. 1 indicates the dominant effects which can be expected on the basis of theoretical considerations. Band filling causes occupation of states with larger in-plane wavevector, k_{xy}. The 12% larger effective mass in the upper subband [5] leads to a negative frequency dispersion (lower part of Fig. 1). Taking into account the density of states, the center of gravity of the intersubband transitions has been calculated as a function of electron concentration. The results are shown in Fig. 2, curve ω_{01}. The predicted redshift with increasing electron concentration has been observed experimentally via single particle Raman scattering [4]. However, the IR absorption is due to collective intersubband plasmon-phonon mode transitions, where the difference in frequency with respect to the single particle transition is given by the depolarization shift [4]. In Fig. 2, curve ω_{01}^{*} was obtained by adding the experimentally observed depolarization shift [4] to the calculated ω_{01} values. The linear relation between the electron concentration and the IR frequency proves to be very useful when studying illumination effects as discussed below.

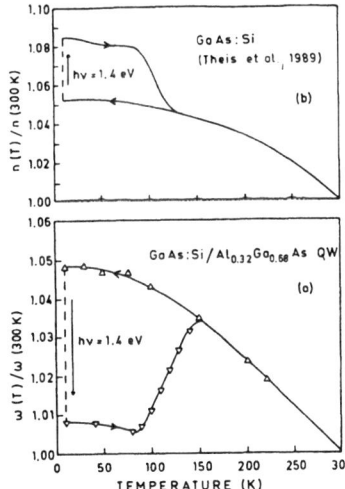

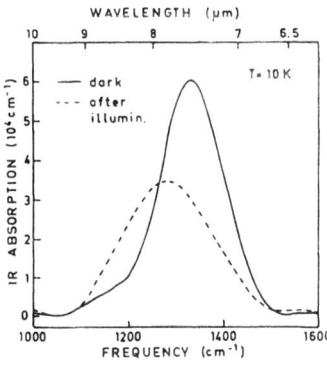

Fig. 3. Change of free electron concentration as a function of temperature and under illumination: (a) for a QW sample doped with 8×10^{18} cm^{-3}, (b) for a very highly doped (1.4×10^{19} cm^{-3}) GaAs MBE-layer containing DX centers (reproduced from [8]).

Fig. 4. Intersubband infrared absorption measured at 10 K in the dark (full line) and after illumination (broken line). The absorption strength is normalized for the case of light polarization perpendicular to the layers.

In agreement with earlier reports [1,5,6], a blueshift and a narrowing of the infrared absorption are observed upon cooling. Both effects can be interpreted by considering the influence of temperature on the Fermi distribution and subband population. For higher temperatures additional states with larger k_{xy} become occupied, corresponding to a broadening and to a redshift of the intersubband transitions. Fig. 3(a) shows the temperature dependence of the IR absorption frequency, demonstrating the predicted redshift. The increase in halfwidth over this temperature range is 50 %.

At temperatures below 70 K and for a Si doping level above 5×10^{18} cm^{-3} a persistent photo-induced decrease in the free electron concentration is observed. This behaviour is shown in Fig. 4 for a sample with 8×10^{18} cm^{-3} Si. The integrated absorption, which is proportional to the free electron concentration [1,3], decreases by 13 % upon illumination. A redshift of the IR absorption peak is also observed, and attributed to the corresponding reduction in depolarization shift (see Fig. 2). This agreement is an independent confirmation for the dependence of ω^{*}_{01} on the electron concentration shown in Fig. 2, which was originally obtained using a set of samples with different doping levels [4]. After turning off the light the effect of illumination is persistent for several hours. However, the samples return to the dark state when the temperature is raised above 70 K. Preliminary experiments indicate regeneration time constants of 2.5×10^{3} sec. and 1.1×10^{3} sec. at 110 K and 120 K, respectively. Assuming an Arrhenius relation an activation energy of 0.1 eV is obtained. The corresponding value for Si DX centers is 0.21 eV [7].

The spectral dependence of the photo-induced changes was studied at a sample temperature of 10 K. The light-induced redshift of the IR absorption peak was found to be linearly related to the decrease in electron density (see Fig. 2). The illumination effect showed an exponential dependence on exposure time, eventually reaching a saturated value. The inverse time constant of this exponential was taken as a measure for the spectral response and was normalized for constant quantum flux. The results are shown by the triangles in Fig. 5; in this log-log plot the spectral dependence can be approximated by a straight line.

It is suggested that the observed light-induced reduction of the free carrier concentration is related to electron trapping at an unknown defect. The reverse effect, namely the persistent light induced detrapping of electrons at low temperature has been reported for DX centers [7,8] (see Fig. 3(b)). For comparison, the spectral dependence of the photoionization of the DX center, as reported in the literature [7], is included in Fig. 5 . A common scale can be used by calibrating the present instrumental setup with a sample showing DX center related effects. The new center observed in the present work is tentatively called "AX", meaning "anti-DX" or "acceptor-like/unknown". In view of the similarities between AX and DX centers demonstrated in Fig. 3 (except for the opposite sign) and in Fig. 5, the AX center is also likely to have a metastable state with large lattice relaxation, as inferred for DX donor centers in n-type III-V [7] and for acceptor centers in p-type II-VI compounds [9].

At present no microscopic model for the AX center is proposed, but the observed behaviour gives several clues. The QW/barrier interfaces play an important role, because the illumination effect is completely absent in the 300 nm thick uniformly Si-doped GaAs layer prepared under the same conditions as the MQW samples. The corresponding 300 nm thick $Al_{0.32}Ga_{0.68}As$ MBE-layer showed the usual DX behaviour, i.e. almost no free electrons in the dark. The electron concentration in the

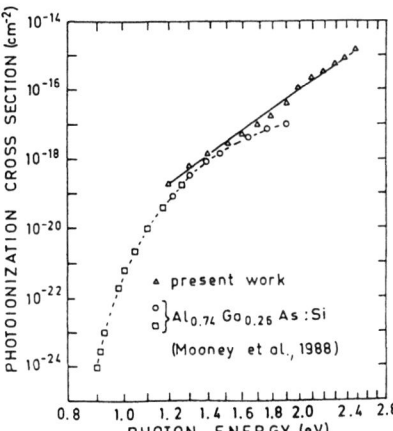

Fig. 5. Spectral dependence of the illumination effects for the QW samples (triangles). For comparison the DX photoionization cross section is reproduced from Ref.[7].

thick layers was monitored both by Hall measurements and by IR absorption due to the plasma resonance as described in the literature [10]. A second clue is given by the results from samples which were Si-doped in the AlGaAs barriers, where a mixture of AX and DX behaviour is observed. For example, the sample with 8×10^{18} cm^{-3} Si demonstrated a redshift of only 1.3 % along with a 9 % increase in integrated infrared absorption intensity. At temperatures above 100 K there is a dynamic balance between light induced transfer to the metastable AX state and thermal regeneration of the "dark" state. Under very strong Ar laser irradiation at 300 K, redshifts sufficient to be useful for modulation of mid-infrared radiation have been observed [11].

Fig. 6 shows the Hall sheet carrier concentration as a function of temperature for five QW doped samples with different doping levels. Below 100 K almost no temperature dependence of the carrier concentration is observed. The experimental points in Fig. 6 were obtained in the dark. The changes induced by illumination at a temperature of 100 K were studied for the sample with 8×10^{18} cm^{-3} Si doping. The 14 % decrease in carrier concentration and the spectral dependence are the same as observed in the IR experiments. The illumination caused a 20 % increase in electron mobility.

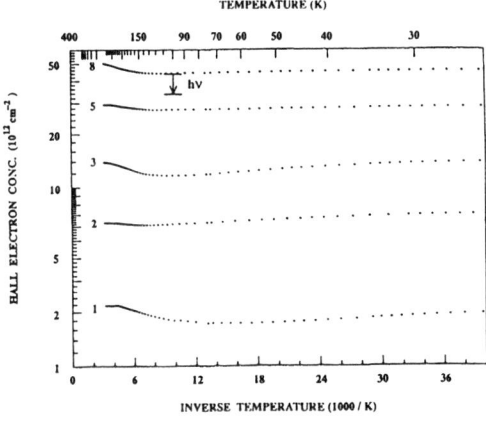

Fig. 6. Hall sheet electron concentration (cm^{-2}) as a function of inverse temperature for five QW samples with different nominal Si doping concentrations indicated by the numbers (10^{18} cm^{-3}).

SUMMARY

The effects of temperature variation on intersubband transitions in GaAs/AlGaAs QW structures have been studied over a wide range of Si doping levels. The observed changes in the linewidth and absolute frequency of the intersubband transitions are adequately modelled by taking into account the different subband effective masses along with band filling effects. At temperatures below 70 K and for Si concentrations exceeding 5×10^{18} cm^{-3}, a persistent photo-induced decrease of the Hall electron concentration and of the integrated intersubband absorption, along with a redshift of the IR absorption peak is observed. An acceptor-like analogue of the well-known DX center is tentatively proposed to account for this photo-effect, although the present effect appears to require the presence of hetero-interfaces and is not observed in uniformly doped GaAs or AlGaAs samples.

ACKNOWLEDGEMENTS

Valuable discussions with J. Wagner, H. Schneider and P. Tasker and technical assistance from H. Biebl, G. Bihlman and H. Thaden are gratefully acknowledged.

REFERENCES

1. L.C. West and S.J. Eglash, Appl. Phys. Lett. **46**, 1156 (1985).
2. B.F. Levine, K.K. Choi, C.G. Bethea, J. Walker, and R.J. Malik, Appl. Phys. Lett. **50**,1092 (1987).
3. J.D. Ralston, H. Ennen, M. Maier, M. Ramsteiner, B. Dischler, P. Koidl, and P. Hiesinger, Mat. Res. Soc. Symp. Proc. Vol. **163**, p. 875 (1990).
4. M. Ramsteiner, J.D. Ralston, P. Koidl, B. Dischler, H. Biebl, J. Wagner, and H. Ennen, J. Appl. Phys. **67**, 3900 (1990).
5. P. von Allmen, M. Berz, G. Petrocelli, F.-K. Reinhart, and G. Harbeke, Semicond. Sci. Technol. **3**, 1211 (1988).
6. B. C. Covington, C.C. Lee, B.H. Hu, H.F. Taylor, and D.C. Streit, Appl. Phys. Lett. **54**, 2145 (1989).
7. P.M. Mooney, J. Appl. Phys. **67**, R 1 (1990).
8. T.N. Theis, T.N. Morgan, B.D. Parker and S.L. Wright, Mater. Sci. Forum **38-41**, 1073 (1989).
9. D.J. Chadi and K.J. Chang, Appl. Phys. Lett. **55**, 575 (1989).
10. D. Kirillov, D. Lin, and Shang-Lin Weng, Appl. Phys. Lett. **55**, 2199 (1989).
11. H. Tsuchiya, Y. Shakuda, and H. Katahama, Surface Sci. **228**, 172 (1990).

PROPERTIES OF GaInAsP ALLOYS INVESTIGATED BY OPTICALLY DETECTED MAGNETIC
RESONANCE TECHNIQUES

C. Wetzel, B.K. Meyer, D. Grützmacher* and P. Omling+

Physikdepartment TU München, E16, James-Franck-Str., D-8046 Garching, FRG
* Institut für Halbleitertechnik, RWTH Aachen, Sommerfeldstr., 5100 Aachen, FRG
+ Department of Solid State Physics, University of Lund, Box 118, S-22100 Lund,
Sweden

1. INTRODUCTION

The quaternary $Ga_xIn_{1-x}As_yP_{1-y}$ semiconductor alloy system has considerable
importance for present day optoelectronic and microwave device applications. For
state of the art high mobility samples grown by metal organic chemical vapor
deposition (MOVPE) there are few experimental techniques which both can asess
band structure related properties (effective mass m^*, g-values of free
electrons) and impurity related properties (luminescence, mobility and
lifetimes). In this paper we compare optical and transport properties of the
quaternary compound $Ga_xIn_{1-x}As_yP_{1-y}$ (x-0.47,y-1; x-0.42,y-0.92; x-0.28,y-0.61;
x-0.12,y-0.34) lattice matched to InP by optically detected magnetic resonance
techniques.

2. EXPERIMENTAL

A. Sample preparation

Growth of the GaInAsP samples was performed in a LP-MOVPE on 2 inch wafers at
913 K [1]. The pressure was kept at 20 mbar. Semi-insulating InP substrate was
buffered by 400 nm InP epilayer (n < $5x10^{14}cm^{-3}$). The thickness of the
quarternary layers were 1000 nm (2000 nm for y-0.92) protected by a caplayer.
One sample is intentionally Zn-doped (p - $3x10^{17}cm^{-3}$).

B. Optical and microwave experiments

Photoluminescence (PL) spectra of the MOVPE layers were recorded at 1.6 K with
the sample mounted in a liquid He cryomagnetsystem (4 Tesla). Excitation was
provided by an Ar ion laser at 514 nm, the luminescence was analyzed by a Spex
1684 double monochromator in connection with a North Coast Ge-detector. The
samples were placed in a cylindrical open microwave-resonator operating in the
TE_{011} mode to study the influence of the microwave electrical/magnetical field
on the PL intensity. The microwave system operates at 24 GHz, as source we used
a Gunn diode amplified by a Hughes travelling wave tube (TWT) giving a maximum
power of 3 W. Complementary investigations were performed in a similar 12 GHz
system.

3. RESULTS AND DISCUSSION

A. Photoluminescence and impact ionisation studies

Fig.1 shows the PL of the quaternary layer GaInAsP with x-0.28 and y-0.61 at 1.6
K. The center of the band is at 0.995 eV with a width of 8 meV. The band gap for
this composition is given by [2]
    $E_g(4 K) - 1.421 - 0.767 y + 0.149 y^2$ (eV) - 1.008 eV
and marked in fig.1. with the arrow. The contributions of free or bound excitons
as well as band acceptor or donor-acceptor transitions to the total line width
are not resolved. In favourable cases they can be deduced by

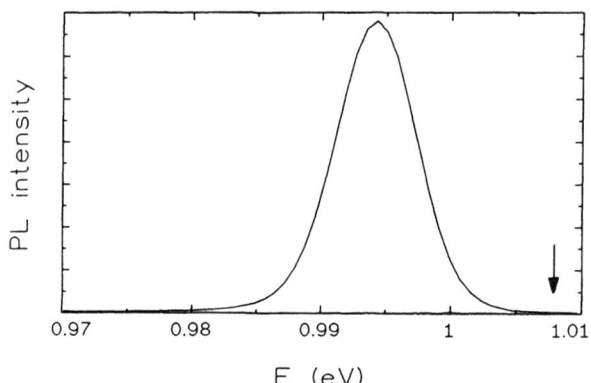

Fig.1: Photoluminescence of the $Ga_{0.28}In_{0.72}As_{0.61}P_{0.39}$ layer at 1.6K. The band is centered at 0.995 eV and has a width of 8 meV. The value of E_g is indicated.

studying the luminescence at higher temperatures. Here we used a different experimental technique taking advantage of the microwave electrical field present in the resonator. From comparative studies on GaAs, CdTe and Si [3,4,5] it is known that hot carriers accelerated by the electric field can impact ionize shallow bound excitons and shallow donors thus reducing their contributions to the photoluminecence and enhancing others e.g. free exciton transitions. The optically detected impact ionisation (ODII) is detected by chopping the microwaves (100 Hz -10 kHz) and monitoring the synchronous changes in the PL intensity. This is shown in fig.2 where a small increase with a maximum at 1.002 eV is observed followed by a rather strong decrease centered at 0.994 eV. The relative changes in the PL intensity, which can be as high as 60 %, depend strongly on the microwave power used which is demonstrated in fig.3. The first onset is around 50 mW and a second threshold is found at 300 mW. The linear dependence at low microwave powers with a threshold at 50 mW suggest that indeed impact ionisation is the responsible mechanism.

The spectral positions of the different recombination mechanisms can roughly

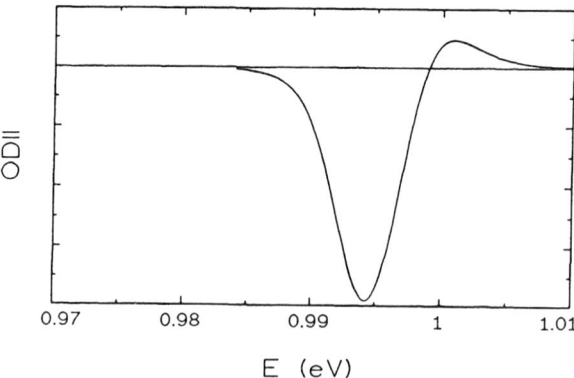

Fig.2: Microwave induced change of the PL (ODII) for the $Ga_{0.28}In_{0.72}As_{0.61}P_{0.39}$ sample at T=1.6K, 24 GHz.

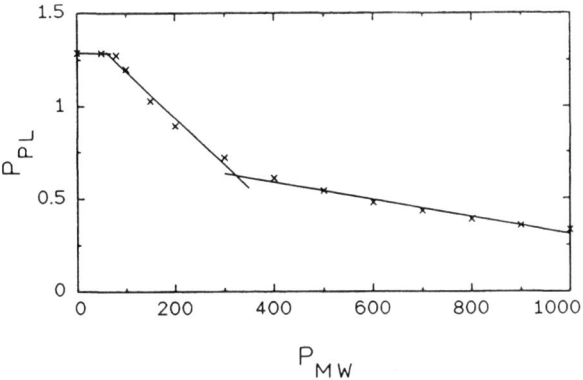

Fig.3: PL intensity as a function of microwave power P_{MW} at 1.6K, 24 GHz (symbols) with drawn lines as a guide for the eyes.

be predicted by calculating the respective binding energies on the basis of the effective mass theory (EMT) and scaling the donor/acceptor bound excitons (see table 1) according to Haynes rule as was done for the ternary InP lattice matched compound $In_{0.53}Ga_{0.47}As$ [6]. The binding energies of the free exciton resp. donor/acceptor is given by

$$E = 13.6 \ [\ (\mu/m_0)/\varepsilon^2] \quad (eV)$$

with $\mu = m_e$ (m_{hh}) the conduction band effective mass for the shallow donor resp. the heavy hole mass for the shallow acceptor, $1/\mu = 1/m_e + 1/2(1/m_{1h} + 1/m_{hh})$ the exciton reduced mass and ε the dielectric constant. The variations of m_e, m_{1h}, m_{hh} and ε with the composition (x,y) were taken from [7].

Recombination process	Emission energy in eV	
(X)	$E_g - E_X$	= 1.0051
(D^0,X)	$E_g - (E_X + \alpha E_D)$	= 1.0047
(D^+,X)	$E_g - (E_D + \alpha E_D)$	= 1.0036
(A_1^0,X)	$E_g - (E_X + \beta E_{A1})$	= 1.0029
(A_2^0,X)	$E_g - (E_X + \beta E_{A2})$	= 1.0019
(D^0,h)	$E_g - E_D$	= 1.0039
(e,A_1^0)	$E_g - E_{A1}$	= 0.987
(e,A_2^0)	$E_g - E_{A2}$	= 0.976
(D^0,A_1^0)	$E_g - E_{A2} - E_D$	= 0.971

Table 1:

The coefficients α and β were taken as 0.1, for the acceptor binding energies E_{A1} and E_{A2} we used the experimental values 23 and 32 meV respectively (see below). The effective mass value for the shallow acceptor is calculated to be 39 meV, it is however known from luminescence studies on $In_xGa_{1-x}As$ that the experimental value is about 10-20 % lower for x=0 and the discrepancy increases with increasing x, probably due to the neglect of central cell effects in the EMT calculation.

When comparing with a Zn-doped sample (p=3*10^{17} cm^{-3}) we see that the luminescence (see fig.4) has an additional band centered at 0.969 ± 0.002 eV,

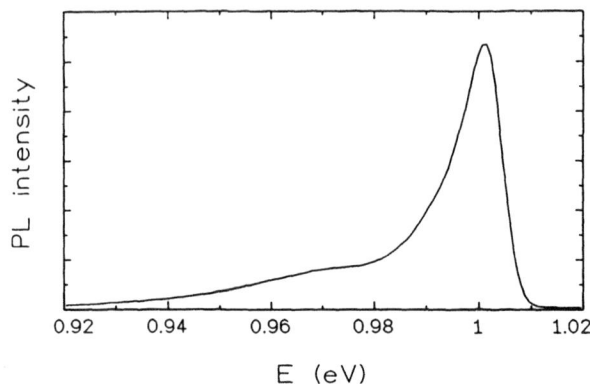

E (eV)

Fig.4: PL of the Zn-doped $Ga_{0.28}In_{0.72}As_{0.61}P_{0.39}$ sample. Compared to the undoped sample an additional band at 0.969 eV is observed.

which we can attribute to the donor-acceptor transition involving Zn with a binding energy of 32 ± 1 meV. The binding energy of the shallow acceptor calculated by EMT is 32.5 meV for InGaAs and 48 meV for InP compared to the experimental values for Zn in InGaAs of 22 meV and 43 meV in InP. A linear interpolation for InGaAsP (between y=0 i.e. InP and y=1 i.e InGaAs) with y=0.61 results in a value for Zn of 32 meV in very good agreement with the experimentally determined value (a similar estimate for C gives 23 meV). Based on this analysis the ODII-spectrum of the undoped sample may then be attributed to an acceptor bound exciton transition (enhancing the PL) and a free to bound transition involving C (quenching the PL). It is interesting to note that Carbon is the dominating residual acceptor in MOVPE grown samples.

B. Optically detected cyclotron resonance (ODCR)

The magnetic field dependence of the ODII signal monitored at 1 eV is shown in fig.5.a for a y=0.61 sample attributed to the electron cyclotron resonance (ODCR) [5,8]. The resonance itself is not resolved thus we are in the limit wt< 1. The measured curve can be fitted by the following expression [8]

$$P \sim \frac{1 + \omega_c^2 \tau^2 + \omega^2 \tau^2}{(1 + \omega_c^2 \tau^2 - \omega^2 \tau^2) + 4\omega^2 \tau^2} \cdot \left(\frac{Ne^2 \tau}{m^*} \right)$$

where w_c is the cyclotron frequency, w is the microwave frequency (24 GHz), t the carrier scattering time, P is the power absorbed, N the number of optically excited carriers and m^* the effective mass. Only for wt>1 can the effective mass be directly deduced from the microwave frequency $w_c = w$ with $w_c = eB/m^*$ the resonance condition. A best fit to the experimental curve gives the values for $m^* = 0.053 \pm 0.002$ (electron mass) and $t = 3 \times 10^{-13}$ s. For comparison the ODCR spectrum for y=0.34 is shown in fig 5.b.

With the values obtained for the electron masses and the scattering times for all three lattice matched alloys it is possible to calculate the mobilities of the samples under the conditions of optical excitation using the expression $\mu = et/m^*$ (see table 2). The Hall mobilities are 3500 at RT and 20000 at 77 K for y=0.61 compared to 16000 cm^2/Vs at 1.6 K. The difference between the low temperature values is not as much as one could expect when comparing with results from other ODCR investigations on III-V binary or ternary alloys. In AlInAs MBE layers [8] the mobility was higher by a factor of 100, similar results were reported on GaAs [4]. Those increases were attributed [4] to the photoneutralisation of shallow ionized impurities by the optically created

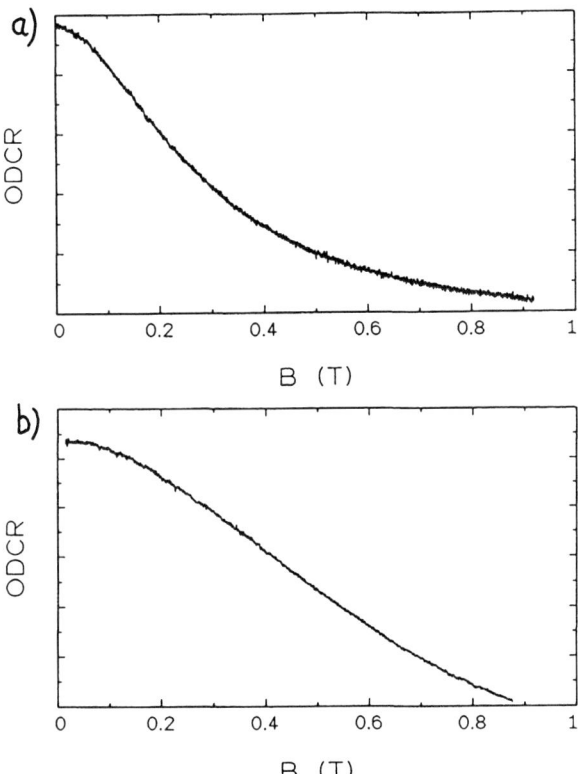

Fig.5: a) ODCR of the undoped $Ga_{0.28}In_{0.72}As_{0.61}P_{0.39}$ sample monotoring the ODII signal at 1 eV. The cyclotron resonance is attributed to the electrons.
b) ODCR of a $Ga_{0.12}In_{0.88}As_{0.34}P_{0.66}$ sample in comparison. The shape of the curve indicates the lower mobility.

carriers. This implies that ionized impurity scattering is the dominant mechanism at low temperatures. However, in the quaternary alloys three scattering mechanism are relavant a) polar optical phonon scattering b) ionized impurity scattering and c) alloy scattering. The available experimental data at 77 K show that with increasing y the mobility increases by a factor 5 to 6 from y=0.31 towards y=0.92 [7]. This enhancement is not found in our experimental results. Our w*t values are in resonable agreement with conventional cyclotron resonance experiments; there the values for different y ranged from 1.7 to 6.4 at maximum using a 118 μm radiation source [9]. Classical cyclotron experiments with and without light excitation are planed to study the alloy dependence of the mobilty at low temperatures in more detail.

$Ga_x In_{1-x} As_y P_{1-y}$	$\tau_{scatt}(ps)$	mobility ODCR $(cm^2/Vs)T = 1.6K$	mobility Hall $(cm^2/Vs)T = 77K/RT$	$\omega\tau$
x=0.12 y=0.34	0.47	13 000	20 000 / 3 300	0.035
x=0.28 y=0.61	0.49	16 000	20 000 / 3 500	0.036
x=0.42 y=0.92	0.75	30 000	45 000 / 8 500	0.056

Table 2:

5. CONCLUSIONS

The optical and transport properties of quaternary InGaAsP alloys were investigated by optically detected magnetic resonance techniques. Making advantage of the microwave electrical field impact ionisation studies were performed, which allow to distinguish between different radiative recombinations not resolved in the PL experiments. The ODCR experiments offer the possibility to investigate the scattering mechanism at low temperatures and to contribute to the understanding of alloy versus ionised impurity scattering in these alloys.

One of us (P.O.) acknowledges support by the Swedish Natural Science Council. We thank J. Kolodzey for helpful discussions and critical reading of the manuscript.

REFERENCES

[1] R. Meyer, D. Grützmacher, H. Jürgensen and P. Balk, J. Cryst. Growth 93 (1988) 285
[2] T.P. Pearsall, L. Eaves and J.C. Portal, J. Appl. Phys. 54 (1983) 1037
[3] E.J. Pakulis and G.A. Northrop, Appl. Phys. Lett. 50 (1987) 1672
[4] R. Romestain and C. Weisbuch, Phys. Rev. Lett. 45 (1980) 2067
[5] F. P. Wang, B. Monemar and M. Ahlström, Phys Rev. B 39 (1989) 11195
[6] K.H. Goetz, D. Bimberg, H. Jürgensen, J. Selder, A.V. Solomonov, G.F. Glinskii and M. Razeghi, J. Appl. Phys. 54 (1983) 4543
[7] GaInAsP Alloy Semiconductors, ed. by T.P. Pearsall, John Wiley (1982)
[8] M.G. Wright, A. Kana'ah, B.C. Cavenett, G.R. Johnson and S.T. Davies, Semicond. Sci. Technology 4 (1989) 590
[9] R.J. Nicholas, S.J. Sessions and J.C. Portal, Appl. Phys. Lett. 37 (1980) 178

A SYSTEMATIC TEM AND RHEED INVESTIGATION OF THE MBE GROWTH OF $IN_x GA_{1-x}$ AS AS A FUNCTION OF COMPOSITION

S.P. EDIRISINGHE*, A.E. STATON-BEVAN*, D.W. PASHLEY*, P. FAWCETT[+], and B.A. JOYCE[+].
*Department of Materials, Imperial College of Science, Technology and Medicine, London, S.W.7 2AZ, U.K.
[+]IRC Semiconductor Materials, The Blackett Laboratory, London, S.W.7 2BZ,U.K.

ABSTRACT

0.15µm epilayers of $In_x Ga_{1-x}$ As grown on GaAs (001) by MBE, having In concentrations in the range $x = 0.05 - 0.30$, have been investigated using RHEED and TEM. RHEED patterns indicate a 2-D growth mode for low In concentrations changing to Stranski-Krastanov growth for $x > 0.30$. TEM showed misfit dislocations for $x > 0.05$ only, which were found to relieve only a small part of the misfit strain. Although threading dislocations were rarely found in the epilayers, dislocations originating at the interface and penetrating the buffer layer were observed for $0.1 < x < 0.25$.

INTRODUCTION

There is considerable interest in $In_x Ga_{1-x}$ As/GaAs semiconductor heterostructures because of their use in optoelectronic device applications[1,2]. In this lattice mismatched system, if the misfit between the substrate (GaAs) and the epilayer ($In_x Ga_{1-x}$ As) is sufficiently small, the first few atomic layers deposited will be strained to match the substrate[3]. However, if the epilayer thickness exceeds a certain value the lattice mismatch will be accommodated partly by strain and partly by the formation of misfit dislocations[3] which cause degradation of the optoelectronic properties[4].

In the investigation described in this paper, TEM and RHEED have been used to study the defect structures in $In_x Ga_{1-x}$ As/GaAs with a systematic variation in composition for a constant epilayer thickness and growth temperature.

EXPERIMENTAL PROCEDURE

The samples consisted of a 0.15µm thick $In_x Ga_{1-x}$ As epilayer grown by MBE over a 0.5µm thick GaAs buffer layer on (001) GaAs substrate at a temperature of 500°C. The layers were grown on a MBE system with extensive facilities for RHEED observations.

The In content (x) was varied between $x = 0.05$ and 0.30. The structural properties of the layers were examined by plan view and cross-section TEM using Jeol 120CX and FX2000 microscopes. Samples for plan-view TEM were prepared by chemical thinning from the substrate side. Cross-sections were prepared by standard techniques involving mechanical polishing and ion milling.

Mat. Res. Soc. Symp. Proc. Vol. 216. ©1991 Materials Research Society

RESULTS AND DISCUSSION

RHEED observations (Figure 1) revealed that the initial growth of the $In_xGa_{1-x}As$ alloy was two-dimensional for the compositions used since oscillations were observed and a streak pattern remained. For $x \leq 0.2$ the growth mode was two-dimensional throughout (Figure 1a). For $\overline{x} = 0.3$, there was a transition to a three-dimensional growth mode after 12 monolayers had been deposited (as measured by RHEED oscillations) (Figures 1b and 1c). As growth continued, the intensity of these features gradually diminished until at the end of the growth a streak pattern was left with a modulation of intensity along the streaks where three-dimensional features had previously been evident (Figure 1d).

TEM cross sections revealed that threading dislocations were rarely present in the $In_xGa_{1-x}As$ epilayer (Figures 2a - 2e). However, in samples where $0.10 < x < 0.25$, dislocations were present in the buffer layer adjacent to the interface (Figures 2b, 2c and 2d) with most of them forming large loops. In samples where $x = 0.05$ and $x = 0.30$ no buffer layer dislocations were observed (Figures 2a and 2e).

In plan view TEM samples no misfit dislocations were observed in the sample where $x = 0.05$ (Figure 3a). An orthogonal array of misfit dislocations along <110> directions was observed in compositions where $x > 0.10$, as shown in Figures 3b and 3c. There is an asymmetry in the dislocation density along the two <110> directions in most of the samples studied. It is well known that misfit dislocations lying along <110> directions in (001) plane (namely α and β) are not chemically equivalent[5].

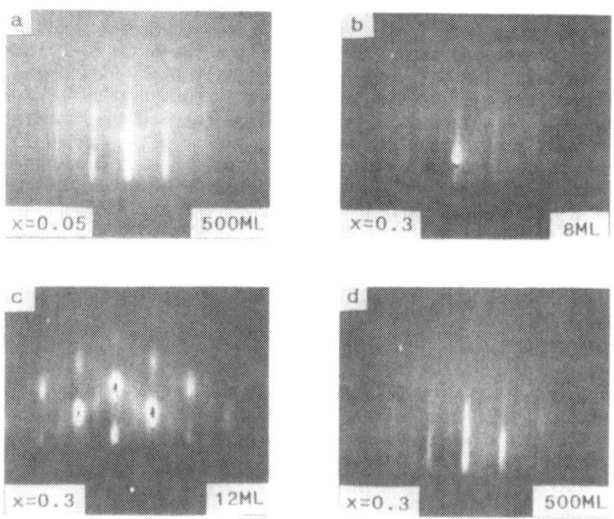

FIGURE 1. RHEED patterns of $In_xGa_{1-x}As$ epilayers on GaAs (001)

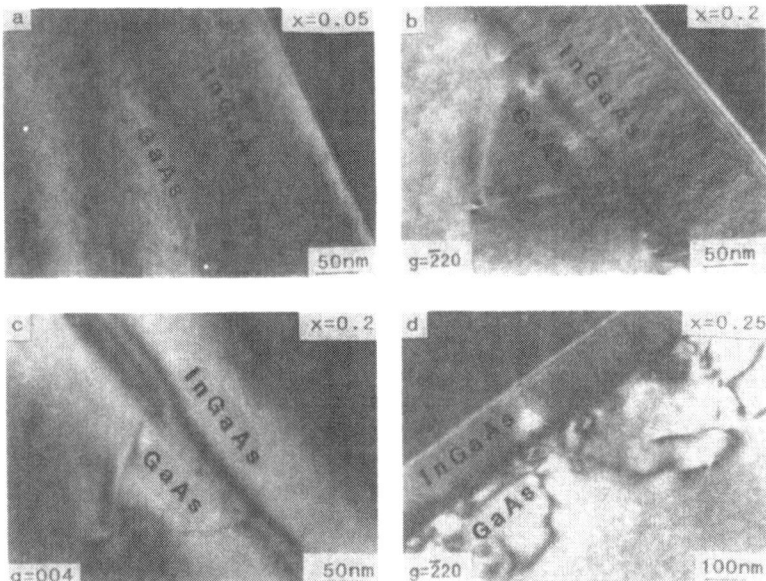

FIGURE 2. HRDF TEM micrographs of cross-sections of $In_xGa_{1-x}As$ epilayer/GaAs buffer interface.

Similar observations have been made in GaInAs/GaAs[6], GaInP and GaAsP/GaAs[7,8] interfaces and attributed to the difference in the mobility of the α and β dislocations.

From dislocation contrast experiments it was observed that misfit dislocations are mainly of the 60° type with burgers vector of the type $\frac{a}{2}$ <110> inclined to the interface plane. When x > 0.25 dislocations became more random and tangled in appearance (Figure 3c). The average dislocation density D along <110> was measured from plan view TEM micrographs. Dislocation density (number/unit length) increases with In composition (x) as shown in Table I. The strain relieved by the formation of dislocations (δ) is equal to b' D where b' is the magnitude of the burgers vector in the plane of the interface. Assuming all the

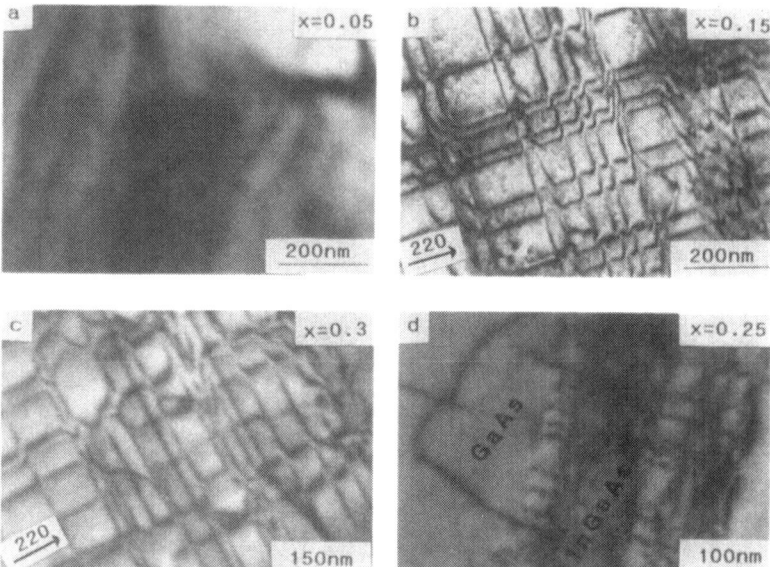

FIGURE 3.   TEM micrographs
(a) plan-view, x = 0.05;   (b) plan-view, x = 0.15
(c) plan-view, x = 0.30;   (d) inclined view of
cross section, x = 0.25.

dislocations present were of the 60° type, $\delta = (1/2\sqrt{2})a$ D, where a is the lattice parameter.   The residual strain $\epsilon = m - \delta$ where m is the lattice mismatch.   The values for m, δ and ϵ are given in Table I.   The average ϵ/m value of the compositions studied is in the range 0.71 - 0.76 and therefore seems to be independent of the composition.

Figure 3d shows an inclined view of a cross section of the sample where x = 0.25.   Misfit and buffer layer dislocations were clearly seen in this sample.

The epilayer and the interface were removed from the plan-view samples to investigate the buffer layer dislocations.   Figure 4 shows the relevant micrographs which illustrate that dislocations 1 and 2 parallel to [110] disappear when g = 220 (Figure 4b), also dislocations 3 and 4 which are parallel to [$\bar{1}$10] disappear when g = $\bar{2}$20 (Figure 4c).   Therefore we can conclude that dislocations 1 and 2 are edge type with burgers vector a/2($\bar{1}$10) and 3 and 4 are edge type with burgers vector a/2(110).   These dislocations could be the result of a reaction between two glissile 60° dislocations with appropriate burgers vectors which are expelled from the interface to form edge dislocations[9].   A few dislocations retain contrast with both <220> type reflections and are of the 60° type (e.g. dislocation 5).

TABLE I. Dislocation Density and Interfacial Strain Values

X	Dislocation density along $<110>cm^{-1}$		Interfacial strains (see text)			
	Theoretical assuming 60°	Measured	m (%)	δ (%)	ε (%)	ε/m (avg.)
0.05	1.7×10^5	-	0.4	-	-	-
0.10	3.6×10^5	-	0.7	-	-	-
0.15	5.4×10^5	1.3×10^5 1.3×10^5	1.1	0.27 0.27	0.83 0.83	0.76
0.20	7.0×10^5	1.4×10^5 2.2×10^5	1.4	0.28 0.45	1.12 0.95	0.74
0.25	8.8×10^5	2.2×10^5 2.5×10^5	1.8	0.44 0.51	1.36 1.29	0.73
0.30	1.0×10^6	2.0×10^5 3.1×10^5	2.2	0.41 0.63	1.79 1.57	0.71

For $0.1 > x > 0.3$, a fine mottled structure aligned parallel to the growth direction was observed when g = $\bar{2}20$ (Figures 2b and 2d). This fine structure goes out of contrast when g = 004 (Figure 2c) which corresponds to the growth direction and therefore this mottled structure has lattice

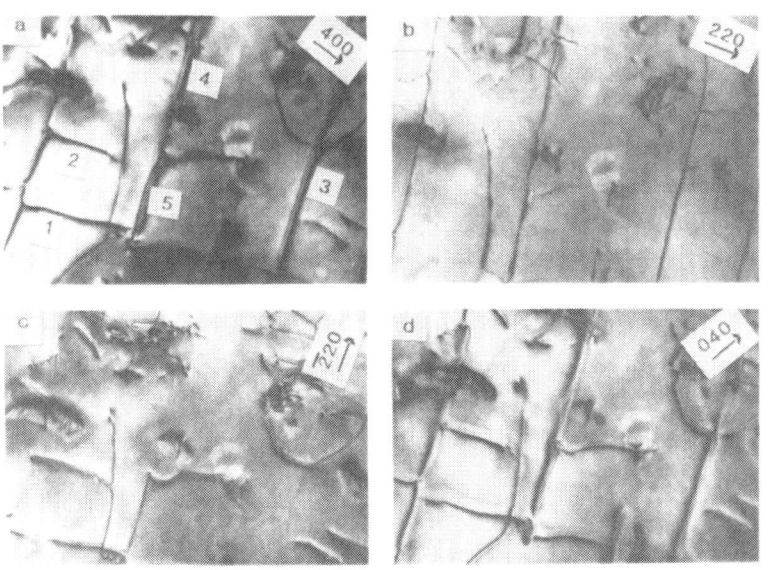

FIGURE 4. TEM micrographs of plan-views (x = 0.20)
(a) g = 400; (b) g = 220; (c) g = $\bar{2}$20; (d) g = 040

strains normal to the growth direction. This could be attributed to spinodal decomposition. Similar observations have been made for VPE grown GaInAsP[10] and MBE grown epilayers[11]. When x > 0.25 only, vertical band-like contrast (Figure 2e) indicated coarser composition modulation.

CONCLUSIONS

I. The mode of MBE growth of $In_xGa_{1-x}As$ at 500°C is 2-D for x < 0.25 and Stranski-Krastanov for x = 0.3.

II. Orthogonal arrays of misfit dislocations, predominantly of the 60° type, are observed for compositions $0.10 < x \leq 0.25$ becoming more random for x = 0.3. These relieve only a small part of the misfit strain.

III. Threading dislocations are rarely observed in the epilayer, however, dislocations originating from the interface penetrate the buffer layer for 0.1 < x < 0.25.

IV. Composition fluctuations are indicated in TEM cross-sections of the $In_xGa_{1-x}As$ epilayer by coarse vertical banding for x > 0.25 and a fine mottled structure for low x values.

ACKNOWLEDGEMENTS

This project was funded by the U.K. Science & Engineering Research Council.

REFERENCES

1. G.C. Osbourn, Phys.Rev.B 27, 5126 (1983)
2. W.D. Laidig, Y.F. Lin and P.J. Caldwell, J.Appl.Phys. 57, 33 (1985)
3. J.W. Matthews and A.E. Blakeslee, J.Cryst. Growth, 27, 118 (1974)
4. T.G. Anderson, Z.G. Chen, V.D. Kulakovskii, A. Uddin and J.T. Vallin, Appl. Phys.Lett. 51, 752 (1987)
5. M.S. Abrahams, J. Blanc and C.J. Buiocchi, Appl.Phys.Lett. 21, 185 (1972)
6. K.L. Kavanagh, M.A. Capano and L.W. Hobbs, J.Appl. Phys. 64, 4843 (1988)
7. G.A. Rozgonyi, P.M. Petroff and M.B. Panish, J.Cryst.Growth 27, 106, (1974)
8. M.S. Abrahams, L.R. Weisberg, C.J. Buiocchi and J. Blanc, J.Mater.Sci. 4, 223 (1969)
9. E.A. Fitzgerald and D.G. Ast, J.Appl.Phys. 63, 693, (1988)
10. S.N.G. Chu, S. Nakahara, K.E. Strege and W.D. Johnston Jr., J.Appl.Phys. 57, 4610 (1985)
11. A.G. Norman and G.R. Booker, in Microscopy of Semiconducting Materials, Eds. A.G. Cullis and D.B. Holt, Inst.Phys., U.K. p.257 (1985).

HIGH-QUALITY InGaAsP CRYSTALS GROWN BY MOVPE USING TBA AND TBP

A. Kuramata, S. Yamazaki, and K. Nakajima
Fujitsu Laboratories Ltd., 10-1 Morinosato-Wakamiya, Atsugi 243-01, Japan

ABSTRACT

TBA and TBP are attractive candidates for group V sources for MOVPE growth from the viewpoint of safety. We studied how the composition of InGaAsP crystals depends on growth conditions, and investigated its electrical and optical properties. The relationship between group V sources and crystals indicates that TBA and TBP decompose into AsH and PH. Since there is no carbon in AsH and PH, carbon contamination in the crystals is expected to be small. Carrier concentrations ranged from 5×10^{14} cm^{-3} to 1.5×10^{15} cm^{-3}. Photoluminescence spectra at 4.2K showed strong band-edge emission with no acceptor-related emission. Based on the electrical and optical properties of the crystals, we conclude that high-quality InGaAsP crystals can be grown using TBA and TBP.

INTRODUCTION

MOVPE growth is very promising for fabricating alloy semiconductor devices because crystals can be grown over a large area and heterointerfaces are sharp; however, arsine (AsH$_3$) and phosphine (PH$_3$), the group V sources commonly used in MOVPE, are dangerous gases. It is important that good alternative group V sources are found to make MOVPE growth safer. This can be done by using group V organic sources because they are liquid and less toxic than arsine and phosphine. Tertiarybutylarsine (TBA) and tertiarybutylphosphine (TBP) are attractive candidates because high-quality GaAs and InP have been grown using these sources [1]-[4]. Duncan et. al. reported InGaAsP crystal growth using these sources for making InGaAsP laser diodes [5]. In these devices, only active layer was grown using TBA and TBP, and lasing properties were not so good. No reports on the quality and the growth mechanism for the InGaAsP crystals grown using these sources are available.

In this report, we discuss InGaAsP crystal growth using TBA and TBP and the characteristics of the resulting crystals. Based on the relationship between the group V sources and crystals, we considered the decomposition products of TBA and TBP. We grew high-quality InGaAsP crystals and investigated their electrical and optical properties.

EXPERIMENT

We grew InGaAsP crystals in a low-pressure vertical reactor. The reactor design was reported on in a previous paper [6]. The total pressure was 76 torr, the growth temperature was 600°C, and the total flow rate was 6 l/min. Hydrogen was used as a carrier gas and trimethylindium (TMI) and triethylgallium (TEG) as the group III sources. Both group III and group V sources were kept in stainless cylinders, where the temperature was controlled by isothermal baths. The growth rate was 2.4 um/h. The V/III ratio was 100 for InP, 80 for InGaAsP, and 15 for InGaAs.

The carrier concentrations of the crystals were measured using capacitance-voltage (C-V) measurement at 300 K with a Polaron etch profiler.

Hall characteristics were measured at 300 K and 77 K using the Van der Paw method. Photoluminescence (PL) spectra were measured at 300 K and 4.2 K using a Kr$^+$ laser with excitation intensity of 1220 W/cm^2. Double crystal X-ray diffraction was used to measure the crystal's lattice constants.

RESULTS AND DISCUSSION

The relationship between the group V sources and the crystals

In ordinary MOVPE growth, group V sources decompose, and the decomposition products react with the crystals. Crystal's quality is affected by the decomposition products that take part in crystallization. There is a high probability that those atoms which combine with group V atom in the decomposition products contaminate the crystals. Since TBA and TBP molecules contain many carbon atoms, it is feared that carbon might be incorporated into the crystals. However, in the decomposition of TBA and TBP, the bond between the carbon and group V atom is believed to dissociate first, because this molecular bond is weakest. The decomposition products do not contain carbon, so it is not incorporated into the crystals.
    The following model was developed based on these ideas. It was assumed that TBA, TBP, TMI, and TEG would decompose into AsH, PH, In, and Ga. We treated these decomposition reactions as first order reactions. In this case, the partial pressure of the sources that remains after decomposition is proportional to the initial partial pressure as follows,

$$p_{TBA} = (1-k_{As})p_{o,TBA} \tag{1}$$

$$p_{TBP} = (1-k_P)p_{o,TBA} \tag{2}$$

$$p_{TMI} = (1-k_{In})p_{o,TMI} \tag{3}$$

$$p_{TEG} = (1-k_{Ga})p_{o,TEG} \tag{4},$$

where p_i is the partial pressure of the element i, $p_{o,i}$ is the initial partial pressure of the element i before decomposition, and k_i is a parameter which represents the decomposition degree. For crystallization, we assumed that the AsH, PH, As$_2$, P$_2$, In, Ga, H$_2$, and InGaAsP crystals are in equilibrium. As$_2$ and P$_2$ were included because we considered that group V elements are thought to have been adsorbed out of the crystals in the shape of these molecules. Reactions and equilibrium constants used in the calculation are listed in table 1 [7]-[11]. To include equilibrium relationships, following equation were used;

Tabel 1. Reactions and equilibrium constant used in the calculation.

Reaction	Equilibrium constant
In(g) + 1/2 P$_2$(g) = InP(s)	ln K$_1$ = -26.82 + 46770 / T
Ga(g) + 1/2 As$_2$(g) = GaAs(s)	ln K$_2$ = -26.99 + 54760 / T
In(g) + 1/2 As$_2$(g) = InAs(s)	ln K$_3$ = -25.71 + 47740 / T
1/2 As$_2$(g) + 1/2 H$_2$(g) = AsH	ln K$_4$ = 1.908 - 17630 / T
1/2 P$_2$(g) + 1/2 H$_2$(g) = PH	ln K$_5$ = 2.652 - 19760 / T

$$a_{InP} \ / \ p_{In} \ p_{P2}^{1/2} = K_1 \tag{5}$$

$$a_{GaAs} \ / \ p_{Ga} \ p_{As2}^{1/2} = K_2 \tag{6}$$

$$a_{InAs} \ / \ p_{In} \ p_{As2}^{1/2} = K_3 \tag{7}$$

$$p_{AsH} \ / \ p_{As2}^{1/2} \ p_{H2}^{1/2} = K_4 \tag{8}$$

$$p_{PH} \ / \ p_{P2}^{1/2} \ p_{H2}^{1/2} = K_5 \tag{9},$$

where a_i is the activity of the element i. The activity of binary crystals in quaternary crystal were calculated using the regular solution model [12]. To keep the mass and the total pressure constant, following equations were used;

$$x = (p_{o,TMI} - p_{TMI} - p_{In}) \ / (p_{o,TMI} + p_{o,TEG} - p_{TMI} - p_{In} - p_{TEG} - p_{Ga}) \tag{10}$$

$$y = (p_{o,TBP} - p_{TBP} - p_{PH} - p_{P2}) \ / (p_{o,TBP} + p_{o,TBA} - p_{TBP} - p_{PH} - p_{P2} - p_{TBA} - p_{AsH} - p_{As2}) \tag{11}$$

$$p_{o,TMI} + p_{o,TEG} - p_{TMI} - p_{In} - p_{TEG} - p_{Ga} = p_{o,TBP} + p_{o,TBA} - p_{TBP} - p_{PH} - p_{P2} - p_{TBA} - p_{AsH} - p_{As2} \tag{12}$$

$$p_{total} = p_{H2} + p_{TMI} + p_{In} + p_{TEG} + p_{Ga} + p_{TBP} + p_{PH} + p_{P2} + p_{TBA} + p_{AsH} + p_{As2} \tag{13},$$

where x is the In composition of the crystal, y is the P composition of the crystal, and p_{total} is the total pressure. We calculated the relationship between the initial partial pressure of the sources and the group V composition of the crystal using this model and compared them with the experimental relationships.

From the calculation, we determined that when the V/III ratio was larger than about 50, the group V composition of the crystal is almost independent of the V/III ratio, and can be determined by the ratio of PH to AsH. Since this relationship results in a nearly straight log-log plot, it can be written as follows,

$$\log(P/As)_{crystal} = A \log(p_{PH}/p_{AsH}) + B = A \log(p_{o,TBP}/p_{o,TBA}) + A \log(k_P/k_{As}) + B \tag{14},$$

where A and B are constants obtained by resolving the simultaneous equations. The calculated value for A was 0.60, and that for B was -0.66. Since the calculation contains no fitting parameters other than k_P and k_{As}, the slope of relationship A is determined by the equilibrium constants. As the decomposition changes, k_P and k_{As} change, and the relationship between the sources and the crystals shifts on the log-log plot, but the slope does not change. Figure 1 shows the calculated and the experimental relationships.

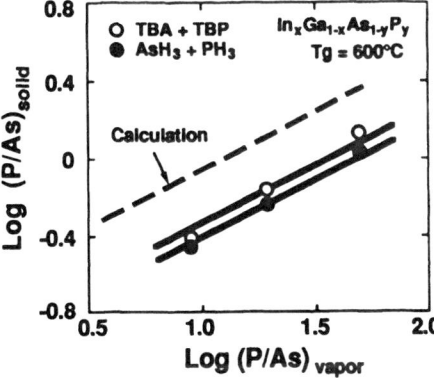

Figure 1. Relationship between group V sources and InGaAsP crystal composition.

The dashed line is calculated for the case where k_P/k_{As} is 1. The experimental relationship of TBA and TBP is almost the same as that of arsine and phosphine, and have the same slope as the calculation. This indicates that our model in which TBA and TBP decomposition products were assumed to be AsH and PH holds for both source combinations. In addition, using the difference between the experimental results and the calculated results, we can determine the value of k_P/k_{As}. The value of k_P/k_{As} is 0.32 for the TBA and TBP combination, and 0.25 for the arsine and phosphine combination.

As mentioned above, the relationship between the sources and the crystals indicate that the TBA and TBP decomposition products are AsH and PH. Since AsH and PH contain no carbon atoms, carbon should not have been incorporated into the crystals. To confirm this, we investigated the electrical and optical properties of the crystals.

Electrical and optical properties of the crystals

Figure 2 shows the carrier concentration of InGaAsP crystals obtained from C-V measurements at room temperature. The composition at the left is InGaAs, and that at the right is InP. The crystal's conductivity was of the n-type. Carrier concentration ranged from 5×10^{14} cm^{-3} to 1.5×10^{15} cm^{-3}. A low carrier concentration was found over the entire composition range.

The Hall characteristics of InP and InGaAs were measured, and compared with those of arsine and phosphine (Table 2). Mobility at 77 K is the most sensitive for the sum of donor and acceptor concentrations. We obtained a mobility of 51,000 cm^2/Vs for InP, which is comparable to 62,000 cm^2/Vs for phosphine. For InGaAs, we obtained a mobility of 44,000 cm^2/Vs, which is better than the 24,000 cm^2/Vs obtained for arsine. The purity of the InP and InGaAs crystals grown with TBA and TBP sources was as high as those grown with arsine and phosphine sources.

Figure 3 shows PL spectra for InGaAsP crystals at 4.2 K. The phosphorus compositions, starting from the left, are 0.55, 0.4, 0.25, and 0. The strong exciton emission observed

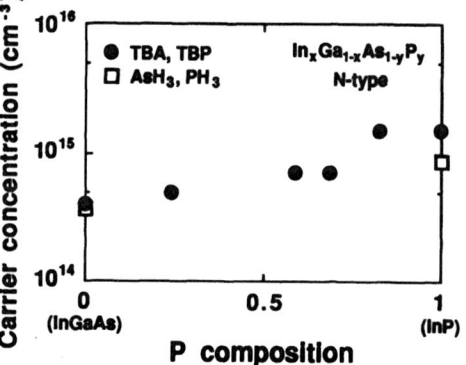

Figure 2. Carrier concentration of InGaAsP crystals at room temperature.

Table 2. Hall characteristics of InP and InGaAs crystal grown using phosphine and arsine, and those grown using TBP and TBA.

Temp. (K)	300		77	
Crystal	n (cm^{-3})	μ (cm^2/Vs)	n (cm^{-3})	μ (cm^2/Vs)
InP (PH₃)	2×10^{15}	3700	9×10^{14}	62000
InP (TBP)	5×10^{14}	4403	3×10^{14}	51000
InGaAs (AsH₃)	8×10^{14}	9600	4×10^{14}	24000
InGaAs (TBA)	3×10^{14}	10000	2×10^{14}	44000

near the band edge for each composition indicates that the optical properties of the crystals are good. Had the crystals contained a large amount of acceptors, we would have seen acceptor-related emission to the right of the band-edge emission. Based on the crystal's electrical and optical properties, we conclude that high-quality InGaAsP crystals can be grown using TBA and TBP.

It is not clear how carbon is ionized in the InGaAsP crystals. The ionized donor and acceptor concentrations calculated using the Hall characteristics for InP crystals were 1.8×10^{15} cm^{-3} and 1.2×10^{15} cm^{-3}. We considered the InGaAsP crystals to be purer than InP crystals because the carrier concentration monotonically decreased as the phosphorous composition decreased. We considered that the donor and acceptor concentrations of InGaAsP crystals were probably less than 1×10^{15} cm^{-3}. This was supported by the PL spectra at 4.2 K, since no acceptor-related emission was observed. We don't know what are donors and what are acceptors in the InGaAsP crystals. Even when all donors and acceptors are carbon, the ionized carbon concentration is less than 2×10^{15} cm^{-3}. We confirmed

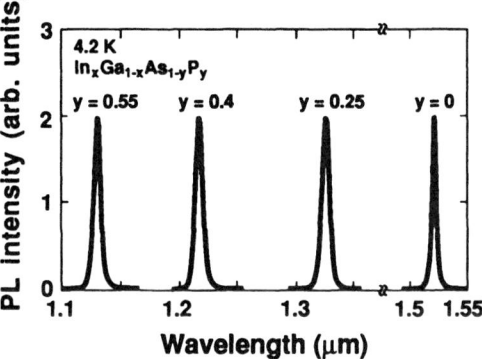

Figure 3. Photoluminescence spectra for InGaAsP crystals at 4.2 K.

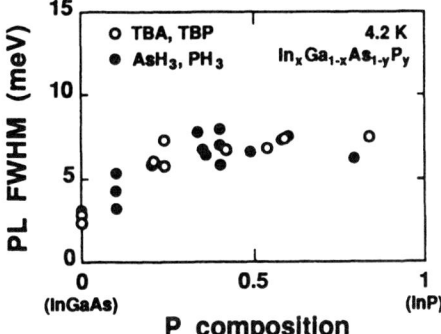

Figure 4. Photoluminescence full width at half maximum (FWHM) for InGaAsP crystals at 4.2 K.

that contamination by the ionized carbon was very small with the TBA and the TBP. In addition, we performed a secondary ion mass spectroscopy (SIMS) measurement of the InP crystals. The detection limit of the carbon was 5×10^{17} cm^{-3}. No carbon was detected by the SIMS measurement. This indicates that the ionized and non-ionized carbon concentration was less than 5×10^{17} cm^{-3}. The detection limit of the SIMS measurement is too high to conclude that non-ionized carbon contamination of the crystals is small. If carbon exists as a non-ionized impurity in the InGaAsP crystals, additional studies are necessary to determine the extent of carbon contamination from TBA and TBP.

Finally, we evaluated the PL full width at 4.2 K and considered the use of TBA and TBP in the fabrication of optical devices. Figure 4 shows that the PL full width depends on the phosphorus composition. The PL full width is about 2 meV for InGaAs and about 6 meV for InGaAsP. These values are small enough for device fabrication and are comparable to the values obtained using arsine and phosphine. We think that we will be able to fabricate good optical devices using TBA and TBP.

CONCLUSION

We found that the relationship of the group V sources and the crystals can be explained using a model which assumes that the decomposition of TBA and TBP produces AsH and PH. The quality of the resulting crystals was expected to be high, since carbon was not included in the decomposition products. Carrier concentration ranged from 5×10^{14} cm^{-3} to 1.5×10^{15} cm^{-3}. PL spectra at 4.2 K showed strong band-edge emission with no acceptor-related emission. Thus, the results of our study on the crystal's electrical and optical properties showed that high-quality InGaAsP crystals can be grown using TBA and TBP.

REFERENCES

1. G. Haacke, S. P. Watkins, and H. burkhard, Appl. Phys. Lett. <u>56</u>, 478 (1990).
2. T. Kikkawa, H. Tanaka, and J. Komeno, J. Appl. Phys. <u>67</u>, 3576 (1990).
3. C. Michel, M. Viscogliosi, J. Baumann, S. Watkins, L. Bunz, and R. Schachter, in OMVPE Workshop, Monterey, California, 1989.
4. A. Kuramata, S. Yamazaki, and K. Nakajima, Ninth symposium Record of Alloy Semiconductor Physics and Electronics, Izunagaoka, 1990, pp. 49.
5. W. J. Duncan, D. M. Baker, M. Harlow, A. English, A. L. Burness, and J. Haigh, Electron. Lett. <u>25</u>, 1603 (1989)
6. A. Kuramata, S. Yamazaki, and K. Nakajima, <u>Inst. Phys. Conf. Ser. 96</u> (Inst. Phys., London, 1989) pp. 113.
7. D. W. Shaw, J. Phys. Chem. Solids <u>36</u>, 118 (1975).
8. A. Koukitu and H. Seki, J. Crystal Growth <u>49</u>, 325 (1980).
9. V. S. Bans and M. Etternberg, J. Phys. Chem. Solids <u>34</u>, 1119 (1973).
10. M. Tirtowidjojo and R. Pollard, J. Crystal Growth <u>77</u>, 200 (1986).
11. in <u>JANAF</u> <u>Thermodynamical</u> <u>Tables</u>, edited by Chase et. al. (Nat. Bur. Standard, 1985).
12. K. Nakajima, in <u>Semiconductors and Semimetals</u> <u>22</u>, edited by W. T. Tsang (Academic Press, Inc., Orlando, 1985) pp 1.

PART VII

General Issues, Materials, Processing and Applications on Long-Wavelength Semiconductors

PHOTOVOLTAIC INFRARED DEVICES IN EPITAXIAL NARROW GAP LEAD CHALCOGENIDES ON SILICON SUBSTRATES

H. Zogg, C. Maissen, J. Masek, T. Hoshino, S. Blunier
AFIF (Arbeitsgemeinschaft für Industrielle Forschung) at Swiss Federal Institute of Technology, ETH Hönggerberg, CH-8093 Zürich, Switzerland

ABSTRACT

We review MBE growth of epitaxial IV-VI layers on Si(111) substrates and fabrication of photovoltaic infrared devices in the layers. Cut-off wavelengths are chemically tailored from 3 μm up to above 12 μm by using PbS, PbTe, $Pb_{1-x}Eu_xSe$ and $Pb_{1-x}Sn_xSe$. An intermediate epitaxial stacked CaF_2-BaF_2 bilayer of 200 nm thickness serves to overcome the large lattice- and thermal expansion mismatch, and device quality IV-VI layers are obtained with layer thicknesses of only 2-4 μm. The layers are untwinned single crystal, exhibit perfectly smooth surfaces with surface defect concentrations down to 10^3 cm^{-2}, and x-ray rocking curve line-width of ≈150 arcsec. Despite the large thermal expansion mismatch, the (111)-oriented layers withstand multiple cooling cycles down to 15K without problems.

Although our IR-device fabrication technique is far from optimized, the sensitivities of our best photovoltaic sensors are comparable to MCT. IV-VI on Si IR sensors have the potential for a low cost technique of large IR focal plane arrays both for the 3-5 μm and 8-12 μm range because of the easy fabrication procedure and because uniformity problems are much less severe in IV-VIs due to the weaker dependence of the band-gap of $Pb_{1-x}Sn_xSe$ on composition x compared to MCT.

INTRODUCTION

Infrared sensors fabricated in narrow band-gap semiconductors exhibit lowest noise at any given operation temperature due to the band-to-band carrier excitation mechanism. While the theoretical noise level is similar in photovoltaic $Hg_{1-x}Cd_xTe$ (MCT) and lead chalcogenide devices [1], most development work is presently performed with MCT. However, lead chalcogenides grown as single crystal layers on silicon substrates offer significant *practical advantages* for the construction of large infrared focal plane arrays which justifies a renewed interest in IV-VI materials for IR-sensor applications. This is because

a) homogeneity problems are much less severe in IV-VIs than in MCT due to the weaker dependence of the band-gap on composition x of e.g. $Pb_{1-x}Sn_xSe$ (LTS) compared to $Hg_{1-x}Cd_xTe$: The bandgap for the binary compounds (x=0) is already ≈0.2 eV for PbSe, while it is as high as 1.6 eV for CdTe [2]. According to the well known dependence of the bandgap vs composition, the cut-off wavelength of a 0.1 eV (12.4 μm cut-off) LTS sensor depends 5 times less on composition x than in a similar sensor fabricated in MCT.

b) MBE of IV-VIs is easy and straight-forward, it is already in routine use for laser fabrication [3]. No difficult to handle high vapour pressure elements are involved. Carrier concentrations in the low 10^{17} cm^{-2} range as needed for optimised photovoltaic sensors are easy to obtain with e.g. an additional Se-flux for p-type doping. The operational parameter range like substrate temperature or background pressure are not critical. Thicknesses of 2-4 μm suffice to obtain device quality layers. When grown on Si wafers, a ≈200 nm thick epitaxial CaF_2-BaF_2 buffer layer ensures single crystal growth with low defect density, high mobilities and low x-ray line widths [4,5].

c) the whole wavelength range of interest between 3 μm and up to above 12 μm is covered by similarly behaving materials like PbS, $PbS_{1-x}Se_x$, $Pb_{1-x}Eu_xSe$, PbTe, $Pb_{1-x}Sn_xSe$ or $Pb_{1-x}Sn_xTe$. Some properties of these materials are listed in table 1.

Mat. Res. Soc. Symp. Proc. Vol. 216. ©1991 Materials Research Society

Material	lattice constant Å	thermal expansion coefficient 10^{-6} /K	band gap energy eV	cut-off wavelength µm
Si	5.431	2.6	1.1	1.1
CaF$_2$	5.46	19.2	>>1	
BaF$_2$	6.20	19.8	>>1	
CdTe	6.48	5.0	1.6	.8
PbS	5.94	20.2	0.42 / 0.37 / 0.31	3.0 (300K) / 3.4 (200K) / 4.0 (77K)
PbTe	6.46	19.8	0.27 / 0.22	4.6 (200K) / 5.6 (77K)
PbS$_{1-x}$Se$_x$ (x=0-1)	5.94-6.12		.42 - .17	3 - 7
Pb$_{1-x}$Eu$_x$Se (x=0--0.02)	6.12		.17 - .42	7 - 3
Pb$_{1-x}$Sn$_x$Se (x=0--0.02)	6.12-6.06		.17 - 0	7 - ∞
Pb$_{1-x}$Sn$_x$Te (x=0--0.4)	6.46-6.40		.3 - 0	4 - ∞

Table 1
Properties of materials for IV-VI narrow gap sensor arrays on fluoride covered Si.

d) MBE growth on Si allows fabrication of heteroepitaxial, but monolithic IR focal plane arrays with signal processing electronics integrated into the Si-substrate.

e) The theoretical sensitivity of LTS is somewhat higher than that of LTT and CMT [1]. However, the difference is not pronounced and the accuracy of the calculations are uncertain to some extent. We prefer LTS instead of LTT mainly because the blocking contact technique for sensor fabrication (see below) works well with this material (but not with LTT), and its permittivity is somewhat lower than in the Te-containing compounds.

One should note that *single crystal* lead-salts are suited for photo*voltaic* devices only, photo*conductive* or MIS-type devices can not be fabricated because of the low carrier density needed for these purposes. The well known photo*conductive* devices fabricated in large quantities in polycrystalline PbS and PbSe exhibit lower ultimate sensitivities and their cut-off wavelength is limited to the SWIR and MWIR range.

Disadvantages of the IV-VI materials are their high permittivities and large thermal expansion coefficients. However, the bandwidth attainable (>100MHz) with photovoltaic IV-VIs is high enough and not a limiting factor for thermal imaging applications. The high thermal expansion coefficients (see table 1) do not impede fabrication of high quality material, as will be shown below. This is because thermally generated strains do not cause cracks in the layer, but relax by some plastic deformation [6-8].

The lattice mismatch between IV-VIs and Si is up to 20% (see table 1). It has not been possible up to now to grow IV-VI materials epitaxially *directly* on Si by any technique. However, by using an appropriate buffer layer, stacked CaF$_2$-BaF$_2$, high quality epitaxial IV-VI layers are obtained. This buffer was chosen because CaF$_2$ which has 0.6% lattice mismatch with Si was successfully grown on Si-substrates by different groups [9-11], and because *bulk* BaF$_2$ was known to be a suitable substrate for epitaxial IV-VI overgrowth and fabrication of photovoltaic IR-sensors in the layers [12].

We found that epitaxy of BaF$_2$ on CaF$_2$ does not pose special problems despite the 14% lattice mismatch [4,5,7]. When grown on Si, the large difference in thermal expansion between fluorides and Si may lead to cracks in the layers. Such cracks have been observed in CaF$_2$ on Si [9-11], but are avoided by using proper growth techniques and clean conditions. It seems that relief of thermally generated strain by dislocation movement is efficient enough to avoid cracking if the movement is not hindered by e.g. contamination. We found that cracking in BaF$_2$ layers grown even under less clean conditions is rather seldom, while CaF$_2$ grown under similar conditions is heavily cracked. This is most probably because the elastic constants of BaF$_2$ are lower than those of CaF$_2$; BaF$_2$ is softer and easier to deform plastically.

In the following, we describe the growth and characterisation of epitaxial narrow gap IV-VI layers on fluoride covered Si(111) and Si(100) substrates, and give some results of IR-sensor arrays fabricated in (111) oriented layers.

GROWTH AND MATERIAL PROPERTIES on Si(111)

The IIa-fluorides grow easiest with (111) orientation because of their low (111)-surface free energy. Growth by MBE is usually performed with solid source material and graphite crucibles. The fluorides evaporate as molecules and do not decompose on sublimation. The orientation of the fluoride lattice with respect to the Si-substrate is type B, i.e. the lattice of the layer is rotated 180° about the [111] surface normal. Typical growth rates are up to a few Å/sec, and usual substrate temperatures 700°C - 750°C for CaF_2(111)-growth [9-11]. These temperatures can be lowered considerably after switching to BaF_2 [4,5]. Growth is 2-d as revealed by streaky RHEED-patterns. Even when changing the beam flux abruptly from CaF_2 to BaF_2, no spots indicative of formation of BaF_2 nuclei are observed in the RHEED pattern: The distance between the streaks decreases within a few monolayer growth time to that corresponding to the bulk lattice constant of BaF_2. In situ rapid thermal anneal cycles (up to ≈1000°C) improve the sharpness of the RHEED-patterns [5], but are not needed for layers to be used as buffers for further growth of IV-VI materials. Best results are obtained when the thickness of the CaF_2 part is kept around 100Å, followed by 1500-3000Å BaF_2.

These temperatures are too high if one would like to grow on a Si-wafer which already contains standard integrated circuits because the Al-metallization can not withstand temperatures above about 450-500°C. We therefore recently attempted to grow at 450°C substrate temperature. It turned out that even such low temperature grown buffer layers can be overgrown with IV-VI materials, and the quality of the $Pb_{1-x}Sn_xSe$ deposited on such a buffer was suitable to fabricate a whole 66 element IR-sensor array (see below) for the LWIR range [13].

IV-VI materials are grown onto the fluoride covered substrates with HWE (hot wall epitaxy) or MBE [14]. HWE suffices for the binary compositions PbTe and PbS, while we prefer MBE for the ternary $Pb_{1-x}Sn_xSe$ because it allows accurate control of the composition x. This is accomplished by adjusting the fluxes of the PbSe and SnSe source. Typical growth rates are again a few Å/sec, substrate temperature is 400°C-450°C, and 2-4 μm thick layers are grown. An additional Se source is used to obtain p-type layers with hole densities in the low 10^{17} cm^{-3} range as needed for device fabrication.

X-ray rocking curves of two samples are shown in fig. 1. The line width in fig. 1a for a 3 μm thick PbTe layer is 180 arc sec. The width for a stack consisting of 2 μm PbSe overgrown by 2 μm $Pb_{1-x}Sn_xSe$ with x≈.05 can be resolved into its two components since the lattice constant for $Pb_{1-x}Sn_xSe$ decreases slightly with increasing x (fig. 1b, we use such stacks for IR-sensor fabrication). The width is ≈150 arc sec for the top $Pb_{1-x}Sn_xSe$ layer, while it is somewhat larger for the underlying PbSe. These line widths are comparable to widths in GaAs layers of similar thickness on Si substrates [15], despite the lattice mismatch is much larger in our case.

The surfaces of the as grown layers are perfectly mirror smooth and with microscopic surface defect densities down to 1000 cm^{-2}, this despite the fact we do not handle the samples under clean room conditions. Slip lines running along the three <110>-directions indicate thermal mismatch strain relief (Fig. 2). These slip lines are the intersections of the {100} glide planes operative in IV-VI materials with the (111)-surface. We never observed cracks in the layers, strain relief is by plastic deformation even in layers with thicknesses above 10 μm.

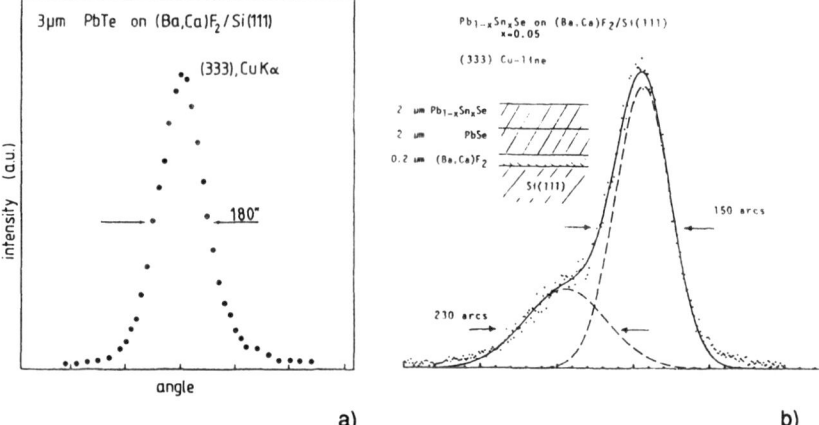

Fig. 1.
X-ray rocking curve of a (a) 3 μm thick PbTe layer (b) PbSe/Pb$_{1-x}$Sn$_x$Se on fluoride covered Si. The larger peak in (b) is from the Pb$_{1-x}$Sn$_x$Se top layer. The dashed lines are fitted curves, the solid line is the sum of the two dashed curves.

Fig. 2.
Nomarski micrograph of a Pb$_{1-x}$Sn$_x$Se on fluoride/Si(111) layer. The directions of the three sets of slip lines resulting from the {100}<110> glide system are indicated.

Strain relaxation

The residual strains in the layers were measured with x-ray diffraction and independently with the Rutherford backscattering angular scan channelling yield technique. Strains are completely relaxed at room temperature for 3 μm thick IV-VI layers, while some residual tensile strain, but much below the elastic limit is observed in thinner layers. Since the lattice mismatch between the IV-VIs and BaF$_2$ is too large for coherent growth, the lattice mismatch strain is near completely relaxed at growth temperature for layer thicknesses of 3 μm. If no relief of the thermally induced strain would occur when cooling down from growth to room temperature (elastic limit), tensile strains as high as 0.8% would be expected from the difference of the thermal expansion of IV-VIs and Si. These strains are much higher than the measured results. We observed plasticity even below room temperature, e.g. strains were near completely relaxed in PbSe layers at cryogenic temperatures. No signs of strain hardening were detected even after many cooling cycles down to 100K.

Since the hardness of IV-VIs depends on the compositions, it will be possible to tailor the amount of plastic deformation occurring on cool-down with intermediate ternary layers like PbSe$_{1-x}$Te$_x$. This material is known to be soft near the binary end-compositions, while it gets more and more brittle with deviation of x from 0 or 1.

Note that the fluoride layers also relief thermally induced strains by plastic deformation

[6,7]. The residual tensile strain in the layers at room temperature was explained by applying Matthews equilibrium model of misfit dislocation formation to the thermal mismatch build-up of strains. However, we found that below room temperature, thermal mismatch strain increases in the BaF_2 layer, but the strain does not reach levels sufficient to cause cracks even when cooling down to 20K.

GROWTH AND MATERIAL PROPERTIES on Si(100)

The MBE growth window for CaF_2 on Si(100) is rather narrow, substrate temperatures must be around 550°C [9-11]. The layers grow 3-d and exhibit a slightly rough surface consisting of pyramids with (111) side faces and ≈200Å basal width. Such a growth mode is also observed for BaF_2(100) grown on CaF_2 on Si(100) [16]. However, if a two temperature step growth procedure is applied with increased substrate temperature after formation of a few 100Å of CaF_2, Morimoto et al [17] showed that CaF_2(100) grows 2-d on Si(100). We found the same behaviour for BaF_2(100) on CaF_2 on Si(100): Streaky RHEED-patterns indicative for 2-d growth were observed, and the surface was terminated with real (100) atomic planes [7]. The behaviour can be explained with a significantly reduced free (100)-surface energy for growth at high temperatures.
Tensile strains induced by the thermal mismatch were found to be relaxed at room temperature similarly as for (111) oriented layers [16].

The (100) surface is the preferred growth plane of the IV-VIs which have the rocksalt crystal structure. We performed some MBE growth experiments with PbSe on fluoride covered Si(100) substrates. The substrate sizes were as usual up to ≈ 3 x 3 cm². It was possible to achieve epitaxy with 2-d growth mode. X-ray rocking curve line width were ≈250 arc sec for 2.5 µm thick layers, slightly larger than in (111) orientation, and smooth surfaces were observed. However, cracks appeared in the (100) oriented layers grown with thicknesses ranging from 1-3 µm. Strain relief is therefore not as easy as in the (111) case. This is most probably because the primary {100}<110>-slip system is not efficient for strain relief, the glide planes are either perpendicular to the surface where they do not contribute to a relaxation, while strain relief with glide *only* in the glide plane parallel to the surface would require dislocation movement over macroscopic distances. For applications with (100) oriented layers, patterning is most probably needed, e.g. an island approach has to be used in IR focal plane arrays with each sensor element fabricated on its own IV-VI island.

It is interesting to note that the primary glide system in IIa-fluorides is the same {100}<110> [19] as for the IV-VIs. However, we did not observe cracks in (100) oriented BaF_2 layers grown on Si(100). A possible explanation why (100) oriented lead salt layers crack on Si(100) while BaF_2 does not might be related to the different thicknesses of the layers or that the secondary slip system operates more easily in fluorides than in IV-VIs.

IR-DEVICE FABRICATION

To demonstrate the device quality of the layers, we fabricated linear arrays using a simple blocking contact technique. An array consists of 66 elements arranged in a staggered way, the areas of the individual sensors are 50 x 100 µm², and the pitch on each side is 100 µm (fig. 3). The substrate orientation is (111). The active areas are defined by evaporated Pb, Pb inverts the surface of p-type material, thereby creating a photovoltaic sensor. A common ohmic contact is formed by vacuum deposited Pt, and an insulator, either vacuum deposited (polycristalline) BaF_2 or a polyimide insulates the fan-out pads to the Si-substrate from the low ohmic lead-chalcogenide layer. The delineations are performed with shadow masks or photolithographic techniques. Illumination is from the backside through the IR-transparent substrate and fluoride buffer. No surface passivation layer is used.

Since most of our device fabrication steps are rather crude, considerable improvement is still possible by optimising the process, or by using diffusions or implantations in order rather to form buried junctions instead of Schottky-type barriers at the surface.

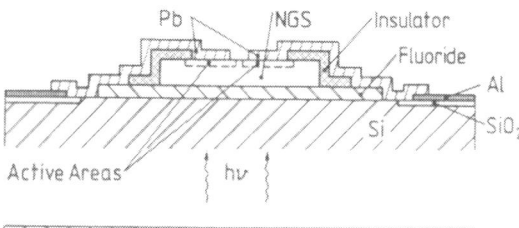

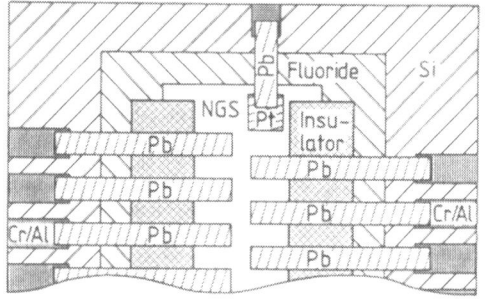

Fig. 3.
Cross section and top view of a bilinear lead-salt-on-Si photovoltaic IR-sensor array. The epitaxial fluoride buffer is 200 nm stacked CaF$_2$-BaF$_2$, the lead salt layer is 2-3 µm thick.

IR-DEVICES ON SILICON

MWIR-range: We have done most work up to now with PbTe grown by HWE [20,21]. Cut-off wavelength is ≈5.5µm at 80K. Measured external quantum efficiencies, which are essentially constant up to the cut-off wavelength apart from interference effects, are above 50% without AR-coating. Nearly every photon, which is not reflected back but penetrates into the PbTe layer therefore contributes to the signal current. Fig. 4 shows the I-V characteristics of an individual sensor of the array with and without illumination. The sensitivities are determined by the noise currents at 0V, which are given by the differential resistances R_0 at zero bias. Normalised with the area A, these values are up to 3000Ωcm^2 at 90K (the lowest temperature achieved in the cryostat used for the measurements). The mean value for the whole array is above 1000 Ωcm^2 (Fig. 5) at this temperature, despite this array contains some mechanically damaged pixels with much lower R_0A values. The temperature dependence of the R_0A values indicates a

Fig. 4.
I-V characteristics of a PbTe-on-Si sensor in the linear array with and without impinging room temperature radiation. Differential resistance R_0 at zero bias at 90K is 40 MΩ, and cut-off wavelength is ≈5.5 µm at this temperature.

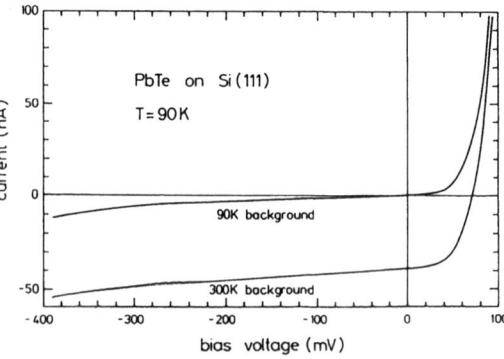

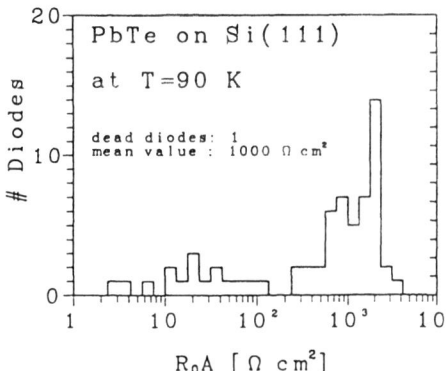

Fig. 5.
Distribution of differential resistance at zero bias times area R_oA values of a linear 66 element PbTe-on-Si array for the MWIR at 90K.

diffusion limited behaviour down to 120K (R_o prop. $\exp(E_g/kT)$), while depletion limited noise is observed below (R_o prop. $\exp(E_g/2kT)$). Calculated detectivities [12] using

$$D^* = \eta q/E_g \cdot (R_oA/4kT)^{1/2}$$

(η is the quantum efficiency, E_g the band gap, T the temperature, q electronic charge and k Boltzmans constant) are up to $D^* = 2 \cdot 10^{12}$ cm $Hz^{1/2}$/W at 90K. This values is as high as that of $Hg_{1-x}Cd_xTe$ diodes with similar cut-off wavelength at this temperature (Since values for comparison are available rather at 77K than at 90K, we extrapolated the 90K values by assuming a temperature dependence as for depletion limited noise for this comparison [22]).

SWIR-range:
These layers were grown with HWE by Dr. Ishida at Shizuoka University, Hamamatsu, Japan on our fluoride covered Si-substrates. Array delineation was again done at Zürich [23]. Fig. 6 shows the distribution of the R_oA-products at 297K, 200K and 85K. Quantum efficiencies are again around 50%, the measured different cut-off wavelengths at the different temperatures are due to the temperature dependence of the bandgap of the IV-VI material. The R_oA-distributions over the array are narrow, indicative of a

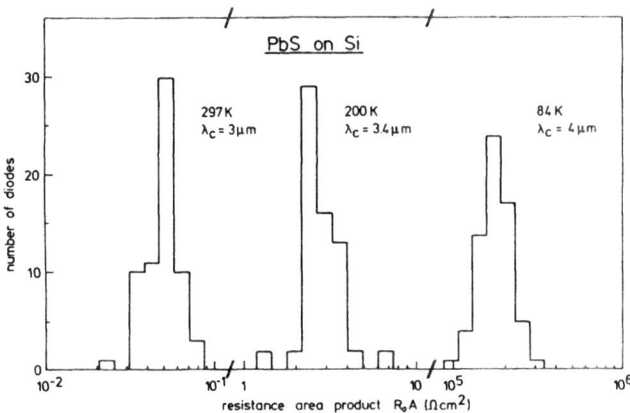

Fig. 6. Distribution of R_oA values of a linear 66 element PbS-on-Si array for the SWIR-range at 297K, 200K and 84K.

uniform material and devices quality. The detectivities as determined from the R_0A-products with the formula above are $3 \cdot 10^9$, $2 \cdot 10^{10}$, and $1 \cdot 10^{13}$ cm $Hz^{1/2}/W$ at 297, 200, and 85K, respectively.

For PbTe and PbS, the bandgap and therefore the cut-off wavelength of the sensors can only be varied by changing the temperature. *Variable* bandgaps with cut-off wavelengths tunable between 3 and 7 μm are obtained with $PbS_{1-x}Se_x$ or $PbEu_{1-x}Se_x$. The blocking Pb-contact technique works with both materials [12,24]. By proper tuning of the band-gap, such arrays with integrated read-out electronics in the Si-substrate might be used at 200K-300K instead of PtSi sensors which need 77K cooling and have low quantum efficiencies.

LWIR-range: $Pb_{1-x}Sn_xSe$ arrays on Si with cut-off wavelength up to 12 μm were fabricated. The first arrays were made in $Pb_{1-x}Sn_xSe$ layers grown by Dr. Lambrecht at Freiburg, BRD, on our fluoride covered Si(111)-substrates [25]. The MBE apparatus used for this purpose was primarily dedicated for laser structure work [3]. Fig. 7 shows the distribution of the R_0A-values of such an array with 11.3 μm cut-off wavelength at 87K ($x \approx 0.065$). The cut-off increases to above 12 μm when cooling to 50K.

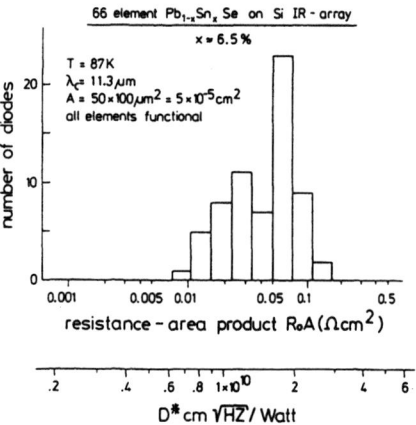

Fig. 7.
Distribution of R_0A values at 87K of a linear 66 element $Pb_{1-x}Sn_xSe$-on-Si (x=0.065) array for the LWIR.

Results of an $Pb_{1-x}Sn_xSe$ array fabricated in a layer grown with our recently completed IV-VI MBE-chamber are shown in fig. 8 and 9 [13]. This layer was grown on a CaF_2-BaF_2 buffer which was deposited at a substrate temperature of 450°C only. The composition chosen was $x \approx 0.05$, leading to about 10.5 μm cut-off wavelength at 77K. Even without substrate rotation, we obtained excellent uniformity over the array. The spectral response measured at 85K (the lowest temperature we achieved in the cryostat used for the measurements) is shown in Fig. 8. The experimentally determined spread of the cut-off wavelengths over the whole array is below ±0.1 μm. The second peak in the response at 8 μm is due to interference in the about 3 μm thick layer.

Quantum efficiencies were determined with a 500K blackbody source and a 8.2 μm cut-on filter. Every second pair of opposite diodes was measured. As shown in Fig. 9, the mean quantum efficiency is 0.59, and the standard deviation over the measured diodes is 0.03. This low spread is of the same order as the spread of the shadow mask used for this delineation. Since no anti-reflection coating was applied, we deduce from the measured high quantum efficiency that nearly every photon which reaches in the $Pb_{1-x}Sn_xSe$ layer contributes to the signal current.

Resistance area products R_0A were up to 1 Ωcm^2 at 77K. This corresponds to a detec-

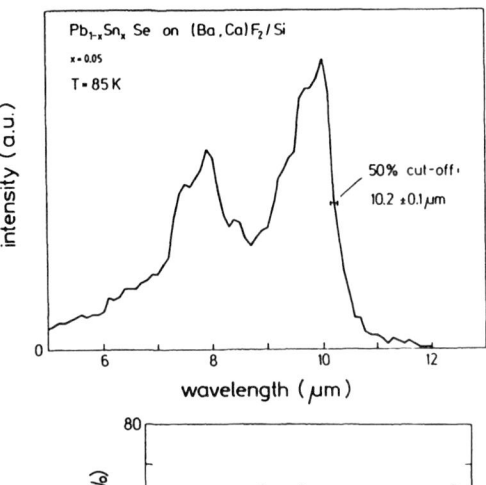

Fig. 8.
Spectral response of a typical sensor of a 66 element LWIR $Pb_{1-x}Sn_xSe$-on-Si (x=0.05) array. The variation of the cut-off wavelength over the whole array is below 0.1 μm as indicated with a bar.

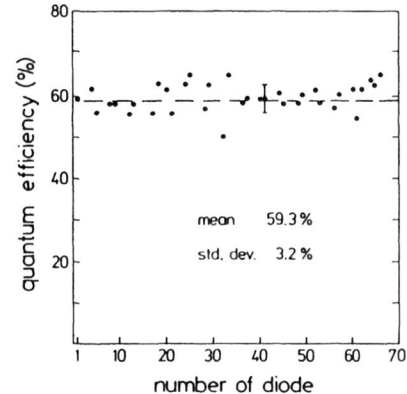

Fig. 9.
Distribution of the measured quantum efficiencies at 90K of the photovoltaic $Pb_{1-x}Sn_xSe$-on-Si array of fig. 8. Every second pair of adjacent sensors was measured.

tivity $D^* = 8 \cdot 10^{10}$ cm Hz$^{1/2}$/W by using the formula above. The temperature dependence of the R_oA-product indicates diffusion limited behaviour down to ≈100K, while depletion noise dominates below [13].

The R_oA-products we achieved in the LWIR range up to now are above the background noise limit for full field of view. Compared to state of the art photovoltaic LWIR $Hg_{1-x}Cd_xTe$ (on CdTe or CdZnTe substrates) with similar cut-off wavelengths at 77K [22], our best R_{oA}-values of the sensors from the two LWIR arrays described here are between a factor 4 to 40 below. Since the detectivity D^* are proportional to the square root of R_oA, the junction noise limited detectivities of our sensors in a strongly reduced field of view are about 2 to 6 times below $Hg_{1-x}Cd_xTe$. However, even the simple Pb blocking contact technique for formation of $Pb_{1-x}Sn_xSe$ sensors is able to reach the state-of-the-art $Hg_{1-x}Cd_xTe$ limit: We have been able to fabricate such sensors on **bulk** BaF$_2$ substrates with 14.5 μm cut-off wavelength long before we turned to the IV-VI-on-Si technique [22].

CONCLUSIONS

We have successfully grown epitaxial IV-VI narrow gap semiconductor layers on Si(111) and fabricated IR-sensor arrays in the layers with cut-off wavelength up to 12 μm. Advantages are the easy fabrication procedure, better homogeneity, and that layers of

2-4 μm thickness only can be used. Compatibility with standard Si-VLSI processing technique can be obtained with a low temperature growth procedure with all material growth steps performed below or at ≈450°C, thus allowing the IR-sensor fabrication to be performed on completely processed and metallized Si-wafers. Despite we have already obtained good sensitivities for the SWIR, MWIR and LWIR range, considerable improvements are still possible by optimizing the device fabrication technique. We believe that especially using buried p-n junctions with a thin wider bandgap cap layer will reduce noise currents due to surface effects. A concept of a possible schematic lay-out of such a focal plane array is depicted in fig. 10.

LWIR monolithic IV–VI on Si IR–sensors

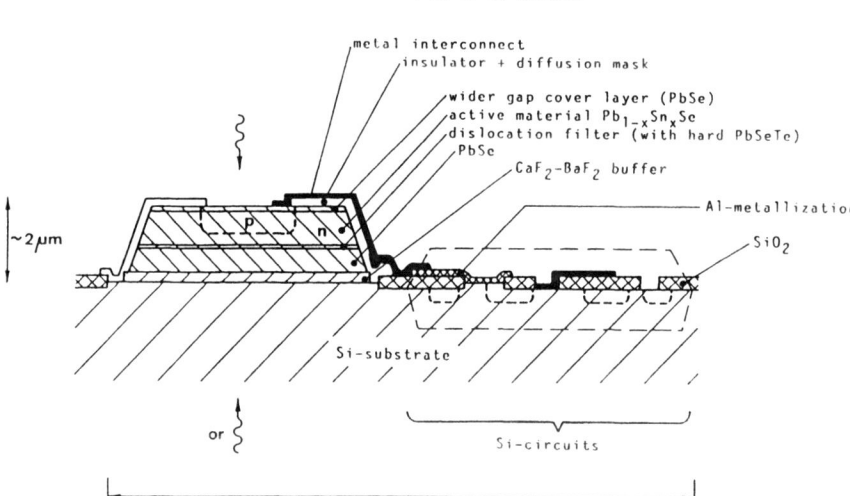

Fig. 10. *Proposed structure of a pixel of a large lead-salt-on-Si infrared focal plane array. The Si-circuits are fabricated with standard processing, and the narrow-gap layer for the IR-sensors is grown* <u>after</u> *completion of the Si-VLSI circuit.*

REFERENCES

[1]   A. Rogalski, Infrared Phys. **28**, 139, 1988; A. Rogalski, J. Piotrowsky, Progress in Quantum Electronics **12**, 88, 1988.
[2]   G. Nimtz, B. Schlicht, R. Dornhaus, Narrow-Gap Semiconductors, Springer Tracts in Modern Physics Vol. 98, 1985.
[3]   M. Tacke, B. Spanger, A. Lambrecht, P.R. Norton, H. Böttner, Appl. Phys. Lett. **53**, 2260, 1988.
[4]   H. Zogg, M. Hüppi, Appl. Phys. Lett. 47, 133, 1985.
[5]   H. Zogg, S. Blunier, J. Masek, J. Electrochem. Soc. **136**, 775, 1989.
[6]   H. Zogg, Appl. Phys. Lett. 49, 933, 1986.
[7]   S. Blunier, H. Zogg, A. Rüegge, Thin Solid Films **184**, 387, 1990.
[8]   H. Zogg, C. Maissen, S. Blunier, J. Masek, V. Meyer, R.E. Pixley, Mat. Res. Soc. Symp. Proc. **198**, ..., 1990.
[9]   J.M. Phillips, Mat. Res. Symp. Proc. **71**, 97, 1986.
[10]   T. Asano, H. Ishiwara, N. Kaifu, Jap. J. Appl. Phys. **22**, 1474, 1983.
[11]   L. J. Schowalter, R.W. Fathauer, CRC Crit. Reviews in Solid State and Mat. Sci. **15**, 367, 1989.

[12]   H. Holloway, Physics of Thin Films, Academic Press, G. Haas, M. H. Francombe ed., Vol. 11, 1980, p. 105.

[13]   T. Hoshino, C. Maissen, H. Zogg, J. Masek, S. Blunier, A.N. Tiwari, S. Teodoropol, W.J. Borer, Infrared Physics Feb. 1991, to be published.

[14]   H. Zogg, P. Maier, P. Norton, Mat. Res. Soc. Symp. Proc. **56**, 253, 1986.

[15]   L. Tapfer, J.R. Martinez, K. Ploog, Semicond. Sci. Technol. **4**, 617, 1988.

[16]   S. Blunier, H. Zogg, H. Weibel, Appl. Phys. Lett. **53**, 1512, 1988.

[17]   Y. Morimoto, S. Sudo, K. Yoneda, Mat. Res. Soc. Symp. Proc. **116**, 413, 1988.

[18]   J.J. Gilman, Acta Met. **7**, 608, 1959.

[19]   A.G. Evans, P.L. Pratt, Phil. Mag. **20**, 1213, 1969.

[20]   C. Maissen, J. Masek, H. Zogg, S. Blunier, Appl. Phys. Lett. **53**, 1608, 1988.

[21]   J. Masek, C. Maissen, H. Zogg, W. Platz, H. Riedel, M. Königer, A. Lambrecht, M. Tacke, Nucl. Instr. & Meths. **A288**, 104, 1990.

[22]   H. Zogg, C. Maissen, J. Masek, S. Blunier, in "Advanced Infrared Detectors and Systems", IEE conference publication No. 321, 1990, p. 36.

[23]   J. Masek, C. Maissen, A. Ishida, H. Zogg, IEEE Electron Dev. Lett. **11**, 12, 1990.

[24]   J. Masek, C. Maissen, H. Zogg, S. Blunier, H. Weibel, A. Lambrecht, B. Spanger, H. Böttner, M. Tacke, J. de Physique **C4** (Suppl. **49**), 587, 1988.

[25]   H. Zogg, C. Maissen, J. Masek, S. Blunier, A. Lambrecht, M. Tacke, Appl. Phys. Lett. **55**, 970, 1989.

PROPERTIES OF GRAPHITE INTERCONNECT CIRCUIT BOARDS WITH
ANISOTROPIC THERMAL EXPANSION

J. Malamas, R.P. Bambha, J.B. Ramsey Jr., W.C. Garrett, and E.G. Kelso,
U.S. Army Center for Night Vision and Electro-Optics, Fort Belvoir, VA
22060-5677
T.A. Hahn, Naval Research Laboratory, Washington, D.C. 20375

ABSTRACT

We report the investigation of an interconnect circuit board (ICB)
with anisotropic thermal expansion for use with bump bonded, indirect
hybrid, scanning focal plane arrays. This ICB is designed to reduce
significantly the thermal stresses on the indium bump bonds during thermal
cycling. Highly oriented pyrolitic graphite (HOPG) was chosen because its
anisotropic thermal expansion meets the criteria for forming an indirect
hybrid ICB using silicon processor circuits and mecury cadmium telluride
detectors. Properties of HOPG influencing its performance as an ICB have
been investigated including thermal expansion, electrical conductivity,
durability, and adherence of electrically insulating thin films.

INTRODUCTION

Solid state focal plane arrays (FPA's) for visible light are
monolithic. FPA's for use in the 8 to 12 micrometer infrared region use
detectors made of mercury cadmium telluride and signal processing circuits
fabricated on silicon. A second generation scanning focal plane array for
this wavelength region contains thousands of detectors which must be
connected to silicon integrated circuits. In an indirect hybrid FPA, the
silicon and mercury cadmium telluride chips are connected by a sapphire
interconnect circuit board (ICB) which contains all the necessary
metalization lines. The chips are bonded to the ICB by pressing together
arrays of indium columns on the corresponding chips and ICB. This
fabrication technology has been developed to the point where it is possible
to form an indirect hybrid FPA consisting of a HgCdTe detector array which
is approximately 17mm long by 2.5mm wide and four silicon integrated
circuit chips each 7 X 9mm.

A difficulty arises because the thermal expansions of silicon,
sapphire, and mercury cadmium telluride are different. Cooling the array
to the operating temperature of 77K places stresses on the indium columns
as illustrated in figure 1. Repeated thermal cycling between room
temperature and 77K first results in high resistance electrical contacts
and eventually in mechanical failure (separation of the semiconductor chip
from the ICB). The use of scanning FPA's in thermal imaging systems
requires a capability to withstand several thousand temperature cycles
before failing.

Mat. Res. Soc. Symp. Proc. Vol. 216. ©1991 Materials Research Society

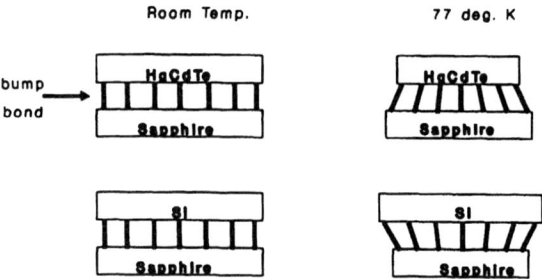

Figure 1. Schematic view of bump bonds strained
by thermal expansion mismatch

The linear thermal expansion, e, is defined as $e(T)=[L(T)-L_O]/L_O$, where L_O is the length of a material at a specified reference temperature, and $L(T)$ is its length at a temperature, T.   Figure 2 shows the thermal expansion for the materials discussed above.   Infrared detectors can be fabricated on bulk $Hg_{1-x}Cd_xTe$ or epitaxially grown material on CdTe.   The difference in thermal expansion for x values between 0.3 and 1.0 is small and may be neglected.   Data for the mercury cadmium telluride curve in figure 2 corresponds to an x value of 0.25 and was obtained from Brice and Capper[1].   Data for plotting the thermal expansion of silicon were obtained from the *American Institute of Physics Handbook*[2]; the data for Al_2O_3 were obtained from *Thermophysical Properties of Matter*[3].   The curves were drawn by fitting tabulated data to cubic polynomials.

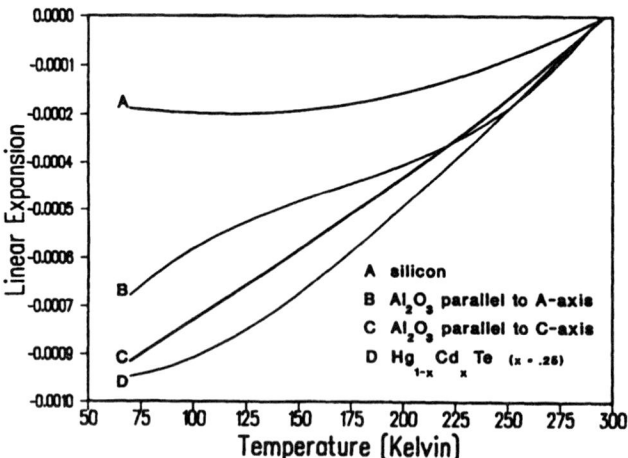

Figure 2. Plot of expansion vs. temperature for
indirect hybrid materials

A new approach to solving the thermal expansion mismatch problem has recently been proposed by Malamas and Bambha[4]. This approach as well as some alternates are described in reference 4. Briefly, the new solution to the problem uses an ICB with anisotropic thermal expansion as illustrated in figure 3. The long narrow rectangle in the center represents the mercury cadmium telluride detector array whose thermal expansion along its length is perfectly matched by the ICB expansion in the vertical direction. Although the thermal expansion in the horizontal direction is not matched, the total difference in expansions is small because the width is only 2.5mm. The four rectangles represent silicon integrated circuits whose thermal expansion in the horizontal direction is equal to that of the ICB. By placing the indium columns in a band whose height is approximately 2.5mm, the difference in expansion in the direction of mismatch is small.

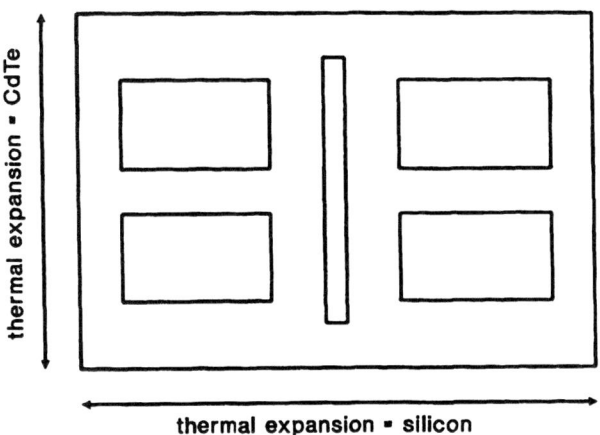

Figure 3. ICB with anisotropic thermal expansion

It has been shown[4] that if a material has a thermal expansion greater than that of silicon along one crystallographic axis and a thermal expansion less than HgCdTe along any axis perpendicular to the first, then that material can be cut so as to produce an ICB with the properties shown in figure 3. Boron nitride, graphite, and tellurium are three materials which meet the required thermal expansion criteria. Boron nitride is not available in large crystalline pieces. A crystal would have to be approximately 3cm in diameter and 1.5cm long in order to produce an ICB, and the cost must be comparable to sapphire. Figure 4 shows the thermal expansion properties of graphite and tellurium as a function of temperature.

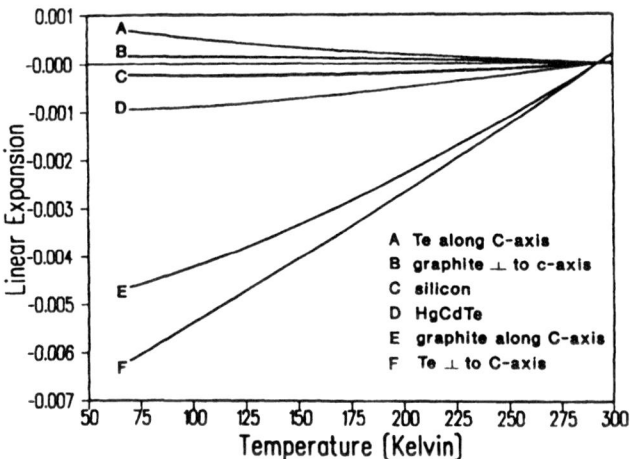

Figure 4. Thermal expansions of tellurium, graphite,
silicon and mercury cadmium telluride

MATERIALS PROCESSING

Attempts were made to grow single crystal tellurium using a modified
Bridgeman method, but the largest dimension of the single crystal grains
contained within the boule were only approximately 1cm. Moreover, the
material was found to be too brittle to be used as an interconnect circuit
board.

Single crystal graphite satisfies the expansion criteria, but it is
not available in sizes required. A polycrystalline form of graphite, known
as highly oriented pyrolytic graphite (HOPG), is known to have properties
similar to the single crystal and is available in large pieces. As the
data presented below show, the thermal expansion of HOPG satisfies the
requirements in the temperature range of interest. The pyrolitic growth
process for carbons involves the deposition of preferentially oriented
carbon crystallites on a suitably prepared substrate in a vacuum furnace.
The as-grown pyrolitic carbon can later be annealed into highly oriented
pyrolitic graphite. The HOPG described here is a form of Pfizer's Pyroid
graphite which was annealed at 2700° C under a c-axis compression of 4000
psi. The material studied was readily cut on a wire saw into 2.5cm X 2.5cm
wafers of 1mm thickness at the desired orientation. Such wafers were able
to withstand polishing and handling without damage. Surfaces of HOPG have
an unctuous quality, and the adhesion of thin films was therefore a matter
of concern. As discussed below, thin film adhesion was not found to be a
problem. Because HOPG is available in the necessary sizes and is
reasonably durable, we chose this material for the initial fabrication
studies of the anisotropic ICB.

The hybrid bonding process is extremely delicate and requires the surface of the ICB to be both flat and smooth to within a fraction of a micrometer. Our attempts to polish the surface of the HOPG wafers left a rough "orange peel" finish with surface feature sizes as large as 3 micrometers. More appropriate polishing techniques must be explored in the future. In the event that polishing to the desired smoothness proves too difficult, the surface roughness can be leveled with a polymer coating. A photoresist layer of 7 micrometers has been spun onto the HOPG surface. The surface roughness was reduced to less than .3 micrometers, and the coating withstood 35 thermal cycles between 77K and room temperature without observable change.

A necessary quality of the ICB is that it electrically isolates the conductors on the surface. Unfortunately, HOPG is a conductor[5]. In order to use graphite as an ICB material it is necessary to coat the surface with an insulating thin film. Silicon oxide (SiO_x) films of approximately 800 angstroms were deposited on graphite substrates using ion beam sputtering. The SiO_x coated substrates were thermal cycled between room temp and 77K by immersion in liquid nitrogen. There were no observed changes in adhesion during the 25 cycles tested. Aluminum contacts were deposited on the SiO_x film and on the graphite substrate. Measurements were made of current vs. voltage. In the range measured, -1V to +1V, corresponding currents were in the picoamp range, varifying the high resistivity of the SiO_x film.

MEASUREMENTS

Thermal expansion measurements were performed on HOPG in the temperature range of 293K to 113K on a differential dilatometer. The thermal expansions satisfy the criteria for fabrication of the anisotropic ICB, and the results were highly reproducible. The c-axis sample was cut parallel to the direction of compression, and two other samples were cut along two arbitrary axes (a and b) perpendicular to the c-axis and to each other. The results are given in table I. Extrapolation of the data to 77K

TABLE I. Thermal Expansion of HOPG (X 10^{-6})

Thermal Expansion of HOPG
(X 10^{-6})

TEMP.($°K$)	A-AXIS	B-AXIS	C-AXIS
293.0	0.0	0.0	0.0
273.0	25.1	20.5	-493.9
253.0	52.2	42.4	-999.7
233.0	78.5	65.8	-1517.0
213.0	106.7	90.6	-2036.8
193.0	135.5	116.5	-2550.3
173.0	164.6	143.4	-3048.8
153.0	193.7	170.9	-3523.4
133.0	222.5	199.0	-3956.3
113.0	250.9	227.3	-4365.8

C-axis direction parallel to direction of compression
A-axis and B-axis perpendicular to c-axis and each other

using a cubic polynomial fit gave thermal expansions of 300 ppm (a-axis), 278 ppm (b-axis), -495 ppm (c-axis). If this data were to be plotted on figure 2, the differences between HOPG and single crystal graphite would not be apparent. As table I reveals, two perpendicular axes in the basal plane were measured, and there is a slight discrepancy. The difference measured is not large enough to appreciably effect the expansion of the resulting ICB, but it worthy of further investigation. If the discrepancy were the result of a slight misalignment of the axes, the samples could be properly aligned with x-ray analysis prior to cutting. In order to obtain an ICB with expansion equal to HgCdTe in one direction and equal to Si in the perpendicular direction the HOPG sample must be cut at 35 deg to the c-axis.

CONCLUSION

The results presented here are evidence that highly oriented pyrolytic graphite is a good candidate material for demonstrating the feasibility of a new concept which utilizes an interconnect circuit board that has anisotropic thermal expansion. Proof of concept will be demonstrated when temperature cycling tests are performed on a graphite ICB containing silicon and HgCdTe chips bonded by the indium hybridization process.

REFERENCES

1. J. Brice and P. Capper, Properties of Mercury Cadmium Telluride (The Institution of Electrical Engineers, London and New York, 1987).

2. D.E. Gray, American Institute of Physics Handbook (McGraw Hill, New York, 1972).

3. Y.S. Touloukian, R.W. Powell, C.Y. Ho, P.G. Klemens, Thermophysical Properties of Matter, Volume 13. (Plenum Publishing Corporation, New York, 1972).

4. J. Malamas and R. Bambha, Meeting of the IRIS Specialty Group on Infrared Materials. National Institute of Standards and Technology, Gaithersburg, MD. 13, 14 August 1990.

5. S. Mrozowski, Phys. Rev. 85, 609(1952)

EXPERIMENTAL METHOD OF DETERMINATION OF STRUCTURAL
CORRELATIONS IN SURFACE LAYERS OF OXIDE GLASS

ZENON BOCHYŃSKI
Non-Crystalline Materials Division, Institute of Physics,
Adam Mickiewicz University, 60-780 Poznań 2, Grunwaldzka 6,
Poland

ABSTRACT

A new method of X-ray diffraction analysis of structural
inhomogeneities in the quartz $/SiO_2/_n$ based inorganic glasses
is presented. The method enables the determination of structural changes occuring in the real nodal lattice in the regions of $10...20$ Å or more as well as substructural changes
in the regions $5...15$ Å comparable to the molecular size of
$SiO_2...SiO_4$. In consequence these changes can be correlated
with approximate nodal lattice models of different degree of
ordering. The applied method provided the possibility of constructing structural models of nodal lattices describing the
surface and inner layers of the real glasses, changes in the
local inhomogeneities as well as boundaries in water-gel
associates.

INTRODUCTION

The structure of the surface and internal layers in a
series of the studied quartz-based oxide glasses /quartz, melted quartz and sodium-silicate/ is first of all determined by
the conformations of the relevant molecules $/SiO_2 ... SiO_4/$.
The idea of looking for local structural correlations in
oxide glasses arose from detail analysis and numerical estimation of structural parameters determined by X-ray diffraction
employing the Fourier of Bessel of analyses of angular intensity distribution.
The results of structural analyses prompted the elabora-
tion and construction of spatial models of nodal lattice which
describe local degrees of ordering and determine the correlated structural parameters of spatial distribution of atoms
/mainly O and Si, partly Na, Ca and Mg/, molecules $/SiO_2...$
$...SiO_4/$, and their groups $/SiO_2/_n ... /SiO_4/_m$.
This work aims at determination of the characteristic
structural parameters of the surface and internal layers in
order to find mean size of local higher symmetry complexes
occuring in the glass.
The aperiodic nodal lattice which can be ascribed to real
inorganic glass includes randomly distributed local regions
/in structural or substructural range/ of mutually ordered
configurations of atoms and molecules. The regions are detectable by X-ray diffraction together with Fourier and/or Bessel
analysis. Structural or substructural regions made of groups
or complexes of atoms and molecules are characterized by a
higher ordering degree and higher local symmetry. The mean size of the analysed regions occuring in real glasses was determined by a stage approximation from the distribution function
of the broad- and low-angle X-ray scattering, applying numerical analysis.

THE OUTLINE OF THEORY

Quantitative analysis of structural correlations in quartz based oxide glasses was performed by a new method which is a modification of the hitherto reparted methods.

In the first stage the total:

$$4 \pi r^2 \sum_{1}^{2 \to 5} \overline{K}_m \cdot \rho_m(r) =$$

and differential

$$4 \pi r^2 \left[\sum_{1}^{2 \to 5} \overline{K}_m \cdot \rho_m(r) - \sum_{1}^{2 \to 5} \overline{K}_m \cdot \rho_{mo} \right] =$$

radial distribution functions are determined for particular effective molecules from the functions of angular distribution of intensities expressed in electron units normalized by the oscillation method applied to the method of field comparison.

In the second stage the courses of the radial distribution functions are determined by sustraction:

$$4 \pi r^2 \left[\sum_{1}^{3 \to 5} \overline{K}_{m(2)} \rho_{m(2)}(r) - \sum_{1}^{2} \overline{K}_{m(1)} \rho_{m(1)}(r) \right] = \quad .$$

They describe the interatomic distances in the analysed atomic and molecular groups /or mixed/ of different degree of local inhomogeneities.

In order to construct a nodal model of spatial lattice of simple systems /in the regions sized 0...10 Å and even to 16 Å/ the one-stage approximation of the effective molecule was used:

$$SiO_2 - \quad - R_2O_3$$
$$SiO_2 - A_2O - R_2O_3$$
$$\cdot \cdot \cdot \cdot \cdot \cdot \cdot \cdot$$

which gave a simplified structure of spatial nodal lattice of A_2O.

To obtain a model of spatial lattice of more complex systems, the multistage approximation of the effective molecule is used:

$$SiO_2 - \quad \quad \quad - R_2O_3$$
$$SiO_2 - A_2O - \quad \quad - R_2O_3$$
$$SiO_2 - A_2O - CO - \quad - R_2O_3$$
$$SiO_2 - A_2O - CO - B_2O - R_2O_3$$
$$\cdot \cdot \cdot \cdot \cdot \cdot \cdot \cdot \cdot \cdot \cdot \cdot \cdot$$

which allows to find a complex structure of spatial nodal lattice of $A_2O-CO-B_2O$.

Direct analysis of the structure of complex glasses composed of m,n,...,t different atoms $/2 \leqslant t \leqslant 6/$ characterized by atom scattering factors $f_m, f_n, ..., f_t$ and the corresponding effective scattering factors $\overline{K}_m = f_m/f_e$, $\underline{K}_n = f_n/f_e$, ..., $K_t = f_t/f_e$ and mean effective scattering factors $\overline{K}_m = 1/S_o \int \overline{K}_m \cdot dS, ...$ we used the integral equation proposed by B.E. Warren, H. Krutter and O. Morningstar:

$$4 \pi r^2 \sum_{m=1}^{t} \overline{K}_m \rho_m(r) = 4 \pi r^2 \rho_{mo} \sum_{m=1}^{t} \overline{K}_m + \frac{2r}{\pi} \int_{S_1}^{S_2} S \cdot i(S) \cdot \sin Sr dS$$

with the appropriate modifications.

EXPERIMENTAL

The structure, physical and partly chemical properties of
a series of quartz glasses $/SiO_2-R_2O_3/$, sodium-silicate glasses $/SiO_2-Na_2O-CaO-MgO-R_2O_3/$ and melted quartz $/SiO_2/$ were
determined.
All these systems are based on quartz $/SiO_2/$ which content varies from ~72 wt % /sodium-silicate glasses/ to
99,90...99,98 wt % /melted quartz/.
The samples to be studied were thin-walled plates
/9,10...0,50 mm thick/, optical plates /1...2 mm thick/ and
thick-walled plates /2...3 mm/ as well as fibres
$/\emptyset$ 0,20...0,60 mm/.
The methods applied were Bragg-Brentano - in reflection,
symmetric preparation - in transmission and low-angle /in
transmission and reflection/ of Debye-Scherrer-Hull.
Broad-angle diffraction patterns were obtained from auto-
matic goniometers HZG 3 and HZG 4 made by Freiberger C. Zeiss
/Jena, Germany/ whereas low-angle diffraction patterns from
the goniometers HZG 4 and KRM 1 /Buryevestnik, Leningrad/.
Stricly monochromatic $K_{\alpha}Cu$ radiation was used of
λ = 0,15418 nm /the monochromatization accuracy $\Delta\lambda$ = $\pm$0,00012
nm/ along with partly monochromatic radiation $K_{\alpha}Mo$
λ = 0,07107 nm $/\Delta\lambda$ = $\pm$0,00021 nm/. Both kinds of radiation
were provided by an effective graphite reflection monochroma-
tor made in our laboratory.

RESULTS

Structural and substructural parameters /= distances,
atomic and electron densities, coordination numbers and so on/
as well as local structural and substructural correlations
/= inhomogeneities in packing and arranqement, mutual orientation and so on/ were determined in the 1st, 2nd and partially
in the 3rd coordination sphere.
A structural init /group/ that is repeated in quartz glasses, melted quartz and in the whole series of silicate based
materials is a SiO_4 tetrahedrons which can be complete or incomplete, more or less distorted but occurs always, independently of the chemical composition.
All SiO_4 tetrahedrons in crystalline quartz are joined
through the corner oxygen atoms. Each of them belongs to 2n
silicon atoms to meet the requirements of the stoichiometric
composition SiO_2.
Thus, each crystal of quartz, in one of its basic crystalline forms: quartz, tridymite, cristobalite, is a gigan-
tic macromolecule $/SiO_2/_n$ composed of only 2 kinds of atoms.
According to experimental results obtained for quartz
glass, melted quartz and some silicate varities $/SiO_2-Na_2O-$
$-CaO-MgO-R_2O_3$, $SiO_2 \cdot xH_2O-R_2O_3/$ the coordination number was
lower varying from 2,86 to 3,58 oxygen ions.
On the other hand, the ratio of ionic radii $r_{Si}/r_O=0,29_3$
allows to ascribe a coordination number of 4 oxygen atoms to
a silicon ion Si^{4+} in the crystal varieties of quartz.
The results of X-ray structural analyses of some exampla-
ry quartz glasses and melted quartz are fiven in figures.
The figures also include the parameters and local structural as well as substructural correlations, and the models of

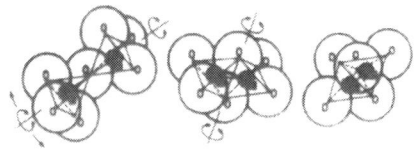

Fig. 1. A simplified scheme of binding between oxygen-silicon tetrahedra: top to top /rotation and deflection/, edge to edge /deflection/ and wall to wall.

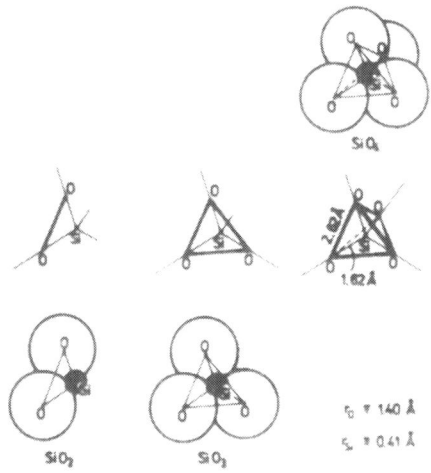

Fig. 2. A skeleton and spatial model of a degraded oxygen--silicon group with 2 /SiO_2/ 3 /SiO_2/ and 4 /SiO_4/ 0-Si bonds, in quartz glass.

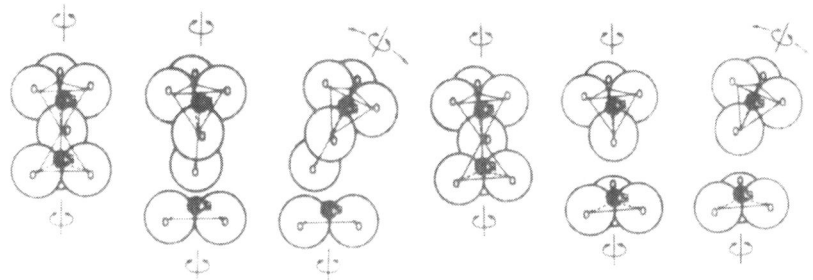

Fig. 3. Structural and substructural local oxygen-silicon arrangement conformations in quartz glass. The stages of trans-formations in quartz glass. The stages of transformation: $Si_2O_7 \longrightarrow SiO_5 + SiO_2$ and $Si_2O_7 \longrightarrow SiO_4 + SiO_3$ in consequence of bond degeneration and group distortion, are illustrated.

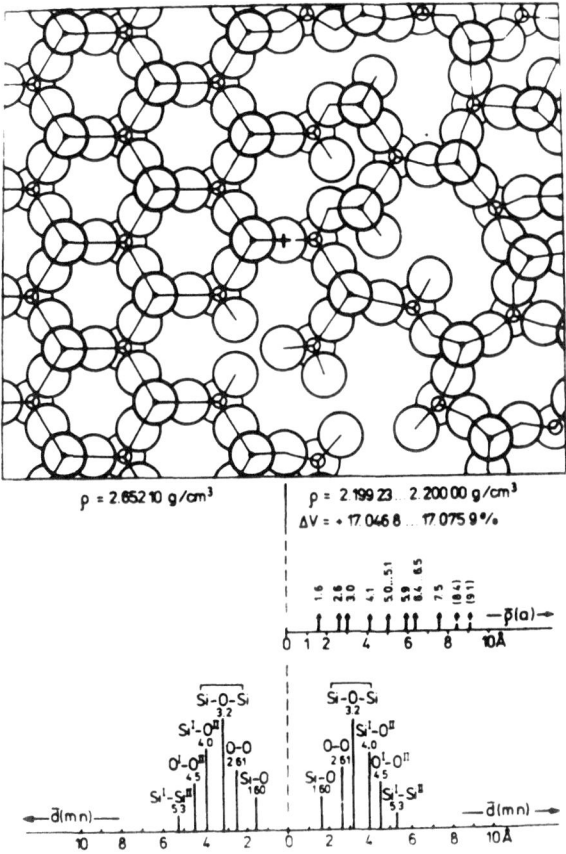

Fig. 4. A two-dimensional projection of oxygen-silicon nodal lattice /the actual distances are given/ in crystalline quartz /tetrahedral conformation/ quartz glass /bond degeneration/ and quartz glass /bond degeneration and group distorsion/.

Fig. 5. A scheme of the measuring setup for X-ray experiment, Bragg-Brentano biaxial goniometer /in reflection and transmission/ uniaxial goniometer /in transmission/ oscillating uniaxial goniometer /in reflection and transmission/.

nodal lattices of different degree of local ordering, which
a detectable by the X-ray diffraction method in the regions
of extreme density.

CONCLUSIONS

The analysis of radial distribution functions and low-
angle distribution functions as well as physical and chemical
parameters indicates the occurrence of structural and substructural regions of pseudocrystalline structure in the bulk
of the quartz glass /melted quartz, sodium-silicate glass/.
Complete SiO_4 tetrahedrons or their degenerated fragments
/SiO_2 ... $SiO_{3.5}$ .../ of different degree of distorsion are
not randomly and homogeneously distributed in the bulk of the
crystals. They are assembled into well distinguishable groups
or clusters.
X-ray structural investigation provides the possibility
to follow the process of transformation from melted quartz to
glass formation. This process involves:

- the disappearance of the ordering typical of crystalline quartz,

- preservation of a certain amount of tetrahedral structures of different degree of reorientation in space,

- the appearence of various oxygen-silicon conformations
/SiO_2 ... SiO_4/ of different coordination numbers of oxygen
/$2 < n < 4$/,

- the stabilization of the structure of the surface and
subsurface layers.

In recapitulation I would like to stress that the above
mentioned stages of glass formation /reorientation and distor-
tions of SiO_4 tetrahedra and partial breaking of bonds between
the tetrahedra/ load to the formation such structural and substructural conformations where local inhomogeneities occur in
the regions taken only by few molecules.

REFERENCES

1. H.P. Klug, L.E. Alexander, X-Ray Diffraction Procedures
   for Polycristalline and Amorphous Materials, John Wiley,
   New York /1966/.
2. A.M. Smoljegowskij, Razwitije prjedstawljenij o strukturje
   silikatow, Nauka, Moskwa, 1979.
3. Z. Bochyński, J. Non-Crystalline Solids, 38–39, 135–140
   /1980/; 46, 405–425 /1982/; 56, 373–374 /1983/.
4. Z. Bochyński, MRS Proc., 61, 135–157 /1986/; 105, 187–190
   /1988/.
5. M.M. Szulc, R.G. Grjebjenszczikow /Red./, Fizika i chimija
   silikatow, Nauka, Leningrad 1987.
6. W.A. Szaragow, Chimichjeskoje wzaimodjejstwije powjerchno-
   sti stjekla z gazami, Sztinca, Kiszinjew, 1988.
7. Z. Bochyński, High-temperature surface corrosion of quartz
   glass, The "UNITECR' 89 CONGRESS", Anhaim, November 1–4,
   1989, California, USA.
8. H. Gleiter, Europhysics New, 20 /9/, 117–136 /1989/.

CHEMICAL DESIGN AND STRUCTURAL CHEMISTRY
OF LWIR OPTICAL MATERIALS

C.K. LOWE-MA, D.O. KIPP, AND T.A. VANDERAH
Chemistry Division, Research Department
Naval Weapons Center, China Lake, CA 93555

ABSTRACT

Some applications for long-wavelength infrared (LWIR) windows require environmental durability and mechanical strength in addition to infrared optical transparency; i.e., the windows must simultaneously serve as optical and as structural ceramics. The requirement of optical transparency at long IR wavelengths eliminates from consideration all ceramics based on oxides and other light-anion compounds, making this a particularly difficult materials problem. The structure-property relationships and chemical rationale used to guide both the screening of known compounds and the synthesis of new compounds likely to possess the desired properties rely on factors such as atomic mass, electronic configuration, coordination number, and crystal structure type.

Our research has included the directed synthesis and characterization of a number of ternary indium sulfides as well as ternary calcium yttrium sulfides. Ternary indium sulfides feature both tetrahedral and octahedral coordination of indium. The crystal structure of $KInS_2$ and its relationship to structures observed in other systems such as AIn_2S_4, A = Ca,Sr,Ba, is described. The crystal structure of CaY_2S_4 along with studies of yttrium-doped CaS are also described. The AIn_2S_4 compounds are more fully described in references [1] and [2].

MATERIALS REQUIREMENTS AND CHEMICAL DESIGN

Compounds that are candidates for use as LWIR window materials must meet a certain set of chemical and physical requirements if they are to serve as both optical and structural ceramics. These requirements are: (a) exhibit near-zero infrared absorption in the 8-12 μm region; (b) be chemically inert, ie. water-, air-, and heat-stable; (c) have a melting point greater than $1000°C$; (d) be mechanically tough; (e) exhibit minimal thermal expansion; and (f) have a band gap greater than 2.0 eV.

Mat. Res. Soc. Symp. Proc. Vol. 216. ©1991 Materials Research Society

How does the synthetic solid-state chemist go about "designing" new compounds and/or identify existing compounds most likely to have, intrinsically, the desired properties? The search for suitable candidate compounds that meet the requirements is guided by the following constraints. (1) The IR transparency of a compound is determined by the masses of the constituent atoms and the nature of the chemical bonding among them. The relatively low fundamental vibrational frequencies that are required for 8-12 μm transparency are favored by weak chemical bonding, heavy atoms, high coordination numbers, and heteropolar ("ionic") bonds. Constituent "ions" should have noble gas or pseudo-noble gas electronic configurations (non-paramagnetic), and, to avoid IR opacity, should be insulators or large bandgap semiconductors. (2) Candidate compounds must be chemically inert; the chemical bonds must be strong enough to resist hydrolysis, oxidation, and decomposition at temperatures many hundreds of degrees above ambient conditions. (3) Candidate compounds must be refractory, with melting points of 1000°C or higher, and hard, but not brittle. Refractory, hard materials have crystal structures featuring strong three-dimensional covalent bonding. (4) As shown by Hazen and Finger [3], thermal expansion is approximately proportional to the ratio of the coordination number of the cation to the product of the cation and anion formal charges.

These constraints show that the factors favoring LWIR transparency tend to conflict with those favoring good structural-mechanical properties. Our compromise approach has been to consider compounds that contain heavy, fully-reduced "anions", such as sulfide and phosphide, and also have non-paramagnetic "cations" heavier than Si, with coordination numbers less than eight but greater than four, and with high formal oxidation states ($\geq 2^+$). Target compounds have crystal structures with strong three-dimensional linkages and exhibit insulating or high-bandgap semiconducting electrical behavior.

After selecting the most appropriate target compounds, the next steps are synthesis and crystal growth of the compounds. Considerable effort has been devoted to the growth of single crystals of both reported and new compositions in order to confirm/determine structural details and correlations with properties. Measurements of intrinsic optical and mechanical properties are best made on quality single crystals.

STRUCTURAL CHEMISTRY OF TERNARY INDIUM SULFIDES

Synthesis

All of the ternary sulfides discussed below have been grown in single-crystal form using eutectic halide fluxes ACl_2-KCl (m.p. 500-700°C), where A is a cation, Ca, Sr, or Ba, in common with the charge. The charges were pre- or unreacted binary sulfide mixtures. Reactions were carred out in graphite crucibles enclosed in evacuated silica ampules. Crystal-growth mixtures were soaked at 1000-1075°C for 4-17 days and slow-cooled at 1-2°/h to the freezing point of the flux. Products were recovered by dissolving the flux with water. Approximate stoichiometries were found by SEM/EDX and determined more accurately by ICP emission analysis.

AIn_2S_4-type Systems (A = Ca,Sr,Ba)

We have encountered a number of sulfides with metal stoichiometries near, but not necessarily equal to 1:2. Unfortunately, many of the phases obtained were insufficiently well-ordered for complete single-crystal structure determinations, although unit cells could be determined from crystals of some of the phases.

Two different ternary calcium-indium-sulfide phases, $Ca_{3.3}In_{6.5}S_{13}$ ($Ca_{1.0}In_{2.0}S_4$) and $Ca_{1.2}In_{1.9}S_4$ ($Ca_{3.9}In_{6.2}S_{13}$), both exhibiting whisker-like morphology, have been obtained. The $Ca_{3.3}In_{6.5}S_{13}$ phase was found to have a monoclinic C-centered unit cell with dimensions a = 37.628(4), b = 3.8360(8), c = 13.722(1) Å, β = 91.66(1)°. From analyses of the X-ray powder diffraction data obtained from crushed crystals, this phase was confirmed to be isostructural with the known compounds $Ca_{3.1}In_{6.6}S_{13}$ [4], $Pb_{3.0}In_{6.7}S_{13}$ [5], and $Sn_{2.5}In_{7.0}S_{13}$ [6]. The structures of these compounds feature infinite ribbons seven-[InS_6]-octahedra in width interconnected to form "stepped layers" throughout the structure. The A^{2+} cations, Ca, Pb, Sn, are accommodated in bicapped trigonal prismatic sites. Crushed whiskers of the $Ca_{1.2}In_{1.9}S_4$ phase yielded an X-ray diffraction pattern distinctly different from that of $Ca_{3.3}In_{6.5}S_{13}$. Although several crystals were examined, all were poorly ordered and of inadequate quality for a single-crystal structure determination. Hence, the structural relationship of this phase to that of $Ca_{3.3}In_{6.5}S_{13}$ is unknown. Further crystal growth experiments and optical transmission studies using an IR microscope are in progress to address this question.

Two different strontium-indium-sulfide phases were also obtained. $Sr_{0.9}In_{2.1}S_4$ exhibits a monoclinic C-centered unit cell with dimensions a = 27.66(1), b = 3.943(2), c = 12.683(7) Å, ß = 94.25(4)°. Comparison of the X-ray powder diffraction pattern of crushed crystals of this phase with that obtained for crushed crystals of the known $SrIn_2S_4$ phase, and also with that of the previously discussed $Ca_{1.2}In_{1.9}S_4$ whiskers, indicates that $Sr_{0.9}In_{2.1}S_4$ is a new phase. Curiously, whiskers of orthorhombic $SrIn_2S_4$ (Fddd, a = 20.892(3), b = 21.123(3), c = 13.017(2) Å) are identical in appearance to those of the unknown phase, even though $SrIn_2S_4$ is isostructural with the plate-like $BaIn_2S_4$ (Fddd, a = 21.808(3), b = 21.654(4), c = 13.107(2) Å). [7]

The structure of orthorhombic $BaIn_2S_4$, and by implication $SrIn_2S_4$, is far different from that observed in the Ca-, Pb-, and Sn-based ternaries previously mentioned. In the $BaIn_2S_4$ orthorhombic structure [8] the indium coordination shifts from octahedral to tetrahedral in a highly covalent, "Zintl-like" $In_2S_4^{2-}$ network. The Sr and Ba cations are accommodated in square antiprismatic sites.

The coordinative versatility of In(III) seems to lead to two classes of ternary sulfides with stoichiometries near, but not necessarily equal to 1:2:4. Both classes adopt framework-type structures, one featuring octahedral, the other, tetrahedral coordination of In by sulfur. The structural class adopted may be driven by the chemistry of the large A-cation. Divalent Ca, Pb, and Sn lead to octahedral $[InS_6]$-based framework; Sr and Ba have tetrahedral $[InS_4]$-based frameworks. Sr may be a "cross-over" A cation. We suspect that the new phase, $Sr_{0.9}In_{2.1}S_4$ may feature an octahedral $[InS_6]$-based framework, in contrast to the tetrahedral In coordination in $SrIn_2S_4$. Confirmation awaits the growth of sufficiently well-ordered $Sr_{0.9}In_{2.1}S_4$ crystals, and/or definitive IR transmission measurements.

$KInS_2-I$ (near-ambient-pressure form)

Light yellow platelets and rods of this compound were obtained as co-products in the growth of the Ca-In-S phases. A single-crystal X-ray structure determination has now been completed [9], confirming the $TlGaSe_2$ structure-type indicated by earlier studies of polycrystalline samples [10].

$KInS_2-I$, the ambient-pressure form of this composition, crystallizes in monoclinic space group $C2/c$; a = 10.981(3),

$b = 10.979(3)$, $c = 15.010(5)$ Å, $\beta = 100.55(2)°$, $Z = 16$. The bonding in $KInS_2$-I is highly covalent and exhibits both two-dimensional and three-dimensional features. The $KInS_2$ structure features tetrahedral coordination of indium. The structure is comprised of layers of vertex-sharing $[In_4S_{10}]$ adamantane-like units built of $[InS_4]$ tetrahedra. The stacking arrangement of these layers are such that the potassium atoms are coordinated by sulfur in distorted trigonal prismatic sites formed between the layers. The trigonal prisms share trigonal faces, creating channels that contain strings of potassium atoms. The In-S layers lend the structure a two-dimensional flavor. However, the "interlayer" potassium-sulfur bonding is strong, as indicated by the normal-to-short K-S bond lengths, providing a significant three-dimensionsal component to the structure. The covalent nature of the interlayer potassium-sulfur bonding is reflected in the non-micaceous morphology of the platelet and rectangular rod-like crystals, as well as the water-stability of the transparent light-yellow crystals.

STRUCTURAL CHEMISTRY OF TERNARY YTTRIUM SULFIDES

CaY$_2$S$_4$

Previous studies of polycrystalline samples by Patrie & Flahaut [11] indicated a $CaFe_2O_4$-related structure with an orthorhombic unit cell. The first crystals of this phase have now been grown in the eutectic halide flux described above by slow thermal cycling. The powder pattern obtained from our crushed crystals matches that which was previously reported (ICDD PDF #17-244) for CaY_2S_4, except for a minor discrepancy at higher angles due to improved resolution.

A full X-ray single-crystal structure determination of CaY_2S_4 has now been completed. The structure differs somewhat from that previously proposed. Although still related to the $CaFe_2O_4$-structure-type, the CaY_2S_4 structure consists of a three-dimensional network of edge- and corner-shared $[YS_6]$ octahedra. Pairs of seven-coordinate sites occupied by Ca atoms are formed by the intersections of "ribbons" that are four-$[YS_6]$-octahedra wide.

Yttrium-substituted CaS, $Ca_{1-x}Y_{2x/3}\square_{x/3}S$

$Ca_{0.68}Y_{0.21}S$ crystals with a simple rock salt unit cell were obtained as a co-product in the CaY_2S_4 experiments. No evidence of superstructure formation was detected on single-crystal precession photographs with long, multiple-day exposure times. The vacancies and metal cations appear to be completely disordered.

Conflicting data exist in the literature on the solubility limit of Y^{3+} in the CaS structure. We have carried out a solid solution study in this system at 1025°C and obtain a solubility limit near the composition $Ca_{0.63}Y_{0.24}S$. These results confirm the earlier report by Tsai & Mechter [12]. The structure tolerates a maximum of about 10% cation vacancies.

ACKNOWLEDGEMENTS

These work was supported by the Office of Naval Research. D.O.K.'s Postdoctoral Fellowship was administered by the American Society for Engineering Education.

REFERENCES

1. D.O. Kipp, C.K. Lowe-Ma, and T.A. Vanderah in Optical Materials: Processing and Science, edited by C. Ortiz and D.B. Poker (Mat. Res. Soc. Symp. Proc. 152, 1989), pp. 63-70.
2. D.O. Kipp, C.K. Lowe-Ma, and T.A. Vanderah, Chem. of Materials 2, 506 (1990).
3. R.M. Hazen and L.W. Finger, Comparative Crystal Chemistry, (John Wiley & Sons, New York, 1982), p. 136.
4. G. Chapuis and A. Niggli, J. Solid State Chem. 5, 126 (1972).
5. D. Ginderow, Acta Crystallogr. B34, 1804 (1978).
6. A. Likforman, S. Jaulmes, and M. Guittard, Acta Crystallogr. C44 424 (1988).
7. P.C. Donohue and J.E. Hanlon, J. Electroshem. Soc. 121 137 (1974).
8. B. Eisenmann, M. Jakowski, W. Klee, and H. Schafer, Rev. Chim. Miner. 20, 255 (1983).
9. C.K. Lowe-Ma, D.O. Kipp, and T.A. Vanderah, J. Solid State Chem. (submitted).
10. H. Schubert and R. Hoppe, Z. Naturforsch. 25b, 886 (1970).
11. M. Patrie and J. Flahaut, C.R. Acad. Sci. Paris 264, 395 (1967).
12. H.L. Tsai and P.J. Mechter, J. Electrochem. Soc. 128 2229 (1981).

PECULIARITIES OF THE FAR-INFRARED REFLECTION SPECTRA OF THE DOPED LEAD-TIN TELLURIDES REVEALING THE PERSISTENT PHOTOCONDUCTIVITY EFFECT

ALEKSANDER I. BELOGOROKHOV[*], IVAN I. IVANCHIK[**], DMITRIY R. KHOKHLOV[**], ZORAN V. POPOVICH[***] AND NEBOJSHA ROMCHEVICH[***]
[*]Institute of Rare Metals, Moscow, USSR
[**]Physics Department, Moscow State University, Moscow 119899, USSR
[***]Institute of Physics, P.O.Box 57, 11001 Belgrade, Yugoslavia

ABSTRACT

We measured the far-infrared reflectivity spectra of PbTe(Ga) and $Pb_{1-x}Sn_xTe(In)$ - materials revealing the persistent photoconductivity effect at the low temperatures $T<T_c$; $T_c \approx 80K$ and 25K, respectively. The reflectivity spectra display the singuliarity nearby the plasmon-phonon minimum. The spectra may be fitted satisfactory by means of the introduction of an additional oscillator into the dispersion relation. The result is interpreted in the framework of the model taking into account the temperature change of the one-electron metastable impurity state position in the configuration-coordinate space. The oscillator corresponds to the two-electron - one-electron state transitions. Some previously unexplained results find the satisfactory interpretation in the framework of the model proposed.

Doping of the lead-tin tellurides with group III impurities leads to the appearance of the persistent photoconductivity effect at the low temperatures $T<T_c$. The characteristic features of this effect are the "integral" photoresponce and the long-term nonexponential conductivity relaxation after the switch-off the light [1,2]. In PbTe(Ga) $T_c \approx 80$ K, in $Pb_{1-x}Sn_xTe(In)$ $T_c \approx 25$ K. Up to now it was beleived that the critical temperature T_c is defined by the relation between the photogenegation and recombination rates and doesn't correspond to the appearance of any peculiarities in the semiconductor energy spectrum.

We measured the infrared reflectivity spectra of $Pb_{0.75}Sn_{0.25}Te(In)$ and PbTe(Ga) in the wavenumbers range (50-500) cm^{-1} at the temperatures (5-300)K. The features of these spectra are affected not only by the free carriers, as in the galvanomagnetic measurements, but are also influenced by the crystalline lattice and the bound carriers. Therefore these measurements may provide some additional information concerning the nature of the persistent photoconductivity effect.

The samples we measured were the $Pb_{0.75}Sn_{0.25}Te(In)$ and PbTe(Ga) monocrystals grown by the modyfied Bridgeman technique and Chokhralski technique, respectively. The tin content x=0.25 in $Pb_{1-x}Sn_xTe(In)$ alloy and the Ga concentration in PbTe(Ga) were chosen in order to provide the Fermi level pinning within the bandgap [2,3]. In this case the reflectivity spectra

features, that are not defined by the free carriers, reveal
themselves clearly.

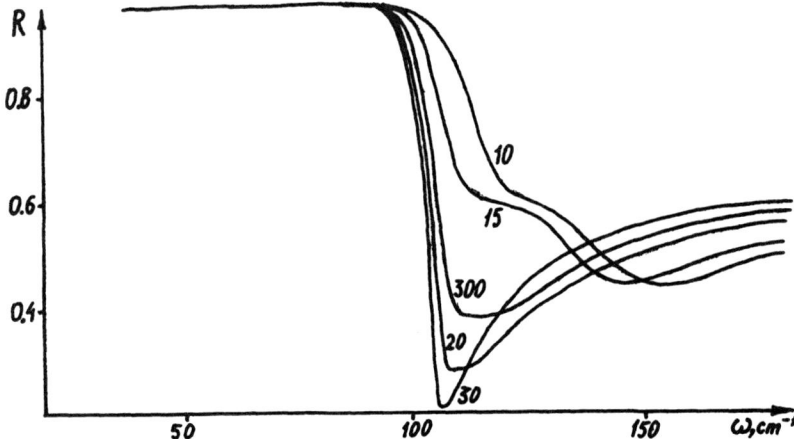

Fig. 1 The reflectivity spectra of $Pb_{0.75}Sn_{0.25}Te(In)$. Figures
near the curves - T in K.

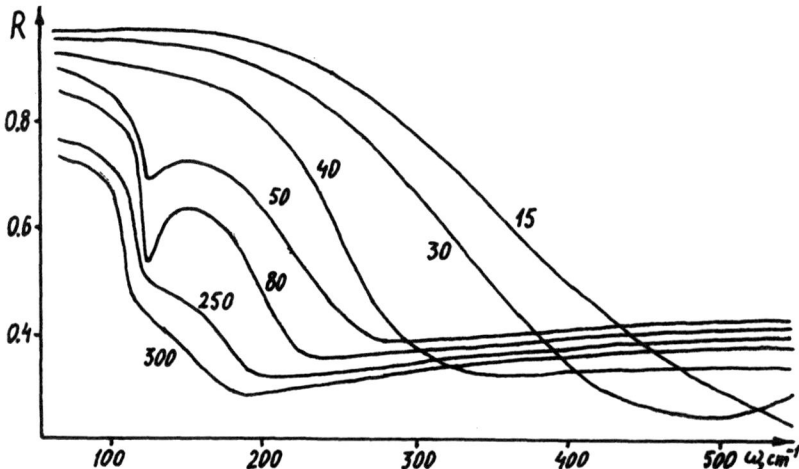

Fig. 2. The reflectivity spectra of PbTe(Ga). Figures near the
curves - T in K.

The reflectivity spectra of the In- and Ga-doped samples
are shown in figs. 1, 2. One can see that for $Pb_{0.75}Sn_{0.25}Te(In)$
the reflectivity minimum observed at $\omega \sim (130-150)\,cm^{-1}$ becomes
more pronounced with the temperature lowering from 300 K to 30
K, then it broadens and shifts to the higher wavenumber values.
At the temperatures T>30 K the spectra have no peculiarities,
whereas at T<25 K an additional structure appears at
$\omega \sim 120\ cm^{-1}$. The structure amplitude increases with the
temperature lowering.

The main features of the Ga-doped material reflectivity spectra are in general analogous, though there exist some considerable differences. The additional structure appears already at T≈250 K. Its position is shifted to the higher wavenumbers with respect to the case of In-doped sample. The structure amplitude strongly changes at the temperatures T~80 K. The reflectivity minimum shift to the higher wavenumber values occurs at T<50 K.

The reflectivity spectra of the lead-tin tellurides in the wavenumber range (50-500) cm^{-1} are usually interpreted using the dispersion relation based on the plasmon-phonon interaction model

$$\varepsilon(\omega) = \varepsilon_\infty \left(1 - \frac{\omega_p^2}{\omega(\omega + i\tau^{-1})} - \frac{\omega_{to}^2 - \omega_{to}^2}{\omega_{to}^2 - \omega^2 - i\gamma_{to}\omega} \right) \qquad (1)$$

where ω_{to}, ω_{lo} and ω_p are the transverse, longitudinal phonon and plasma frequency, respectively; γ_{to} is the phonon damping factor, τ is the free carrier relaxation time and ε_∞ is the high frequency dielectric constant. The second term in Eq. (1) is the free carrier and the third – the lattice vibration contribution to the dielectric constant. The spectra of the In-doped sample at the temperatures T>30 K may be fitted satisfactory using Eq. (1).

The plasmon-phonon minimum shift to the higher wavenumbers both in $Pb_{1-x}Sn_xTe(In)$ and PbTe(Ga) at the low temperatures is obviously due to the persistent photoconductivity effect appearance at $T<T_c$≈25 K in the In-doped alloy and at $T<T_c$≈80 K in PbTe(Ga), respectively. This shift becomes evident at the temperatures somewhat lower than T_c because if ω_p is much lower than ω_{lo} the effect is hidden.

The low-temperature plasmon-phonon minimum broadening is due to the more fine effect – unhomogeneous spacial distribution of the photogenerated free carriers [4].

In the In-doped sample the spectra at the temperatures below 25 K cannot be fitted within the framework of the ordinary dispersion relation taking into account the plasmon and the phonon modes. An additional oscillator of the type $\omega_{loc}^2/(\omega_0^2 - \omega^2 - i\omega\gamma)$ should be introduced into the dispersion relation (1) in order to obtain the best fit. The temperature dependencies of the oscillator frequency ω_0 and relative

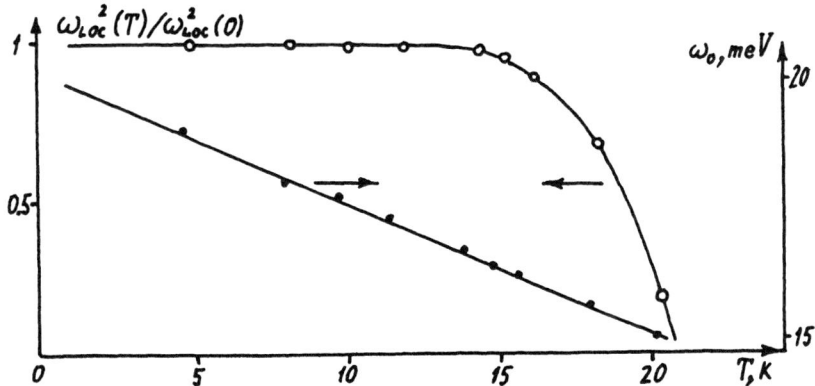

Fig. 3. The ω_0 and $\omega_{loc}^2(T)/\omega_{loc}^2(0)$ temperature dependencies in $Pb_{0.75}Sn_xTe(In)$.

strength $\omega_{loc}^2(T)/\omega_{loc}^2(0)$ are shown in fig. 3. One can see that ω_0 slowly rizes with the temperature lowering, and ω_{loc} tends to zero at $T \to 21K$.

The galvanomagnetic measurements show that the persistent photoconductivity effect appears in this alloy at the temperatures $T<21K$. The resistivity temperature dependence measured in darkness has a low-temperature activation part. The corresponding activation energy is $E_a \sim 25meV$.

The question arises, what is the origin of the additional oscillator we observe in the reflectivity spectra. To our opinion the nature of effect is defined by the optical transitions from the ground two-electron states that pin the Fermi level to the metastable one-electron local level revealing in some other effects, as the negative photoconductivity [5] or giant negative magnetoresistance [6]. This metastable state is separated from the extended and the two-electron localized states by the barriers in the configuration-coordinate space [5,6].

The energy of the metastable state is close to the conduction band bottom in the alloy with $x=0.25$ [6]. Our data confirm this result: the ω_0 value extrapolated to the zero temperature is close to the conductivity low-temperature activation energy E_a.

The ω_{loc} value corresponds to the transitions probability. Our results show that the optical transitions between the ground and the metastable states occur only at $T<21K$. To the other hand, the persistent photoconductivity effect takes place

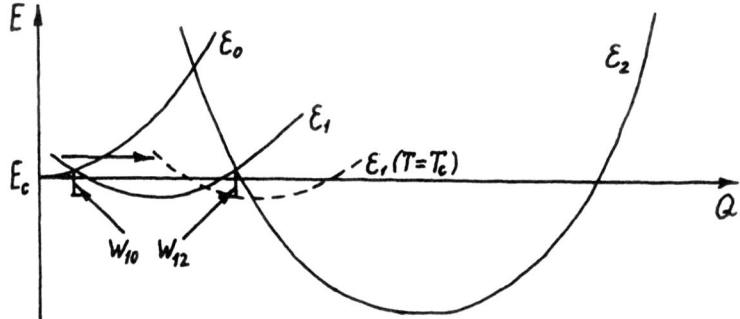

Fig. 4. The configuration-coordinate diagram of the system in the one-electron approximation. E_c - the conduction band bottom. $\mathcal{E}_n$ $(n=0,1,2)$ curves correspond to the states with n localized electrons. These states are separated by the barriers W_{10} and W_{12}. The dashed curve shows the $\mathcal{E}_1$ state position at $T=T_c$.

also at the temperatures $T<T_c=21K$. To our opinion this cannot be simply a coincidence. We propose the following model.

The configuration-coordinate diagram of the system is shown in fig. 4 [5,6]. The curve $\mathcal{E}_0$ corresponds to the extended electron state, $\mathcal{E}_2$ and $\mathcal{E}_1$ - to the ground and metastable localized states, respectively. The diagram at the zero temperature is shown by the solid curves. One can see that all of the states are separated by a barrier. The temperature rising leads to the shift of the curve corresponding to the metastable state to the right side of the diagram. At the temperature $T=T_c$ the metastable state minimum crosses the ground state curve (fig. 4, dashed curve), and the barrier separating these states disappears. If $T>T_c$ the metastable state cannot be occupied because there exist a state with the lower energy and the same configuration coordinate. So the temperature T_c is defined by the appearance of the metastable local electronic states and not by the experiment conditions.

In PbTe(Ga) the persistent photoconductivity appears at $T<T_c \approx 80$ K. The drastic change in the local oscillator strength also corresponds to this temperature. In contrary to the case of $Pb_{1-x}Sn_xTe(In)$ ω_{loc} isn't zero at $T>T_c$. This fact is probably due to the high dispersion of the one-electron state positions in the configuration-coordinate diagram of PbTe(Ga).

Some of the previously unexplained results find the natural interpretation in the framework of the model we proposed. It was observed in [5] that at the temperatures $T>T_c$ the photoconductivity relaxation in $Pb_{1-x}Sn_xTe(In)$ is exponential whereas at $T<T_c$ both the photogeneration and the relaxation processes are very complex. Moreover at $T<T_c$ the relaxation rate depends on the level of the initial photoexcitation [1]. If there exist two local levels separated by a barrier with each other and with the extended electron states these results are quite easy to explain: the relaxation process reflects the photogenerated electron redistribution between these states, and the conductivity relaxation rate depends on the occupancy of each state.

So the results we present allow to conclude that the difference of the system states at the temperatures above and below T_c is the absence or the presence of a second local metastable level.

REFERENCES

1. Akimov B. A., Brandt N. B., Klimonskiy S. O., Ryabova L. I. and Khokhlov D. R. Phys. Lett. A **88A** 483 (1982).
2. Akimov B. A., Brandt N. B., Gas'kov A. M., Zlomanov V. P., Ryabova L. I. and Khokhlov D. R. Sov. Phys. Semicond. **17** 53 (1983).
3. Akimov B. A., Ryabova L. I., Yatsenko O. B. and Chudinov S. M. Sov. Phys. Semicond. **13** 441 (1979).
4. Romchevich N., Popovich Z., Khokhlov D., Nikorich A. V. and Konig W. Infrared Phys. (in press).
5. Zasavitskiy I. I., Matveenko A. V., Matsonashvili B. N. and Trofimov V. T. Sov. Phys. Semicond. **20** 135 (1986).
6. Akimov B. A., Nikorich A. B., Khokhlov D. R. and Chesnokov S. N. Fiz. Tekh. Polupr. **23** 668 (1989).

STRUCTURE AND PROPERTIES OF WN$_x$ THIN FILM ON GaAs SUBSTRATE

DALI MAO, WEILI YU[*] AND DONGLIANG LIN(T.L.LIN)
Shanghai Jiao Tong University, Department of Materials Science
and Engineering, Shanghai 200030, P.R.China
*Materials Engineering, Auburn University, Auburn, AL36849, USA

ABSTRACT

In this paper, the effect of sputtering parameters on the interfacial reaction and the electronic properties of the WN$_x$ system were reported.

The report showed that W and W$_2$N phases were observed in WN$_x$ film on GaAs substrate annealed at 800-900°C, in which no WN phase was found and W$_2$N was the stable phase. Throughout the WN$_x$ film, W,N distributed uniformly. There was no interdiffusion between WN$_x$ films and GaAs substrate annealed at 800°C. However, at 900°C, there was some N in-diffusion to substrate, but no Ga or As out-diffusion. The electrical resistivity ρ of WN$_x$ films increased with increasing nitrogen partial pressure Γ. All the samples with $\Gamma < 0.2$ showed the ρ below 200 $\mu\Omega cm$. By the I/V measurement, the Schottky Barrier Height was obtained with the value: $\phi_B = 0.93ev$, n=1.30 for WN$_x$/n-GaAs contact that annealed at 850°C, 15 minutes.

INTRODUCTION

Among the most promising GaAs devices are self-aligned GaAs Field Effect Transistors(FET)[1]. WN$_x$ film is one prospective gate material for GaAs self-aligned FET's[2,3]. It has been reported that the WN$_x$ film has low electrical resistivity and a thermally stable high Schottky Barrier Height(SBH) at WN$_x$/GaAs interface. However, the properties of WN$_x$ film and its interface with the semiconductor depend on the chemical composition and structure of the film, which in turn a function of the deposition conditions. In the previous paper[4], the effect of the mixing ratio of the reaction gases and other sputtering parameters, and the annealing environment on the WN$_x$ films deposited on Si and GaAs by rf reactive sputtering have been studied. In this paper, we reported the effect of these factors on the interfacial reaction and the electronic properties of the WN$_x$/GaAs.

EXPERIMENTAL

WN$_x$ films were deposited on cleaned GaAs substrates by rf reactive sputtering. A SPF-210B rf sputtering system was used with a pure metallic W target and Ar and N$_2$ gas sources. The chamber was pumped down to about 2×10^{-7} torr before the sputtering gases introduced. The sputtering parameters such as effective power, working pressure, nitrogen partial pressure Γ were controlled. After deposition, the wafers were annealed in flowing N$_2$ at different temperature.

The thickness of thin films was measured by Talystep profilometer and the structure of the films was determined by XRD(Rigaku D/Max IIIA). Detailed depth profile information of WN$_x$ films and interface contacts with substrate were obtained

from AES profiling technique using a PHI 610 Scanning Auger
spectroscope.

Electrical resistivity measurements were carried out on the
STZ-1 digital four-point probe. After deposition of WN_x films
on (100) epitaxial GaAs substrates (dopant Si, $n-10^{16}$ cm^{-3}),
Schottky electrodes of 100 μm in diameter were fabricated.
Alloyed ohmic contacts of AuGeNi were formed on the back side of
the wafer after annealing. Barrier heights and ideality
factors of $WN_x/n-GaAs$ contacts were determined from Schottky
diode forward-current characteristics.

RESULTS AND DISCUSSION

Structure

As reported previously[4], with the increase of the partial
pressure ratio of nitrogen gas or the decrease of the working
pressure, deposition rate of WN_x film decreased before
approaching saturation. With the effective power of 300w, the
WN_x films formed at high working pressure($\approx 5 \times 10^{-2}$ torr)
consisted of W, WN or W_2N phases depending on the nitrogen
partial pressure ratio, whereas the films formed at low working
pressure($\approx 3 \times 10^{-3}$ torr) were usually amorphous. Annealing in a
flowing N_2 caused the crystallization of the amorphous films,
which mainly consisted of $W+W_2N$. However, annealing in H_2 gas
caused severe loss of nitrogen in the film, in which single W
phase formed eventually.

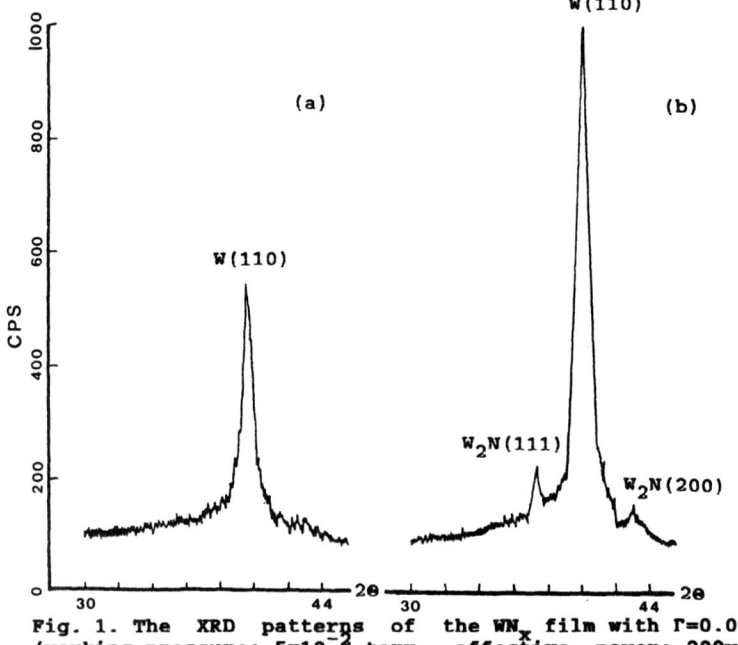

Fig. 1. The XRD patterns of the WN_x film with $\Gamma=0.05$
(working pressure: 5×10^{-2} torr, effective power: 200w)
(a)as-deposited,(b)after annealed at 800°C, 20 minutes
in flowing N_2.

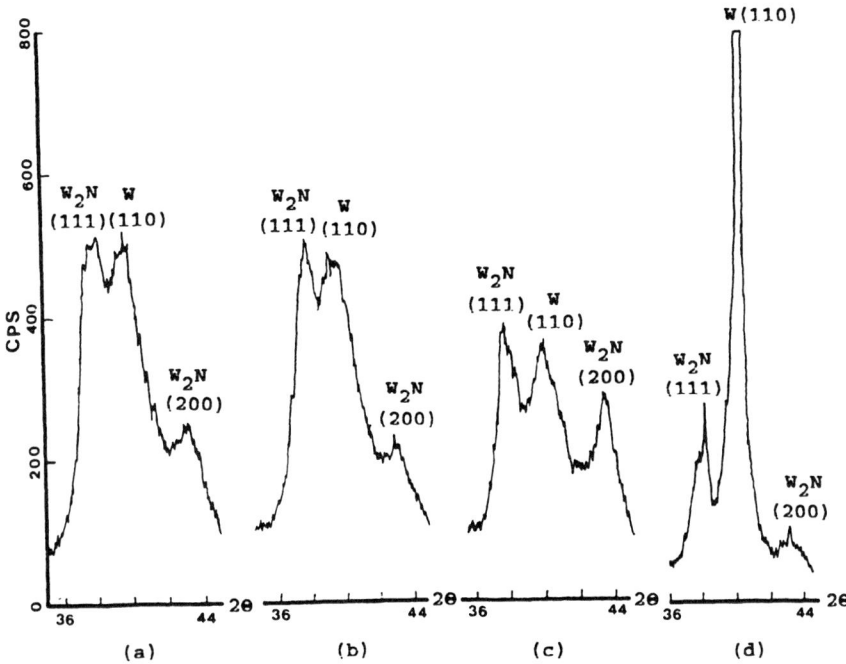

Fig.2. The XRD patterns of WN$_x$ film with Γ=0.1 (working pressure:5x10^{-2} torr, effective power: 200w). (a)as-deposited; (b)after annealed at 600°C, 20 minutes in flowing N$_2$;(c)after annealed at 800°C,20 minutes in flowing N$_2$;(d)after annealed at 900°C,20 minutes in flowing N$_2$.

Fig.1 and Fig.2 were the results of XRD for the specimens that sputtered in the working pressure of 5x10^{-2} torr with the effective power of 200w. There was only W peak for as-deposited film(Fig.1a), but W+W$_2$N phases for films after annealed at 800°C, 20 minutes in flowing N$_2$(Fig.1b). The reason might be that the N concentration in as-deposited film was very low, which was only about less than 4 at% shown in AES, and the N atoms maybe only solved in W but did not form the nitride. However even for such low N content, W$_2$N phase was found in the annealed film. Fig.2 also showed the temperature effects for the samples with Γ=0.1. In the as-deposited film, there was a mixture of W+W$_2$N phases and amorphous-like phase(Fig.2a), which was revealed by a halo in electron diffraction pattern. Annealing at 600°C did not change the XRD spectrum(Fig.2b), which indicated the amorphous phase did not crystallize. After annealing at 800°C and 900°C, the spectrum showed only the crystalline W$_2$N and W phases existed(Fig.2c,2d). The appearance of stronger peak of W phase and weaker peak of W$_2$N phase in films annealed at 900°C indicated N loss, which was confirmed by AES to be N in-diffusion to the substrate. In this film, the N content, which was about 12 at% analysed in AES, was higher than in the films with Γ=0.05. However, WN phase was never found and W$_2$N is a stable phase. These different phenomena may be due to the different sputtering situation.

Interfacial reaction of WN_x/GaAs system

AES depth profile technique was used to analyse the interfacial contact between WN_x film and GaAs substrate. In Fig.3 and Fig.4, the AES results of the annealed WN_x/GaAs samples with Γ of 0.05 and 0.1 were shown together with that of as-deposited samples. From Fig.3, it can be seen that the spectrum did not change greatly after annealing at 800°C for 20 minutes. There was no interdiffusion between WN_x thin film and GaAs substrate. The similar result was also obtained from the samples with $\Gamma=0.1$ annealed at 800°C, but there was slight change for samples annealed at 900°C. In Fig.4. though there was no out-diffusion of Ga or As, some N in-diffusion to the substrate was found. However, all the results have shown that WN_x film was uniform and chould prevent the out-diffusion of Ga or As at 800°C. Even at 900°C, WN_x film still had high stability and was a good diffusion barrier for GaAs except a little N in-diffusion to the substrate.

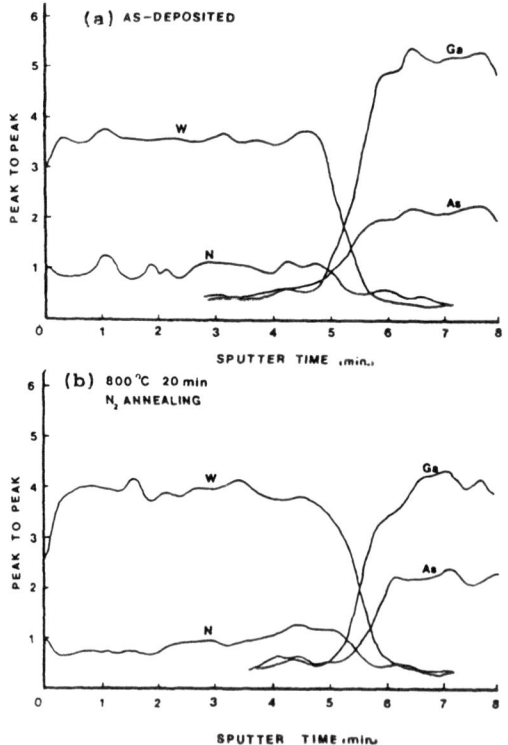

Fig.3 The AES depth profile analysis of WN_x/GaAs samples with $\Gamma=0.05$. (a)as-deposited;(b)annealed at 800°C, 20 min. in flowing N_2.

Electrical properties of WN_x thin film

The electrical resistivity p of WN_x is one of the most important properties for a self-aligned gate in MESFETs. Table I showed the electrical resistivity p of WN_x film at different annealing temperature obtained by STZ-1 four point probe. The p of WN_x film with high Γ was higher than that with low Γ. The p of WN_x film with $\Gamma=0.05$ after annealing at $800^{\circ}C$ was slight less than that before annealing due to the fact that W phase dominated in the film. While $\Gamma=0.1$, the high p was related to the W_2N+W phase. In the films that annealed at lower temperatures than $900^{\circ}C$, the phase did not change greatly and AES showed that there was no change in N content. When annealed at $900^{\circ}C$, as shown in XRD and AES, there was some N in-diffusion to the substrate in the film and comparatively the W_2N phase was less but W phase was more at the surface, thus the p was decreased

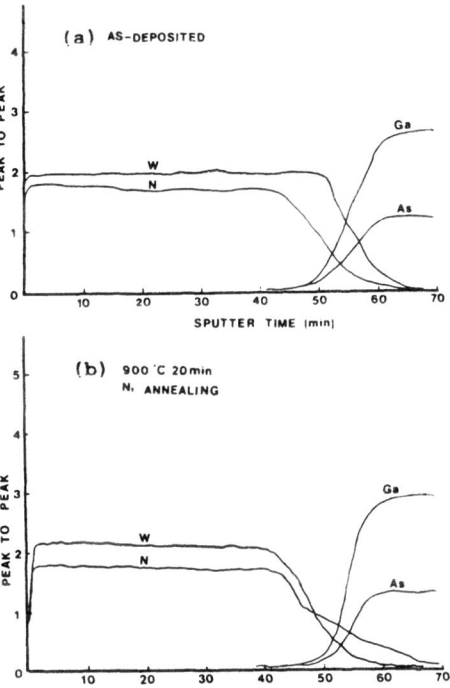

Fig.4 The AES depth profile analysis of WN_x/GaAs samples with $\Gamma=0.1$. (a) as-deposited; (b) annealed at $900^{\circ}C$,20 min. in flowing N_2.

rapidly. There were also some p measurements for other Γ. The values of electrical resistivity for as-deposited and annealed samples with $\Gamma<0.2$ were below 200 $\mu\Omega cm$. In order to investigate the Schottky Barrier Height and ideality factor n, WN_x films about 4000 A thick were sputtered on (100) epitaxial GaAs wafers with Si dopant of about 1×10^{16} cm^{-3}. The sputtering parameters were as follows: background pressure: 2×10^{-7} torr, working pressure: 5×10^{-2} torr, effective power: 200w, Γ: 0.05. After annealed at $600^{\circ}C$, $850^{\circ}C$, 15 minutes in flowing N_2, the WN_x/n-GaAs contact regions were photoengraved with 100 μm in diameter. The WN_x film was etched by a chemical dry etching method. Then ohmic contacts on their back surface were formed with AuGeNi alloy. The results of ϕ_B and n for different temperature calculated using the theory of thermal electron emission[5] were listed in Table II.

Table I. The p measurements of WN_x film

Temp.($^{\circ}C$)	as-deposited	600	800	900
p($\Gamma=0.05$)$\mu\Omega cm$	73		66	
p($\Gamma=0.1$)$\mu\Omega cm$	126	125	117	41

Table II. The SBH and n results of
WN_x/n-GaAs contract

Temp.($^\circ$C)	as-deposited	600	850
ϕ_B(eV)	0.61	0.63	0.93
n	1.18	1.20	1.30

CONCLUSION

The results can be summarized as follows:

[1] Sputtering parameters such as effective power, working pressure, and nitrogen partial pressure Γ, greatly affected the composition, interfacial reaction and electrical properties of WN_x films.

[2] By XRD, the mixed phase of W and W_2N was observed in WN_x films deposited on GaAs substrate and annealed up to T=800–900°C. There was no WN phase existed. W_2N phase was the stable phase in WN_x film.

[3] AES depth profile revealed that the distribution of W, N in WN_x films was uniform throughout the film, When annealed at 800°C, there was no interdiffusion between WN_x film and GaAs substrate. However at 900°C, there was some N in-diffusion to substrate, but no Ga or As out-diffusion. WN_x was a good diffusion barrier for GaAs.

[4] The electrical resistivity ρ of as-deposited WN_x films increased with increasing Γ. When Γ=0.05, ρ was about 80 $\mu\Omega$cm, when Γ=0.1, ρ was about 130 $\mu\Omega$cm. All the as-deposited and annealed samples with Γ<0.2, showed the ρ below 200 $\mu\Omega$cm. The SBH characteristic was obtained by I/V measurements for the samples and showed the good thermal stability with ϕ_B=0.93eV, n=1.30 annealed at 850°C, 15 minutes.

ACKNOWLEDGEMENT

The research project was supported by The National Natural Scientific Fundation of China.

REFERENCE

[1]. N.Yakoyama, T.Mimura, M.Fukuta and H.Ishidawa, 1981 IEEE International Solid-State Circuits Conference, p218.

[2]. H. Yamagishi, Jpn. J. Appl. Phys. 23, L895 (1984).

[3]. H.Yamagishi and Yamamoto, Jpn. J. Appl. Phys. 26, 122 (1987)

[4]. Dongliang Lin, Benda Yan and Weili Yu, Proc. Mater. Res. Soc. Symp., Materials Research Society, Pittsburgh, PA, 1990, Vol.187.

[5]. S.M.Sze, Phys. Semiconductor Device (John Wiley and Sons, 2nd Edition, 1981). Chap. 5.

NEAR IR EMISSIONS FROM Er, Tm, AND Pr IMPLANTED GaAs AND AlGaAs

G.S. POMRENKE,* Y.K. YEO,** and R.L. HENGEHOLD**
*Air Force Office of Scientific Research, Bolling Air Force Base, Washington, D.C. 20332
**Air Force Institute of Technology, Wright-Patterson Air Force Base, OH 45433

ABSTRACT

Optical properties of Er, Tm, and Pr ion implanted GaAs and $Al_xGa_{1-x}As$ were investigated using photoluminescence. For Er, the characteristic sharp emissions around 1.54 μm were observed at sample temperatures as high as 296 K, which are due to the transitions between the weakly crystal-field-split spin-orbit levels $^4I_{13/2}$ and $^4I_{15/2}$ of Er^{3+} ($4f^{11}$); the strongest luminescence signal was from $Al_{0.4}Ga_{0.6}As$. For Tm, the sharp 4f-emissions were observed between 1.22 to 1.33 μm at sample temperatures up to 200 K, and are due to the transitions between levels $^3H_5$ and $^3H_6$ of Tm^{3+}($4f^{12}$). For Pr, three sets of 4f-emissions were observed at 1.05, 1.35, and 1.6 μm due to transitions between levels $^1G_4-^3H_4$, $^1G_4-^3H_5$, and $^3F_3-^3H_4$ of Pr^{3+}($4f^2$), respectively, and the highest temperature at which sharp emissions were observed was around 175 K.

INTRODUCTION

Rare-earth (RE) doping of III-V compound semiconductors has been considered attractive due to possible application to opto-electronic and photonic devices. The RE elements form generally trivalent ions in a crystal and have abundant $4f^n$ excited states covering the range from ultra violet to visible to infrared, and thus the emissions could have a spectral range of interest to integrated optics, fiber optics, and various detectors. In particular, the 1.23 and 1.54 μm spectral regions, which are very important to present silica based fiber optics, are obtainable from the 4f-intracenter transitions of rare earth ions in III-V semiconductors. The 1.54 μm wavelength corresponds to the minimum attenuation and the 1.3 μm wavelength to the minimum dispersion in silica based fiber optics. Of the various rare-earth ions which have been incorporated into the III-V semiconductors, erbium (Er) appears to have the greatest technological potential, because several sharp and strong emissions were observed around 1.54 μm. Since the initial investigations by Ushakov et al [1] and Ennen et al [2], Er has been successfully introduced into III-V semiconductors not only through implantation [3,4], but also through growth by LPE into GaAs [5], MOCVD into InP [6], and MBE into GaAs [7] and AlGaAs [8]. Also, devices have been produced in the form of GaAs:Er pn-diodes [9,10]. The first definitive spectra of thulium (Tm) related emissions around 1.2 μm in III-V semiconductors were reported only recently [4]. The photoluminescence studies of praseodymium (Pr) incorporated into III-V semiconductors has up until now remained essentially uninvestigated. The only studies reported thus far have been Pr doped GaP by Kasatkin et al [11], a study of Pr implantation into GaP by Gippius et al [12], and further studies on the luminescence of Pr^{3+} in GaAs, InP, and GaP by Pomrenke et al [4].

In this investigation, we carried out a photoluminescence (PL) study of Er^{3+} by examining deep high energy (1 MeV) implants into semi-insulating GaAs, AlAs, and $Al_xGa_{1-x}As$ having different Al mole fractions. In addition, PL investigations of the 4f-intracenter transition emissions of Tm^{3+} and Pr^{3+} in GaAs were studied in order to better understand the luminescence centers responsible for the emissions. We also performed a PL study on sample temperature dependence. From the technological point of view, it is of great interest to observe the rare earth emission at high temperatures (especially room temperature).

EXPERIMENTS

The samples used for this study were semi-insulating Cr-doped GaAs or LEC grown undoped GaAs, and MOCVD grown $Al_xGa_{1-x}As$. The samples were carefully cleaned and room temperature implantation was carried out at ion energies of 380, 390, and 1000 keV for Pr, Tm, and Er, respectively, with a dose of $5X10^{13}$ cm^{-2}. After implantation the samples were proximity cap annealed, face down on a Si:P wafer in a conventional annealing furnace in a flowing forming gas atmosphere. Annealing temperatures were chosen to be close to the optimum annealing temperatures as previously established [13]. The luminescence excitation source for Er doped samples was an air cooled argon ion laser at 514.5 nm, and that for Tm and Pr doped samples a krypton ion laser at 647.1 nm and a power of 200 mW. The luminescence was dispersed by a 1200 gr/mm grating blazed at 1.0 μm for Tm, and a 600 gr/mm grating blazed at 1.6 μm for Er and Pr using a Spex 1702 3/4-m Czerny-Turner spectrometer. Signal detection was obtained with a liquid nitrogen cooled Ge detector.

RESULTS AND DISCUSSION

Figure 1 depicts the photoluminescence spectra of Er implanted Cr-doped semi-insulating (SI) GaAs, AlAs, and $Al_xGa_{1-x}As$ at different Al mole fractions; $x = 0.1$, 0.2, 0.3, and 0.4. These materials have bandgaps ranging from 1.52 eV for GaAs to 2.24 eV for AlAs at 4 K. The samples were annealed at 680 °C for 10 min. The sample emission temperature is 7 K, the resolution is approximately 3 meV, and irradiance is 600 mW/cm^2. Seen in the figure are the characteristic Er^{3+} emissions from 1.53 to 1.67 μm with the strongest signal occuring around 1.54 μm. At least up to seven emission lines may be identified for each spectrum. These emissions are due to the transitions between the weakly crystal-field-split spin-orbit levels $^4I_{13/2}$ and $^4I_{15/2}$ of Er^{3+} ($4f^{11}$). The near-edge emissions due to free-to-bound and donor-acceptor pair transitions were also observed for several samples, although these are not shown here. In Fig. 1, aside from possibly AlAs, there are no shifts in peak energy positions or significant changes in the spectral characteristics of the Er^{3+} emissions

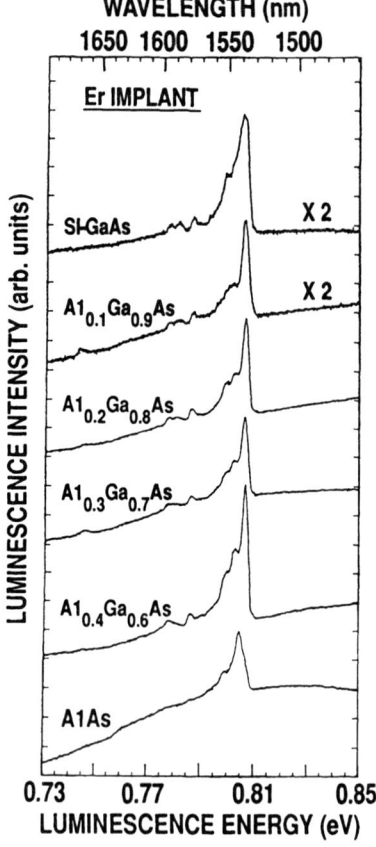

Fig. 1. Photoluminescence spectra of Er-implanted into GaAs, AlAs, and AlGaAs with different Al mole fractions.

from material to material, i.e., with variation of the semiconductor bandgaps, which seems to indicate that the crystal-field has only a minor effect on 4f levels of Er³⁺ in these materials. These observations agree with the previous investigations of Ennen et al [2] and more recently by Rochaix et al [14]. The photoluminescence intensity is generally stronger for the AlGaAs samples than for the GaAs sample. The strongest signal was obtained for $Al_{0.4}Ga_{0.6}As$ at 1.537 μm with a full-width at half maximum (fwhm) of around 0.4 meV at 7 K. However, this must be interpreted with caution because the penetration depth differs from one semicondutor material to another, depending upon the laser excitation line. That is, the penetration depth of the 514.5 nm laser line is greater in $Al_{0.4}Ga_{0.6}As$ as compared to GaAs, and hence the probe beam can access more of the Er centers in the implanted layer. It has been found that for the same Er-implanted GaAs sample under identical emission conditions, the spectra from the 647.1 nm excited layer were twice as intense than those from the 514.5 nm excited layer. This also demonstrates the desirability of implanting at high energy, e.g., 1 MeV to create a much deeper RE doped layer.

Figure 2 shows the results of the 4f intra-transition emissions for 1 MeV, Er-implanted $Al_{0.4}Ga_{0.6}As$ measured at the sample temperatures of 7 K and 270 K. Although the luminescence signal is reduced by a factor of ten at the higher temperature of 270 K, the signal is seen even at room temperature (296 K). The presence of a high energy shoulder in the 270 K spectrum could be a result of the development of 'hot' lines at the higher temperatures. From a technological point of view, it is of great interest that the Er-specific emission is seen near room temperature.

The Tm³⁺ photoluminescence emissions in GaAs were investigated as a function of sample temperature from 6 to 200 K to determine the nature and behavior of the Tm-related emissions. The implanted samples were encapsulated with Si_3N_4 layers prior to the proximity annealing at 750 °C for 10 min. The spectra in Fig. 3 show a dominant, sharp peak around 1.01 eV (approximately 1.22 μm) with some low energy emission structure. The emissions extending from 0.93 to 1.03 eV are assigned to the transitions between the weakly crystal-field-split spin-orbit levels of $^3H_4$ and $^3H_6$ of Tm³⁺ $(4f^{12})$. The main 1233 nm line in GaAs shows doublet structure with the proper system resolution, and each doublet component has a measured full-width at half-maximum of better than 0.15 meV. These temperature dependent studies show that the 4f-emissions may be seen up to a temperature of 200 K. The luminescence peak intensity remains relatively unchanged from 6 to 42 K, and the intensity decreases gradually as the emission temperature increases to 166 K, after which the intensity falls off such that it shows only a very weak presence at 200 K. It is also shown that the peak energy positions or relative intensities among the various emission peaks are unchanged at all sample temperatures. Quite obvious, however, is the development of a high energy 'hot' line at 1.211 μm at temperatures above 42 K. This is due to thermal

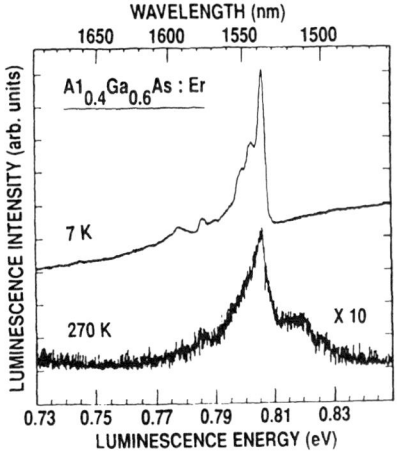

Fig. 2. Temperature dependence of the Er-specific emission in $Al_{0.4}Ga_{0.6}As$

excitation involving the higher energy terms of the specific crystal-field-split spin-orbit levels. At the same resolution settings, the sharpness (1.4 meV at fwhm) of the main 1233 nm emission line is maintained up to 166 K. Higher emission temperatures might be possible with improved techniques of incorporating Tm into GaAs such as through MOCVD or MBE growth.

The praseodymium doped semiconductors, specifically III-V semiconductors, have remained relatively unexplored. The results of our study of 4f-intracenter emissions for Pr^{3+} in GaAs are shown in Fig. 4. The samples were annealed at 750 °C for 10 min and the emission was measured at a sample temperature of approximately 6 K. The trivalent Pr has an electron configuration of $4f^2$ with a $^3H_4$ ground state. Three sets of sharp lines are evident around 1.05, 1.35, and 1.6 μm. The emissions around 1.05 μm as shown in the upper spectrum in Fig. 4 are the result of transitions between the spin-orbit levels $^1G_4$ and the ground state $^3H_4$ of Pr^{3+}. The lower spectra of Fig. 4 identifies the Pr^{3+} emission at around 1.35 μm as due to transitions from the spin-orbit level $^1G_4$ to the first excited level $^3H_5$ of Pr^{3+} and the luminescence around 1.6 μm as due to transitions between the spin-orbit levels $^3F_3$ and $^3H_4$ of Pr^{3+}. The $^1G_4$-$^3H_5$ transitions are the most intense. Due to slightly different experimental conditions, the intensities of the $^1G_4$-$^3H_5$ emissions are not at the same intensity for the two spectra shown in the upper and lower figures; however, these emissions can be used to compare the relative intensities of all the emission lines in the 0.73 to 1.55 eV spectral range. High resolution investigations of the strongest Pr^{3+} emission at 1382 nm show the full-width at half-maximum to be less than 0.9 meV. Fig. 4 also identifies the free-to-bound (FB) near-edge emissions and the bandgap E_g of the GaAs at 4 K. A strong transition metal (Cu) emission can be seen at 1.36 eV along with its lower energy phonon replicas. The lower spectra show broad background bands which could be associated with deep levels and defect complexes as a result of ion implantation damage.

Further investigations of sample temperature dependent studies were performed on Pr implanted undoped LEC grown semi-insulating GaAs, and the results are shown in Fig. 5. The luminescence results indicate that the 1.35 and 1.6 μm emissions have their own unique temperature dependent

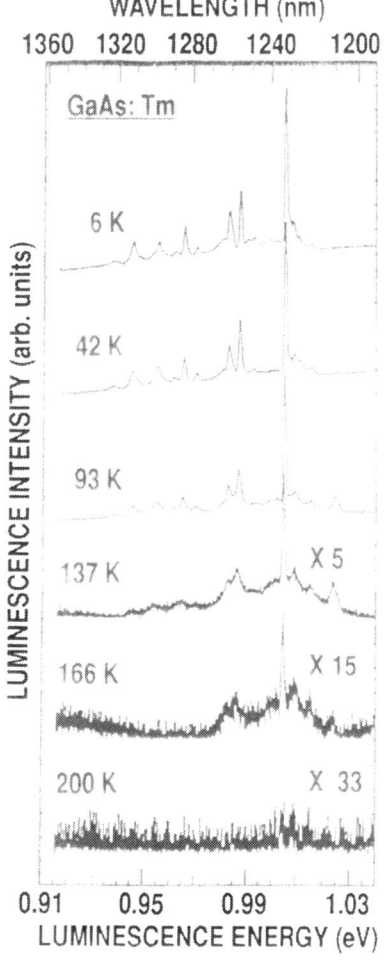

Fig. 3. Tm^{3+} emissions in GaAs as a function of sample temperature.

behavior. The 1.35 μm Pr³⁺ emissions essentially disappear at around 107 K, whereas the 1.6 μm Pr³⁺ emissions become quenched at around 174 K. The intense broad band, however, also masks the Pr-related emissions such that they can not effectively be seen. Due to the low intensity of the 1.05 μm emission line, temperature dependent studies were not performed. Aside from the differences in the above mentioned temperature dependence between the 1.35 and 1.6 μm emissions, the relative intensities of various emission lines also change with temperature, specifically for those lines marked by 'up-pointing' arrows in Fig. 5. Based on these emission

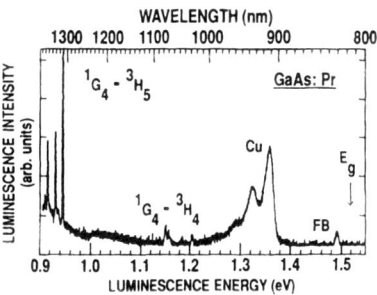

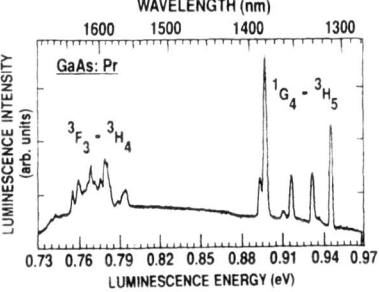

Fig. 4. Transitions between three sets of spin-orbit levels of Pr³⁺ implanted GaAs.

temperature dependent studies for GaAs:Pr, it can be concluded that the rare earth luminescent centers may occupy either substitutional lattice sites and/or interstitial sites. Some of the centers may also possibly be due to the radiation damage from implantation. At and near the optimum anneal temperature for GaAs:Pr, there is essentially no evidence of near-edge emissions, hinting at the presence of possible competing processes that favor the excitation of the rare earths versus the free-to-bound or donor-acceptor recombinations. For this Pr-implanted sample, evidence of 'hot' lines was not seen.

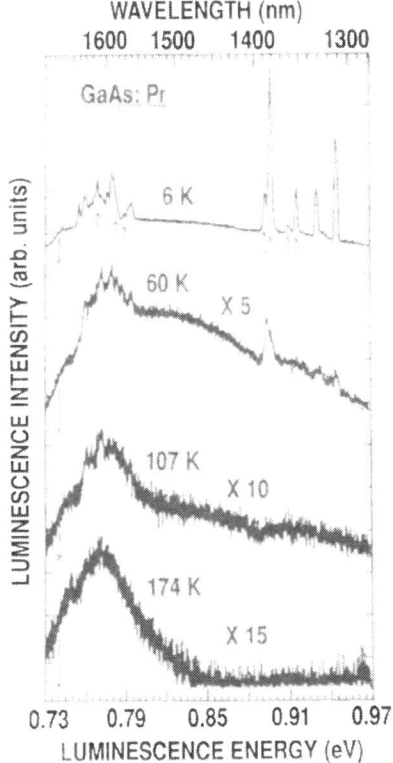

Fig. 5. Intracenter emissions of Pr³⁺ in GaAs as a function of sample temperature.

SUMMARY

In conclusion, photoluminescence investigations were performed on implanted ions of Er, Tm, and Pr in GaAs and Er in AlₓGa₁₋ₓAs and AlAs. The characteristic 1.54 μm Er emissions have been demonstrated for GaAs, AlAs, and $Al_xGa_{1-x}As$ having different mole fractions, with the strongest signal for $Al_{0.4}Ga_{0.6}As$. The observations of Er^{3+} emissions in the numerous semiconductors, emissions at room temperature, and the reported fabrication of Er-doped light-emitting diodes support the idea that Er doped semiconductors can have substantial technological importance. The characteristic sharp 4f intra transitions between the weakly crystal-field-split spin orbit levels $^3H_5$ and $^3H_6$ of Tm^{3+} are identified around 1.22 μm from the 390 keV Tm implant in GaAs, with the strongest emission line observed even at a temperature of 200 K. The three sets of characteristic intra 4f-shell luminescence spectra of Pr^{3+} were observed from the 380 keV Pr-implanted GaAs. These emissions at 1.05, 1.35, and 1.6 μm are assigned to the transitions between the crystal-field split spin-orbit levels $^1G_4-^3H_4$, $^1G_4-^3H_5$, and $^3F_3-^3H_4$ of Pr^{3+}, respectively, with some emission lines seen up to the temperature of 174 K. The successful identification of many rare-earth 4f intra-transition emission lines ranging from 1.02 to 1.64 μm from the Er, Tm, and Pr in GaAs and Er in $Al_xGa_{1-x}As$ allows for potential opto-electronic applications for these rare-earth/semiconductor systems.

REFERENCES

1. V.V. Ushakov, A.A. Gippius, V.A. Dravin, and A.V. Spitsyn, Sov. Phys. Semicond. 16, 723 (1982).
2. H. Ennen, U. Kaufmann, G. Pomrenke, J. Schneider, J. Windscheif, and A. Axmann, J. Cryst. Growth 64, 165 (1983).
3. P.N. Favennec, H. L'Haridon, M. Salvi, D. Moutonnet, and Y. Le Guillou, Electron. Lett. 25, 719 (1989).
4. G.S. Pomrenke, R.L. Hengehold, and Y.K. Yeo, in GaAs and Related Compounds, Karuizawa, Japan, 1989; Inst. Phys. Conf. Ser. No. 106 (Institute of Physics, Bristol, UK, 1990), pp. 339-344.
5. F. Bantien, E. Bauser, and J. Weber, J. Appl. Phys. 61, 2803 (1987).
6. K. Takahei, K. Uwai, and H. Nakagome, J. Lumin. 40 & 41, 901 (1988).
7. R.S. Smith, H.D. Mueller, H. Ennen, P. Wennekers, and M. Maier, Appl. Phys. Lett. 50, 49 (1986).
8. P. Galtier, M.N. Charasse, J. Chazelas, A.M. Huber, C. Grattepain, J. Siejka, and J.P. Hirtz, in GaAs and Related Compounds, Atlanta, USA, 1988; Inst. Phys. Conf. Ser. No. 96 (Institute of Physics, Bristol, UK, 1989), pp. 61-64.
9. P.S. Whitney, K. Uwai, H. Nakagome, K. Takahei, Electron. Lett. 24, 741 (1988).
10. A. Rolland, A. Le Corre, P.N. Favennec, M. Gauneau, B. Lambert, D. LeCrosnier, H. L'Haridon, D. Moutonnet, and C. Rochaix, Electron. Lett. 24, 957 (1988).
11. V.A. Kasatkin, F.P. Kesamanly, and B.E. Samorukov, Sov. Phys. Semicond. 15, 352 (1981); V.A. Kasatkin, ibid., 19, 1174 (1985).
12. A.A. Gippius, V.S. Vavilov, V.V. Ushakov, V.M. Konnov, N.A. Rzakuliev, S.A. Kazarian, A.A. Shirokov, and V.N. Jakimkin, Proc. 14th Int. Conf. Defects in Semicod. ICDS-14, edited by H.G. von Bardeleben, Materials Sciences Forum Vol. 10-12, 1195 (1986).
13. G.S. Pomrenke, H. Ennen, and W. Haydl, J. Appl. Phys. 59, 601 (1986).
14. C. Rochaix, A. Rolland, P.N. Favennec, B. Lambert, A. Le Corre, H. L'Haridon, and M. Salvi, Jpan. J. Appl. Phys. 27, L2348 (1988).

ELECTRON MOBILITY IN N-TYPE EPITAXIAL ZnSe

M. Vaziri* and R. Reifenberger**
* University of Michigan - Flint; Department of Physics and Engineering; Flint MI 48502;
U.S.A.
** Purdue University; Department of Physics; W. Lafayette, IN 47907; U.S.A.

ABSTRACT

An analysis of the temperature dependent mobility in lightly doped ZnSe epitaxial layers grown on a semi-insulating GaAs substrate by Molecular Beam Epitaxy is reported.
Our results indicate that the temperature dependence of the mobility is in poor agreement with calculated values based on typical phonon and ionized impurity scattering
mechanisms. Good agreement between theory and experimental data can be obtained by
including a scattering term associated with the space-charge region surrounding defects.

INTRODUCTION

The electrical and optical characterization of MBE-grown Ga-doped ZnSe was reported
earlier.[1] From an analysis of these transport measurements, the room temperature mobility was found to be substantially lower than the expected value of 600 cm^2/V − s.[2]
Furthermore, the temperature dependence of the mobility was in poor agreement with calculations based on typical phonon and ionized-impurity scattering mechanisms. Similar
observations have been reported by other investigators.[3,4,5] However no suitable model
has been proposed to explain these discrepancies. Van Houten et al.[5] have reported
photo-Hall measurements in ZnSe/GaAs heterostructures and have concluded that space-
charge surrounding defects is a main source of mobility reduction at room temperature.
However, they did not discuss how space-charge scattering would influence the temperature dependence of the mobility in these epilayers. This topic is the subject of this study.

EXPERIMENTAL RESULTS

We report on the temperature dependence of mobility in several ZnSe epitaxial layers
grown on semi-insulating GaAs by Molecular Beam Epitaxy. The growth conditions for
these samples has been reported elsewhere.[1,6] All measurements were performed with
the Van der Pauw configuration. The Hall coefficients were measured at a magnetic field
strength of 0.65 Tesla. Two calibrated thermometers, one adjacent to the sample and the
other mounted inside a copper block (for temperature control), were used to measure the
temperature to better than ±0.1 K.

Fig. (1) shows the carrier concentration as a function of temperature for five selected
samples designated by the letters (a)-(e). To extract the carrier concentration from the
Hall measurements, the thickness of the each sample was corrected for the depletion
width. The Hall factor was assumed equal to unity. It is clear from Figure 1 that sample
(e) - with a room temperature carrier concentration of ∼ 10^{17}cm^{-3} - behaves differently
than the others. The minimum in the carrier concentration near 40K for this sample is
an indication of impurity conduction. This sample will not be discussed further in this
paper.

The solid lines through the experimental data in Fig. 1 for samples (a) through (d) are
the results of least-squares fit to the standard expression for the carrier concentration vs.
temperature for a non-degenerate semiconductor (see Ref. [1]) and gives an estimate for
the activation energy, E_D, the donor concentration, N_D, and the acceptor concentration

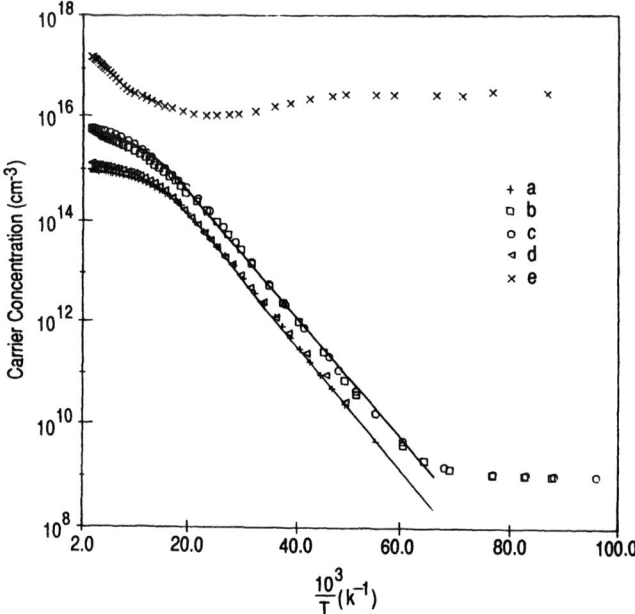

Figure 1: The temperature dependence of the carrier concentration in epitaxial ZnSe. Five samples labeled (a)-(e) were investigated.

N_A. The fitting parameters and the room temperature mobility for each sample are listed in Table I.

Sample	Thickness (μm)	n_{300} (10^{15}cm^{-3})	μ_{300} (cm^2/V.s)	N_D (10^{15}cm^{-3})	N_A (10^{15}cm^{-3})	N_A/N_D
			Summary of Results			
a	4.3	1.0	382	2.5	1.6	.64
b	2.7	5.2	424	6.5	3.3	.51
c	3.8	6.3	381	10.0	5.0	.50
d	1.9	1.3	390	3.0	1.8	.60
e	2.7	150.0	395			

Table I: Fitting parameters obtained from least square fit to data shown in Fig. 1. The activation energy for these samples was $E_D = 21.5 \pm 1$ mev. The sample thickness, the room temperature carrier concentration and mobility, as well as the compensation ratio are also given.

Fig. (2) shows the temperature dependence of the Hall mobility for these epitaxial layers. The room temperature mobility of these samples, irrespective of their doping level and thickness, is around 400 cm^2/V $-$ s. This is considerably lower than the expected theoretical value of 600 cm^2/V $-$ s. Also, the solid line shows the mobility (referred to as $\mu_{lattice}$ below) for sample (a) calculated by combining the polar optical, deformation-potential, piezoelectric and ionized impurity scattering using Matthiessen's rule. It is evident that poor agreement exists between the experimental results and the calculated mobility. In addition, the solid line can be regard as the calculated mobility for sample (d), since n vs. 1/T for both samples are almost identical. This observation indicates that for thinner samples the discrepancy between theory and experiment has increased significantly. This suggests that a new scattering term which is sensitive to epilayer thickness must be included in the theory.

As pointed out in Ref. [5], there is evidence that the space charge region surrounding defects is responsible for mobility reduction in these materials. Briefly, the scattering of conduction electrons by a space charge region depends on the relative size of the electron mean free path, ℓ, in dislocation free material to the radius, r, of the space charge region. For the case when $\ell > r$, the space charge region acts like a scattering center and the coulombic interaction between the conduction electrons and the charge center becomes important. The suitable model to describe this case is the one proposed by Weisberg [7] in which the mobility associated with space charge scattering is given by

$$\mu_{sc} = \frac{e}{N_{defect}\pi r^2(2mK_BT)^{1/2}}. \tag{1}$$

The effect of this space charge contribution to the total mobility can be written as

$$\mu_{total} = \frac{\mu_{sc}\,\mu_{lattice}}{\mu_{sc} + \mu_{lattice}}. \tag{2}$$

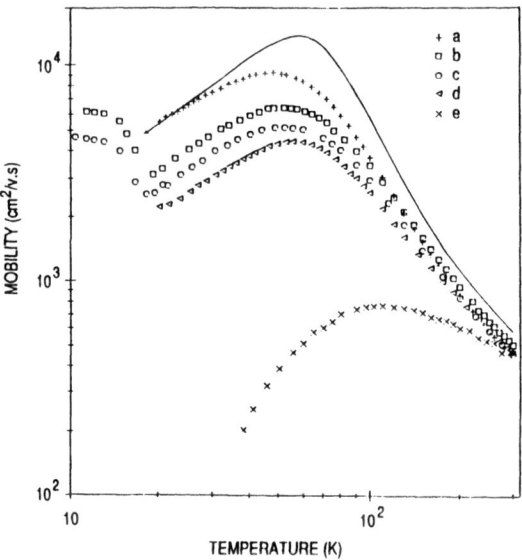

Figure 2: Temperature dependence of the mobility of the five samples shown in Fig. 1. The solid line is the calculated mobility using the model discussed in the text. The origin of the increase in the mobility for samples (b) and (c) at temperatures below 20 K is not presently understood.

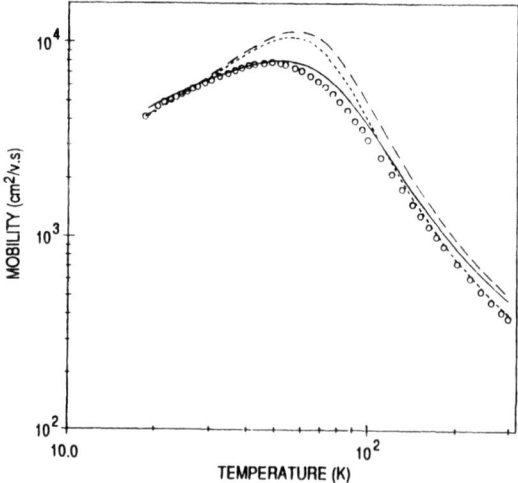

Figure 3: Mobility as a function of temperature for sample (a). The dashed curve is the result expected when phonon and ionized impurity scattering mechanisms are included (i.e. μ_{lattice}). The dotted curve is from Eq. 3 in the text. The solid line is the result of Eq. 2.

For the case $\ell < r$, the space charge region begins to distort the current flow and acts like macroscopic voids inside the sample. Under these conditions, the space charge contribution to the mobility acts more like a reduction factor rather than a scattering mechanism. If ϵ is the fraction of missing volume, the carrier mobility can be written[8]

$$\mu_{total} = g(\epsilon) \times \mu_{lattice} \tag{3}$$

where $g(\epsilon)$ is given in Ref. [9]. For $\epsilon < 0.5$, $g(\epsilon)$ can be approximated by $\frac{1}{1+\epsilon}$.

For a defect free sample, the mean free path of a conduction electron is about 15 nm using an effective mass of $0.17m_o$ and a room temperature mobility of 600 cm^2/V $-$ s. This value is much smaller than the Debye screening length of lightly doped ZnSe which is around 120 nm. As the temperature is lowered, the mean free path can easily exceed a few hundred nanometers. As a result, the Weisberg formula will be applicable at lower temperatures.

Using the above argument, we have fit the temperature dependence of the mobility for the thickest available sample (a). The data from the thickest sample was chosen for this fitting procedure because the effects of the depletion layer will be minimized in determining the donor and acceptor concentrations. In addition, the effects of interface scattering at the GaAs/ZnSe interface will be reduced for a thick sample. The result of such a fitting procedure is given in Fig. (3).

Since $\ell < r$ at room temperature, Eq. (3) was used to fit the high temperature range of the data. In this case ϵ was used as a fitting parameter and a best value of $\epsilon \sim 0.23$ was obtained. It is clear that this model provides a reasonable fit to the data down to ~ 120 K. To fit the low temperature portion of the data where $\ell > r$, Eq. (2) was used. The calculated mobility based on this equation is given by the solid line in Fig. 3 and shows a reasonable fit to the data at temperatures below ~ 100K.

It is important to make a few remarks regarding the procedure used to calculate the mobility based on Eq. (2). First, the value of ϵ obtained from fitting the room temperature data was also used to estimate the fraction of the acceptor centers associated with the space charge region by asserting that $N_{defect} = \epsilon N_A$. Second, the acceptor concentration used in the Brooks-Herring formula was replaced by a new value which was $(1 - \epsilon)N_A$. Third, the space-charge radius, r, was used as a fitting parameter and assumed to be temperature independent. This assumption gives a value of 50 nm for the space-charge radius which is consistent with the assumptions underlying the data analysis.

We have shown that the temperature dependence of the mobility in ZnSe/GaAs heterostructures can be described by invoking space-charge scattering associated with defects in bulk epilayers. Further studies are planned to calculate the mobility by solving the Boltzmann transport equation and including scattering in which the space-charge radius will be temperature dependent.

ACKNOWLEDGEMENTS

This work was partially supported by DARPA/URI Contract Number 218-25015 and AFSOR-89-0438. The authors would like to thank Prof. R. L. Gunshor and Dr. M. Kobayashi for providing samples and many helpful discussions.

REFERENCES

1. M. Vaziri, et al. J. Vac. Sci. Technol. B7, 253(1989).
2. H. E. Ruda, J. Appl. Phys. 59, 1220(1986).
3. K. Morimoto, J. Appl. Phys. 64, 4951(1988).
4. T. Yao, J. Cryst. Growth 72, 31(1985).
5. H. Van Houten, et al. J. Appl. Phys. 66, 3047(1989).
6. R. L. Gunshor, et al. IEEE J. Quantum Electron. 24, 1744(1988).
7. L. R. Weisberg, J. Appl. Phys., 33, 1817(1962).
8. R. A. Logan, et al. J. Appl. Phys. 30, 885(1959).
9. W.T. Read, Tr. Phil. Mag. 46, 111(1955).

TEM STUDIES OF CD:ZN:TE-BASED II-VI SUPERLATTICES AND EPITAXIAL LAYERS

P.D. BROWN, H. KELLY, P.A. CLIFTON, J.T. MULLINS[‡], M.Y. SIMMONS,
K. DUROSE, A.W. BRINKMAN, T.D. GOLDING[*†] AND J. DINAN[*]

Applied Physics Group, School of Engineering and Applied Science, Science Site, South
Road, University of Durham, Durham, DH1 3LE, U.K.
* C2NVEO, Fort Belvoir, Virginia, USA.
‡ Now at Dept. of Engineering, Osaka University, Osaka 565, Japan.
† Now at Space Vacuum Epitaxy Centre, University of Houston, Texas, USA.

ABSTRACT

A TEM study is presented charting the development of a MOVPE growth process for the
deposition of CdTe//ZnTe superlattices. In addition, MBE grown (Cd,Zn)Te//CdTe
superlattices deposited onto GaAs and InSb substrates are compared.

INTRODUCTION

There has been considerable investment in the demonstration and development of a
wide range of II-VI materials systems containing zinc in recent years. For example,
CdTe//ZnTe superlattices are of interest because of their promising optoelectronic properties [1,2], and further for their potential use as buffer layers for the growth of (Hg,Zn)Te
[3] and HgTe//ZnTe superlattices [4,5] onto highly lattice mismatched GaAs substrates.
Similarly, the ternary alloy (Cd,Zn)Te may be used as a buffer layer for the growth of
(Hg,Cd)Te onto GaAs or GaAs/Si substrates [6], or alternatively we feel that graded
(Cd,Zn)Te may be of use within the novel p-i-n ZnTe/(Cd,Zn)Te/CdS solar cell structure. (Cd,Zn)Te//CdTe superlattices tailored to have the same free standing lattice
parameter as $Cd_{0.96}Zn_{0.04}Te$ may also provide a suitable buffer layer for the growth of
$Hg_{0.08}Cd_{0.02}Te$ onto GaAs [7].

The first demonstration of the MBE growth of the CdTe//ZnTe superlattice system by
Monfroy et al [8] was soon repeated by workers using the MOVPE growth technique
[9-11]. To complement the studies presented by Clifton et al [11] and Mullins et al [12],
TEM investigations made during the development of a MOVPE process for the growth
of CdTe//ZnTe superlattices will be reported on here. The feasibility of reducing the
density of threading dislocations by use of II-VI superlattice structures has also recently
been considered [13,14]. To continue a study of the (Cd,Zn)Te//CdTe system presented
in our earlier paper [14], comparison is made of $Cd_{0.92}Zn_{0.08}Te$ //CdTe superlattices
deposited using MBE onto $Cd_{0.96}Zn_{0.04}Te$ buffer layers on {100} oriented GaAs and
InSb substrates.

EXPERIMENTAL

The growth conditions for each of the MOVPE and MBE grown samples considered in
this paper here have been previously described [7,11,12]. The MOVPE growth conditions are summarised in Table 1. One important aspect of the growth of the CdTe//ZnTe
superlattice is that the binaries were found to have widely different growth rates which
made it necessary to establish longer growth times for ZnTe than CdTe within the
MOVPE growth cycle (see Table 1).

TABLE 1. MOVPE GROWTH CONDITIONS

Fig	Sample	Heat Clean	T_{growth}	Dilution	Cd mole/min	Zn mole/min	Te mole/min	No. Pds.	Growth Time	thickness	Growth Rate/Comment
1	ZnTe/{100} GaAs	10min/555°C	350°C	5000sccm	-	1.64×10^{-4}	5.2×10^{-6}	-	1hr	0.08μm	2.8Ås^{-1}
2	ZnTe/{100} GaAs	10min/550°C	300°C	5000sccm	-	1.64×10^{-4}	5.55×10^{-6}	-	1hr	0.10μm	0.28Ås^{-1}
3	CdTe/{100} GaAs	10min/550°C	300°C	5000sccm	1.64×10^{-4}	-	5.55×10^{-6}	-	1hr	1.43μm	4.0Ås^{-1}
4	ZnTe/{$\bar{1}\bar{1}\bar{1}$}BGaAs	10min/550°C	300°C	5000sccm	-	1.64×10^{-4}	5.55×10^{-6}	-	1hr	0.14μm	0.28Ås^{-1}
5	CTZT/{100} GaSb	10min/530°C	395°C	8000sccm	6.29×10^{-5}	6.73×10^{-5}	6.18×10^{-5}	120	2:15†	0.4μm	no Zn
6	CTZT/{100} GaAs	10min/550°C	300°C	5000sccm	1.64×10^{-4}	4.1×10^{-4}	5.05×10^{-6}	3×30	2:16,3:24,4:32†	0.3μm	$\sim Cd_{0.67}Zn_{0.33}Te$
7	CTZT/{100} GaAs	5min/325°C	325°C	7000sccm	1.43×10^{-4}	3.28×10^{-4}	4.96×10^{-6}	100	2:40†	0.65μm	65Å period

All samples were grown using DMCd, DMZn and DIPTe except for sample 5 in which case DETe was used as the Te-alkyl.

Bubbler temperatures: DMZn(-12°C), DMCd(0°C), DIPTe(22°C), DETe(30°C).

† CdTe:ZnTe growth times (in seconds) within the deposition cycle.

The (Cd,Zn)Te//CdTe samples were grown in a Varian 360 MBE machine. 50 period Cd$_{0.92}$Zn$_{0.08}$Te //CdTe superlattices sandwiched between Cd$_{0.96}$Zn$_{0.04}$Te buffer and capping layers were deposited onto (001) oriented GaAs and InSb substrates respectively. Prior to growth, the GaAs substrates were heated to 610°C to desorb the native oxide, following which Zn and CdTe cells were opened to establish a {100} epilayer growth orientation [15]. Growth then proceeded at a temperature of 250°C. For the case of growth onto InSb, the substrates were first maintained at a temperature of 410°C and irradiated in an Sb$_4$ flux to remove the native oxide, and then a 0.1μm InSb buffer layer followed by a 0.5μm Cd$_{0.96}$Zn$_{0.04}$Te buffer layer was grown. The (Cd,Zn)Te//CdTe superlattice structure was then grown at 270°C.

TEM samples were prepared either in cross-section or in plan view using conventional techniques, with iodine reactive ion sputtering being used for the final stage of sample preparation [16].

RESULTS AND DISCUSSION

MOVPE grown CdTe, ZnTe and CdTe//ZnTe

As a prerequisite to the growth of CdTe//ZnTe superlattices, conditions for the deposition of the respective binary compounds were first established. Figure 1 illustrates the high threading dislocation content within epitaxial (001)ZnTe grown on (001) oriented GaAs at 350°C. Use of DIPTe as the Te-precursor allowed growth of ZnTe to be performed as low as 300°C. Figure 2 shows that a 0.1μm thick epilayer had formed after one hour of growth on a (001) GaAs substrate and selected area diffraction patterns demonstrated that the layer was fully relaxed as one would expect for this layer thickness. The array of features at the interface correspond to the interfacial array of misfit dislocations. The average separation of these features was determined to be 46±2Å (using a catalase crystal to calibrate the magnification) and this compares closely with the expected misfit dislocation separation of ≈ 50Å [17] for full relaxation via the formation of pure edge $\frac{1}{2} < 110 >$ dislocations. A micrograph similar to Fig. 1 showing epitaxial (001) CdTe/GaAs grown at 300°C is shown in Fig. 3. The respective growth rates of CdTe and ZnTe at 300°C were determined to be 4.0Ås^{-1} and 0.28Ås^{-1}, which suggests that a growth time ratio of the order of 1:14 would be necessary for the deposition of CdTe:ZnTe superlattices by MOVPE. The growth of ZnTe under identical conditions as Sample 2 at 300°C onto a {$\bar{1}\bar{1}\bar{1}$}B GaAs substrate resulted in the formation of a 0.14μm

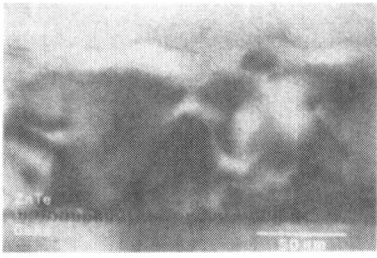

Figure 1. ZnTe/{100} GaAs grown at 350°C (g=220).
Figure 2. ZnTe/{100} GaAs grown at 300°C (g=220).

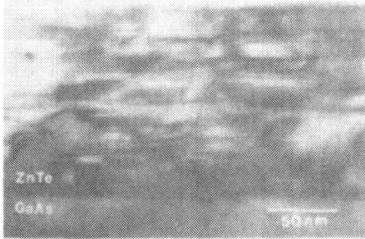

Figure 3. CdTe/{100} GaAs grown at 300°C (g=220).
Figure 4. ZnTe/{1̄1̄1̄}B GaAs grown at 300°C (g=111).

thick epilayer which exhibited a large density of twin lamellae lying parallel to the epilayer/substrate interface (Fig. 4). These features are similar to defects commonly found in epitaxial {111} CdTe [18].

One early attempt to grow a CdTe//ZnTe superlattice at 395°C onto (001)GaSb using DETe as the Te-precursor and growth times of 2 and 15 seconds for CdTe and ZnTe respectively produced a layer which gave no evidence of multilayer contrast as shown in Fig. 5. The lack of Zn (i.e. undetected using EDX) suggested that the layer was just CdTe and that a residual amount of the Cd-alkyl was acting to inhibit the deposition of ZnTe which indicated the need for longer flushing times between each growth cycle. Alternatively, it is possible that there is a delay in the nucleation of ZnTe on CdTe and this in turn would necessitate a longer growth time for ZnTe within the growth cycle. The highly complex interfacial structure may be related to a rough GaSb substrate surface used for growth. GaSb is recognised as being a difficult substrate material to prepare for epitaxy. Regions of {111} oriented twinned CdTe were also observed within this sample.

An attempt to grow three different multilayer structures each of 30 periods at 300°C to investigate the effect of using increasing growth time ratios of 2:16, 3:24 and 4:32 seconds for CdTe:ZnTe led to the production of a highly faulted layer in which three distinct zones could be distinguished (Fig. 6). Although no evidence of a multilayer

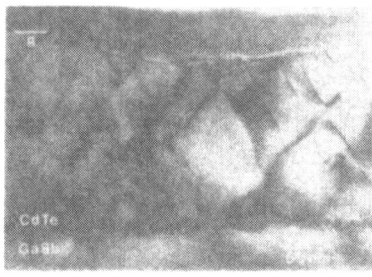

Figure 5. Attempt to grow CdTe//ZnTe/{100} GaSb at 395°C (see text). No zinc was present within the layer. (g=220)

Figure 6. Attempt to grow three CdTe//ZnTe stacks onto /{100} GaAs at 300°C. Zinc was present within the layer but no multilayer structure was detected. (g=400)

Figure 7. CdTe//ZnTe/{100} GaAs grown at 325°C. Superlattice period = 65Å. (g=400)

Figure 8. Diffraction pattern taken from the same sample along a < 110 > zone axis showing superlattice spots.

structure (predicted to have a period between 10Å and 20Å) was detected, EDX demonstrated the presence of Zn, suggesting that a multilayer deposition process coupled with the very low growth rates associated with low temperature growth allows the formation of a ternary alloy. Indeed, diffraction patterns contained a single spot pattern from the epilayer which indicated a volume averaged composition of $\approx Cd_{0.87}Zn_{0.13}Te$ in this instance.

The most encouraging result was obtained at 325°C using CdTe:ZnTe growth times of 2:40 seconds. This led to the production of a regular superlattice structure with a period of 65Å as shown in Fig. 7. This periodicity was confirmed by measurement of the separation of the associated superlattice spots formed in < 110 > diffraction patterns (Fig. 8). The composite epilayer was relaxed with respect to the (001) GaAs substrate and hence exhibited a high density of threading dislocations.

MBE grown (Cd,Zn)Te//CdTe

Figure 9 shows a schematic of the intended $Cd_{0.92}Zn_{0.08}Te$ //CdTe superlattice to be grown on (001) GaAs. [110] and [1$\bar{1}$0] projections of this system presented previously [14] demonstrated a periodicity of 330Å and showed that the epilayer was highly faulted

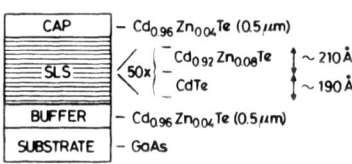
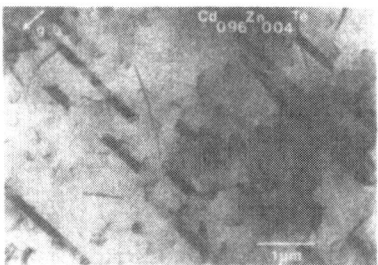

CAP	— Cd$_{0.96}$Zn$_{0.04}$Te (0.5 μm)
SLS	$\langle 50x \rangle \begin{cases} \text{Cd}_{0.92}\text{Zn}_{0.08}\text{Te} & \updownarrow \sim 210\text{Å} \\ \text{CdTe} & \updownarrow \sim 190\text{Å} \end{cases}$
BUFFER	— Cd$_{0.96}$Zn$_{0.04}$Te (0.5 μm)
SUBSTRATE	— GaAs

Figure 9. Proposed CdTe//Cd$_{0.92}$Zn$_{0.08}$Te superlattice structure to be grown by MBE.
Figure 10. Plan view through the Cd$_{0.96}$Zn$_{0.04}$Te capping layer for superlattice grown on GaAs showing anisotropic distribution of microtwins (g=220).

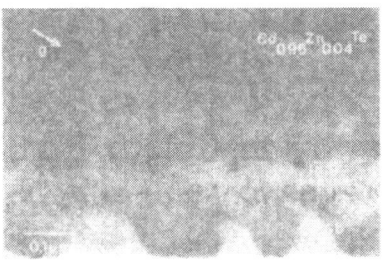
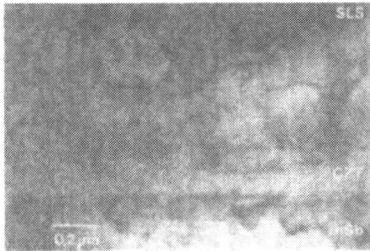

Figure 11. Buffer layer region of a CdTe/Cd$_{0.92}$Zn$_{0.08}$Te superlattice structure grown onto InSb showing precipitates at the InSb/InSb and Cd$_{0.96}$Zn$_{0.04}$Te /InSb interfaces (g=111).
Figure 12. Lower magnification shot of the same sample showing the bending of dislocations into the plane of the superlattice (g=220).

and characterised by a strong anisotropic distribution of microtwins. This anisotropy is more clearly illustrated by Figure 10 which shows an [001] plan view projection through the Cd$_{0.96}$Zn$_{0.04}$Te capping layer. The microtwins intersecting the surface give rise to fringe contrast similar to overlapping stacking faults [19] and this may be related to a reduction in the thickness of the microtwins as they approach the sample surface. These defects were oriented predominantly along the [1$\bar{1}$0] direction. It is noted that this anisotropic defect distribution may best be explained in terms of stacking errors on {111} planes of growth facets as proposed by Pirouz et al [20], rather than by a stress induced deformation process as previously suggested [14]. The surface defect density within this layer was estimated to be 5×10^8cm^{-2}.

Considerable improvement in the epilayer defect microstructure was obtained by growth of an identical structure onto closely lattice matched InSb substrates as shown in Figures 11 and 12. X-ray characterisation confirmed the improved epilayer defect microstructure of layers grown on InSb substrates [7]. It is apparent that a band of precipitates had formed at the InSb/InSb interface, along with similar but slightly larger features at the Cd$_{0.96}$Zn$_{0.04}$Te /InSb interface. Such features have been shown to be In rich [21]. More significantly, however, is a strong dislocation filtering effect within the superlattice structure itself, with a much higher incidence of dislocations bending over into the plane of the superlattice as compared with layers grown on GaAs. It is interesting to note that the majority of the threading dislocations appear to nucleate within the buffer layer, rather than at the buffer layer/InSb interface which would suggest that they are formed during growth following interruptions to the growth surface, rather than being strain induced defects.

ACKNOWLEDGEMENTS

The author would like to acknowledge SERC for financial support under grant No. F05855.

REFERENCES

1. R.H. Miles, G.Y. Wu, M.B. Johnson, T.C. McGill, J.P. Faurie and S. Sivananthan, Appl. Phys. Lett. 48 1383 (1986)

2. H. Mathieu, J. Allegre, A. Chatt, P. Lefebvre and J.P. Faurie, Phys. Rev. B38 7740 (1988)

3. S. Sivananthan, X. Chu, M. Boukerche and J.P. Faurie, Appl. Phys. Lett. 47 1291 (1985)

4. P.A. Clifton, J.T. Mullins, P.D. Brown, N. Lovergine, A.W. Brinkman and J. Woods, J. Crystal Growth 99 468 (1990)

5. J.T. Mullins, P.A. Clifton, P.D. Brown, D.O. Hall and A.W. Brinkman, Mat. Res. Soc. Proc. 161 375 (1990)

6. H.-J. Kleebe, W.J. Hamilton, W.L. Ahlgren, S.M. Johnson and M. Rühle, Mat. Res. Soc. Proc. 161 63 (1990)

7. T.D. Golding, S.B. Qadri and J.H. Dinan, J. Vacuum Sci. Technol. A7 616 (1989)

8. G. Monfroy, S. Sivananthan, X. Chu, J.P. Faurie, R.D. Knox and J.L. Staudermann, Appl. Phys. Lett. 49 152 (1986)

9. D.W. Kisker, P.H. Fuoss, J.J. Krajewski, P.M. Amirtharaj, S. Nakahara and J. Menendez, J. Crystal Growth 86 210 (1988)

10. H. Shtrikman, A. Raizman, M. Oron and D. Eger, Mater. Lett. 5 345 (1987)

11. P.A. Clifton, J.T. Mullins, P.D. Brown, G.J. Russell, A.W. Brinkman and J. Woods, J. Crystal Growth 93 726 (1988)

12. J.T. Mullins, P.A. Clifton, P.D. Brown, A.W. Brinkman and J. Woods, J. Crystal Growth 101 100 (1990)

13. J. Petruzzello, D. Olego, X. Chu and J.P. Faurie, J. Appl. Phys. 66 2980 (1989)

14. P.D. Brown, T.D. Golding, G.J. Russell, J.H. Dinan and J. Woods, Inst. Phys. Conf. Ser. No. 100 357 (1989)

15. L.A. Kolodziejski, R.L. Gunshor, N. Otsuka, X.C. Zhang, S.K. Chang and A.V. Nurmikko, Appl. Phys. Lett. 47 882 (1985)

16. N.G. Chew and A.G. Cullis, Ultramicroscopy 23 175 (1987)

17. G. Feuillet, L. Di Cioccio, A. Million, J. Cibert and S. Tararenko, Inst. Phys. Conf. Ser. No. 87 135 (1987)

18. P.D. Brown, J.E. Hails, G.J. Russell and J. Woods, Appl. Phys. Lett. 50 1144 (1987)

19. *ElectronMicroscopyofThinCrystals*, ed. P.B. Hirsch, A. Howie, R.B. Nicholson, D.W. Pashley and M.J. Whelan, pub. Butterworths, 1965, p242.

20. P. Pirouz, F. Ernst and T.T. Cheng, Mat. Res. Soc. Proc. 116 57 (1988)

21. S. Wood, J. Greggi Jr., R.F.C. Farrow, W.J. Takei, F.A. Shirland and A.J. Noreika, J. Appl. Phys. 55 4225 (1984)

II–VI SEMICONDUCTOR HETEROSTRUCTURES: POINTERS TO APPROPRIATE ANALYSIS TECHNIQUES

ERICA G. BITHELL
University of Cambridge, Department of Materials Science and Metallurgy, Pembroke Street, Cambridge, CB2 3QZ, UK.

ABSTRACT

Approximate methods are used, for a variety of II–VI semiconductor alloys, to estimate the sensitivity to composition change of quantitative transmission electron microscopy techniques which have proved successful in characterising III–V heterostructures. It is shown that bright field thickness fringe matching at the [001] axis is likely to prove relatively more successful than 200 dark field intensity measurement for many alloy systems. It is also noted that alternative methods would be necessary if quantitative characterisation of (Mn,Zn) compounds were required.

TRANSMISSION ELECTRON MICROSCOPY OF II–VI HETEROSTRUCTURES

The relatively recent application to II–VI materials of growth technologies (e.g. molecular beam epitaxy) capable of producing very high quality epilayers has provided a route for the manufacture of heterostructures in a wide variety of alloys. It is to be expected that the same characterisation requirements will arise (and indeed have already arisen) as in the case of III–V semiconductors. That is, it must be possible to measure alloy compositions with high accuracy (generally with x better than ± 0.05 for an alloy $A_x B_{1-x} C$) and at a spatial resolution comparable with the atomic layer spacings. In this respect transmission electron microscopy (TEM) has proved outstandingly successful for the III–V systems and merits exploration as a quantitative characterisation technique for the II–VI materials. It is already widely used for qualitative characterisation and admittedly it is clear that the problems are more serious than for the III–Vs, both because of the increased ion beam damage which occurs during TEM sample preparation of II–VIs and also because of the higher susceptibility to beam damage in the electron microscope. To some extent both of these problems can be overcome, the first by the use of chemical or chemically assisted [1] polishing methods and the second by the use of lower microscope accelerating voltages. An assessment of the sensitivity to composition change of a range of TEM techniques which have proved successful for the III–Vs is therefore worthwhile, in order to discover which are likely to be successful for the II–VI alloys and to identify those alloy systems for which further technique development will be required. This kind of general overview has already been performed for the III–Vs, and the reader is referred to [2] for detail of the methods used, which are summarised only briefly here.

Three specific characterisation techniques are considered, these being composition measurement using (i) kinematic line shifts in convergent beam electron diffraction (CBED) patterns [3], (ii) 200 dark field intensity measurement [4], and (iii) [001] bright field thickness fringe matching [5]. These methods are representative of the three parameters most easily measured by TEM: changes in the lattice parameter, in a single, sensitive Fourier component of the crystal potential and in the interference between several Fourier components of the potential. The aim here is to use approximate methods to estimate the composition sensitivity of these approaches, rather than to perform a detailed analysis.

Convergent beam electron diffraction

The angular radius of the first Laue zone at the [001] axis of a sphalerite–type semiconductor is approximately

$$\theta \approx \sqrt{\frac{\lambda}{2a}} \, .$$

(1)

If Vegard's Law is assumed to apply to the lattice parameter of an alloy $A_x B_{1-x} C$ then

Mat. Res. Soc. Symp. Proc. Vol. 216. ©1991 Materials Research Society

$$\left|\frac{d\theta}{dx}\right| \approx (a_{AC} - a_{BC}) \sqrt{\frac{\lambda}{8[a_{AC}x + a_{BC}(1-x)]^3}} \ .$$

(2)

The position of a deficit line in the zero order disc of a pattern can be measured to approximately $\pm 0.05 nm^{-1}$ (i.e. half a typical linewidth). Thus x can be measured to $\pm \Delta x$, where

$$\Delta x \approx \frac{0.05}{a_{AC} - a_{BC}} \sqrt{8\lambda[a_{AC}x + a_{BC}(1-x)]^3} \ ,$$

(3)

with λ being measured in nanometres.

200 dark field intensity measurement

For a sphalerite–type material of composition $A_xB_{1-x}C$ and atomic scattering factors f_A, f_B and f_C the kinematic approximation for the intensity of the 200 reflection is

$$I \propto [xf_A + (1-x)f_B - f_C]^2 \ .$$

(4)

In practice an intensity ratio to some standard is measured. This both avoids the difficulties of making absolute measurements of intensity and also allows limited cancellation of the effects of thickness and inelastic scattering [4]. The remaining uncertainties (due chiefly to inelastic scattering) are equivalent to being able to measure an intensity ratio with an accuracy of about $\pm 10\%$ which, if anything, overestimates the likely errors. The sensitivity to composition change can thus (see [2]) be estimated as

$$\Delta x \approx \frac{0.1[xf_A + (1-x)f_B - f_C]}{2(f_A - f_B)} \ .$$

(5)

[001] bright field thickness fringe matching

This technique for composition measurement relies upon the interference of the {200}, {400} and {220} Fourier components of the crystal potential, as seen in bright field images obtained at the [001] zone axis [5]. Unfortunately, many–beam Bloch wave calculations must be performed in order to match experimental thickness fringes and thus to measure alloy compositions: whilst this is reasonable as a characterisation procedure, it is not feasible as an approach to a general assessment of compositional sensitivity for a large number of alloy systems. However the high symmetry of the diffraction condition permits the problem to be reduced to an analytic solution for the {200}/{400}/{220} combination, with higher beams introduced as perturbations [6]. Although this approach is inadequate for matching experimental images, it does produce a very rapid means of assessing general composition sensitivity. Briefly, the criterion adopted is that two thickness fringe profiles differ significantly if the root mean square difference in their intensity levels up to 100nm thickness exceeds 5% of the incident intensity. The sensitivity to composition changes is then given as $x \pm \Delta x$ where Δx is *half* the change required to produce a distinguishably different profile.

Application to II–VI ternary alloys

Figures 1, 2 and 3 show diagrammatically the estimated sensitivity of composition measurement by convergent beam diffraction, 200 dark field intensity measurement and [001] bright field thickness fringe matching respectively, using equations (3) and (5) and the methods described in [6]. Lattice parameter versus bandgap diagrams are a concise way of displaying the composition sensitivity information and these have been constructed assuming Vegard's Law and linear variation of bandgaps with composition: the II–VI ternary alloys have not been explored as widely as the III–Vs, and whilst data are available for some systems relating the lattice parameter and bandgap to the composition, this is by no means universally true. Ternary alloy

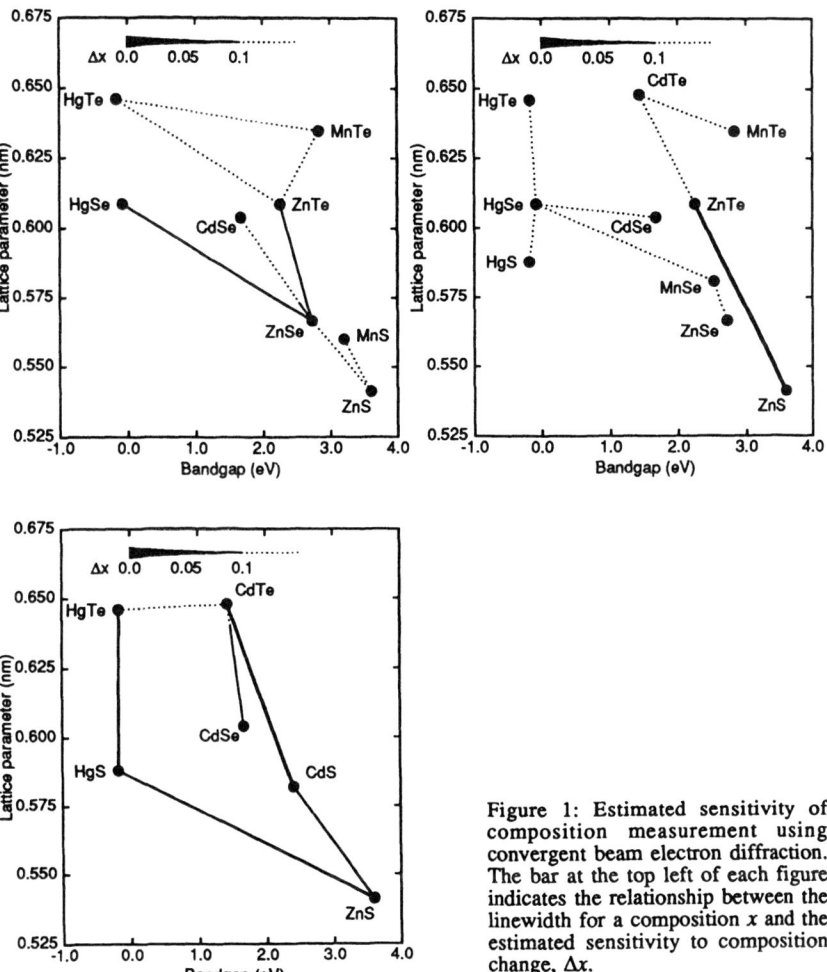

Figure 1: Estimated sensitivity of composition measurement using convergent beam electron diffraction. The bar at the top left of each figure indicates the relationship between the linewidth for a composition x and the estimated sensitivity to composition change, Δx.

combinations of the group IIb elements Zn, Cd and Hg with the group VIa elements S, Se and Te have been considered: some Mn compounds have also been included. The values of lattice parameter and bandgap indicated for the binary compounds are those which pertain to the sphalerite structures at room temperature (although these are not always the stable structures) and were obtained from a similar figure in [7]. The ternary combinations shown are those for which at least one end–member forms the sphalerite structure at room temperature (i.e. CdTe, ZnS, ZnSe, ZnTe, HgSe, HgTe [8]). For these systems at least limited solid solution would be expected and in many cases has been achieved.

All the calculations have been performed for incident electron energies of 100keV: although in some instances (e.g. CBED) higher sensitivity can be achieved at higher voltages, it appears that this would be largely impractical due to the tendency of these materials to suffer beam damage. It is of course possible that for this reason some of the alloys included here will prove entirely unsuitable for TEM examination. No attempt has been made to include any effects arising from coherent strain between an epilayer and its substrate: whilst this severely complicates the approach to quantification it appears unlikely that strain effects will seriously

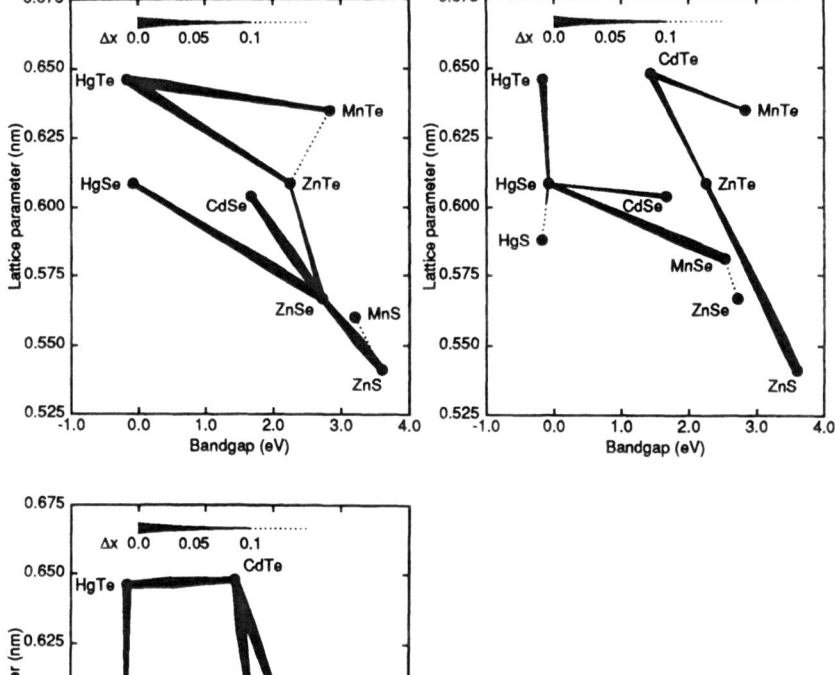

Figure 2: Estimated sensitivity of composition measurement using 200 dark field intensities. Certain alloy systems (e.g. $Cd_xHg_{1-x}Te$) contain points at which Δx is apparently zero: this arises because of the form of equation (5) when the scattering factor for the 200 reflection is exactly zero. It is unlikely that such arbitrarily high accuracy could be achieved in practice.

modify the overall sensitivity to composition.

DISCUSSION AND CONCLUSIONS

The relative effectiveness of the various techniques examined is closely connected with the magnitude of the 200 Fourier component of the potential. If the two atomic species which mutually substitute bracket the third in atomic number, then at some composition the average atomic scattering factors of the atoms on the group II and group VI sites are equal, and the 200 component of the potential is exactly zero. Close to this composition, the *fractional* change in intensity of a 200 dark field image for a given change in composition is at its highest (equation (5)). Thus composition measurement using 200 dark field intensities is most successful when the absolute intensities are low (as is the case for the III–V semiconductor $Al_xGa_{1-x}As$). This is precisely the condition for which [001] bright field thickness fringe matching is insensitive. The bright field image is normally formed by the combination of three 'extinction distances',

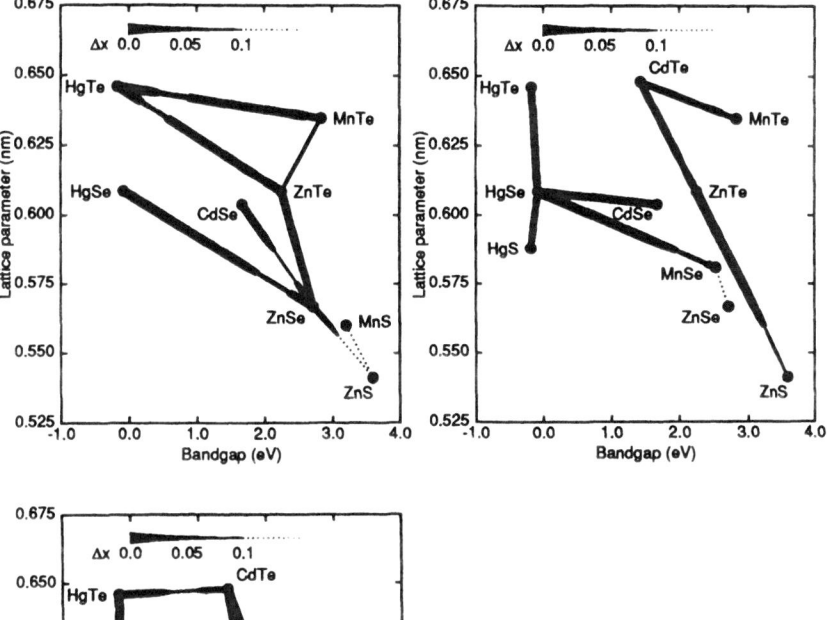

Figure 3: Estimated accuracy of composition measurement using [001] bright field thickness fringe matching. A number of alloy systems (e.g. $Zn_xCd_{1-x}Se$) contain a well-defined region where the sensitivity to composition change is rather low. This is not an artefact of the calculation method but a real effect which arises when two of the top three branches of the [001] dispersion surface cross: it is closely associated with the scattering factor for 200 passing through zero (see figure 2).

each of which originates in the interference of two out of the top three Bloch states. However when the atomic scattering factors are equal for the group II and group VI species, the top two states are degenerate, because each is localised on only one of the two sites. The image then consists of a single frequency of intensity oscillation with thickness and is relatively insensitive to composition change. Thus it may be concluded that the two imaging techniques are genuinely complementary, and this is reflected in the summary given in Table I of the relative success of the different methods. Whilst composition measurement by CBED is not a particularly sensitive method, it does have the merit that a given pattern is uniquely associated with a given lattice parameter (and hence composition). It can therefore be used to distinguish the 'double-valued' cases which arise on either side of a zero in the 200 component of the potential: in dark field, identical image intensities occur at two differing compositions.

Overall it can be said that bright field thickness fringe matching is likely to be the most successful technique for the II–VI alloys rather more frequently than it is for the III–V systems (c.f. [2]). Zeros in the value of the 200 component of the potential as a function of composition are relatively less common in the II–VI alloys than is the case for the III–V systems because a

TABLE I

Summary of sensitivity to composition change of convergent beam electron diffraction, 200 dark field intensities and [001] bright field thickness fringes. $\sqrt{\sqrt{}}$: composition sensitivity of ±0.05 or better over the entire alloy system. $\sqrt{}$: ±0.05 or better over part of the alloy system. ×: ±0.05 to ±0.1 over the entire alloy system. ××: ±0.1 or worse over some or all of the alloy system.

Alloy	CBED	200DF	[001]BF	Alloy	CBED	200DF	[001]BF
$Mn_xZn_{1-x}S$	××	×	×	$Zn_xCd_{1-x}Se$	××	$\sqrt{\sqrt{}}$	$\sqrt{}$
$Mn_xZn_{1-x}Se$	××	×	×	$Zn_xHg_{1-x}Se$	×	$\sqrt{\sqrt{}}$	$\sqrt{\sqrt{}}$
$Mn_xHg_{1-x}Se$	××	$\sqrt{\sqrt{}}$	$\sqrt{\sqrt{}}$	$Zn_xCd_{1-x}Te$	××	$\sqrt{}$	$\sqrt{\sqrt{}}$
$Mn_xZn_{1-x}Te$	××	×	×	$Zn_xHg_{1-x}Te$	××	$\sqrt{\sqrt{}}$	$\sqrt{\sqrt{}}$
$Mn_xCd_{1-x}Te$	××	$\sqrt{}$	$\sqrt{\sqrt{}}$	CdS_yTe_{1-y}	×	$\sqrt{\sqrt{}}$	$\sqrt{\sqrt{}}$
$Mn_xHg_{1-x}Te$	××	$\sqrt{\sqrt{}}$	$\sqrt{\sqrt{}}$	$CdSe_yTe_{1-y}$	××	$\sqrt{\sqrt{}}$	$\sqrt{}$
ZnS_ySe_{1-y}	××	$\sqrt{\sqrt{}}$	$\sqrt{}$	$Cd_xHg_{1-x}Se$	××	$\sqrt{}$	$\sqrt{\sqrt{}}$
ZnS_yTe_{1-y}	×	$\sqrt{\sqrt{}}$	$\sqrt{}$	$Cd_xHg_{1-x}Te$	××	$\sqrt{\sqrt{}}$	$\sqrt{}$
$Zn_xCd_{1-x}S$	×	$\sqrt{}$	$\sqrt{\sqrt{}}$	HgS_ySe_{1-y}	××	$\sqrt{}$	$\sqrt{\sqrt{}}$
$Zn_xHg_{1-x}S$	×	$\sqrt{}$	$\sqrt{\sqrt{}}$	HgS_yTe_{1-y}	×	$\sqrt{}$	$\sqrt{\sqrt{}}$
$ZnSe_yTe_{1-y}$	×	$\sqrt{}$	$\sqrt{\sqrt{}}$	$HgSe_yTe_{1-y}$	××	$\sqrt{}$	$\sqrt{\sqrt{}}$

smaller proportion of the alloys consist of two elements of one group bracketting a single element of the other group (as in $Cd_xHg_{1-x}Te$): composition measurement using the 200 dark field intensity is thus relatively less effective. It is worth noting that a further advantage of thickness fringe matching for the II–VI alloys is that the specimen preparation method used – cleaving a wedge – minimises mechanical and ion beam damage to the material.

Finally, attention should be drawn to the (Mn,Zn) compounds, for which no successful technique has been identified. Being rather close in atomic number, the substitution of Mn for Zn does not produce very significant changes in the crystal potential. It seems unlikely therefore that any method relying on TEM imaging and/or diffraction will be effective, and other methods (e.g. spectroscopy) would have to be sought.

ACKNOWLEDGEMENTS

The author is grateful to the Royal Society for financial support and to Professor D. Hull for the provision of laboratory facilities.

REFERENCES

1. N.G. Chew and A.G. Cullis, Appl. Phys. Lett. 44, 142 (1984).
2. E.G. Bithell and W.M. Stobbs, J. Appl. Phys. in press (1991).
3. P.M. Jones, G.M. Rackham and J.W. Steeds, Proc. Roy. Soc. Lond. A354, 197 (1977).
4. E.G. Bithell and W.M. Stobbs, Phil. Mag. A60, 39 (1989).
5. H. Kakibayashi and F. Nagata, Jap. J. Appl. Phys. 25, 1644 (1986).
6. E.G. Bithell and W.M. Stobbs, Phil. Mag. in press (1990).
7. J.F. Schetzina, N.C. Giles, S. Hwang and R.L. Harper in Growth and Optical Properties of Wide–Gap II–VI Low–Dimensional Semiconductors, edited by T.C. Mcgill, C.M. Sotomayor Torres and W. Gebhardt (NATO ASI Series: Plenum Press, New York, 1988), p. 129.
8. M. Hansen, Constitution of Binary Alloys (McGraw–Hill, New York, 1958); also First Supplement (R.P. Elliot, 1965) and Second Supplement (F.A. Shunk, 1969).

ORIENTATION DEPENDENCE OF THE STABILITY OF STRAINED Sn EPILAYERS GROWN ON $Cd_{0.8}Zn_{0.2}Te$ SUBSTRATES

DAVID W. NILES AND HARTMUT HÖCHST
Synchrotron Radiation Center, University of Wisconsin-Madison, 3731 Schneider Drive, Stoughton, WI 53589

ABSTRACT

The strain energy density in epitaxial overlayers depends strongly on the substrate's surface orientation. The affect of a growth orientation dependent strain energy contribution to the phase stability was studied by growing overlayers of metastable, semiconducting α-Sn films on the (100) and (111) surfaces of $Cd_{0.8}Zn_{0.2}Te$ substrates. On the (100) surface, the α-phase of Sn remains stable for coverages up to and beyond 125 ML. On the (111) surface though, the Sn overlayer shows metallic behavior from a partial α→β phase separation for coverages lower than 100 ML. Static calculations using the stress-strain relationship bear out these experimental results by showing that the strain energy density for the (111) growth direction is 38% larger than for the (100) direction.

INTRODUCTION

Pseudomorphic Growth of Strained α-Sn Films

The modification of the electronic properties in strained, epitaxial interlayers is an important additional tool for fine-tuning the performances of novel, specially tailored semiconductor heterostructure devices [1]. A large variety of structures may be grown as thin, strained heteroepitaxial films with molecular beam epitaxy (MBE) by simply offering substrates which are slightly lattice mismatched to the bulk lattice constant of the overlayer material. The strain obtained by sandwiching thin epilayers between materials whose lattice constant is a few percent smaller or larger is conceivably higher than what is achievable by applying mechanical pressure to a bulk sample, and will profoundly influence the electronic band structure [2]. However, the incorporation of lattice strain is limited to a critical thickness beyond which misfit dislocation formation starts to relieve the accumulated epilayer strain energy.

In previous papers, we showed that metastable semiconducting α-Sn films of high electronic quality can be grown up to $\sim 10^4$ Å by MBE on lattice matched substrates such as CdTe and InSb [3,4]. The present paper reports on the growth conditions and stability of strained α-Sn films grown on $Cd_{0.8}Zn_{0.2}Te$ substrates. This ternary alloy substrate offers a lattice constant which is 1.25% smaller than that of α-Sn and forces the overlayer to grow under compressive strain. One should keep in mind that for bulk Sn the stable phase is the metallic β-phase, a body centered tetragonal crystal structure with D_{4h} point group symmetry, while the semiconducting α-Sn phase, which has the diamond crystal structure, is stable only below 13.2 C. Even though thin epitaxial films of α-Sn can be grown at room temperature and are substrate stabilized with an α→β phase transition temperature well above 100°C, small amounts of strain may destabilize the metastable α-phase and trigger a transition into the thermodynamic stable β-phase [5].

We studied the growth of Sn on $Cd_{0.8}Zn_{0.2}Te$ substrates in both the (100) and (111) directions in order to explore the effect of the built-in strain on the stability of the metastable α-Sn phase. Our goal was to learn whether the α-phase would be stable in the presence of strain, and whether different surface orientations influenced the phase stability. The premise relies on the fact that the strain energy density is an orientation dependent quantity. For pseudomorphic growth, the in-plane lattice constant of the overlayer must match the lattice constant of the substrate. The only free strain parameter in the overlayer is the amount of relaxation perpendicular to the surface, which one may calculate by using the stress-strain relationship with the stress perpendicular to the surface set to zero. [6]

Strain Energy

Knowing the lattice strain, one may calculate the strain energy density in the overlayer using the equation

$$E = \frac{1}{2} \sum_{ij} C_{ij} \, \varepsilon_i \, \varepsilon_j,$$

where C_{ij} represents the elastic coefficients and ε_i, the strain components. Taking the fractional lattice mismatch between the overlayer and the substrate to be $\delta = (a-a_0)/a_0$, one can show that the strain energy density for the (100) orientation is

$$E(100) = \left(1 + \frac{2\,C_{12}}{C_{11}} \right)\left(1 - \frac{C_{12}}{C_{11}} \right) C_{11}\, \delta^2,$$

while for the (111) growth direction the strain energy density is given by

$$E(111) = \frac{6\,C_{44}\,(\,C_{11} + 2\,C_{12}\,)}{(\,C_{11} + 2\,C_{12} + 4\,C_{44}\,)}\, \delta^2.$$

Using the elastic and lattice constants from Table I, we find that the strain energy density for the (100) orientation is 0.459 δ^2 (eV/Å^3), but it increases by 38 % to 0.635 δ^2 (eV/Å^3) for the (111) orientation.

Table I

$C_{11} = 6.90 \times 10^{11}$ dyne/cm^2	a(CdTe) = 6.49 Å	E(100) = 0.459 δ^2 (eV/Å^3)
$C_{12} = 2.93 \times 10^{11}$ dyne/cm^2	a(ZnTe) = 6.10 Å	E(110) = 0.600 δ^2 (eV/Å^3)
$C_{44} = 3.62 \times 10^{11}$ dyne/cm^2	a(α–Sn) = 6.49 Å	E(111) = 0.635 δ^2 (eV/Å^3)
	a(Cd$_{0.8}$Zn$_{0.2}$Te) = 6.41 Å	

EXPERIMENTAL RESULTS
Sn deposition on Cd$_{0.8}$Zn$_{0.2}$Te(100)

Figures 1 and 2 show photoemission spectra of the valence band and the Sn4d core levels for the growth of Sn on Cd$_{0.8}$Zn$_{0.2}$Te(100). For this orientation,the photoemission spectra and the RHEED pattern indicate that Sn grows in the α–phase, although with the tetragonal distortion associated with the 1.25% compressive strain in the (100) plane. From the separation of RHEED reflexes, we determine the in-plane lattice constant. A quantitative RHEED analysis shows no relaxation of the lattice constant of Sn overlayers of 125ML thickness, indicating that the in-plane lattice constant for the Sn film is equal to the lattice constant of the Cd$_{0.8}$Zn$_{0.2}$Te(100) substrate. Therefore, the α–Sn film compresses in the growth plane by 1.25%, and according to the stress-strain relationship must expand by 2.02% in the perpendicular direction.

The absence of a Fermi level emission in the valence band spectra of Figure 1 indicates that the films consist only of the α-phase and do not contain precipitates of the metallic β-phase. Note that the Fermi level is the reference energy for these spectra. The sharp peak labeled A in the 0 ML spectrum originates from a direct transition from the split-off valence band into the G = (+2, 0, 0) 2 π / a free electron like final state band. The peak disperses with photon energy and with emission angle. We shall report on the band mapping analysis of the Cd$_{0.8}$Zn$_{0.2}$Te(100) crystalline alloys in another paper. Peak B is a density of states feature resulting from the flat portions of the split-off band in the vicinity of the X$_5$ point in the Brillouin zone.

The 63 ML and 125 ML Sn spectra are nearly identical. The sharp feature labeled C is again a direct transition from the split-off band into the G = (+2, 0, 0) 2 π / a free electron like final state band. The fact that we observe this dispersive transition is exceptionally strong evidence for a high degree of crystallinity, since even a small degree of disorder greatly reduces

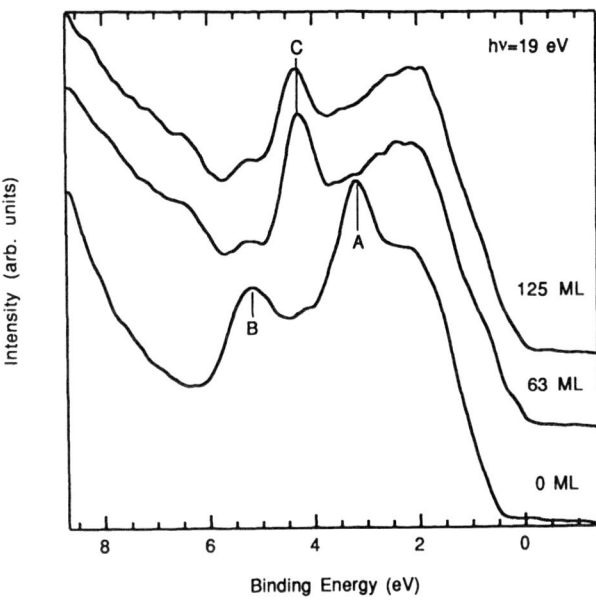

Fig. 1: Normal emission valence band spectra of Sn overlayers grown on $Cd_{0.8}Zn_{0.2}Te(100)$

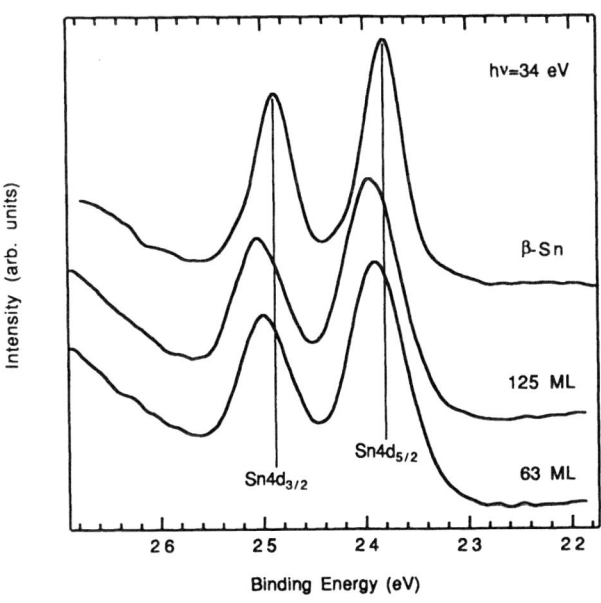

Fig. 2: Sn 4d core spectra of Sn overlayers grown on $Cd_{0.8}Zn_{0.2}Te(100)$

direct transitions and may even eliminate them altogether.

The Sn4d core emission also shows that the Sn film grown on the (100) surface is in the α-phase. The top spectrum of Fig. 2 is from β-Sn which forms on the sample clips during deposition of the Sn films. The two spectra at the bottom are from the sample with Sn coverages of 63 and 125 ML. The Sn4d core emission of the Sn epilayer is 0.12 eV higher in binding energy and 0.2 eV (50%) broader than the spectra from the metallic β-Sn measured at the substrate clips. The observed increase in binding energy and width are typical fingerprints of α-Sn.

Sn deposition on $Cd_{0.8}Zn_{0.2}Te(111)$

Sn growth on the (111) oriented substrate is in contradistinction with the α-Sn formation on the (100) substrate. The valence band spectra of Figure 3 indicate the formation of β-Sn on this surface, although the initial 20 to 30 MLs have a significant amount of α-Sn. Note that at a coverage of 37 ML Sn, the Fermi level is well pronounced, and becomes even stronger for the 87 ML and 150 ML thick films. The presence of Fermi level emission evince the metallic behavior of the Sn film. We also do not see any strong direct transitions for any of the film thicknesses, indicating a film of poorer crystalline quality compared to the (100) orientation.

The Sn4d core emission corroborates the presence of the β–phase at the 150 ML coverage. The Sn4d core emission from the 37 ML and 87 ML coverages matches the emission seen from α-Sn, in apparent contradiction with the metallic appearance of the valence band spectra. However, the Fermi level emission is a much more sensitive measurement for the presence of β-Sn than the Sn4d core level emission. We can resolve this apparent contradiction by assuming that the initial layers of Sn are in the α-phase, and that a mixture of α- and β-Sn starts to form at a coverage between the 16 ML and 37 ML spectra. At higher coverages, the phase separation is also very obvious in the RHEED pattern. Fig. 5 shows the intensity profile of the (00) and (01) RHEED reflex along the [11-2] azimuth. The build up of an additional RHEED

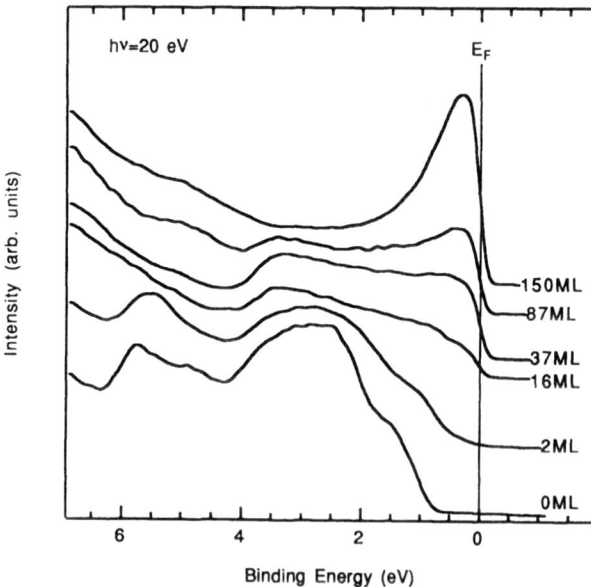

Fig. 3: Normal emission valence band spectra of Sn overlayers grown on $Cd_{0.8}Zn_{0.2}Te(111)$

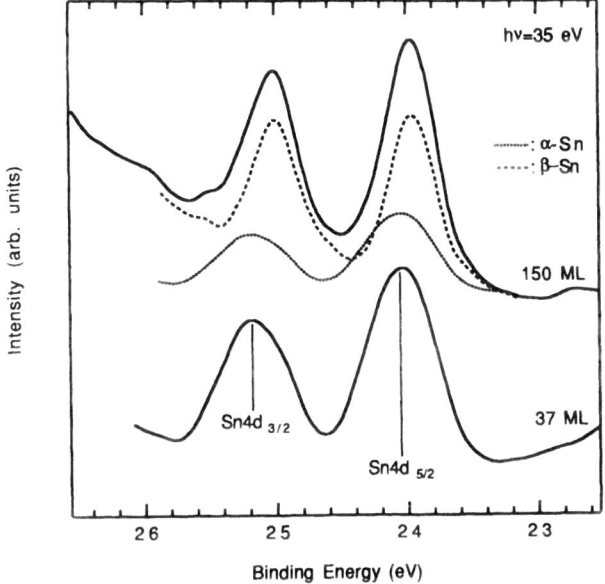

Fig. 4: Sn 4d core spectra of Sn overlayers grown on $Cd_{0.8}Zn_{0.2}Te(100)$

reflex with increasing Sn coverage indicates the formation of ordered β-Sn crystallites which are intermixed or floating on top of the initial α-Sn film.

CONCLUSIONS

Our growth study shows that strain is a contributing factor to the instability of metastable Sn epilayers. Since the strain energy density is an orientation dependent quantity, lattice mismatched pseudomorphic growth is different for different surfaces. However, the observed differences in the surface orientation dependence in growth of a-Sn films can not solely be related to orientation dependent strain distribution in the strained overlayer. For the (111) growth direction, we observe the inception of β-Sn formation at a coverage of ~37 ML (100 Å along [111]) . If the build up of strain energy were the only cause for the phase separation in the Sn overlayer, then one would expect the same breakdown to occur for the (100) growth at a thickness of 138Å = (0.635/0.459)x100 Å, at which the equivalent amount of strain energy accumulates in a (100) oriented epilayer. The fact that the (100) oriented Sn film remains in the α–phase up to a coverage of at least 125ML (~200Å), which is double the estimated critical thickness, indicates that additional parameters such as the interfacial bond orientation and growth kinetics must also be important parameters in the growth process of metastable epilayers.

444

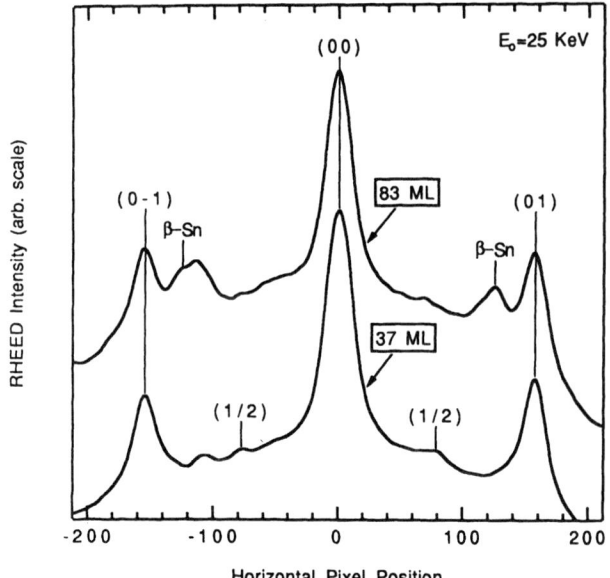

Fig. 5: RHEED intensity profile of Sn overlayers grown on $Cd_{0.8}Zn_{0.2}Te(111)$

REFERENCES

1. C. Mailhoit and D.L. Smith, Critical Review in Solid State and Material Sciences, CRC **16**, 131 (1990)
2. C. Van de Walle, Phys. Rev. B **39**, 1871 (1989).
3. R.C. Bowman Jr., P.M. Adams, M.A. Engelhardt and H.Höchst, J. Vac. Sci. Technol. A 8, 1577 (1990) and references therein
4. C.A. Hoffman, J.R. Meyer, R.J. Wagner, F.J. Bartoli, M.A. Engelhardt and H. Höchst, Phys. Rev. B **40**, 11693 (1989)
5. J. Menéndez and H.Höchst, Thin Solid Films **111**, 375 (1984).
6. J.F. Nye, *Physical Properties of Crystals* (Clarendon, Oxford 1969).

TRAPS IN ZnS/ANODIC SULFIDE FILM
ON MERCURY CADMIUM TELLURIDE

S. Hikida, N. Kajihara, and Y. Miyamoto

Fujitsu Laboratories Ltd., 10-1 Morinosato-Wakamiya, Atsugi, 243-01, Japan.

ABSTRACT

We studied the influence of visible light on the ZnS/anodic sulfide/HgCdTe interface.

We measured flatband voltage of metal-insulator-semiconductor diodes with semitransparent electrodes at 77 K. In the dark (i.e., without visible light). The flatband voltage of an metal-insulator-semiconductor diode is 0 V. We scanned visible light from 400 to 800 nm and measured the flatband voltage using the photocapacitance measurement. After an initial positive shift, the flatband voltage moved negative by as the wavelength shortened. We studied the relationship between the flatband voltage shift and photon energy using Fowler plots. Results suggest one electron trap level 1.5 eV above the valence band and two hole trap levels 1.7 and 2.3 eV below the conduction band.

ZnS/anodic sulfide films on mercury cadmium telluride, studied using photoluminescence spectroscopy, showed broad photoluminescence peaks at 1.5, 1.7, and 2.3 eV, and the edge emission of the anodic sulfide.

These results suggest that photoionization of the electron trap and two hole traps in the ZnS/anodic sulfide film causes the flatband shift.

INTRODUCTION

Mercury cadmium telluride (MCT) is a promising material for infrared sensors, because it is a narrow band gap semiconductor, and its band gap varies with composition. To get a high-performance sensor using MCT requires an excellent passivation film due to this narrow band gap.

ZnS/anodic sulfide film on MCT is one of the most promising passivants, due to its low fixed-charge density and low surface-state density[1, 2]. We found that the flatband voltage of a metal-insulator-semiconductor(MIS) diode shifts in visible light. We thought the shift was due to traps ionized by visible light. We studied these traps by measuring the photocapacitance to distinguish the types of traps. Using photoluminescence, we confirmed the existence in the film. The purpose of our paper is to clarify the behavior of the photoionized traps in ZnS/anodic sulfide film on MCT.

EXPERIMENT

Our experiment sample was the metal-insulator-semiconductor (MIS) diode in Figure 1. The p-$Hg_{0.8}Cd_{0.2}Te$ ($Na = 1 \times 10^{16}$ cm^{-3}) crystal was grown by liquid-phase epitaxy on CdTe. The anodic sulfide film was electrochemically grown on p-$Hg_{0.8}Cd_{0.2}Te$ in a solution of Na_2S with ethylene glycol using a platinum counter electrode. Growth was for 45 min and current density at the surface of the p-$Hg_{0.8}Cd_{0.2}Te$ was 100 $\mu A/cm^2$. The anodic sulfide film was 30 nm thick. After this film was grown on MCT, 800 nm of ZnS film was evaporated. A chromate gate electrode 20 nm thick was used so that visible light illuminated the ZnS/anodic sulfide film through the gate electrode.

Using a monochrometer with a tungsten lamp (3200 K), we measured the deviation in the capacitance of the MIS diode at 77 K scanning the illumination light from 400 nm (3.1 eV) to 800 nm (1.6 eV). The MIS diode's capacitance was measured at 1 MHz. The gate bias voltage was made 0 V to exclude the influence of the electric field to charge emission/capture rate. During illumination, traps in the insulator were ionized by emitting or capturing charges. These ionized traps changed the band condition at the semiconductor's surface, thuschanging the diode's capacitance. Measuring the capacitance-voltage (C-V) curve in the absence of visible light illumination (i. e. in the dark), we calculated flatband voltage shifts from the deviation in the diode capacitance using the C-V curve in the dark. The flatband voltage shift depends on the illumination time, and the shift eventually saturates. The measured flatband voltage shifts are saturated.

We prepared a sample for photoluminescence measurement like the MIS diode but without a gate electrode, and measured the photoluminescence spectra of the ZnS/anodic sulfide film on MCT at 77 K. The exciting light source was a 325-nm He-Cd laser.

RESULT AND DISCUSSION

Figure 2 shows the relationship between the flatband shift and incident photon energy. Where incident energy is between 1.5 and 2.0 eV, the flatband voltage moves positively. Where incident photon energy exceeds 2.0 eV, the flatband voltage moves negatively. This behavior of indicates that at least two

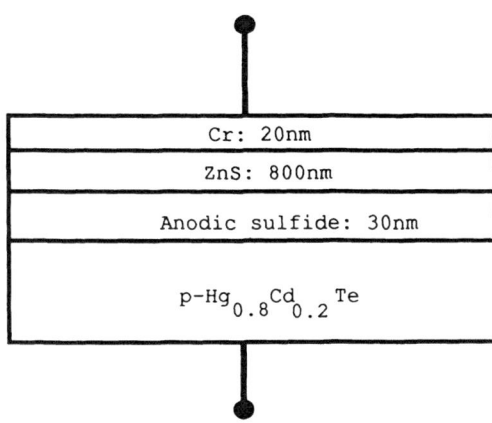

Fig. 1 MIS diode for photocapacitance measurement

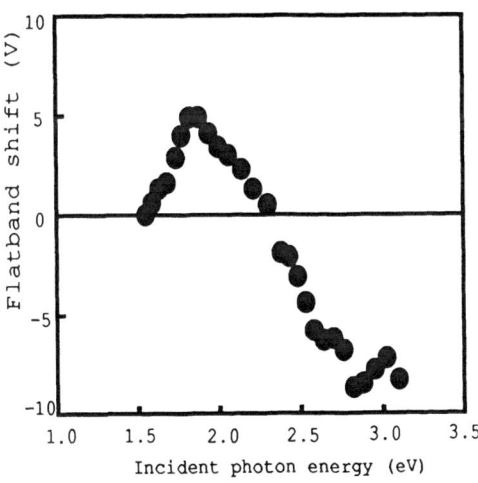

Fig. 2 Spectral responce of flatband voltage shifts

types of traps exist in the film and are ionized by visible light. Generally, such a photoionization trap response is expressed using a Fowler plot formula[3] :.

$$R = C(h\nu-\emptyset)^2 \tag{1}$$

where R is the response, $h\nu$ the incident photon energy, $\emptyset$ the trap level, and C a constant. Based on this relationship, we analyzed the flatband voltage shift caused by visible light (Figure 3). The response corresponds to flatband voltage shifts normalized by the incident photon number.

Where incident photon energy is between 1.5 and 1.7 eV, the response appears dominated by one photoionized trap which shifts the flatband positively. In this region, the response moves positively. This means that photoionized traps include an electron trap which increases the negative charge in the film. Based on the Fowler plot, the response is proportional to the square of the difference between incident photon energy and trap depth. This relationship is shown by the upper broken curve in Figure 3.

Where incident photon energy exceeds 1.7 eV, the response moves negatively. The response is dominated by two types of photoionized traps, meaning the other trap is a hole trap increases the positive charge in the film. Here, the response synthesized by both traps is expressed by a sum of above formula(1),

$$R = C_1(h\nu-\emptyset_1)^2 + C_2(h\nu-\emptyset_2)^2 , \tag{2}$$

where $h\nu > \emptyset_2 > \emptyset_1$, $C_1>0$, $C_2<0$.

One electron trap's behavior has been already given by initial fitting. The relationship between the other hole trap and incident photon energy is easily obtained by subtracting $C_1(h\nu-\emptyset_1)^2$ from normalized data. The relationship between the response of the other hole trap and incident photon energy is obtained as shown by lower broken curve in the figure. Similarly, experiment data on the response shown by solid curve can be broken down into three curves (not shown here).

Figure 4 shows the results of Fowler plot fitting using the least squares method. Three types of traps exist in the ZnS/anodic sulfide film. Of these, trap A is an electron trap 1.5 eV above the valence band. Traps B and C are hole traps 1.7 and 2.3 eV below the conduction band.

To verify the nature of these traps, we also measured the photoluminescence spectra of the ZnS/anodic sulfide film (Figure 5). Broad peaks can be observed. In these peaks, the emission at 2.6 eV is the edge emission of the film, because this film is high CdS content[2, 4]. We also observed emissions

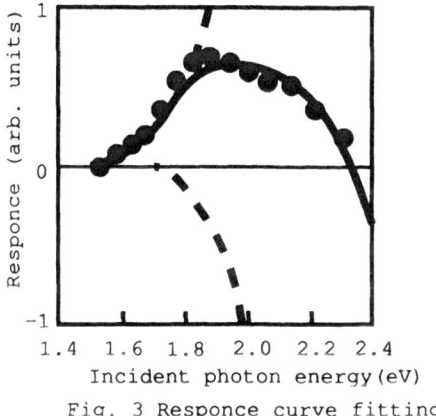

Fig. 3 Responce curve fitting

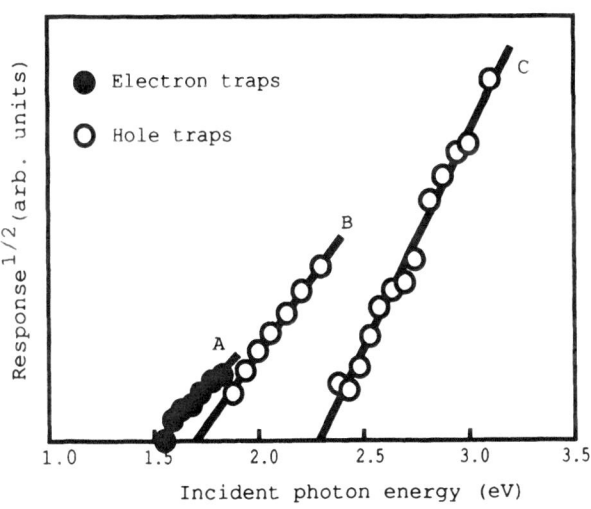

Fig. 4 Results of curve fitting for Fowler plots

450

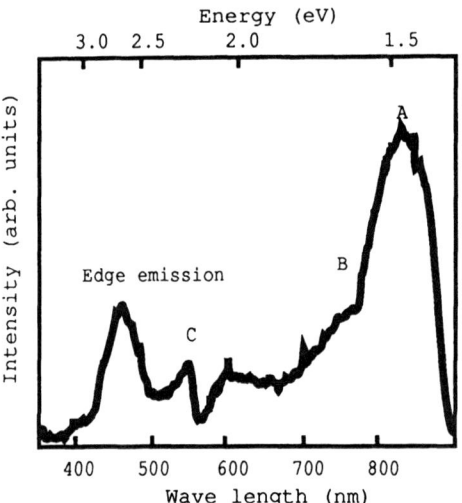

Fig. 5 Photoluminescence spectra
of ZnS/anodic sulfide film on MCT

at 1.5, 1.7, and 2.3 eV. These peaks are consistent with the energy levels of traps A, B, and C obtained from the photocapacitance measurement.

CONCLUSIONS

We studied the photoionized traps in ZnS/anodic sulfide film on MCT. These traps, when photoionized by visible light, shift move the MIS diode's flatband voltage. Using photocapacitance measurement, we found that one electron trap is 1.5 eV above the valence band and two hole traps are 1.7 and 2.3 eV below the conduction band. The response of photoionized traps for incident photon energy can be broken down based on the Fowler plot formula. The photoluminescence spectra of the ZnS/anodic sulfide film on MCT is consistent with the energy levels obtained from the photocapacitance measurement.

REFERRENCES

1. Y. Nemirovsky, L. Burstein, and I. Kidron, J. Appl. Phys. 58, 366 (1985)

2. Y. Nemirovsky, L. Burstein, Appl. Phys. Lett. 44, 443 (1984)

3. C. R. Crowell, S. M. Sze, and W. G. Spitzer, Appl. Phys. Lett. 4, 91 (1964)

4. T. Ipposhi, K. Takita, K. Murakami, K. Masuda, H. Kudo, and S. Seki, J. Appl. Phys. 63, 132(1988)

IMPLANT ISOLATION OF DEVICE STRUCTURES CONTAINING BURIED, HIGHLY-DOPED LAYERS

S. J. PEARTON, F. REN, L. A. D'ASARO, W. S. HOBSON, T. R. FULLOWAN, J. LOTHIAN, C. R. ABERNATHY, R. F. KOPF AND J.-M. KUO
AT&T Bell Laboratories, Murray Hill, NJ 07974

ABSTRACT

The formation of high resistivity ($> 10^7$ $\Omega/\square$) regions in GaAs-AlGaAs HBT and SEED structures by oxygen and hydrogen ion implantation is described. Multiple energy implants in the dose range 10^{13} cm^{-3} (for O$^+$) and 10^{15} cm^{-2} (for H$^+$), followed by annealing around 500°C are necessary to isolate structures ~2 µm thick. In each case, the evolution of the sheet resistance of the implanted material with annealing is consistent with a reduction in hopping probabilities of trapped carriers between deep level states for temperatures up to ~600°C, followed by significant annealing of these deep levels. A comparison of the relative thermal stability of O$^+$ or H$^+$ ion implant-isolated p$^+$ material is given. Small geometry (2×9 µm^2) HBTs exhibiting current gain of 44 and cut-off frequency f_T as high as 45 GHz are demonstrated using implant isolation.

INTRODUCTION

Implant isolation has the obvious advantage over mesa etching that it retains the planarity of the surface, and, in general, it also intrudes less under the mask edges. Wet etching is isotropic, so that undercutting at mask edges can be quite severe when a deep mesa must be etched. This undercutting can be eliminated by using dry-etching techniques, but the non-planarity of the surface is still a problem and causes difficulties in subsequent resist application and metal step coverage. Implant isolation also intrudes under mask edges because of lateral straggle of the implanted ions, although this straggle is typically about half of the projected range. For very high-speed applications where the achievement of very low parasitic capacitances is necessary, mesa etching may be superior because the capacitance of a given area of semiconductor is larger than the capacitance of the same area of air. For most isolation purposes, however, implant isolation is very effective because the resistivity of the bombarded material is often larger than that of the semi-insulating substrate itself, leading to lower leakage currents through the material.

In particular there is interest in the use of implantation to isolate relatively thick (~2 µm) structures containing several different semiconductors and also to selectively alter the conductivity of a buried layer within a device structure without affecting the overlying layers. In this paper we detail the application of implant isolation to Heterojunction Bipolar Transistors (HBTs) and structures typical of those used for the optical switches known as Self-Electro-Optic Devices (SEEDs).

EXPERIMENTAL

A typical HBT structure is shown in the upper part of Fig. 1. Most of our work has been performed using material grown in a barrel-type metalorganic chemical vapor deposition (MOCVD) reactor operating at atmospheric pressure, although similar results have been obtained with material grown by molecular beam epitaxy (MBE). In Fig. 1 the n$^+$-GaAs doping level is ~3×10^{18} cm^{-3}, the n-GaAs is 5×10^{16} cm^{-3}, and the n-AlGaAs is 5×10^{17} cm^{-3}. Silicon is the n-type dopant in each case. The p-type dopant

(Zn in MOCVD, Be in MBE) level was $5 \times 10^{18} - 2 \times 10^{19}$ cm^{-3}. To achieve an approximately uniform damage level throughout this HBT structure, a multiple energy oxygen and hydrogen ion implant scheme was used, as shown in the lower part of Fig. 1. We have shown the ion profiles assuming Pearson IV distributions, which appear to describe most ion profiles in GaAs. The implant condition in this case are O$^+$ at 40 keV (8×10^{12} cm^{-2}), 100 keV (6×10^{12} cm^{-2}), 200 keV (7×10^{12} cm^{-2}), 250 keV (8×10^{12} cm^{-2}), 300 keV (8×10^{12} cm^{-2}), and 400 keV (8×10^{12} cm^{-2}), followed by H$^+$ at 100 keV (10^{15} cm^{-2}), 150 keV (10^{15} cm^{-2}), and 200 keV (10^{15} cm^{-2}). We prefer to use oxygen ions for isolation whenever possible because of their relatively high efficiency for removing carriers (~50−100 carriers per implanted ion depending on the ion energy) compared to hydrogen ions (~3−5 carriers per ion) and because oxygen ions have a relatively lower ratio of lateral straggle to projected range. The maximum energy available from most conventional ion implanters is 400 keV by using doubly ionized beams in a 200 keV machine. Since the projected range (R_p) of a 400 keV O$^+$ ion is not great enough to isolate our HBT structure, further implants of protons are used to isolate the n$^+$ subcollector region. Higher ion doses are used for the protons in order to compensate for their lower carrier removal rate. We emphasize that it is really the damage distribution created by the ion stopping process that is important, rather than the ion profile itself. The maximum in the damage distribution occurs at ~0.75 R_p for oxygen ions, but for protons is essentially coincident with the ion profile [1]. Protons have been widely used for isolating GaAs-AlGaAs laser [2]. It is important to note that in Fig. 1 we have chosen higher ion doses to compensate the highly doped n$^+$-GaAs layers.

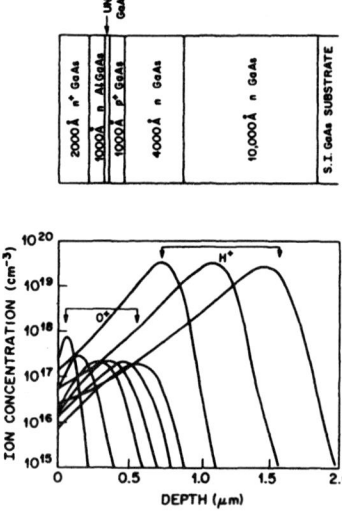

Fig. 1.   Typical GaAs-AlGaAs HBT structure (at top) and ion profiles of O$^+$ and H$^+$ ion implanted into the structure (at bottom). The O$^+$ implant parameters are 8×10^{12} cm^{-2}, 40 keV + 6×10^{12} cm^{-2}, 100 keV + 7×10^{12} cm^{-2}, 200 keV + 8×10^{12} cm^{-2}, 250 keV + 8×10^{12} cm^{-2}, 300 keV + 8×10^{12} cm^{-2}, 400 keV, while the H$^+$ ions were implanted at 100, 150, and 200 keV, each at a dose of 10^{15} cm^{-2}.

To examine the effect of using single energy oxygen or proton implants to create a buried damage profile centered in the collector layer, a structure consisting of half of the HBT in Figure 1 was grown. In that case we eliminated the AlGaAs emitter layer and GaAs cap layer. Finally, to investigate the isolation of p^+ $Al_{0.11}Ga_{0.89}As$ of the type used in SEED structures, we grew 5000 Å thick layers either directly on GaAs, or buried by ~5000 Å of undoped GaAs.

RESULTS AND DISCUSSION

(a) HBT Isolation

Figure 2 shows the sheet resistance obtained from a transmission line measurement (TLM) on the implanted HBT structure, as a function of both ion dose and post-implant annealing temperature. Ion doses were varied by a factor of 4 about central values chosen based on our previous experience with implant isolation of GaAs-based device [3]. The unimplanted sheet resistance of the structure was ~50 $\Omega/\square$.

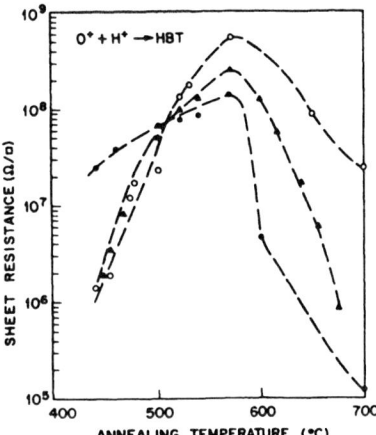

Fig. 2. Sheet resistance of HBT material implanted with multiple energy oxygen and hydrogen ions, as a function of post-implant annealing temperature (10 s anneals). The implant conditions for the data labeled (•) are O^+, 4×10^{12} cm^{-2} (40 keV), 3×10^{12} cm^{-3} (100 keV), 3.5×10^{12} cm^{-2} (200 keV), 4×10^{12} cm^{-2} (250, 300, and 400 keV), H^+ 5×10^{14} cm^{-2} (100, 150, and 200 keV). The doses for the data labeled ($\triangle$) and ($\bigcirc$) are two and four times higher, respectively, at each energy.

The main features of the data in Fig. 2 can be explained as follows. First, the sheet resistance of the structure is increased by several orders of magnitude through the introduction of deep level states that trap both the conduction electrons in the n-type GaAs and AlGaAs, and the holes in the p-type GaAs. Hall measurements show that this compensated material has an extremely low carrier mobility (~1 cm^{2}V^{-1}s^{-1}) relative to the unimplanted values (~1600 cm^{2}V^{-1}s^{-1} in the n$^+$-GaAs capping layer). This low mobility conductivity σ shows a T^{-1} dependence near room temperature, according to

$$\sigma \propto \sigma_0 \exp(T) .$$

This is a characteristic signature of hopping conduction in which the trapped carriers hop from one damage site to another [4,5], i.e., intrastate transitions between neighboring defect sites. Upon annealing to produce the maximum resistivity in the implanted material, this activation energy increases to midgap values (0.72 − 0.78 eV). Indeed as the annealing temperature is increased the sheet resistance increases as some of the damage is annealed out and the hopping probabilities decrease. The resistance reaches a maximum for annealing temperatures between 570 and 650°C, depending on the ion doses, at which point the trap density falls below the carrier concentration and carriers are returned to their respective bands. The resistance then shows a sharp decrease back toward its preimplanted value.

We have found that similar result can be obtained by replacing the proton implants by an MeV oxygen ion implant [6]. Figure 3 shows the sheet resistance of HBT structure somewhat thinner than the one shown in Fig. 1 [a total thickness of 1.2 μm because of a narrower (4000 Å) n^+-GaAs subcollector] obtained from Van der Pauw-Hall measurements after multiple energy oxygen implantation. The projected range of a 1 MeV O^+ ions is ~0.96 μm with a straggle of ~0.22 μm, and this is sufficient to compensate the n^+-GaAs subcollector region. The evolution of the sheet resistance with post-implant annealing temperature has the same form for O^+ + H implantation or O^+ only implantation.

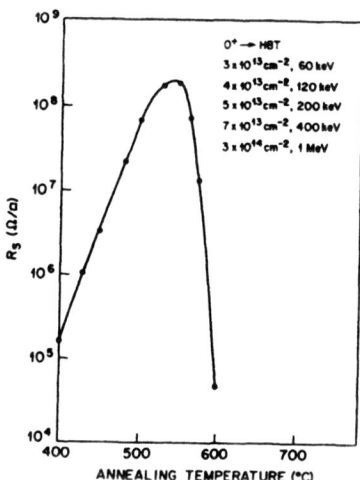

Fig. 3.   Sheet resistance of HBT structure implanted with multiple energy (60, 120, 200, 400 and 1000 keV) oxygen ions, as a function of post-implant annealing temperature (10 s anneals).

High performance HBT device utilizing implant isolation in a self-aligned process were fabricated. For 2×9 μm² devices, gains of 44 were obtained with e-c breakdown of 10V. Isolation currents of 8 μA · mm⁻¹ were measured, with f_T values of 45 GHz at a bias of V_{CE} = 2V and a collector current of 7.6 mA.

(b) C_{BC} Reduction

The results of 1MHz C-V measurements on the H^+ or O^+ implanted half-HBT structure are shown in Figure 4. The as-implanted samples showed high resistance behavior, as measured by probes placed on the surface. Since alloying of the ohmic contacts was required in the processing sequence in any case and maximum resistivities are usually observed in ion-bombarded GaAs for post-implant anneals at $\geq 500°C$, we investigated post-implant anneals at 500 or 550°C for 10 sec. A proton dose of 5×10^{14} cm^{-2} is required to fully deplete the collector layer when a post-implant anneal of 550°C for 10 sec is used. Note that in the unimplanted control sample the zero bias capacitance was 88 pF - this is reduced to ~1.3 pF as a result of the appropriate implant dose. Similar data for oxygen implants into our half-HBT structure are also shown in Figure 2. As expected, the oxygen ions are much more effective in creating damage-related compensation in the collector, and even doses of 10^{12} cm^{-2} are enough to cause full depletion.

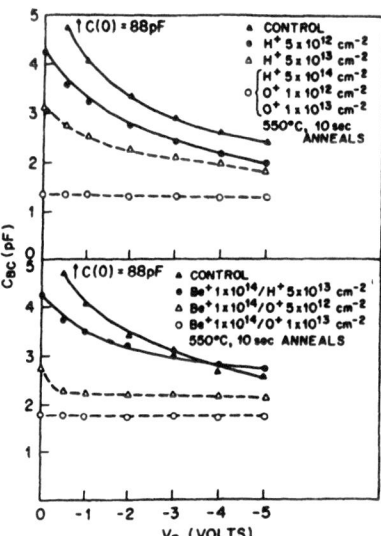

Fig. 4.   Capacitance-voltage characteristics for H^+ or O^+ implanted (top) or Be^+/H^+ or Be^+/O^+ implanted (bottom) p^+nn^+ structures, as a function of ion dose. All anneals were 550°C for 10 sec.

(c)   Isolation of p^+ AlGaAs and GaAs

SEED structures contain 5000Å thick $Al_{0.11}Ga_{0.89}As$ layers. Figure 5 shows the sheet resistance of such a layer as a function of annealing temperature after multiple-energy F implantation of two differents dose levels. The evolution of the resistivity is similar to that of n^+ AlGaAs, indicating that both electron and hole traps are created by the ion bombardment. The same type of data for buried p^+ GaAs layers after multiple energy proton implants is shown in Figure 6. As expected, the peak resistivity occurs at higher temperatures for higher doses.

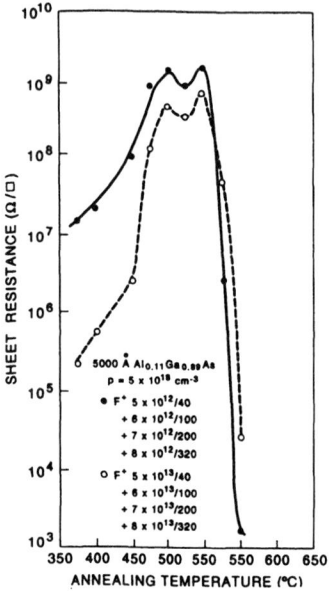

Fig. 5.   Sheet resistance of $p^+ Al_{0.11} Ga_{0.89} As$ implanted with multiple energy F^+ ions, as a function of post-implant annealing temperature.

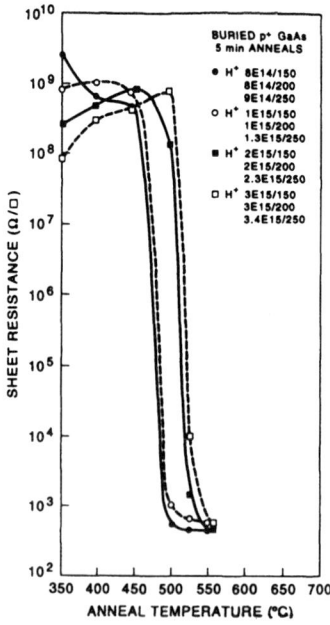

Fig. 6.   Sheet resistance of $p^+ GaAs$ implanted with multiple energy protons, as a function of ion dose and post-implant annealing temperature.

CONCLUSIONS

We have demonstrated the formation of high-resistivity regions in GaAs-AlGaAs heterojunction by multiple energy oxygen and hydrogen ion implantation. Optimum isolation between neighboring HBT devices is obtained after post-implant annealing in the range 575 – 650°C. The evolution of the sheet resistance of the implanted material with annealing can be explained by changes in carrier hopping probabilities as annealing of damage-related defects occurs. Similar results are obtained for p^+ GaAs and AlGaAs, and we have also shown that single O^+ or H^+ implants can be used to selectively reduce the conductivity of a buried layer in an HBT.

REFERENCES

[1] Calculated from a Transport of Ions in Matter (TRIM) program-see J. P. Biersack and L. G. Haggmark, Nucl. Instrum. Methods **174**, 257 (1980).

[2] See, for example J. C. Dyment, J. C. North, B. I. Miller, and J. E. Ripper, Proc. IEEE **60**, 726 (1972).

[3] K. T. Short and S. J. Pearton, J. Electrochem. Soc. **135**, 2835 (1988).

[4] N. F. Mott, Philos. Mag. **19**, 835 (1969); J. Non-Cryst. Solids, **1**, 1 (1968).

[5] M. Cohen, H. Fritsche, and S. Ovshinsky, Phys. Rev. Lett. **22**, 1065 (1969).

[6] F. Ren, S. J. Pearton, W. S. Hobson, T. R. Fullowan, J. Lothian and A. W. Yanof, Appl. Phys. Lett. **56** 860 (1990).

IN SITU SPECTROSCOPIC ELLIPSOMETRY
FOR REAL TIME SEMICONDUCTOR GROWTH MONITOR[+]

BLAINE JOHS*, DUANE MEYER*, GERALD COONEY*, HUADE YAO**, PAUL G. SNYDER**, JOHN A. WOOLLAM**, JOHN EDWARDS*** AND GEORGE MARACAS***
*J.A. Woollam Co., 315 South 9th, Suite 22, Lincoln, NE 68508
**Center for Microelectronic and Optical Materials Research, and Department of Electrical Engineering, University of Nebraska, Lincoln, NE 68588-0511
***Center for Solid State Electronics Research, Arizona State University, Tempe, AZ 85287
[+]Research supported by DARPA Contract DAAH01-89-C-0357, NASA Lewis Grant NAG-3-154, and DARPA/URI Contract N00014-89-J-3120.

ABSTRACT

A modular spectroscopic ellipsometer for in situ and ex situ materials analysis is described, and results for in situ MBE growth of GaAs/AlGaAs are reported.

EXPERIMENTAL

We have designed, built, and operated an in situ spectroscopic ellipsometer (SE) covering the spectral range from 1.25 eV to 4.5 eV. It is of the fixed-polarizer rotating-analyzer design which minimizes alignment and residual polarization problems, and has a chopper to minimize the ambient light difficulties normally associated with rotating analyzer designs. Furthermore, it is modular thus permitting interchangeable use between in situ and variable angle ex situ operation. The apparatus is compact, rugged, and unobstructive in the busy MBE environment.
Ellipsometry measures the complex reflectance ratio

$$\rho = r_p/r_s = \tan \psi \exp i \; \Delta$$

where r_p and r_s are the complex Fresnel reflection coefficients for p- and s-polarized light respectively. The data are frequently reported in terms of ψ and Δ over a given spectral range and particular angle of incidence. Both ψ and Δ data are used to determine optical constants, surface roughness, layer thickness, etc., using a model dependent regression analysis.
We report here on the use of this SE for the acquisition and analysis of data from warm up, oxide blow-off, GaAs surface smoothing, and AlGaAs crystal growth by molecular beam epitaxy.
These experiments were performed in a Vacuum Generators V80-H MBE growth chamber with specially designed ellipsometry ports. Growths of GaAs and AlGaAs was performed with solid Ga and Al and cracked arsine as a group V species.

RESULTS

Figure 1 displays ψ and Δ ellipsometric data vs. time, taken at a wavelength of 5000 Å during a typical MBE growth run. A GaAs wafer was slowly heated in the growth chamber, until at roughly 30 minutes the

Complete Growth Sequence at 5000A

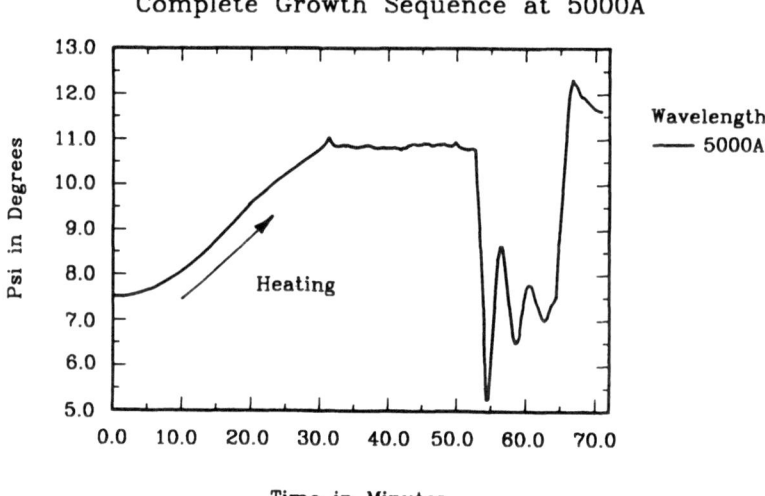

Psi in Degrees

Heating

Wavelength
—— 5000A

Time in Minutes

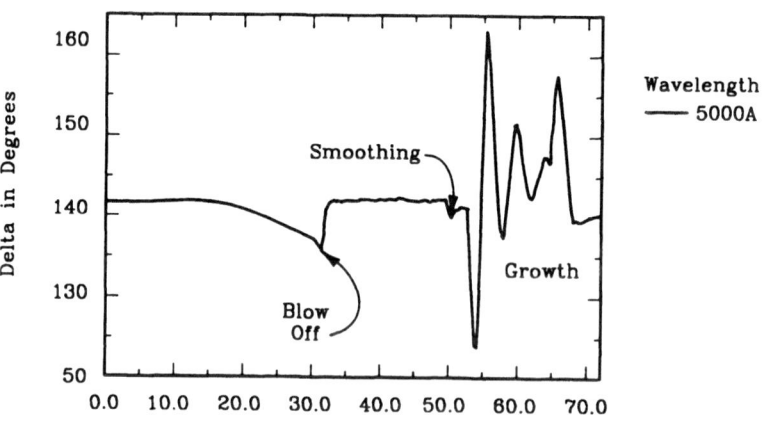

Delta in Degrees

Smoothing

Blow
Off

Growth

Wavelength
—— 5000A

Time in Minutes

AlGaAs/GaAs Layer Growth

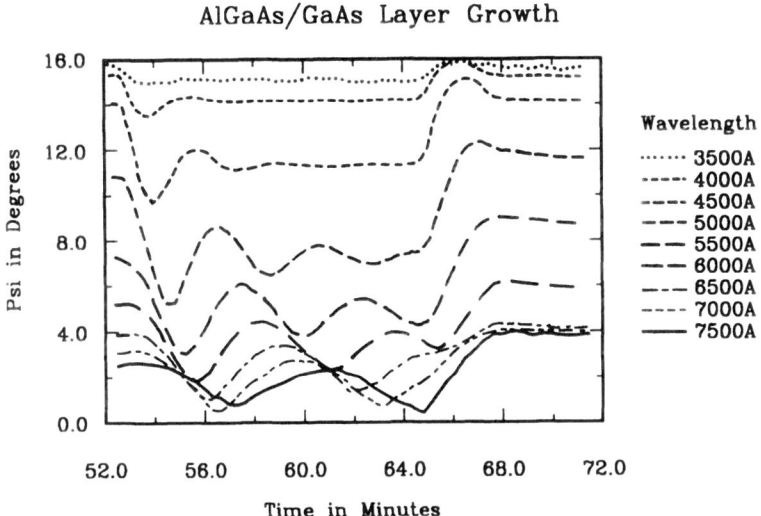

Psi in Degrees

Time in Minutes

Wavelength
...... 3500A
- - - 4000A
-- - 4500A
— — 5000A
— — 5500A
— — 6000A
— - — 6500A
- - - 7000A
——— 7500A

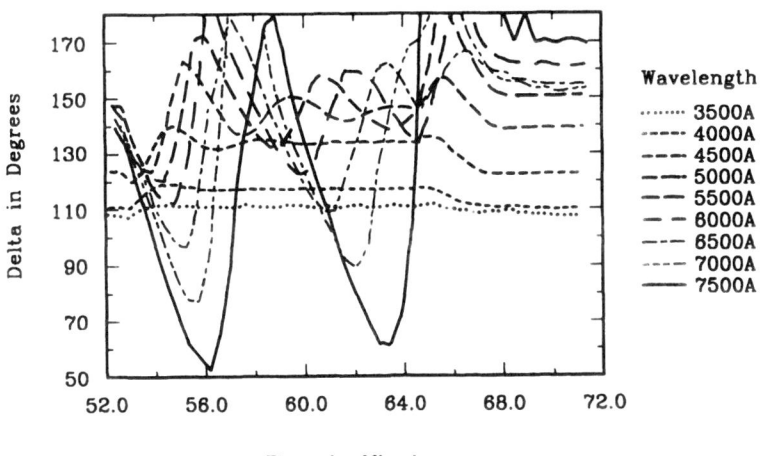

Delta in Degrees

Time in Minutes

Wavelength
...... 3500A
- - - 4000A
-- - 4500A
— — 5000A
— — 5500A
— — 6000A
— - — 6500A
- - - 7000A
——— 7500A

Temp=479C, Oxide=10.7A thick, Angle=75.07
MSE=0.11

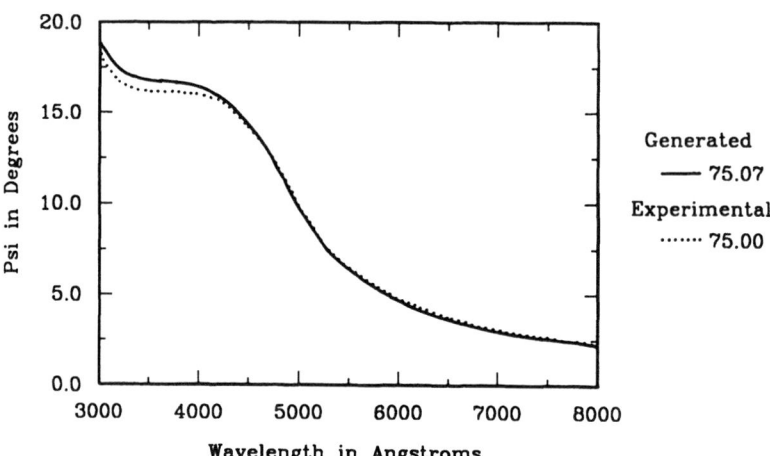

Generated
— 75.07
Experimental
······· 75.00

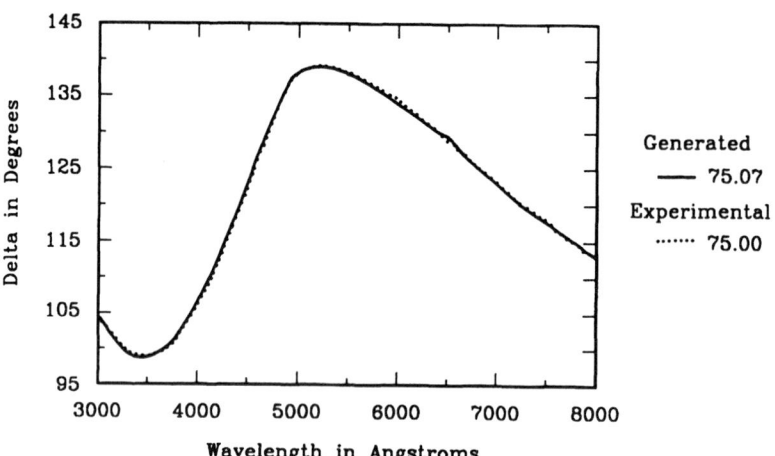

Generated
— 75.07
Experimental
······· 75.00

natural oxide sublimated, as evidenced by the abrupt changes in the ψ and Δ curves. This event was simultaneously monitored and observed using RHEED.

At roughly 50 minutes the Ga shutter was opened and surface smoothing resulted from the growth of a few monolayers of GaAs. This is shown very clearly in Figure 1 and was confirmed by the appearance of discernable streaks in the RHEED pattern.

At 52 minutes the growth of AlGaAs was initiated, and the ellipsometric data taken simultaneously at nine wavelengths are shown in Figure 2. Notice the strong spectral dependence. For example, at 3500 Å there is little variation in data during growth compared with the huge variation seen in the 5000 Å to 7500 Å range. These data show that AlGaAs growth can be monitored with good sensitivity to layer thickness.

Post-deposition SE analysis showed that the final structure grown was GaAs (substrate)/$Al_{0.25}Ga_{0.75}As$ (2133 Å)/GaAs (379 Å).

The optical constants of GaAs were independently measured as a function of temperature (see companion MRS paper). A simple linear interpolation scheme was used in the present experiments to determine the surface temperature. As an example, complete spectra were taken for several temperatures below the oxide blow-off. Fits were made to the data, in which oxide thickness, angle of incidence, and temperature were the variables. A typical example is shown in Figure 3. Only the psi data at short wavelength shows any difference from the model calculation, and this is explainable as being due to slight surface imperfections. The results of three such fits are summarized:

Temperature	Oxide Thickness	Angle
58°C	12.1 Å	75.03
227°C	12.8 Å	75.04
479°C	10.7 Å	75.07

Notice that the oxide thickness and angle of incidence are nearly constant. Thus in situ SE may also be used to determine surface temperature.

CONCLUSIONS

A modular rotating analyzer spectroscopic ellipsometer capable of both in situ and ex situ operation has been used for monitoring details of the in situ MBE growth of a GaAs/AlGaAs structure. RHEED observations of oxide blow-off and surface smoothing complimented and were in agreement with the ellipsometric measurements.

THERMOELECTRIC POWER IN QUANTUM CONFINED BISMUTH
UNDER CLASSICALLY LARGE MAGNETIC FIELD

KAMAKHYA P. GHATAK[*] AND S. N. BISWAS[*]

* Department of Electronics & Tele-communication Engineering, University
of Jadavpur, Calcutta - 700032, INDIA.
** Department of Electronics & Tele-communication Engineering, B.E.
College, Shibpur, Howrah, West Bengal, INDIA.

ABSTRACT

In this paper we studied the thermoelectric power under classically
large magnetic field (TPM) in quantum wells (QWs), quantum well wires
(QWWS) and quantum dots (QDs) of Bi by formulating the respective
electron dispersion laws. The TPM increases with increasing
film thickness in an oscillatory manner in all the cases. The TPM in QD
is greatest and the least for quantum wells respectively. The theoretical
results are in agreement with the experimental observations as reported
elsewhere.

INTRODUCTION

In recent years with the advent of MBE, FLL, MOCVD and other
experimental techniques, low-dimensional microstructures having quantum
confinement in 1,2 and 3 dimensions such as quantum wells (QWs), quantum
well wires (QWWs) and quantum dots (QDs) find extensive applications in
QW lasers. EFTs, high speed digital networks, optical modulators and
other devices [1,2,3]. Microstructures based on various materials are
currently being studied becuase of the enhancement of carrier mobility
[4]. It appears from the literature that the thermoelectric power under
classically large magnetic field (TPM) in quantum confined Bi
microstructures has yet to be investigated. The TPM gives information
about the band structure, the density-of-states function and can be
related to the Einstein relation [5]. The importance of studying the
electronic properties of Bi has been given in [6]. We shall use McClure
and Choi model which fits the data for a large number of
magneto-oscillatory and resonance experiments.

THEORETICAL BACKGROUND

The energy spectrum of the conduction electrons in bulk specimens
of Bi can be expressed in accordance with McClure and Choi model as [6]

$$E(1 + (E/E_g)) = (\hbar^2 k_x^2 / 2m_1) + (\hbar^2 k_y^2 / 2m_2) + (\hbar^2 k_z^2 / 2m_3) +$$

$$(\hbar^2 k_y^2 E / 2E_g m_2)(1 - (m_2/m_2')) + (p_y^4 / 4m_2 m_2' Eg) - (\hbar^4 k_x^2 K_y^2).$$

$$(4m_1 m_2 E_g)^{-1} - (\hbar^4 K_y^2 K_z^2 / 4m_2 m_3 E_g) \tag{1}$$

where the notations are defined in [6]. Therefore the dispersion law in
size quantized Bi with the direction of size quantization is along k_z
direction in the presence of a large magnetic field B_y along k_y axis can
be written as

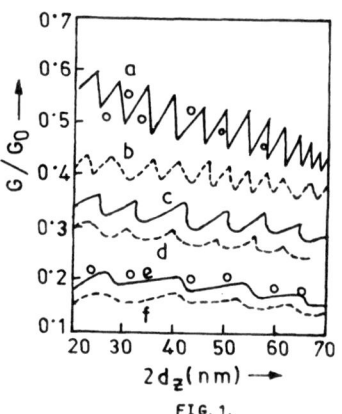

FIG. 1.

Fig. 1. Plot of G/G_o versus $2d_z$ for Bi in accordance with (a) QD($2d_x$ = $2d_y$ = 40 nm, B=1 Tesla, n_o = 10^{20} m^{-3} and T=4.2K) (b) same with $E_g \rightarrow \alpha$ (c) QWW ($2d_x$ = 40nm, B=1 Tesla, n_o = 10^{14} m^{-1} and T=4.2K) (d) same with $E_g \rightarrow \alpha$ (e) QW (B = 1 Tesla; n_o = 10^{15} m^{-2} and T = 4.2K) (f) Same with $E_g \rightarrow \alpha$.

$$L_+ \ (d_1^{\ 2} - L_+^{\ 2})^{\frac{1}{2}} - L_- \ (d_1^{\ 2} - L_-^{\ 2})^{\frac{1}{2}} + d_1^{\ 2} \ [\sin^{-1} (L_+/d_1) - \sin^{-1} (L_-/d_1)]$$

$$= 2a_1 \hbar n_z \tag{2}$$

where $L_+ = (b_1/2a_1) \pm (a_1 d_z)$, $d_1^{\ 2} = c_1 + (b_1^{\ 2}/4a_1^{\ 2})$, $a_1^{\ 2} = (m_3/m_1) \ e^2 B_y^{\ 2}$,

$b_1 = (m_3/m_1) \ 2\hbar k_x \ eB_y$, $C_1 = [(1/2m_3) - (\hbar^2 k_y^{\ 2}/4m_2 m_3 E_g)]^{-1} \ [E + \alpha E^2 - \hbar^2 k_x^{\ 2}]$

$((1/2m_1) - (\hbar^2 k_y^{\ 2}/4m_1 m_2 E_g)) - \hbar^2 k_y^{\ 2} \ ((1/2m_2) + (E/2m_2 E_g)) \ (1-(m_2/m_2')) \ +$

$(\hbar^2 k_y^{\ 2}/4m_2 m_2' E_g) \ ]$, $2d_z$ is the width of the film in the Z-direction with

the quantum number n_z. Putting $k_y = n_y \pi/2d_y$ and $k_x = \pi n_x/2d_x$ seperately in equation (2) we get two different dispersion laws of QWWs respectively. Putting both the said values of k_y and k_x simultaneously in equation (2) we get the dispersion law of QDs of Bi. The TPM for the present case is given by [7] $G = H_o/en_o$ (3)

where n_o and H_o are corresponding electron concentration and entropy respectively.

RESULTS AND DISCUSSION

Using the appropriate equations together with the parameters as given in [6] we have plotted the normalized TPM in all the cases of Bi as functions of film thickness. The circular plots exhibit the experimental results as given elsewhere [8]. The TPM increases with increasing film thickness in an oscillatory manner though the rates of oscillations are completely band structure dependent. The TPM in QD is greatest and least for QWs respectively. Our analysis is valid for holes with proper change in band parameters. By putting $E_g \rightarrow \infty$, equation (1) converts into the ellipsoidal parabolic energy bands. From figure 1 we can compare among the model of Bi. Finally it may be noted that though the many body effects should be considered along with a self consistent procedure, this simplified analysis exhibits the basic features of the TPM in quantum confined Bi and the agreement between the simplified theoretical analysis and the experimental results are rather prominent.

1. D.R. Scifres, C. Lindstrom, R.D. Rurnham, W. Streifer, T.L. Paoli, Electron Letts. 19, 170 (1983).
2. P.M. Salomon, Proc. IEEE, 70, 439 (1982).
3. S. Tarucha and H. Okamoto. Appl. Phys. Letts. 48, 1(1986).
4. N.T. Linch, Festkorperprobleme, 23, 227 (1983).
5. K.P. Ghatak and M. Mondal, J. App. Phys. 66, 3056 (1989).
6. B. Mitra and K.P. Ghatak, Physica Scripta, 40, 776 (1989).
7. I.M. Tsidilkovski, Band structure of semiconductors, Pergamon Press, Oxford, 1982.
8. M. Atali and K. Kakanov, J. Exp. and Theor. Phys. 109, 120 (1990).

ON THE PHOTOEMISSION FROM TERNARY AND QUATERNARY ALLOYS SYSTEMS
UNDER STRONG MAGNETIC QUANTIZATION

KAMAKHYA P. GHATAK*

*Department of Electronics & Tele-communication Engineering, University
of Jadavpur, Calcutta - 700032, INDIA.

ABSTRACT

We study the photomission from ternary and quanternary alloys systems
by formulating the respective expressions considering the spin &
broadening of Landau levels. It is found taking n-$Hg_{1-x}Cd_{x}Te$ and
$In_{1-x}Ga_{x}As_{y}P_{1-y}$ lattice matched to InP as examples of ternary and
quaternary alloys respectively that the photoemission exhibits oscillatory
magnetic field dependence and increases with increasing electron
concentration in spiky manner for both the cases. The numerical magnitudes
of photoemission in ternary alloys are greater than that of the quaternary
systems and the theoretical formulations are in agreement with the
experimental observations as reported elsewhere.

INTRODUCTION

In recent years there has been intensive interest in studying the
various electronic properties of ternary and quaternary alloys systems
because of their importance in device technology. The ternary alloys final
extensive applications as optoelectronic materials [1], photovoltaic
detector arrays [2], infrared detectors [3] and other systems. The
quaternary alloy has received attention as a possible material for the
fabrication of heterojunction lasers [4], light-emitting diodes [5],
avalanche photo-diodes [6], transferred-electron devices [7], etc. Though
considerable work has already been done, nevertheless it appears from
the literature that the photoemission from such alloys has yet to be
studied under strong magnetic quantization by considering spin and
broadening of Landau levels into account. In what follows, we shall study
the magnetic field and doping dependence of the magneto-photo emission,
taking $nHg_{1-x}Cd_{x}Te$ and $In_{1-x}Ga_{x}As_{y}P_{1-y}$ lattice matched to InP as examples
of ternary and quanternary alloys respectively.

THEORETICAL BACKGROUND

The magneto-dispersion relation of the conduction band electrons can
be expressed [3] as

$$A(E) = \hbar^2 k_z^2/2m^* + (n + \tfrac{1}{2}) \hbar w_o \pm \tfrac{1}{2} g^*(E) \text{ B.C.} \tag{1}$$

where $A(E) = [E (E + E_g) (E + E_g + D) (E_g + (2/3)D)] [(E_g + D). E_g (E + E_g + (2/3) D)]^{-1}$, E is the electron energy as measured from the edge of
the conduction band in the absence of any quantization, E_g is band gap,
D is the spin-orbit splitting parameter of the valence band, $\hbar$ is Dirac's

constant, m^* is effective electron mass at the edge of the conduction band, $w_o = eB/m^*$, e is electron charge, B is the quantizing magnetic field along z-direction, C is Bohr magnetron, n(=0, 1, 2, ...) is Landau quantum number, $g^*(E) = (2/3)D\ g_o/(E + E_g + (2/3)D)$ and g_o is the magnetude of the effective g factor at the edge of the conduction band. Thus the magneto-photocurrent density can be expressed as

$$J_B = (ea_o Bk_B T/h^2)\ t\ (\ln\ (1 + a_1^2 + 2a_1\ Cos\ a_2)^{\frac{1}{2}}) \qquad (2)$$

where a_o is the probability of photon absorption, k_B is Boltzmann constant, T is temperature, $t = \sum_{n \geq 0}$, $a_1 = \exp((E_F - E')/k_B T)$, E_F is the Fermi energy in the present case, $E' = W - h\nu + E_n$, W is electron affinity, $h\nu$ is the energy of the incident photon, E_n is obtained by putting $E = E_n$ and $k_z = 0$ in equation (1), $a_2 = L/k_B T$, L is the broadening parameter and is given by $L=\pi k_B T_D$ [8] and T_D is Dingle temperature. It appears, then, that, the evaluation of J_B from equation (2) as a function of n_o requires an expression of electron statistics which can, in turn, be written using equation (1) as

$$n_o = [eB\ (2m^*)^{\frac{1}{2}}/2\pi^2 h^2]\ t\ (p_1 + p_2) \qquad (3)$$

where p_1 = Real part of $(A\ (E_o) - (n + \frac{1}{2})\ h\ w_o \pm \frac{1}{2}g^*\ (E_o)\ BC)^{\frac{1}{2}}$, $E_o = E + iL$, $i = (-1)^{\frac{1}{2}}$, $p_2 = \sum_{r=1}^{} N(p_1)$, $N = 2(k_B T)^{2r}\ (1 - 2^{1-2r})\ \phi\ (2r)$, r = 1,2,.....S and $\phi(2r)$ is the Zeta function of order 2r. Under the condition of non-degeneracy and in the absence of magnetic field, equation (2) assumes the well-known form [9] for parabolic energy bands as

$$J_o = (4\ m^*\ a_o\ ek_B^2\ T^2/h^3)\ \exp\ ((k_B T)^{-1}\ (h\nu - q)) \qquad (4)$$

whre q is the work function.

RESULTS AND DISCUSSION

Using equation (2) and (3) and parameters [3] $E_g = [-0.\ 3.3 + 1.73x + 5.6.10^{-4}\ (1 - 2x)T + (0.25x^4)]$ eV, $m^* = 3E_g/4P^2$, $P^2 = (h^2/2m_o)\ (18 + 3x)$, D = 0.9eV, T_D = 3.2K, T = 4.2K, $h\nu$ = 3.1eV, a_o = 1, B = 1 Tesla, g_o = 51 and W = 3.5eV for n-$Hg_{1-x}Cd_x Te$ and D = $(0.114 + 0.26y - 0.02y^2)$eV, $E_g = (1.337 - 0.73y + 0.13y^2)$eV and $m^* = (0.080 - 0.039y)\ m_o$ for $In_x Ga_{1-x} As_y P_{1-y}$ lattice matched to InP [10, 11] we have plotted J_B/J_o as functions of n_o and B in accordance with three-band Kane model, two-band Kane model, and that of parabolic energy bands for both alloys as shown in Figs. 1 and 2 respectively where the circular plots exhibit [12] the experimental datas. It appears from bothy the figures that the photo current oscillates both with n_o and 1/B due to SdH effect. The numerical magnitude of the photo-emission is ternary alloys are greater than that of the quaternary alloys. Finally, we wish to note that though in a more rigorous treatment the many body effects, the escape and penetration depths and the nonuniformity of the light should be taken into account, nevertheless the experimental results and the theoretical formulations are in close agreement with each other.

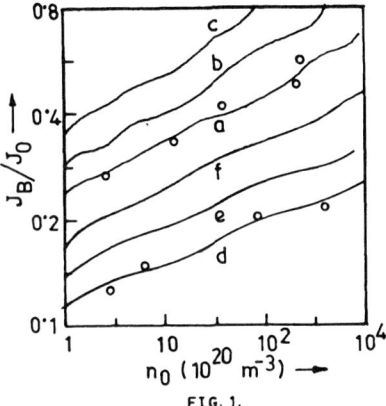

FIG. 1.

Fig.1. Plot of J_B/J_O in n-$Hg_{1-x}Cd_xTe$ as function of n_O in accordance with (a) three-band Kane model, (b) two-band Kane model and (c) parabolic energy bands. The plots d,e and f exhibit the same dependences for $In_{1-x}Ga_xAs_yP_{1-y}$ lattice matched to InP respectively. The circular plots exhibit the experimental datas [x = y = 0.2].

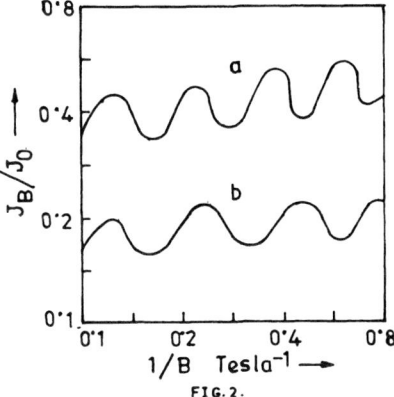

FIG. 2.

Fig.2. Plots of J_B/J_O as functions of $1/B$ for (a) n-$Hg_{1-x}Cd_xTe$ and (b) $In_{1-x}Ga_xAs_yP_{1-y}$ Lattice matched to InP. We have used the three-band Kane model for both the cases. ($n_O = 10^{20}$ m^{-3})

1. P.Y. Lu, C.H. Wang, C.M. Williams, S.N.G. Chu, C.M. Stiles, Appl. Phys. Letts. 49, 1372 (1986).
2. E. Weiss and N. Mainzer, J. Vac. Sci. Tech. 7A, 391 (1989).
3. R. Dornhaus and G. Nimtz, Springer tracts in modern Physics 78, 1 (1976).
4. T. Yamato, K. Sakai, S. Akiba, Y. Suematsu, IEEE J. QE14, 95 (1978)
5. T.P. Pearsall, B.I. Miller, R.J. Capik, Appl. Phys. Letts. 28, 499 (1976).
6. G.E. Hurwitz and J.J. Hsieh, Appl. Phys. Letts. 32, 487 (1978).
7. M.A. Littlejohn, J.R. HYauser, T.H. Glisson, Appl. Phys. Letts. 30, 242 (1977).
8. A.J. Ponomarev, G.A. Potapov, G. I. Kharu, I.M. Tsidilkovskii, Sov. Phys. Semicond. 13, 502 (1979).
9. R.K. Pathria, Statistical Mechanics, (Pergamon Press, London, 1977) p. 242.
10. P.M. Laufer, F.H. Pollak, R.E. Nahary, M.A. Pollack, Solid State Communication, 36, 419 (1980).
11. R.J. Nicholas, S.J. Sessions, J.C. Portal, Appl. Phys. Letts. 37, 178 (1980).
12. M. Elisov and T. Gelmont, J. Exp. Theo. Phys. 22, 101 (1990).

LONG-WAVELENGTH GERMANIUM PHOTODETECTORS BY ION IMPLANTATION

I.C. Wu[1,2], J.W. Beeman[1], P.N. Luke[1], W.L. Hansen[1], and E.E. Haller[1,2]
Lawrence Berkeley Laboratory[1] and Department of Materials Science and Mineral Engineering[2] ,University of California, Berkeley, CA 94720, USA

ABSTRACT

Extrinsic far-infrared photoconductivity in thin high-purity germanium wafers implanted with multiple-energy boron ions has been investigated. Initial results from Fourier transform spectrometer(FTS) measurements have demonstrated that photodetectors fabricated from this material have an extended long-wavelength threshold near 192μm. Due to the high-purity substrate, the ability to block the hopping conduction in the implanted IR-active layer yields dark currents of less than 100 electrons/sec at temperatures below 1.3K under an operating bias of up to 70mV. Optimum peak responsivity and noise equivalent power(NEP) for these sensitive detectors are 0.9A/W and 5×10^{-16} W/Hz$^{1/2}$ at 99μm, respectively. The dependence of the performance of devices on the residual donor concentration in the implanted layer will be discussed.

INTRODUCTION

Because of the potential technological importance of high performance far infrared detectors for space-born telescope applications, the development of far infrared Ge detectors using a two-layer structure based on the Blocked-Impurity-Band(BIB) concept is being extensively studied[1-5]. In contrast to normal photoconductors[1,3], BIB photodetectors have a number of desirable features: 1. extension of the cutoff wavelength limit from the dopant photoconductive onset to longer wavelengths, 2. a decrease in interference from cosmic radiation because of volume reduction, and 3. a reduction in the noise due to fixed unity gain.

These improvements have been realized to a large extent with Si BIB detectors for wavelengths up to 30 μm[6,7]. Furthermore, these novel devices have the capability of achieving background limited performance at the very low infrared levels encountered in space-based observations[8]. In addition, two-dimensional monolithic BIB detector arrays have been fabricated using IC process technologies[6,7].

Conventional BIB detectors consist of a thin, undoped intrinsic blocking layer grown homo-epitaxially on an intermediately doped substrate. Incident photons absorbed in this intermediately doped region generate free carriers which pass through the pure layer, and are collected as photocurrent when the device is properly biased. The ionized impurity state travels via successive bound carriers to the opposite electrode. The intrinsic layer serves to block the large currents which would arise from hopping conduction within the impurity band in the intermediately doped layer. The device derives improved performance from the fact that the presence of the blocking layers makes it possible to dope the IR-active layer more heavily than in conventional photoconductive detectors, without the corresponding increase in dark current due to hopping conduction. The higher doping also allows a reduction in device volume without compromising quantum efficiency of the device.

In view of considerable success of Si BIB detectors, it appears worthwhile to try to develope Ge detectors with similar structure. Because of the much smaller ionization energies of shallow impurities in Ge as compared with Si, Ge BIB detectors are expected to have a longer wavelength cutoff around 200 μm and perhaps beyond. A Ge BIB detector of conventional design consists of a pure Ge epitaxial layer which is grown on an intermediately doped IR-active substrate. Early devices fabricated by chemical vapor deposition (CVD)[9] indeed showed an extended long-wavelength threshold at 190 μm and had a promising peak responsivity of 5 A/W. However, data on Ge BIB devices which exhibit such characteristics as well as low dark currents (< 100 e⁻/s) and NEP[10,11] which are necessary for future satellite born applications have not been published so far.

In this paper, we present a new approach of fabricating two layer Ge far-IR photoconductors which show BIB device characteristics. In this new approach, the IR active layer is formed in an ultra-pure Ge substrate using ion implantation. The high-purity substrate serves as the blocking layer. Since the substrate can be selected from very high quality bulk crystals(| N_a-N_d | < 2x10¹⁰ cm⁻³)[12,13], its ability to block hopping conduction is assured. The thickness and dopant concentration of the IR-active layer can be controlled by varying ion energies and doses. Since complete analysis of BIB device physics and applications have been published elsewhere[1-5], we will confine ourselves to describing the processes required to make the new structure and to their characterization. Preliminary experimental results of the discrete far-IR Ge detectors will be presented and discussed.

EXPERIMENTAL DETAILS

The starting material for these devices investigated in the present work was prepared from ultra-pure Ge with extremely low impurity concentration (about 2x10¹⁰ cm⁻³) and essentially perfect crystalline quality[13]. Wafers of 200 μm thickness were cut from large boules with a diamond saw, then lapped and polish etched to remove all mechanical damage. Boron ions were implanted using several energies to approximate a flat impurity concentration profile of 3 x 10¹⁶ cm⁻³ from the surface to a depth of 0.5 μm . The wafers were then annealed for one hour at 400°C in argon to remove any implantation damage and to fully activate the boron[14]. Boron was used for implantation doping because it is the shallowest elemental acceptor in Ge and it is active as implanted at room temperature. These two characteristics guarantee the longest wavelength response and circumvent very high temperature annealing which might introduce other impurities into the wafer. Furthermore, boron is the lowest mass acceptor thus reaching the greatest implantation depth for a given ion energy.

A schematic diagram of a prototype ion implantation Ge BIB device described in this work is demonstrated in Fig. 1. Square samples of 1x1 or 2x2 mm² were cut from an implanted and annealed wafer with a dicing saw. The small samples were mounted with the implanted side down onto a glass slide using wax. The device was then further thinned by chemically etching the unimplanted side to reach a final sample thickness of 50 μm. The sides were

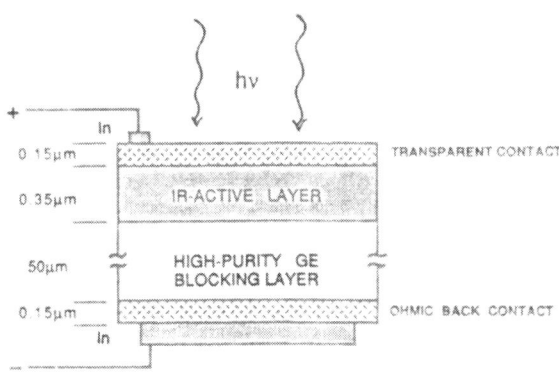

XBL 9010-3359

Fig.1 Schematic cross section of a ion implanted Ge far-IR detector.

also etched to remove any residual saw damage, which could contribute to excess surface leakage currents[8]. A final implantation of B+ ions formed a degenerately doped p++ contact on the top face without any post-implant annealing. We chose an implant dose of 2×10^{13} B+cm^{-2} at an energy of 25 keV. This produce ohmic, semi-transparent contact layers.[15]

Finally, a strip of copper was attached to the bottom surface of the detector using pure indium solder. The specimens were then mounted on copper heatsinks in an integrating cavity for photocurrent response and NEP tests. A copper wire was pressure-welded to the top surface by using a pure indium pad. Measurements of dark current were obtained using an integrating JFET amplifier JF-4 (Infrared Labs.) which has a read noise of less than 20 electrons.[16]

EXPERIMENTAL RESULTS AND DISCUSSIONS

Various detector figures of merit, including spectral response, responsivity, noise equivalent power, and dark current, were measured from several specimen. Because all the devices showed very uniform response, we chose to restrict ourselves to reporting results from just one of these specimen. The two spectral responses of this detector under different bias are shown in Fig.2. It was obtained using a Fourier transform spectrometer (FTS). The periodic variation in the current response resulted from the Fabry-Perot interference between the front and the back surfaces of the detector. When the applied bias was varied from 10 to 25 mV with the detector operating at a temperature of 1.5K, the long-wavelength cutoff moved from 140 to 192 μm, significantly beyond the 120 μm onset of a standard boron-doped Ge photoconductors[17]. Such a shift does not occur in the very pure blocking layer. It must be due to photoexcitation in the moderately doped ion-implanted layer.

BIB device structure can be compared to a metal-insulator-semiconductor (MIS) photodiode. A positive bias applied to the detector surface contact (Fig. 1) will deplete the moderately doped detection layer of ionized acceptors to some depth. The positively charged minority donors constitute the space charge. By solving Poission's equation, we find that the width of depletion layer W relates to the applied bias V_a and minority donor concentration N_d as follows:

$$W = \sqrt{\frac{\varepsilon_0 \varepsilon \ V_a}{e \ N_d} + t^2} \ - t$$

where ε is the dielectric constant of Ge, t the thickness of blocking layer, and e the electronic charge.   It is desirable to increase the applied bias to widen of the depletion layer W from the tail of the implant profile into the intermediately doped plateau so as to extend the cutoffs to the longest possible wavelengths and also to increase the quantum efficiency of the detector.   However, for our ion implanted BIB detectors, no further increase in both cutoff wavelength and photoresponse were observed when the bias voltage exceeded   25mV.   Beyond 25mV, the detector noise level increased sharply with the increase of the applied bias.

It should be mentioned that photocurrent could be also measured under reverse bias operation.   This response has been found to be due to the implanted contact on the back side of the device as shown in Fig.1.   However, the spectral response (Fig.2) generated from the back single-implant contact is different from that the front multiple-implant side.   The former always lacks a sharp long wavelength threshold and shows a lower value of the peak responsivity at about 60 μm.   For a conventional extrinsic photoconductor, the spectral response is completely symmetric with respect to applied bias polarity.   However, for these implanted detectors, the asymmetry of performance in regard to bias polarity is a strong indication that the spectral response   (Fig.2) of the detector comes primarily from the front multiple-implant layer.

In Fig.3, we show three figures of merit of this device.   The absolute responsivity and NEP values at 99 μm as a function of bias were tested at a photon background of 1×10^8 photons/cm^2·s and at a typical photon chopping frequency of 23 Hz with a low noise, trans-impedance preamplifier and a spectrum analyzer.   Dark currents at this test temperature of 1.3K were found to be independent of the bias up to 40 mV.   Operated at 24 mV, the detector has a NEP of 5×10^{-16} W/Hz$^{1/2}$ which is only about an order of magnitude higher than the value of background limited performance (BLIP).   Scaled to the measured responsivity at 99 μm, the corresponding peak responsivity in Fig.2 is   expected to be about 0.9 A/W at 120 μm.

The thickness of the detection layer for these devices is determined by two parameters, N_d and t.   For instance, with an applied bias of 25 mV, a value of t=50 μm, and an assumed $N_d = 1 \times 10^{11}$ cm^{-3}, the depletion width W has a value of 4.2μm.   This is eight times as thick as the range of 150 keV B$^+$ ions, i.e., the device is fully depleted except the electrode layer.   But, for $N_d = 1 \times 10^{12}$ cm^{-3}, with

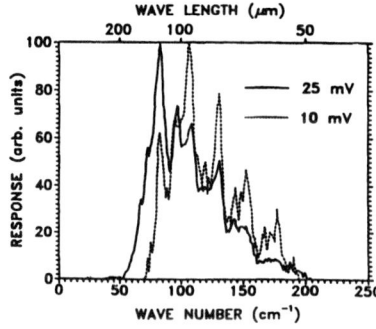

Fig.2   Spectral response of a boron-implanted Ge detector under two different bias 10 and 25 mV.

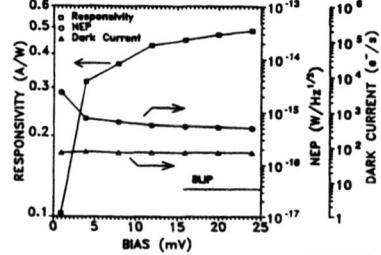

Fig.3   Responsivity and NEP of low-energy implanted Ge detector active layer depth = 0.5 μm, [B] = 3x10^{16}/cm^3 , narrow band filter at 99 μm, chopping frequency = 23Hz.

all other parameters as above, W would be 0.44 μm which is close to depth distribution of 150 keV B$^+$ ions. Therefore, the determination of N_d in the IR-active layer is crucial for device development. Due to the fact that carrier freeze-out does not occur in the intermediately doped layer, both the actual value of N_d and therefore W in the implanted IR-active layer are not known and are still under investigation. We have used variable temperature Hall effect measurements to determine donor concentrations on several low dose B-implanted layers([B]=1x10^{15} cm^{-3}). These low dose layers were deliberately utilized to avoid hopping conductivity in the implanted layer at low temperatures. Our preliminary Hall data indicate that 150 keV boron as-implanted layers are highly compensated with donor concentrations in the 1x10^{14} cm^{-3} range. Post-implant annealing steps for these implanted layers have been found to be necessary to lower donor concentrations to the 1x10^{13} cm^{-3} range, which is still higher than the expected value of $N_d < $ 1x10^{12} cm^{-3}. As already mentioned, the lower donor concentration values lead to an increased depletion region thickness, therefore, if the residual donor concentration N_d can be reduced to lower values ($N_d \sim$ 1x10^{11} cm^{-3}) via optimization of the fabrication processes, then the use of MeV implantation to produce IR-active layer several microns in thickness to improve the detector sensitivity become very desirable.

Finally, low dark current is a key factor which promises the detector performance in low background condition. The far-IR space-based astronomy application requirements for dark current are less than 100 electrons/sec/pixel with a typical pixel size of less than 0.5x0.5 mm^2. As shown in Figs.3 and 4, even lower dark currents, about 70 e$^-$/s, have been achieved with this discrete Ge device 1x1 mm^2 in size at operating temperatures below 1.3K and a bias below 70 mV. The temperature dependence of dark currents under different bias conditions is shown in Fig.4. Indeed, at a temperature of 1.25K, the value of the dark current did not exceed 100 e$^-$/s up to an applied bias of 80 mV. The breakdown voltage in the dark for this operating temperature was about 90 mV.

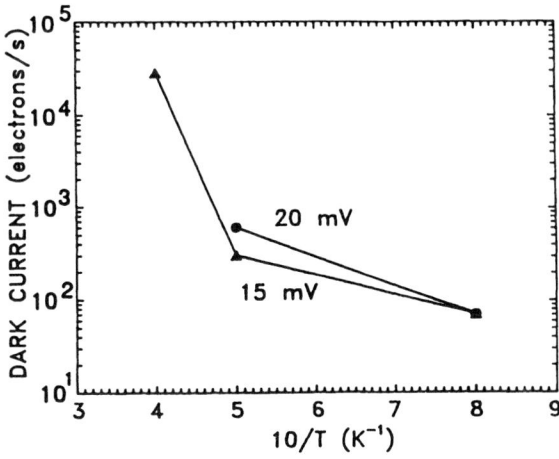

XBL 9010-3360

Fig.4   Dark currents as a function of temperature. At 1.25K, the values of dark currents are less than 70 e$^-$/s over the applied voltages from 1 to 70 mV.

CONCLUSION

We have fabricated the first far-infrared Ge detectors based on the BIB concept using ion implantation technique. These boron-implanted Ge BIB detectors exhibit operating characteristics compatible with requirements for low background applications. Device parameters such as low dark currents, acceptable sensitivity and extended wavelength threshold, demonstrate that ion-implanted Ge far-IR detectors offer promise for use in astrophysics instrumention. Further studies are aimed at fully understanding the device operation. In particular, the origin of the residual donor concentration in the ion-implanted IR-active layer needs to be determined. If the lower N_d value can be achieved in boron implanted layers, the use of high-voltage ion implantation to increase the active layer thickness will also be examined in order to optimize the detector performance.

This work was supported by NASA Contract No. W-14606 under Interagency Agreement with the Director's Office of Energy Research, Office of Health and Environmental Research, U.S. Department of Energy under Contract Numbers DE-A03-76SF00098 and DE-AC03-76SF00515.

REFERENCES

1   See articles in " Proc. 3rd IR Det. Techn. Workshop, NASA Tech. Memo 102209", C. McCreight, ed., (1989).
2. M.D.Petroff and M.G. Stapelbroek, U.S. Patent No. 4-568-960(4 February 1986).
3. V. Hadek, J. Farhoomand, C.A. Beichman, D.M.Watson and M.D. Jack, Appl. Phys. Lett. 46, 403(1985).
4. F. Szmulowicz and F. Madarz, J. Appl. Phys. 62,2533(1987).
5. M.G. Stapelbroek, M.D. Petroff, J.J. Speer, D.D. Arington, and C. Sayre, IRIS Specialty Group in IR Detectors, Boulder, CO, August 4, 1983.
6. D.B. Reynolds, D.H. Seib, S.B. Statson, T. Herter, N. Rowlands, and J. Schoenwald, IEEE Trans, NS 30, 857(1989).
7. T. Herter, N.Rowlands, S.V.W. Beckwith, and G.E. Gull, " Proc. 3rd IR Det. Techn. Workshop, NASA Tech. Memo 102209", C. McCreight, ed., 427(1989).
8. E.E. Haller, M. R. Hueschen, and P. L. Richards, Appl. Phys. Lett., 34,495(1979).
9. D.M. Watson and J.E. Huffman, Appl. Phys. Lett. 52 1602(1988).
10. C.S. Rossington, Ph.D. dissertation, University of California, Berkeley, 1988.
11. I.C.Wu (unpublished work).
12. E.E. Haller, W.L.Hansen , and F.S. Goulding, Adv. in Phys. 30, No.1, 93(1981).
13. W.L.Hansen and E.E. Haller, Mat. Res. Soc. Symp. Proc. 16, 1(1983).
14. K.S. Jones and E.E. Haller, J. Appl. Phys. 61, 2469(1987).
15. V.Hadek, D.M. Watson, C.A. Beichman, and M.D. Jack, Phys. Rev. B 31, 630(1985).
16. Infrared Labs. Inc. Product Brochure, April, 1989, p.30.
17. H. Shenker, W. J. Moore, and E. M. Swiggard, J. Appl. Phys., 35, 2965(1964).

PHASE RELATIONS AND BULK CRYSTAL GROWTH IN THE SYSTEM
CuInTe$_2$-MnIn$_2$Te$_4$

James T. Kelliher and Klaus J. Bachmann,
North Carolina State University, Department of Materials Science and
Engineering,Box 7919, Raleigh, NC 27695-7919.

ABSTRACT:

Recently manganese substituted I-III-VI$_2$ compounds have been investigated.
Their magnetic properties are similar to the paramagnetic transistion metal ion
substituted II-VI materials, but their crystal structure is non-cubic resulting in
anisotropies in their physico-chemical properties that do not exist in the
zincblende structure diluted magnetic semiconductors (DMS) materials. In this
paper, we report the phase relations on the pseudobinary cut CuInTe$_2$-
MnIn$_2$Te$_4$ based on x-ray diffraction and differential thermal analysis data. The
range of chalcopyrite structure alloys of composition Cu$_{1-x}$Mn$_x$□$_x$In$_2$Te$_4$ is
limited to x<0.52 due to an eutectic at x=0.74, T= 734°C. A second eutectic
exists at x = 0.97, T = 737.5°C. A heretofore unknown congruently melting
compound exists in between the two eutectics at x= 0.85, T$_m$ = 760°C. Also, the
thus far unknown melting point of the compound MnIn$_2$Te$_4$ was determined
(740°C). Based on the analyses of the first to freeze parts of directionally
solidified melts in this range of liquidus compositions by inductively coupled
plasma emission spectroscopy the solidus is constructed.

INTRODUCTION:

In the past decade, diluted magnetic semiconductors (DMS) have
received considerable attention because of their interesting physical
properties, e.g. magnetic polaron formation, giant Faraday rotation and other
unusual properties[1-3]. Thus far, research in this field has focused on II-VI
compounds that have been converted into DMS materials by substituting
paramagnetic transition metal ions into the diamagnetic cation sublattice.
Recently the substitution of paramagnetic Mn^{2+} ions into the lattice of I-III-VI$_2$
compounds has been reported[4-6], extending the exploration of DMS materials
to a new class of compounds. In particular, magnetic susceptibility measurements have shown that the substitution of Mn^{2+} in CuInTe$_2$ can be
restricted to the Cu sublattice. Since Cu exists in CuInTe$_2$ as a monovalent
cation, each substitution event must entail the formation of a vacancy for
charge compensation. Therefore, the general composition of substitutional
alloys is Cu$_{2-2x}$Mn$_x$ $_x$In$_2$Te$_4$ extending from pure CuInTe$_2$ (x =0) to MnIn$_2$Te$_4$ (x
=1). The phase relations on this pseudobinary are the topic of this paper,
including the determination of the range of solid solutions, intermediate phases,
and the solidus-liquidus separation. This information is needed to guide experiments concerning bulk crystal growth of the DMS alloys from the melt.
Single crystals of these alloys are essential for the measurements of their
fundamental optical, electrical and magnetic properties.

One of the endpoints of the pseudobinary understudy, CuInTe$_2$,
crystallizes in the chalcopyrite structure with lattice parameters of a = 6.195 Å
and c = 12.39 Å[7]. The reported melting point of CuInTe$_2$ is 780 °C[8].

The other endpoint of the pseudobinary phase diagram is MnIn$_2$Te$_4$,
which was first synthesized in 1975 by Range and Hubner[9]. The crystal
structure of this AB$_2$C$_4$, compound semiconductor was determined to be

Mat. Res. Soc. Symp. Proc. Vol. 216. ©1991 Materials Research Society

tetrahedrally coordinated, representing a defect chalcopyrite structure which contains one vacancy per A atom. The cations are randomly distributed over the occupied sites, but the vacancies are ordered. The lattice parameters for $MnIn_2Te_4$ prepared from the pure elements at 800°C were determined to be a=6.191 Å and c= 12.382 Å[9].

EXPERIMENTAL:

Melts of $CuInTe_2$ and $MnIn_2Te_4$ were synthesized from the elements, and single crystals were produced by gradient freezing. The copper, indium, and tellurium had a purity of 6N's, and the manganese was 4N pure. $CuInTe_2$ was grown using a two zone furnace with the tellurium pressure controlled to ≈ 100 mtorr at 838°C[10]. $MnIn_2Te_4$ was grown in a single zone furnace after the procedure described by Range and Hubner[9]. Portions of the first to freeze ends of these crystals and mixtures thereof corresponding to selected compositions on the $(Cu_2In_2Te_4)_{1-x}$-$(MnIn_2Te_4)_x$ pseudobinary were examined with DTA and x-ray diffraction.

The DTA experiments were initiated by heating the samples, sealed in evacuated fused silia ampoules, at a rate of 2 °C/min to 950°C. The samples are held at 950°C for 3 to 6 hours and is cooled to 550°C at a rate of 2°C/min. After the sample annealled overnight at 550°C, DTA data were collected on runs consisting of ramping to 850°C at 2°C/min, holding for 15 minutes, and then furnace cooling to 550°C. At least 3 independent runs were performed for each composition. Various samples for x-ray analysis were prepared in a similar fashion as described for the DTA.

In addition crystals of three intermediate compositions on the pseudobinary at x = 0.2, x = 0.47, and x = 0.85, were grown using two zone gradient freezing. Chemical analysis by Inductively Coupled Plasma (ICP) emission spectroscopy, was preformed on the first to freeze ends of these crystals to determine the solidus-liquidus separation. For more detailed information the reader is referred to reference [11].

RESULTS:

Based on the data, the phase diagram on the $(Cu_2In_2Te_4)_{1-x}$-$(MnIn_2Te_4)_x$ pseudobinary cut can be constructed and is presented in figure 1. The melting point of $CuInTe_2$ was found to be 780°C with a polymorphic transition at 667°C concurring with the reported work[8,12,13]. The liquidus starts from the $CuInTe_2$ endpoint with a shallow slope which joins smoothly to an almost linear liquidus line with negative slope between x $(MnIn_2Te_4)$ = 0.2 to 0.6. No polymorphic transitions are observed in the DTA traces up to x $(MnIn_2Te_4)$ = 0.61. At this composition, a second transition at 731°C is observed in addition to the thermal arrest at the liquidus. In the composition range $0.74 \leq x \leq 0.85$, the liquidus line has a positive slope. Extrapolation of the negative and positive portions of the liquidus in the range $0 \leq x \leq 0.86$ reveals an eutectic at x =0.74 corresponding to a temperature of ≈734°C. This agrees within the experimental error with the second transition temperature observed at x = 0.61. Also at x =0.78 and x =0.81 thermal arrests are observed below the liquidus at 725 and 735°C, respectively. This agrees within a ±1 % error with the temperature associated with the eutectic at x = 0.74. In view of the difficulty in locating the small heat effects of the eutectic transitions, this is a reasonable error. Therefore, we place the

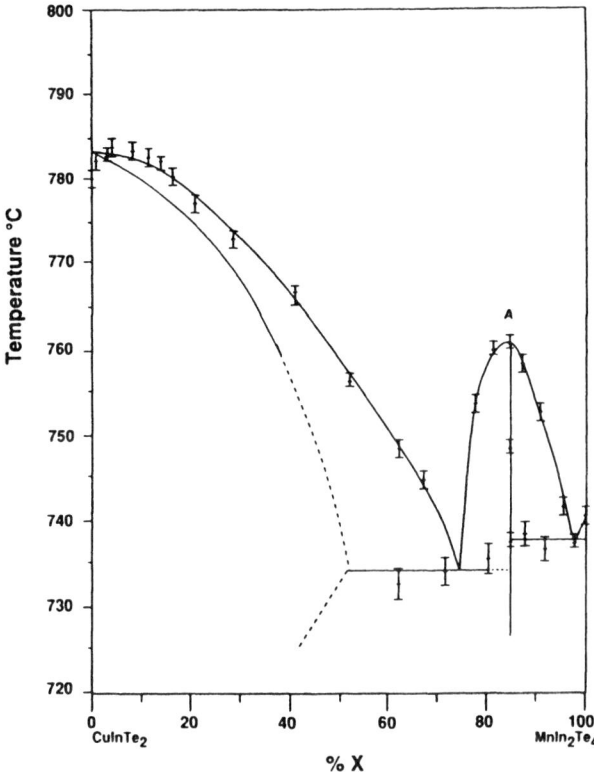

Figure 1: $CuInTe_2$-$MnIn_2Te_4$ pseudobinary phase diagram. Solid lines supported by experimental data (see reference [11] for complete data set), dashed lines shown for completeness, but not secured by experimental data.

eutectic temperature at 734°C which is indicated in the phase diagram of figure 1 as a solid horizontal line. A maximum is found in the liquidus near to x = 0.85 with an associated congruent melting point of 760°C corresponding to a heretofore unknown compound. From the liquidus temperatures observed at x = 0.86, 0.91 and 0.975, it follows that beyond this composition the liquidus slope is negative up to a eutectic close to the $MnIn_2Te_4$ endpoint of the pseudobinary. From the liquidus data, we place this eutectic at x = 0.97 corresponding to a temperature of 737.5, just below the melting point of $MnIn_2Te_4$, which is 740°C. Second thermal arrests below the liquidus were observed for x = 0.86 and x = 0.91 at 739°C and 736°C, respectively. Therefore we place the eutectic at x = 0.97 at a temperature of 737.5°C as represented by a solid horizontal line in the phase diagram.

The interpretation of the maximum in the liquidus near to x = 0.85 as a congruent melting point is further corroborated by inductively coupled plasma emission analyses (ICP) of samples taken from different parts of a crystal grown at this composition revealing the segregation behavior. This analysis shows that a crystal grown at average melt composition x=0.85 freezes out with the same uniform composition within the experimental error of ICP analysis which is of the order of ± 1%. The structure of the compound $Cu_{0.3}Mn_{0.85}\square_{0.85}In_2Te_4$ is presently unknown, but similar to the structure of $MnIn_2Te_4$ as indicated by the similarity of the x-ray powder patterns. ICP analyses of the first to freeze parts of directionally solidified liquids in the range $x\leq 0.52$ establishes a relatively narrow solidus-liquidus separation for $CuInTe_2$ rich compositions. However the error in the ICP analyses results in an uncertainty in the determination of the tie lines connecting the liquidus line of the pseudobinary cut with endpoints on the solidus surface which within the error limits of the analyses may be outside the pseudobinary cut. Indeed a tendency towards Cu-rich and Te-deficient solidus compositions was observed, but is within the experimental error of the ICP analysis. Although the solidus line constructed from these data in the range $0\leq x\leq 0.3$ does not necessarily imply tie lines to this solidus from pseudobinary liquids, the error is small enough to make the pseudobinary phase diagram shown in figure 1 a reasonable guide for the crystal growth in the $CuInTe_2$ - $MnIn_2Te_4$ system. The following conclusions may be drawn:

1: The range of solid solutions in the system $Cu_{2-2x}Mn_xIn_2Te_4$ extends well beyond the limit (x=0.12) utilized in reference [4].
2: The growth of single crystals of Mn^{2+}-rich alloys (x > 0.5) is precluded by the eutectic at x = 0.74.
3: A quaternary congruently melting compound exists at close to x = 0.85.
4. The melting points of this compound and the heretofore unknown melting point of $MnIn_2Te_4$ are 760°C and 740°C, respectively.

ACKNOWLEDGMENTS:

This work has been supported by NASA grant NAG 1-1100.

REFERENCES:

1 E.L. Nagaev, *Physics of Magnetic Semiconductors* MIR Publishers, Moscow,1983.
2 J.K. Furdyna and J. Kossut *Semiconductors and Semimetals* vol 25, Academic Press, Inc. San Diego, 1988.
3 For general review of dilute magnetic semiconductors compounds see *Material Research Society Symposia Proceedings*, 89, R.I. Aggrawal, J.K. Furdyna and S. von Molner, eds. Material Research Society, Pittsburgh, PA, 1987.
4 L.-J. Lin, J.H. Wernick, N. Tabatabaie, G.W. Hull and B. Meagher, *Appl. Phys. Lett.* 51 (24) 2051, 1987.
5 L.-J. Lin, N. Tabatabaie, J.H. Wernick, G.W. Hull and B. Meagher, *J. Electronic Mater.*, vol 17, no. 4, 321, 1988.
6 J.-R. Gong, H. Neff and K.J. Bachmann, *J. Electronic Mater.*, vol 17, no. 5, 361, 1988.
7 H. Hahn, G. Frank, W. Klingler, A. Meyer and G. Storger, *Z.Anorg. Chem.* 271,153,1953.
8 V.P. Zhuze, V.M. Sergeeva and E.L. Shtrum, *Soviet Phys.- Tech. Phys.*, 3, 1925, 1958.
9 K.-J. Range and H.J. Hubner, *Z. Naturforch. Teil B*, 30, 145, 1975.
10 J. van den Boomgaard, *Philips Res. Rep.* 10, 319, 1955.
11 J.T. Kelliher, Masters Thesis, Department of Materials Science and Engineering, North Carolina State University, 1990.
12 L.S. Palatnik and E.I. Rogacheva, *Soviet Physics Doklady*, 12, 503, 1967.
13 S.M. Zalar, *Metallurgy of Semiconductor Materials* , pp263-283, AIME, Interscience Publishing Co., New York 1962.

MICRO-ELLIPSOMETRIC MEASUREMENTS OF STRAINED LAYER STRUCTURES

YI-MING XIONG AND PAUL G. SNYDER
University of Nebraska, Center for Microelectronic and Optical Materials
Research, and Dept. of Electrical Engineering, Lincoln, NE 68588-0511.

ABSTRACT

Ellipsometric measurements are extremely sensitive (on the monolayer scale) to layer thicknesses and surface or interfacial roughness. However, lateral resolution is poor, because of the need for a highly collimated ($\leq 0.05°$) and therefore large diameter (typically 1 mm) optical beam. Many materials are inhomogeneous on a smaller lateral scale, in which case conventional ellipsometry simply measures some average value. Focusing the beam leads, for an incoherent light source, to an uncollimated beam and a consequential spread in the angle of incidence, which in turn degrades the sensitivity. We describe here a technique for focusing a coherent, Gaussian laser beam to a minimum spot size of about $30\mu m$ x $130\mu m$, while maintaining a negligible effective spread in angle of incidence. The technique was used to measure semi-insulating (S.I.) GaAs, laser annealed ion implanted GaAs, and $In_x Ga_{1-x} As$ strained layer structures.

INTRODUCTION

Variable Angle Spectroscopic Ellipsometry (VASE) and fixed angle SE are powerful techniques for characterizing a broad spectrum of materials and multilayer structures. A few examples include determination of layer thicknesses (to within a few Angstroms) and composition in $GaAs/Al_x Ga_{1-x} As$ structures [1]; characterization of multi-dielectric and polysilicon layers [2]; and in-situ characterization of compositionally modulated III-V growth [3]. Homogeneity studies are also possible [4], although here the lateral scale of resolution is on the order of millimeters.

Higher lateral resolution can be achieved with beam focusing [5]. However, focusing an incoherent light source results in an uncollimated beam, with a continuum of incidence angles. As the focusing length decreases, the focused spot size decreases, but the spread in incidence angle becomes correspondingly larger. Because the measured ellipsometric parameters are strongly angle-dependent, this angular spread degrades the sensitivity of the data, and complicates the data analysis.

We describe measurements using a coherent laser source, which take advantage of Gaussian beam properties to achieve both a small focused spot at the sample, and negligible effective angular spread at the focal point. A four wavelength helium-neon laser was used to reduce model parameter correlations.

EXPERIMENTAL

The ellipsometry apparatus is shown schematically in Figure 1. A moveable steering mirror guides the focused laser beam into a conventional rotating analyzer ellipsometer. The laser has four selectable wavelengths: 594.1, 604.0, 611.9, and 632.8 nm. A spatial filter, beam expander, and diffraction-limited focusing optics are attached directly to the laser. For comparison, measurements at the same wavelengths and angles were made using a well-collimated beam (< 0.05° angular spread), from a xenon arc lamp attached to a half-meter spectrometer.

We review here the properties of a focused TEM_{00} mode (spherical Gaussian beam), and their application to micro-ellipsometry (ME). For a beam diameter of A and focal length D, the minimum spot waist size (radius) is as follows [6]:

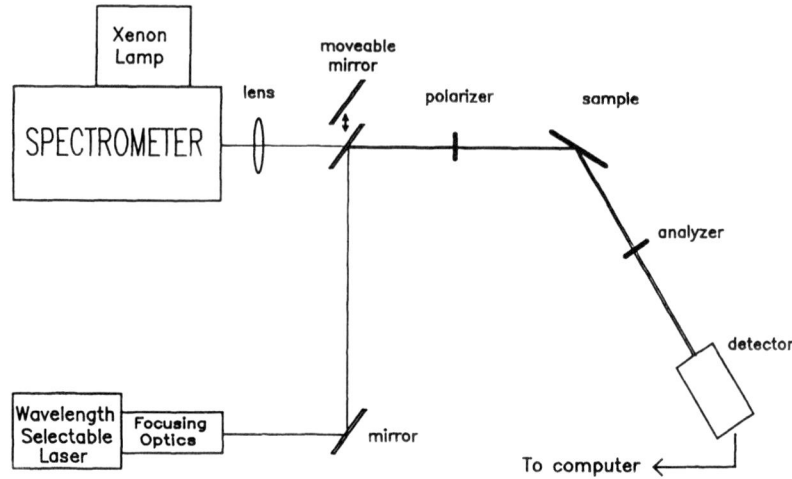

Figure 1. Schematic of experiment, which allows either a conventionally collimated incoherent beam, or a focused laser beam to be used as the light source for a rotating-analyzer ellipsometer.

$$W_o = \frac{\lambda D}{\pi A} \tag{1}$$

For $\lambda = 632.8$ nm, $D = 2$ m, and $A = 2.5$ cm, we have $W_o = 16\mu m$, or a beam diameter of about $32\mu m$.

The intersection of beam and sample produces an elliptical spot on the surface, with minor axis length W_o and major axis length $4W_o$ (for $75°$ angle of incidence). Thus the spot size in our case is $32\mu m$ x $128\mu m$. The confocal parameter b is the distance over which the beam area increases by a factor of two:

$$b = \frac{\pi W_o^2}{\lambda} \simeq 1300\mu m \tag{2}$$

Therefore the beam remains highly collimated over the $130\mu m$ length of the spot. More important is the radius of curvature R of the beam front, as this determines the spread in angle of incidence. R is infinite at the waist minimum (focal point), and equal to $\pm b^2/2W_o$ at the extreme ends of the spot. The corresponding maximum variation in angle of incidence is

$$\delta\phi \simeq \frac{2W_o^2}{b^2} = \frac{2\lambda^2}{\pi^2 W_o^2} \simeq 0.02° \tag{3}$$

This is less than the divergence of the collimated incoherent beam from the spectrometer. Farther away from the focal point, the angular spread approaches the geometric value of $0.72°$. Equation (3) shows the inverse-square relationship between $\delta\phi$ and W_o. For example, reducing W_o by a factor of two (by using a shorter focal length) would increase $\delta\phi$ by a factor of four.

Conventional SE [7] and VASE [1] have been described in detail elsewhere. We only repeat here that 1) the ellipsometric parameters ψ and Δ are defined as [8]

$$\tan\psi \ e^{i\Delta} = \frac{R_p}{R_s} \tag{4}$$

where R_p and R_s are the total complex reflection coefficients of the sample, for light polarized parallel to (p) and perpendicular to (s) the plane of incidence; and 2) the ψ and Δ data measured at each wavelength and angle of incidence are numerically fitted, using a multilayer model and varying designated model parameters to minimize the mean square difference between measured and calculated values:

$$MSE = \frac{1}{N} \sum_{i=1}^{N} [\ (\ \psi_i^m - \psi_i^c \ )^2 + (\ \Delta_i^m - \Delta_i^c \ )^2 \ ] \tag{5}$$

where N is the total number of measurements.

RESULTS

A direct comparison between conventional and micro-ellipsometry was made, by measuring a S.I. GaAs surface. In this case, where the sample is laterally uniform over a large area, the conventional and microellipsometry measurements should be the same. The two sets of measurements are shown in Figure 2. We attribute the small, nearly constant difference between the two Δ sets to a very small (< 0.05°) difference in the angle of incidence. The model used to fit these data was a thin layer of GaAs oxide, on a GaAs substrate. Solving for the oxide thickness, both data sets gave nearly the same result (to within 0.05 nm): 2.2 nm. The MSE for both fitted data sets was also nearly the same. This demonstrates that ME produces the same results as the conventionally collimated beam, for a laterally uniform sample.

The second sample was GaAs, whose surface was uniformly irradiated with Si ions, 180 keV, to a total fluence of $1 \times 10^{16}/cm^2$. The partially amorphized surface was then laser annealed with a single dye laser pulse, at 0.84 J/cm^2, and a Gaussian beam profile. Laser annealing produced a typically 300μm diameter, crystalline region, with very heavy n^+ doping (>$3 \times 10^{19}/cm^3$) [9]. ME measurements, made with the spot centered within the annealed area, are shown in Figure 3. Also shown are conventional measurements, made with the 1mm x 4mm ellipsometer beam spot overlapping the much smaller annealed spot. The very large difference between the two data sets is due to the difference in beam sizes, and corresponding sampled areas.

The ME data were fit very well using a simple model, consisting of an n^+ GaAs substrate (whose dielectric function was measured conventionally, from a separate sample), with a native oxide. The oxide thickness was varied for a best-fit. In addition, the angle of incidence was allowed to vary in the fit, to determine whether a significant error had occured in alignment. The result was an oxide thickness of 2.4 nm, an angle of incidence of 74.97° (nominal angle was 75°), and a very low MSE of 1×10^{-5} (indicating a very good fit). By contrast, the conventional data set could not be fit assuming a crystalline GaAs substrate.

Finally, a strained layer sample containing $In_xGa_{1-x}As$ was measured. The layer structure was: 10nm $n^+In_{.53}Ga_{.47}As$ / 20nm intrinsic $In_{.52}Al_{.48}As$ / 15nm n^+ $In_{.52}Al_{.48}As$ / 5nm intrinsic $In_{.52}Al_{.48}As$ / d intrinsic $In_xGa_{1-x}As$ / 40nm intrinsic $In_{.53}Ga_{.47}As$ / 400nm intrinsic $In_{.52}Al_{.48}As$ / 30 period 3nm InGaAs, 3nm InAlAs superlattice / S.I. InP substrate, with strained layer thickness d=10nm, and composition x=0.7. Figure 4 shows the comparison between ME and conventional measurements. Here again, a constant offset between the two Δ set, due to a slight difference in the incidence angle,

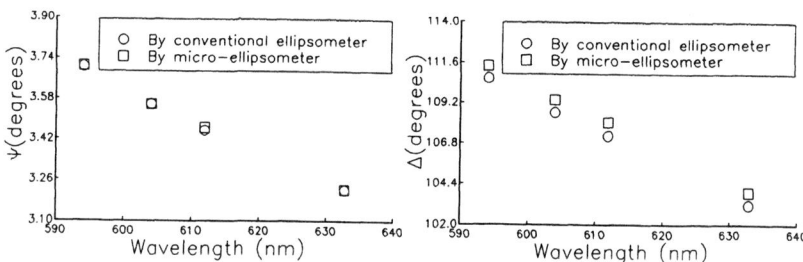

Figure 2. Measured data for semi—insulating GaAs.

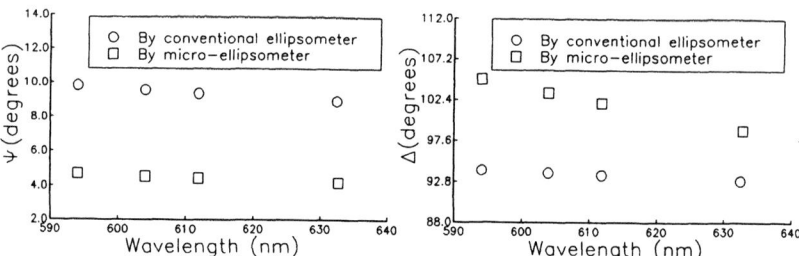

Figure 3. Measured data at annealed spot on GaAs.

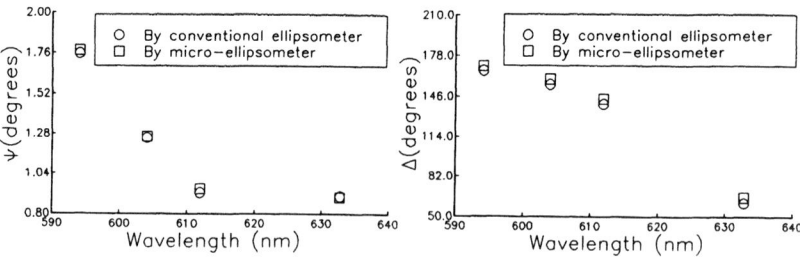

Figure 4. Measured data for In(0.7)Ga(0.3)As.

is observed. These data were not analyzed, because of their limited number with respect to the number of model parameters. Nevertheless, the ME measurements have the advantage of sensitivity to lateral inhomogeneities across the sample, with 0.1 mm or better resolution.

CONCLUSIONS

Conventional spectroscopic ellipsometry, with a highly collimated, large diameter, incoherent optical beam, achieves excellent depth resolution but poor lateral resolution. Focusing the incoherent beam to a smaller spot size spoils the beam collimation, and therefore the sensitivity of the measurements. We have demonstrated that a coherent Gaussian beam may be focused to the order of 100μm or less, with no reduction of the effective beam collimation. Direct comparisons were made between conventional and micro-ellipsometry, on three different samples, to confirm the theoretical predictions and the usefulness of the technique.

ACKNOWLEDGEMENTS

This work was supported by NASA Lewis Research Grant NAG-3-154. The laser-annealed GaAs sample was provided by Dr. Huade Yao (University of Nebraska, Department of Electrical Engineering).

REFERENCES

1. P.G. Snyder, M.C. Rost, G.H. Bu-Abbud, J.A. Woollam, and S.A. Alterovitz, J. Appl. Phys. 60, 3293 (1986).

2. Yi-Ming Xiong, P.G. Snyder, J.A. Woollam, E.R. Krosche, Y.Strausser, MRS Symp. Proc. for 1990 Spring Meeting (in press).

3. D.E. Aspnes, W.E. Quinn, and S. Gregory, Appl. Phys. Lett. 56, 2569 (1990).

4. S.A. Alterovitz, P.G. Snyder, K.G. Merkel, J.A. Woollam, D.C. Radulescu, and L.F. Eastman, J. Appl. Phys. 63, 5081 (1988).

5. M. Erman and J.B. Theeten, J. Appl. Phys. 60, 859 (1986).

6. A.E. Siegman, An Introduction to Lasers and Masers (McGraw-Hill, New York, 1971), p. 304.

7. D.E. Aspnes and A.A. Studna, Appl. Optics 14, 220 (1975).

8. R.M.A. Azzam and N.M. Bashara, Ellipsometry and Polarized Light (North Holland, Amsterdam, 1977).

9. H. Yao and A. Compaan, Appl. Phys. Lett. 57, 147 (1990).

II-VI / III-V STRAINED-LAYER SUPERLATTICE MATERIALS AND DEVICES

GAN FUXI   WANG HAILONG   CUI JIE
Shanghai Institute of Optics & Fine Mechanics, Academia Sinica,
P.O.Box 800-216, Shanghai 201800, P.R.China

ABSTRACT

The optical and spectroscopy properties of II-VI wide gap semiconductor SLSs have been studied. We report the first observation of the pulse compression in ZnS-ZnSe MQW and the folded LA phonon modes in ZnS-ZnSe SLS; The confined LO_n phonon modes in ZnSe-ZnTe SLS are observed and the critical thickness of ZnSe-ZnTe SLS are determined, for the first time, by Raman scattering.

INTRODUCTION

II-VI wide bandgap semiconductor strained-layer superlattices(SLSs) are potential materials for application to optoelectronic devices on the short-wavelength visible region. Because they are of the direct wide energy bandgap, high electro-optic coefficient, wide range of transmission and continuously tunable refractive index.

High quality ZnSe-ZnTe,ZnS-ZnSe and ZnTe-ZnS SLSs have been grown by MBE[1,2], MOCVD[3] and MOMBE[4], etc. Structural and optical properties such as X-ray diffraction[2-4], photoluminescence (PL)[1,3-5], Raman scattering[6,7] have been studied by various groups.

In this paper we report the optical and spectroscopy property studies on ZnS-ZnSe and ZnSe-ZnTe SLSs grown on GaAs and InP by MBE. The photoluminescence spectra and Raman spectra are discussed. The characteristics of a F-P cavity type ZnS-ZnSe multiple quantum well(MQW) optical bistable device are measured.

EXPERIMENTS

The ZnS-ZnSe SLS and MQW are grown on (001) GaAs substrates with ZnSe buffer or ZnS cladding layers, the ZnSe-ZnTe SLSs are grown on (001) InP. The samples are grown at 320-350℃ by MBE, X-ray diffraction measurements combining the growth rates of monolayers determined by RHEED oscillation periods had been used to determine the thicknesses of individual layers.

RESULTS AND DISCUSSION

PL spectra of ZnS-ZnSe and ZnSe-ZnTe SLSs

The photoluminescence measurements of the ZnS-ZnSe SLSs were carried out at 77K. The N₂ laser(337nm) was used as an excitation source. Fig.1 shows the excitonic emission spectra, the sharp emission lines located at 415nm and 440nm which were shorter than the value of the ZnSe band-edge(442nm) were attributed to the recombination of electron-hole pairs between the quantized energy levels in the ZnSe well layers.The value of the full width at half maximum (FWHM) of the peak is 50mev, which is smaller than the values reported previously.

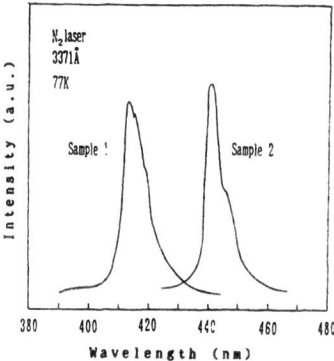

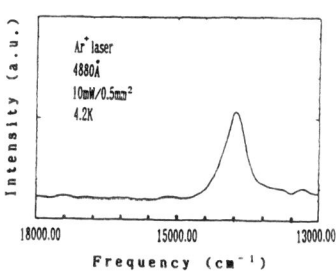

Fig.1 Photoluminescence spectrum
of ZnS–ZnSe samples

Fig.2 A typical PL spectrum at 4.2K
for a ZnSe–ZnTe SLS sample

The PL of ZnSe–ZnTe SLS with period of 60A were carried out at 4.2K excited by Ar^+-ion laser 488nm line. The emission peak located at 675nm. FWHM of the peak is 62.3mev,as shows in Fig.2. As we know, the ZnSe–ZnTe SLS belongs to type II, electrons and holes are separted into two different constituent materials. Though there are some reports on PL spectra and the band–structure calculations[8], the nature of the PL from this system is not well understood.

Raman spectra of ZnSe–ZnTe and ZnS–ZnSe SLSs

The Raman spectra were excited in a backscattering geometry along [001] direction using the 488nm line of an Ar^+-ion laser, the laser power on the sample was 70–100mW, see Ref.[9].

Besides the folding and the confinement effects of SL, for strained–layer superlattice, one has to consider the effect of elastic strains induced by lattice mismatch on the phonon frequencies. This is particularly important for II–VI wide gap compounds ZnSe, ZnTe and ZnS because of the large lattice mismatch between the individual layers.

In typical Raman spectra of the ZnSe–ZnTe SLSs,ZnSe–like and ZnTe–like LO phonons are observed. The LO modes frequencies are shifted as compared with the bulk materials.Fig.3shows the relationship between the shifts of LO frequencies and the thicknesses of ZnSe–ZnTe SLS layers.All the samples have the equal thicknesses for two individual layers,but have different periods. The solid lines show the measured shifts including the red shifts due to confinement and the shifts induced by the elasic strain; the dash lines are the shifts induced by the elastic strain. We use the methods described in Ref.[10]to calculate the red shifts of confined modes in the absence of strain using the dispersion curves of phonons of ZnSe and ZnTe given in Ref.[11,12]. As we see, when the individual layer thickness exceeds 40A, the shifts are almost constant and comparatively small, but when the thickness is smaller than 40A, the shifts changed sharply. For the sample with layer thickness of 10A, the shifts are in better agreement with the calculated values[6] assuming that the two individual layers are commensurate. Therefore,the sample is in a commensurate configuration, and the large elastic strain is

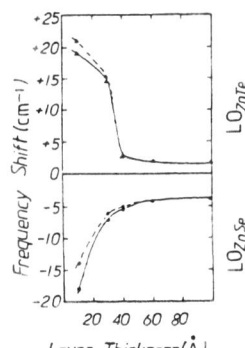

Fig.3 The relationship between the shifts of LO phonon modes frequencies and the individual layer thicknesses. Solid lines are the measured values that including the strain induced shifts and the red shifts due to confinement, deshed lines are strain induced shifts.

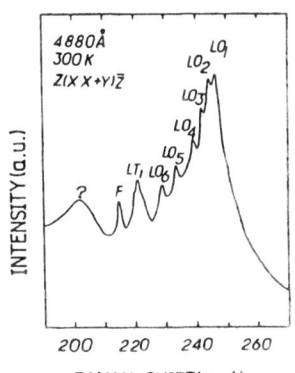

Fig.4 Confined LOm phonon modes in ZnSe layer of the ZnSe-ZnTe SLS sample(27A,29A). LT1 is confined LO mode in ZnTe layer, the F peak is attributed to folded LO mode and the ? band to interface vibration or delocalized optical phonon X=(110),Y=(110),Z=(001).

accommodated in the layers instead of forming misfit defects. But this is not true for the samples which individual layers exceed 40A. For these samples,there are no large elastic strains in the layers, because of small shifts of phonon frequencies. These samples should be in free-standing configurations (or incommensurate growth). We estimate that the critical thickness for ZnSe-ZnTe SLS with the 7.3% lattice mismatch is 40A.

Fig.4 shows the Raman spectrum from the ZnSe-ZnTe(27A,29A) SLS. Six confined LOₘ phonon modes in the ZnSe layers have been observed. We have calculated the confined phonon frequencies by considering the red shifts due to confinement and the shifts induced by elastic strain. Table I lists the calculated and measured frequencies of the confined LOₘ modes in ZnSe layer. The shifts induced by strain[13] were calculated using the parameters in Ref.[14]. The lattice constants of ZnSe and ZnTe are 5.6687A and 6.104A,respectively.The lattice mismatch is as large as 7.3% which is the source of tensile strain in the ZnSe layers and compressive strain in the ZnTe layers in (100)plane. The tensile strain induces the red shifts 6.5cm^{-1} of confined LOₘ frequencies in ZnSe layer, and it is much larger than the red shift due to confinement, 0.5cm^{-1}(for m=1 mode). The compressive strain in the ZnTe layer induces a blue shift as large as 15cm^{-1} which is greater than the red shift due to confinement, 0.3cm^{-1} (for m=1 mode). Thus the two groups of phonon dispersion curves under the strain field would overlap in the region about 210-220 cm^{-1}.Therefore,the overlap exists between the ZnTe-like confined LOₘ mode and the confined LOₘ mode in ZnSe layers with the m≥7. The mode labeled by F must be identified as due to scattering by the folded LO mode, and our calculations show that the frequency of mode F is equal to the m=8 confined LOₘ mode in the ZnSe slab and the m=4 confined LO₄ mode in the ZnTe slab. The frequencies of two modes which have the same symmetry of A₁ are in the overlapping range and the modes can propagate throughout the entire superlattice. The mode labeled by the question mark would be an interface mode or a delocalized mode.

Table I The measured and calculated frequencies (cm^{-1}) of confined LO_m modes in ZnSe layer

m	1	2	3	4	5	6	7
ω_m	246	244	242.5	239	233.5	228.5	
ω_{cal}	246	244.5	242.5	239.2	234.3	229	222

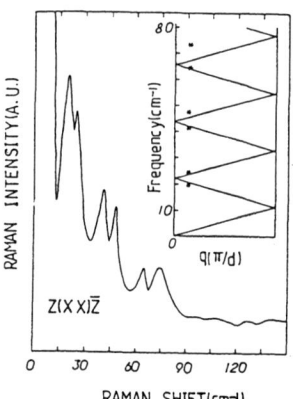

Fig.5 Low-frequency region Raman spectrum for the ZnS-ZnSe(26A,29A) SLS sample, which shows three folded LA doublets.

In Raman spectra of the ZnS-ZnSe SLSs, we have observed the ZnSe-like LO phonon mode and the weak ZnS-like LO phonon mode, we also observed the very weak ZnSe-like TO phonon mode.

Fig.5 shows the Raman spectrum of the folded LA modes for the ZnS-ZnSe(26A,29A) SLS sample. In the region 10-90cm^{-1}, three doublets were seen at room temperature near 22cm^{-1},44cm^{-1} and 67 cm^{-1}, the separation between the components of the doublet is about 5cm^{-1}. The inset of the figure is a calculated dispersion curve for the LA phonons using the Rytov model with the strain-free parameters[14]. The disagreement with the experiment may come from the neglect of the strain effects on the phonon dispersion curves. As we know,in the[001] direction,the ZnSe layers are under tensile strain,while the ZnS layers are under compressive strain.Therefore, the dispersion curve of ZnSe should shift down while that of ZnS should shift up. This effect broadens the peaks and makes the separation of the components of the doublet is larger than the calculated value.

ZnS-ZnSe multiple quantum well(MQW) optical bistable devices

The bistablities in Fabry-Perot(F-P) etalons constructed by Ⅲ-Ⅴ compounds SL have been reported earlier[15,16], but as for zinc-chalcogenides,the work reported is limited to bulk and film materials[17-19] by now. We have, for the first time, observed the optical bistablity in a MBE-grown ZnS-ZnSe MQW Fabry-Perot

etalon.The etalon was fabricated as follows. After the substrate
had been mechanically polished down, the MQW wafer was cleaved
to a certain size and stuck on a polished thin sapphire slice by
epoxy resin(the sample and the sapphire were in optical contact)
Using photoresistant or nail polish as mask, a window less than
1mm diameter was left on the GaAs substrate.The window first was
etched down by a $H_2SO_4+H_2O_2+H_2O$ (1:8:1 by volume) solution, and
then etched by a flowing selective etchant, NH_4OH+H_2O (1:20 by
volume)which has a very small etching rate for ZnSe[20]. A He-Ne
laser was used to monitor the etching process, and the etching
was stopped as soon as light could pass through the window.
In this way, a MQW film with a flat surface and a thickness of
$1\sim2\mu m$ was obtained on a sapphire support. The two faces of the
ZnS-ZnSe MQW film naturally formed a Fabry-Perot etalon.

    The third harmonic beam of a YAG laser was used as a exci-
ting source for bistablity measurements. A photo diode(PD) and
a storage oscilloscope were used for detecting and displaying
the signals.We observed that when the input pulse width is 12ns,
the output pulse width was compressed to about 6ns, nearly half
that of the input pulse width, as shows in Fig.6.

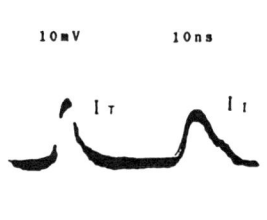

10mV   10ns

I_T I_I

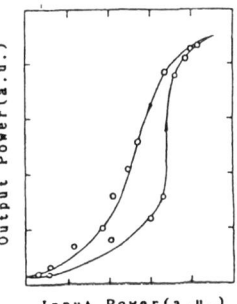

Output Power(a.u.)

Input Power(a.u.)

Fig.6 Pulse compression in a ZnS-ZnSe
MQW etalon with input pulse width of
about 12ns.

Fig.7 Hysteresis loop of
the ZnS-ZnSe MQW OBD.

    From the input and output pulse waveform, the hysteresis
loop of the ZnS-ZnSe MQW Fabry-Perot type bistable device was
obtained, as shown in Fig.7. We note that this device has apparent bistablity. In the experiment, the intensities required for
bistable operation are$\sim 0.1W/(\mu m)^2$ and the switching time are
~ 10ns. The wavelength of the exciting light is far from the
excitonic reconant peak of ZnS-ZnSe MQWs, so the nonlinear
refractive index mechanism may not be excitonic. Perhaps the
dominant mechanism is thermal.

CONCLUSION

    In conclusion, we have grown ZnS-ZnSe and ZnSe-ZnTe SLSs on
GaAs and InP substrates by MBE. In PL measurements, we observed
strong quantum effect. By Raman scattering, we have determined
that the critical thickness of ZnSe-ZnTe SLS is about 40A. For
the firet time, we have observed the confined LO_a mode in ZnSe-
ZnTe SLS and the folded LA modes in ZnS-ZnSe SLS. We have
fabricated the ZnS-ZnSe MQW etalon and observed the pulse
compression effect and obtained the hysteresis loop.

References

1. M.Kobayashi, R.Kimura, M.Konagai and K.Takahashi, J. Crystal Growth **81**, 495(1987).
2. T.Karasawa, K.Ohkawa and T.Mitsuyu, J. Crystal Growth **95**,547 (1989).
3. T. Yokogawa, M. Ogura and T.Kajiwara, Appl. Phys. Lett., **49**, 1702(1986).
4. A. Taike, N.Teraguchi, M. Konagai and K.Takahashi, Japan. J. Appl. Phys. **26**, L989(1987).
5. M. Kobayash, N. Mino, H. Katagiri, R. Kimura, M. Konagai and K. Takahash, J. Appl. Phys. **60**, 773(1986).
6. S.Nakashima, Y.Nakakura, H. Fujiyasu and K. Mochizuki, Appl. Phys. Lett. **46**, 236(1986).
7. S.Nakashima, A.Wada, H.Fujiyasu, M.Aoki and H.Yang, J. Appl. Phys. **62**, 2009(1987).
8. Y. Rajakarunanayake, R.H.Miles, G.Y.Wu,and T.C.McGill, Phys. Rev. B **37**, 10212(1988).
9. Z.P.Wang, D.S.Jiang and K. Ploog, Solid State Commun.**65**, 661 (1988).
10. D. J. Olego, K.Shahzed, D.A.Cammack and H.Cornelissen, Phys. Rev. B **38**, 5554(1988).
11. N. Vagelatos, D. Wehe and J. S.King, J. Chem. Phys. **60**, 3613 (1974).
12. B. Hennion, F. Moussa, G.Pepy and K.Kunc, Phys.Lett.**36A**, 376 (1971).
13. F. Cerdeira, C.J.Buchenaure, F.H.Pollak and M.Cardona, Phys. Rev. B **5**, 580(1972).
14. Landolt-Bornstein Tables, edited by O.Madelung, H.Schulz and H.Weiss(Springer,Berlin, 1982), Vol.Ⅲ-17b.
15. H.M.Gibbs, S.S.Tarng, J.L.Jewell, D.A.Weinberger, K.Tai,A.C. Gossard,S.L.McCall,A.Passner,and W.Wiegmann, Appl.Phys.Lett. **41**, 221(1982).
16. S.S.Tarng, H.M.Gibbs,J.L.Jewell,N.Peyghambarian,A.C.Gossard, T.Venkatesan, and W.Wiefmann, Appl.Phys.Lett. **44**,360(1984).
17. M.R.Taghizadeh, I.Janossy, and S.D.Smith,Appl.Phys.Lett. **46**, 331(1985).
18. A.K.Kar,and B.S.Wherrett, J. Opt. Soc. Am. B**3**,345(1986).
19. B. G. Kim, E. Garmire, N.Shibata, and S.Zembutsu, Appl.Phys. Lett. **51**, 475(1987).
20. F.Kitagawa, T.Mishima, and K.Takahashi, J. Electrochem. Soc. **127**, 937(1980).

LOCAL PHONON HEATING AND RECOMBINATION-ENHANCED DEFECT ANNEALING

V. N. ABAKUMOV, A. A. PAKHOMOV AND I. N. YASSIEVICH
A. F. Ioffe Physico-Technical Institute, Leningrad, U.S.S.R.

ABSTRACT

A detailed study of local phonon heating due to multiphonon recombination process is presented. Nonequilibrium distribution of defects by vibration energy is calculated taking into account the dependence of capture and emission rates on the defect energy. It is shown that multiphonon generation and recombination of electron-hole pairs stimulate diffusion of defects in energy space. The results are used to calculate the rate of recombination-enhanced defect annealing.

The process of nonradiative recombination of electron-hole pairs is accompanied by energy transfer to the lattice of the order of forbidden gap $\mathcal{E}_g$. If the recombination process is controlled by electron-local vibration interaction this energy is accumulated in the defect vicinity. The forbidden gap width is usually of the order of 1 eV, i.e. of the order of the binding energy of defects in the lattice. On these grounds, the accumulation of such large energy in the defect vicinity is expected to lead to reconstruction of the bonds and thus to induce defect reactions (for example, defect jumps and Frenkel pair recombination).

As early as 1961, Seitz [1] pointed out that this fact is important for understanding of degradation of semiconductor devices and defect annealing. The first extensive theoretical investigation of this problem was given by Wheeks et al. [2] who draw attention to essentially nonequilibrium character of defect distribution by vibration energy when the multiphonon recombination of carriers via defect takes place.

In this work we present a detailed study of the relevant nonequilibrium distribution.

Contrary to the work [2] we take into account a strong dependence of carrier capture and emission rates on the defect energy. Besides that, the value of the carrier energy plays an important part in the energy balance of highly excited defects. We found that strongly excited defects emit hot carriers with mean energy highly exceeding the thermal energy. On the contrary, defects capture predominantly cold (thermal) carriers as these are thermalized very quickly. This process results in a specific "electronic" mechanism of energy relaxation of defect vibrations, when the concentrations of free carriers are sufficiently large.

We have also shown that simultaneous action of recombination and generation of electron-hole pairs via highly excited defects induces a special kind of defects diffusion in energy space with a step $\mathcal{E}_g$. As a result, in a semiconductor with high concentrations of excess electrons and holes the high-energy tail of defect distribution by vibration energy falls out with the energy increase considerably slower than that in the equilibrium case. In our case the high-energy tail may be characterized by an effective temperature $T^* = \mathcal{E}_g / [k \log(N_c N_v / n p)]$ where n,p are the electron and hole concentrations, N_c, N_v are the densities of states in the conduction and the valence bands.

The distribution function found was used to calculate the enhancement coefficient η that is the ratio of annealing rates in the presence of excess carriers and without them.

1. Kinetic equations

The distribution of defects by vibration energy E should be characterized by two functions $f_1(E)$, $f_2(E)$ depending on the charge state of the defect. Let state 1 correspond to defect with a bound electron and state 2 to defect without an electron. The functions f_1, f_2 are normalized according to conditions

$$\int_0^\infty f_1(E) \rho_1(E) dE = f \quad , \quad \int_0^\infty f_2(E) \rho_2(E) dE = 1 - f \qquad (1.1)$$

where f is the population number for the recombination centre, $\rho_{1,2}(E)$ is the density of vibronic states.

The kinetic equations may be written in the form

$$\begin{cases} \dfrac{\partial N_1(E,t)}{\partial t} = n C_n(E - \mathcal{E}_{\tau c}) N_2(E - \mathcal{E}_{\tau c}) - [p C_p(E) + e_n(E)] N_1(E) + \\ \qquad + \int_0^\infty d\varepsilon\, W_p(E + \mathcal{E}_{\tau v} + \varepsilon; \varepsilon) N_2(E + \mathcal{E}_{\tau v} + \varepsilon) + \\ \qquad\qquad + \dfrac{\partial}{\partial E}\left\{ \dfrac{E}{\tau_L}\left[N_1(E) + kT \rho_1(E) \dfrac{\partial}{\partial E} \dfrac{N_1(E)}{\rho_1(E)} \right] \right\} \qquad (1.2a) \\[2ex] \dfrac{\partial N_2(E,t)}{\partial t} = p C_p(E - \mathcal{E}_{\tau v}) N_1(E - \mathcal{E}_{\tau v}) - [n C_n(E) + e_p(E)] N_2(E) + \\ \qquad + \int_0^\infty d\varepsilon\, W_n(E + \mathcal{E}_{\tau c} + \varepsilon; \varepsilon) N_1(E + \mathcal{E}_{\tau c} + \varepsilon) + \\ \qquad\qquad + \dfrac{\partial}{\partial E}\left\{ \dfrac{E}{\tau_L}\left[N_2(E) + kT \rho_2(E) \dfrac{\partial}{\partial E} \dfrac{N_2(E)}{\rho_2(E)} \right] \right\} \qquad (1.2b) \end{cases}$$

Here $N_{1,2}(E) = N f_{1,2}(E) \cdot \rho_{1,2}(E)$ is the number of defects in state 1 or 2 with an energy E in a volume unit, N is the defect concentration, $C_{n,p}(E)$ the capture rates of electrons and holes by a defect with vibronic energy E, $e_{n,p}(E)$ are the emission rates of electrons and holes by such a defect, $W_{n,p}(E; \varepsilon)$ are the probabilities of emission of electron or hole with kinetic energy ε by a defect with vibronic energy E. These probabilities are connected with emission rates

$$e_{n,p}(E) = \int_0^\infty d\varepsilon\, W_{n,p}(E; \varepsilon) \qquad (1.3)$$

The terms containing derivatives in (1.2) describe energy relaxation of local vibrations via interaction with thermostat (i.e. the lattice) in a Fokker-Planck approximation, τ_L is the energy relaxation time determined by a local phonon decay into lattice phonons. This approximation is justified when the energy losses occur in small steps and is actually valid when the vibration energy E is significantly larger than the energy of a local phonon $\hbar\omega$. While obtaining (1.2) we neglected the kinetic energy of the carrier being captured by a defect ($\varepsilon \sim kT_e$) compared to the vibration energy E.

The boundary conditions to the system (1.2) are

$$\left[N_i(E) + kT\rho_i(E) \frac{\partial}{\partial E} \frac{N_i(E)}{\rho_i(E)} \right]_{E=0,\infty} = 0 \qquad (1.4)$$

which correspond to zero flux in the energy space at $E = 0$ and $E = \infty$

In one-mode approximation $\rho_{1,2}(E) = 1/\hbar\omega_{1,2}$, where $\omega_{1,2}$ is a local vibration frequency in state 1, 2.

In a multi-mode case the functions $f_{1,2}$ represent the distributions of defects by total energy of vibrations, i.e. E is the sum energy of all vibration modes of the defect. This description is valid in case of energy redistribution between the modes being faster than the total energy change. Here the density of states depends essentially on energy E.

If the capture and the emission of carriers are controlled by interaction with a single mode of local vibrations the emission rate may be represented by relation

$$W_s(E;\varepsilon) = \int_0^E dE' \, W_0(E';\varepsilon) \frac{dK_s(E,E')}{dE'} \qquad (1.5)$$

where $W_0(E';\varepsilon)$ is the emission probability of carrier with energy ε when the relevant mode has accumulated energy E'.

To calculate the probability $W_0(E',\varepsilon)$ one can use the results of the theory of multiphonon transitions obtained in the framework of one-mode approximation. The function $K_s(E,E')$ is a probability for the revelant mode to accumulate energy greater than E', when the total vibration energy is E. In a model of s identical oscillators [2] :

$$K_s(E,E') = \rho_s(E-E')/\rho_s(E) \; , \quad \rho_s(E) = \frac{1}{(\hbar\omega)^s} \frac{E^{s-1}}{(s-1)!} \qquad (1.6)$$

2. Weak heating

Consider the situation when the energy supply to the system of local vibrations determined by the capture of electrons and holes is considerably less than the rate of energy relaxation due to interaction with the lattice:

$$nC_n(E), \; pC_p(E) \ll \tau_L^{-1} \qquad (2.1)$$

In this case we can solve the system (1.2) using iterative procedure. We shall obtain the steady state solution $N_{1,2}(E)$. In the zeroth approximation we may neglect the recombination and generation fluxes in (1.2). Then we arrive at the equilibrium energy distributions $f_{1,2}^{(0)}(E) = A_{1,2}\exp\{-E/kT\}$

To make the next step, i.e. to find the first approximation results, we substitute $f_{1,2}^{(0)}(E)$ into the recombination and generation terms in (1.2). If the concentrations of excess carriers are considerably greater than their equilibrium values we may disregard the terms responsible for thermal emission of electrons and holes by defect in (1.2). It is well known from the theory of multiphonon transitions (cf. [4,5]) that the capture coefficients depend essentially on the defect vibration energy. In the framework of classical description of the defect movement characteristic energy values $E = \varepsilon_{2c}, \varepsilon_{2v}$ exist, which determine the threshold energies for multiphonon transitions. For energy values below the threshold only tunnel transitions are possible. In this energy range the coefficients $C_{n,p}(\;)$ are sharply increasing (actually exponential) functions of vibration energy. As usually we have $\varepsilon_{2c,v} \gg kT$ the capture is controlled by tunneling at some optimal energies

$E = E_{oc,v}(T)$ which correspond to a sharp maximum of the product $C(E) f^{(o)}(E)$. Thus we can represent these in the form

$$C_n(E) f_2^{(o)}(E) \rho_2(E) \approx (1-f) \langle C_n \rangle \delta(E - E_{oc}(T))$$

$$C_p(E) f_1^{(o)}(E) \rho_1(E) \approx f \langle C_p \rangle \delta(E - E_{ov}(T)) \qquad (2.2)$$

where $\langle C_{n,p} \rangle$ are the mean equilibrium values of the capture coefficients for electrons and holes. Now the system (1.2) has splitted to two independent equations which can be easily solved. For $f_1^{(1)}(E)$ we get

$$f_1^{(1)}(E) = A_1 \begin{cases} \exp\{-E/kT\} \quad ; \; E \leq E_{ov} \\ \exp\{-E/kT\} + \dfrac{1}{A_1} \dfrac{\nu_R \tau_L}{kT} \displaystyle\int_{E_{ov}}^{E} \exp\left[\dfrac{E'-E}{kT}\right] \dfrac{dE}{E \rho_1(E)} \; ; \; E_{ov} \leq E < \mathcal{E}_{TC} + E_{oc} \\ \exp\{-E/kT\} + \dfrac{1}{A_1} \dfrac{\nu_R \tau_L \exp\{-(E - E_{oc} - \mathcal{E}_{TC})/kT\}}{(E_{oc} + \mathcal{E}_{TC}) \rho_1(E_{oc} + \mathcal{E}_{TC})} \quad E \geq \mathcal{E}_{TC} + E_{oc} \end{cases} \qquad (2.3)$$

where ν_R is the rate of multiphonon recombination via defect:

$$\nu_R = n \langle C_n \rangle p \langle C_p \rangle / (n \langle C_n \rangle + p \langle C_p \rangle)$$

The constant A_1 is determined from the normalization condition (1.1). An explicit form of $f_2^{(1)}(E)$ may be found similarly.

Thus, the distribution functions have nonequilibrium parts in form of an energy plateau extending up to $E_{oc} + \mathcal{E}_{TC}$ for $f_1(E)$ and to $E_{ov} + \mathcal{E}_{TV}$ for $f_2(E)$.

3. Strong heating

Now we shall discuss the opposite case when the carrier concentrations are so large that an inequality opposite to (2.1) takes place (for high vibration energy E).

In this case the main role in a formation of high-energy tails of the distribution functions is played by capture and emission processes.

The change of energy distribution due to electron capture and emission by defects may be presented as:

$$\frac{\partial N_1(E,t)}{\partial t} = n C_n(E - \mathcal{E}_{TC}) N_2(E - \mathcal{E}_{TC}) - e_n(E) N_1(E)$$

$$\frac{\partial N_2(E - \mathcal{E}_{TC}, t)}{\partial t} = \int_0^\infty W_n(E+\varepsilon; \varepsilon) N_1(E+\varepsilon) d\varepsilon - n C_n(E - \mathcal{E}_{TC}) N_2(E - \mathcal{E}_{TC}) \qquad (3.1)$$

It was shown in [3,4] that the characteristic energy $\mathcal{E}_m(E)$ of the escaping electron in the process of multiphonon emission satisfies the condition $kT_e \ll \mathcal{E}_m(E) \ll E$. This fact allows to present the integral term in (3.1) as

$$\int_0^\infty W_n(E+\varepsilon; \varepsilon) N_1(E+\varepsilon) d\varepsilon \approx e_n(E) N_1(E) + \frac{d}{dE}\left\{ N_1(E) \int_0^\infty \varepsilon W_n(E; \varepsilon) d\varepsilon \right\} \qquad (3.2)$$

Taking into account only the first term in (3.2) we have from (3.1) that the total population of vibration states $\tilde{N}(E) = = N_1(E) + N_2(E - \mathcal{E}_{TC})$ remains unaltered ($\partial \tilde{N}(E)/\partial t = 0$) and a dynamical balance between $N_1(E)$ and $N_2(E - \mathcal{E}_{TC})$:

$$e_n(E) N_1(E) = n C_n(E - \mathcal{E}_{TC}) N_2(E - \mathcal{E}_{TC}) \qquad (3.3)$$

is established after time $\tau_n' \sim e_n^{-1}$. If we take into account also the second term in (3.2) we obtain

$$\frac{\partial \tilde{N}(E,t)}{\partial t} = \frac{\partial}{\partial E}\left[\frac{E}{\tau_e} \tilde{N}(E,t) \right] \qquad (3.4)$$

where "electronic" relaxation time $\tau_e(E)$ is determined by

$$\frac{1}{\tau_e(E)} \approx \frac{n C_n(E - \mathcal{E}_{TC})}{e_n(E)} \frac{1}{E} \int_0^\infty \varepsilon W_n(E; \varepsilon) d\varepsilon = \gamma \frac{\mathcal{E}_m(E)}{E} n C_n(E - \mathcal{E}_{TC}) \qquad (3.5)$$

Here γ is a numerical factor of the order of unit (see [3,4]). Estimates show that for highly excited defects ($E > \varepsilon_{Tc} + \varepsilon_{2c}$) "electronic" relaxation mechanism is expected to be dominant when the carrier concentration exceeds 10^{18} cm^{-3} .

Similarly, the hole capture and emission processes establish a dynamical balance

$$N_1(E + \varepsilon_{TV}) e_p(E + \varepsilon_{TV}) = p \, c_p(E) N_1(E)$$

(3.6)

after time $\tau_p' \sim e_p^{-1}$

From eqs. (3.3) and (3.6) we get

$$N_{1,2}(E + \varepsilon_g) = \frac{np}{N_c N_v} N_{1,2}(E)$$

(3.7)

Here we have used the relations

$$e_n(E + \varepsilon_{Tc}) = N_c \, c_n(E) \quad ; \quad e_p(E + \varepsilon_{TV}) = N_v \, c_p(E)$$

(3.8)

that are valid if the kinetic energies of electrons and holes are neglected.

Thus recombination and generation processes via defects lead to effective diffusion of defects in energy space with a step ε_g . As a result the high-energy tail of energy distribution of defects has the form:

$$f_{1,2}(E) \sim \exp\left\{-E / kT^*\right\} \quad ; \quad T^* = \frac{\varepsilon_g}{k \log(N_c N_v / np)}$$

(3.9)

The effective temperature introduced T^* characterizes the energy distribution in the range of large vibration energy. When the concentration of nonequilibrium carriers is high, the effective temperature may be of the order of several thousands degrees.

We shall find also the distribution functions f_1 (E) and f_2 (E) in case of heavily doped p-type semiconductor when

$$p \, c_p(E) \gg \tau_L^{-1} \quad ; \quad n \, c_n(E) \ll \tau_L^{-1}$$

(3.10)

Here the relation (3.6) remains valid, and to find f (E) one can use the procedure described in section 2. However, in this case we should replace the lattice time τ_L by a total relaxation time defined by

$$\tau_*^{-1} = \tau_L^{-1} + \tau_p^{-1}(E)$$

(3.11)

where τ_p is the "electronic" relaxation time of defects due to emission of hot holes and capture of cold (thermalized) ones. This time is determined by

$$\tau_p^{-1}(E) = \gamma' E^{-1} \varepsilon_m(E + \varepsilon_{TV}) \, p \, c_p(E)$$

(3.12)

where γ' is a numerical coefficient of the order of unit.

Finally we obtain

$$f_1(E) \rho_1(E) = n \langle c_n \rangle \tau^*(E) \cdot E^{-1} \quad ; \quad \varepsilon_{2v} \leqslant E < \varepsilon_{Tc} + \varepsilon_{oc}$$

$$f_2(E) \rho_2(E) = \frac{p \, c_p(E - \varepsilon_{TV})}{e_p(E)} \frac{n \langle c_n \rangle}{E - \varepsilon_{TV}} \tau^*(E - \varepsilon_{TV}) \quad ; \quad \varepsilon_{2v} + \varepsilon_{TV} \leqslant E < \varepsilon_g + \varepsilon_{oc}$$

(3.13)

Thus, the function $f_2(E)$ corresponding to a defect deproved of an electron has a nonequilibrium tail which extends up to $E = \varepsilon_g$.

4. Recombination-enhanced annealing

The rate of defect reactions (for example, defect annealing) is proportional to a frequency of defect jumps to neighbour equilibrium positions ν . According to [2] this frequency is determined by

$$\nu = \nu_0 \int_{E_a}^{\infty} dE \left\{ f_1(E) \rho_{s,1}(E) K_{s,1}(E; E_a) + f_2(E) \rho_{s,2}(E) K_{s,2}(E, E_a) \right\}$$

(4.1)

where E_a is the activation energy of defect reaction.

We define the enhancement coefficient of defect annealing η as a ratio of annealing rates in the presence of excess carriers and without them $\eta = \nu^*/\nu$

The main contribution to the coefficient η comes from that distribution function $f_{1,2}$ (E) which has a longer high-energy nonequilibrium tail. The difference of f_1 and f_2 functions is due to difference of energies supplied to the system of local vibrations in a process of electron or hole capture.

Let the thermal emission energy for electrons be larger than that for holes ($\mathcal{E}_{Tc} > \mathcal{E}_{Tv}$), then for weak heating the coefficient η is determined by contribution of f_1 function. If $E_a < \mathcal{E}_{Tc}$ the defect reactions become athermal and for the coefficient η we get

$$\eta = \nu_R \tau_L \exp\left\{\frac{E_a}{kT}\right\} \int_{E_a}^{\mathcal{E}_{Tc} + \mathcal{E}_{ac}} dE \frac{(E - E_a)^{s-1}}{E^s} \qquad (4.2)$$

For $E_a > \mathcal{E}_{Tc} + \mathcal{E}_{ac}$ the calculation gives

$$\eta = (s-1)! \ \nu_R \tau_L \left(\frac{kT}{\mathcal{E}_{Tc}}\right)^s \exp\left\{\frac{\mathcal{E}_{Tc} + \mathcal{E}_{ac}}{kT}\right\} \qquad (4.3)$$

Our result differs from that found in [2] by an additional activation energy $\mathcal{E}_{ac}(T)$ that depends, in general, on temperature and also by a strong temperature dependence of preexponential factor.

We present also the coefficient η for heavily doped p-type semiconductor when f_2 function has a nonequilibrium plateau extending up to energies $E \simeq \mathcal{E}_g$. For $E_a < \mathcal{E}_g$ we have

$$\eta \approx \frac{n\rho \langle c_n \rangle}{N_v} \int_{E_a}^{\mathcal{E}_g} \tau^*(E) \frac{(E - E_a)^{s-1}}{E^{s-1}} \frac{dE}{E - \mathcal{E}_{Tv}} \qquad (4.4)$$

For high concentration of holes when the energy relaxation of the local vibrations is determined by a "hole" time $\tau_p \sim \rho^{-1}$ (see (3.12), (3.13)) the coefficient becomes independent of the hole concentration.

In [5] a saturation of recombination-enhanced annealing of defects in GaAs as a function of injection hole current in n-region of p-n junction under high injection level (p>>n) was observed. We believe that this observation is connected with the fact previously mentioned. A detailed discussion of this experimental result will be presented in [6].

References

1. F. Seitz, J. Koehler, in Solid State Physics, edited by F. Seitz, P. Turnbull (Acad. Press, N.Y., 1956) V. 2, p. 307.

2. J.P. Wheeks, J. Tully, L.C. Kimmerling, Phys. Rev. B 12, 3286 (1975).

3. V.N. Abakumov, A.A. Pakhomov, I.N. Yassievich, Fiz. Tv. Tela, 31, 145 (1989).

4. V.N. Abakumov, A.A. Pakhomov, M.K. Sheinkman, I.N. Yassievich, Fiz. Tekhn. Polupr. 23, 2232 (1989).

5. D. Stievenard, J.C. Bourgoin, Phys. Rev. B 33, 8140 (1986).

6. V.N. Abakumov, V.I. Perel, I.N. Yassievich, Nonradiative Recombination in Semiconductors (Elsevier, Amsterdam, 1991)

Gaas QUANTUM WELL INFRARED PHOTODETECTORS GROWN BY OMVPE

W. S. HOBSON*, A. ZUSSMAN*, B. F. LEVINE*, S. J. PEARTON*, V. SWAMINATHAN**, AND L. C. LUTHER**
*AT&T Bell Laboratories, Murray Hill, New Jersey 07974
**AT&T Bell Laboratories, Breinigsville, PA 18031

ABSTRACT

We have grown, fabricated, and measured GaAs quantum well infrared photodetectors (QWIPs) using organometallic vapor phase epitaxy (OMVPE). The epitaxial layers were characterized by electrochemical capacitance-voltage profiling, double-crystal X-ray diffraction, cathodoluminescence, and infrared absorption. Dark current, responsivity spectra, and detectivity were measured for the QWIP devices. The performance of these QWIPs was comparable to detectors grown using MBE. This is of importance since OMVPE has advantages for wafer throughout and cost.

INTRODUCTION

There has been considerable recent interest in the use of GaAs quantum well infrared photodetectors (QWIPs) for long-wavelength device applications [1]. This is based in part on the existence of a mature GaAs/AlGaAs growth technology which allows the reproducible deposition of ultra-thin quantum wells with excellent uniformity. To date, the vast majority of work on QWIPs has utilized material grown by molecular beam epitaxy (MBE). There has been only one demonstration, to our knowledge, of the growth of such structures by organometallic vapor phase epitaxy (OMVPE) [2]. These workers reported high-performance QWIPs at peak wavelengths (λ_p) from 6.7 µm to 10 µm for a given set of growth parameters. Here, we provide a detailed material and device characterization of QWIPs grown by OMVPE as a function of growth temperature (T_G), and compare the results with our previous work on MBE-grown devices [1]. We find that high-performance QWIPs can be grown over the range of growth temperatures explored ($T_G \sim 650-700°C$), in contrast to MBE where the optimization of T_G is more critical [3]. Since OMVPE gives comparable device performance to MBE, while exhibiting a greater tolerance to variations in the growth parameters, it is anticipated that OMVPE can provide a robust means for producing these structures and thus may exhibit advantages for increased wafer throughput and reduced cost.

EXPERIMENTAL

The QWIP device structures were grown by atmospheric-pressure OMVPE in a custom reactor similar to that described by Matsumota et al. [4]. Trimethylgallium (TMGa), trimethylaluminum (TMAl), and arsine (AsH_3) were used as the source chemicals, and disilane was utilized for n-type (Si) doping. A high AsH_3-to-(TMGa+TMAl) mole fraction ratio was employed in order to minimize carbon incorporation in the AlGaAs. The growth rate for the GaAs was 0.9 µmh^{-1}, and T_G was varied from 625 to 700°C.

Device structures were grown on undoped, semi-insulating GaAs substrates oriented 2° off (100) toward the nearest (110). The absorption region consists of 25 periods of a

$40\overset{\circ}{A}$ GaAs/$500\overset{\circ}{A}$ AlGaAs superlattice (SL). Only the center $20\overset{\circ}{A}$ of the GaAs quantum well was Si doped ($\sim 0.5 - 1.0 \times 10^{18}$ cm^{-3}) in order to avoid the possibility of forming interface states which could assist tunneling through the AlGaAs barriers. GaAs contact layers (0.5 μm thick, n $\sim 1 \times 10^{18}$ cm^{-3}) adjoin the top and bottom of the absorption region. The operating principles of the device have been described in detail previously [1]. In brief, absorption of infrared radiation promotes electrons bound in the single confined state within the QW into the conduction band continuum, which results in a collected photocurrent under applied bias. The primary figure of merit for device performance is the detectivity, D^*, which is defined as the ratio of the collected photocurrent per incident power density to the noise current, i_n.

The QWIP device layers were characterized by electrochemical capacitance-voltage profiling (ECV), double-crystal X-ray diffraction (XRD), cathodoluminescence (CL), and infrared absorption. Dark current, responsivity spectra, and D^* were measured for the QWIP devices. Device fabrication has been described previously [5]. Details of the measurements and calculations will be presented below.

RESULTS AND DISCUSSION

The QWIP material characteristics are summarized in Table I for samples grown at four different temperatures. XRD measurements were used to determine the SL period and the Al composition. All of the samples exhibited 4th order satellite peaks, indicating a well defined SL structure. The Al composition was intentionally decreased for samples C and D. It is seen that the variation in growth rate over the complete temperature range was quite small with a total variation of less than ±3% from the average SL period for the four samples.

Table I. QWIP Material Characteristics

Sample	T_G (°C)	Period ($\overset{\circ}{A}$)	Al(%)	CL Int. (a.u.)	λ_{max}, (FWHM) (nm) (meV)
A	625	557	34	10	723 (13)
B	650	575	34	4	726 (16.5)
C	675	586	33	4	729 (19.1)
D	700	558	33	3	724 (16.6)

CL spectra were taken at 6°K using an electron beam energy of 20 keV and current of 10^{-8} A. The range of the excitation was estimated to be 2.5 μm and the excitation density $\sim 1.4 \times 10^{15}$ pairs/cm^3 assuming a lifetime of 1 nsec. Identical excitation conditions and sample geometry were used in order to directly compare the relative intensity of the 4 samples. To first order, this should provide an indication of the optical quality of the QWs. The accuracy of the measurement is estimated to be better than a factor of two. It is seen that the QWIP sample grown at lowest temperature

(T_G = 625°C) has the strongest CL intensity, although the variation among the samples was small. The CL intensities obtained from MOCVD-grown QWIPs were typically an order of magnitude stronger than MBE-grown structures. We also calculated the QW thickness based on the measured peak emission wavelength (λ_{max}) and assuming a 60/40 conduction band offset, Al = 34% or 33% and undoped QWs. The results for samples A through D, respectively, were 28, 29, 29, and 28Å. The apparent discrepancy between the calculated and intended QW thickness is not fully understood at present but may be related to the fact that the donor atoms are spatially localized and due to the extrinsic nature of the luminescence. An examination of the full width at half maxima (FWHM) of the CL spectra for MOCVD QWIPs shows values ranging from 13 to 19.1 meV. The MBE structures exhibit more narrow linewidths (~9 meV) for uniformily doped QWs. To be noted is the excellent reproducibility of the QW thickness for different T_G.

ECV measurements of the carrier concentration were done for samples done B, C and D. Sample A exhibited weak IR absorption, which was assumed to be due to low doping in the QWs. The carrier concentration of the top contact layer was 4.5×10^{17}, 7.5×10^{17}, and 1.0×10^{18} cm^{-3} for B, C, and D, respectively, and for the SL region was 2.6×10^{16}, 6.5×10^{16}, and 6.5×10^{16} cm^{-3}. The actual doping in the wells is obtained by multiplying these values by the barrier-well duty cycle.

The bound state to extended continuum absorption was measured at room-temperature using a Perkin-Elmer Fourier Transform Infrared spectrometer as described previously [6]. The spectrum of the absorption coefficient α for 45° incident excitation (sample C) is given in Figure 1. A broad spectrum ($\Delta h\nu$ ~ 100 meV, λ_p = 8.15 μm) is consistent with the extended nature of the excited state. The spectral dependence of the responsivity, R_p, is shown in Fig. 2 for QWIP devices with 200 μm mesas fabricated from sample C. The bias voltage was −3V and the device temperature was 10°K. A maximum in R_p of 0.1 A/W was obtained at λ_p (8.15 μm). The dark current-voltage (I_d-V) characteristic for sample C is given in Figure 3 for T = 77°K. Asymmetry is observed between the reverse-and forward-biased I_d-V curves. The origin of this asymmetry is not known at present but may reflect the inequivalence of the GaAs/AlGaAs and the AlGaAs/GaAs interface as observed in MBE-grown material.

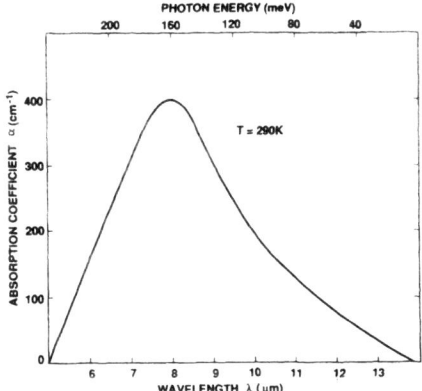

Fig. 1.   Infrared absorption spectrum for QWIP sample C measured at T = 290K.

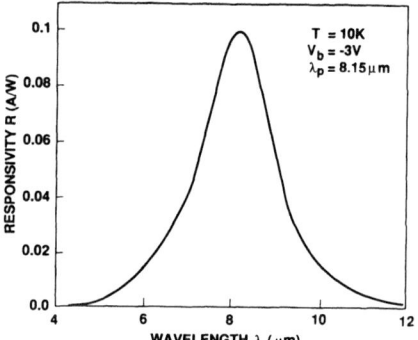

Fig. 2.   Responsivity spectrum for QWIP sample C measured at T = 10K.

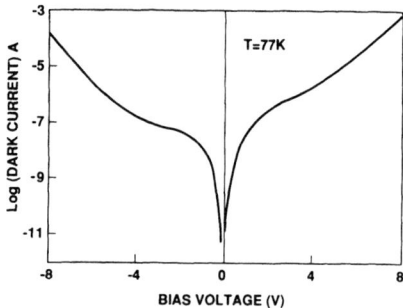

Fig. 3.   Dark current-voltage characteristic for QWIP sample C measured at
T = 77K.

A summary of the QWIP device characteristics is given in Table II. Difficulty in making ohmic contact to the bottom GaAs contact layer prohibited devices from being made for sample A, due to the low doping of this sample, as discussed above. The quantum efficiency, η, was calculated from the peak absorption coefficient, α, from the expression $\eta = 1/2(1 - e^{-2\alpha\ell})$, where ℓ is the length of the SL region. λ_{co} is the cut-off wavelength and corresponds to the long-wavelength responsivity half-height. The peak responsivity, R_p, was calculated from $R_p = I_p/P$ where I_p is the measured peak photocurrent and P is the optical power of the blackbody source. The peak wavelength detectivity, $D^*_{\lambda_p}$, at 77°K is calculated from $D^*_{\lambda_p} = R_p\sqrt{A\Delta f}/i_n$, where A is the active device area, Δf is the bandwidth and i_n is the noise current. The noise current is in turn calculated from the expression $i_n = (4qI_D g\Delta f)^{1/2}$ where q is the electronic charge, I_D is the measured dark current at the operating voltage (-2V), and g is the gain obtained from $g = R_p\,hV/q\eta$ where hv is the peak energy. The calculated $D^*_{\lambda_p}$ values are found to be comparable to those of MBE-grown QWIPs. While the values of R_p are low, I_D is also low. Both R_p and I_D scale with the doping density in the QW.

Table II. QWIP Device Characteristics

Sample	α (cm^{-1})	η (%)	λ_p (μ_m)	λ_{co} (μ_m)	R_p (A/W)	$D^*_{\lambda_p}$ (77 K) cm $\cdot$ Hz$^{1/2}$ $\cdot$ W^{-1}
B	320	4.5	7.5	9.1	0.033	2.4×10^{10} (@-2V)
C	400	5.6	7.9	8.8	0.082	1.8×10^{10} (@-2V)
D	300	4.2	8.4	9.4	0.16	1.4×10^{10} (@-2V)

SUMMARY

We have examined QWIP device performance for OMVPE-grown structures as a function of T_G. High-performance devices are obtained for all growth conditions and suggests that OMVPE is suitable for producing such structures in a reliable manner.

REFERENCES

[1] B. F. Levine, C. G. Bethea, G. Hasnain, V. O. Shen, E. Pelve, R. R. Abbott, and S. J. Hseih, Appl. Phys. Lett. 56, 851 (1990) and references therein.

[2] Z. F. Paska, J. Y. Anderson, and L. Lundqvist, Paper presented at the 5th International Conference on Metalorganic Vapor Phase Epitaxy, 18-21 June, 1990; J. Y. Anderson and G. Landgren, GaAs and Related Compounds, Karuuizawa, Japan, 1989, Inst. Phys. Conf. Ser. 106, 731 (1990).

[3] M. J. Asom and R. E. Leibenguth, (unpublished results).

[4] K. Matsumoto, K. Itoh, T. Tabuchi, and R. Tsunoda, J. Cryst. Growth 77, 151 (1986).

[5] B. F. Levine, K. K. Choi, C. G. Bethea, J. Walker, and R. J. Malik, Appl. Phys. Lett. 50, 1092 (1987).

[6] G. Hasnain, B. F. Levine, C. G. Bethea, R. R. Abbott, and S. J. Hsieh, J. Appl. Phys. 67, 4361 (1990).

ION MILLING DAMAGE IN InP AND GaAs

S. J. PEARTON, U. K. CHAKRABARTI AND A. P. PERLEY
AT&T Bell Laboratories, Murray Hill, NJ 07974

ABSTRACT

Near-surface damage created by Ar^+ ion milling in InP and GaAs was characterized by capacitance-voltage, current-voltage, photoluminescence, ion channeling and transmission electron microscopy. We find no evidence of amorphous layer formation in either material even for Ar^+ ion energies of 800 eV. Low ion energies (200 eV) create thin (≤ 100 Å) damaged regions which can be removed by annealing at 500°C. Higher ion energies (≤ 500 eV) create more thermally stable damaged layers which actually show higher backscattering yields after 500°C annealing. Heating to 800°C is required to restore the near-surface crystallinity, although a layer of extended defects forms in GaAs after such a treatment. No dislocations are observed in InP after this type of annealing. The electrical characteristics of both InP and GaAs after ion milling at ≥ 500 eV cannot be restored by annealing, and it is necessary to remove the damaged surface by wet chemical etching. For the same Ar^+ ion energies the damaged layers are deeper for InP than for GaAs-after 500 eV ion milling at 45° incidence angle, removal of ~485 Å and ~650 Å from GaAs and InP respectively restores the initial current-voltage characteristics of simple Schottky diodes.

INTRODUCTION

Ion beam etching techniques are widely used for mesa-isolation of GaAs electronic and photonic devices [1-9]. The energetic ion bombardment causes the formation of deep level traps which compensate the carrier concentration at depths up to 2000Å from the surface. When ion energies above 400 eV are used, the apparent net electron concentration may actually increase [7], and low temperature annealing often worsens the electrical properties of the material [7] Pang et al. [3] showed the damage depth is shallower with low energy and heavy ion species. Little is known about ion damage to InP surfaces. In this paper we report a comparison of the Ar^+ ion milling characteristics of both InP with those of GaAs.

EXPERIMENTAL

The ion milling was performed in a Technics Micro Ion Mill (Model MIM TLA 20) using an Ar^+ ion beam (200-800 eV) whose angle of incidence was varied between 0-75° from the vertical. The samples were mounted on a water-cooled plate held at 10°C, and the chamber evacuated to ~3×10^{-6} Torr using mechanical and cryogenic pumps. Typical Ar^+ ion current densities were 0.5-0.8 mA cm^{-2}. A variety of doped and semi-insulating GaAs and InP substrates, the former with back AuGeNi ohmic contacts, were used in this work.

RESULTS AND DISCUSSION

The average mill rate normalized to the Ar^+ ion beam current for InP and GaAs as a function of Ar^+ ion energy at 45° incidence angle and a temperature of 10°C is shown in Figure 1. Within experimental error (±10%) in each case the etch rate depends linearly on the Ar^+ ion energy, with slopes of ~0.4 ($Å \cdot min^{-1} \cdot cm^{-2} \cdot eV^{-1}$) and 0.8 respectively for GaAs and InP. For GaAs this appears to be consistent with some previously reported data [1,16,17] in that a doubling of the ion energy leads to an increase in mill rate of ~20%. For InP the mill rate increases more rapidly than for GaAs. Our measured mill rates correspond to sputtering yields for InP of 0.30 atoms per ion at 200 eV Ar^+ ion energy, 0.42 atoms per ion at 500 keV and 0.80 at 800 keV when the Ar ion currents are factored in. For GaAs the sputtering yield is 0.35 atoms per ion at 200 eV and 0.54 atoms per ion at 800 eV Ar^+ ion energy.

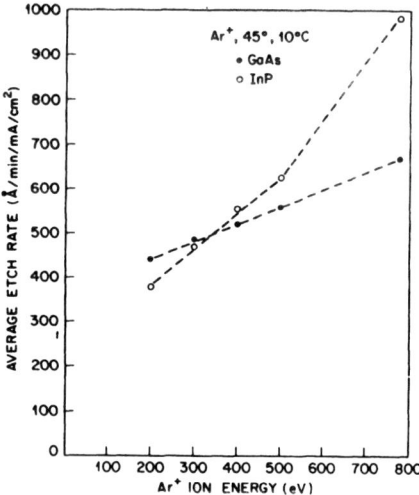

Fig. 1.   Average etch rate normalized to ion current for Ar^+ ion milling of InP and GaAs, as a function of ion energy. The beam incidence angle was 45°, and the sample temperature 10°C.

The experimentally observed mill rate dependence on the angle of incidence of the Ar ion beam for both materials at 500 eV Ar^+ ion energy is shown in Figure 2. The mill rate increases with increasing incidence angle up to ~60° where the mill rate is ~40-50% higher than at normal incidence. The mill rate then decreases for high beam incidence angles, as expected from linear cascade theory [9]. For InP the sputtering yields were 0.36, 0.45, 0.49 and 0.26 atoms per ion respectively at 0°, 45°, 60° and 75° angles of incidence. For GaAs the corresponding values were 0.31, 0.33, 0.39, 0.44 and 0.23 atoms per ion at 0°, 15°, 45°, 60° and 75° incidence angles. Both materials show a similar dependence of the ion milling rate, with the InP removal rate being faster at all incidence angles for this Ar^+ ion energy of 500 eV.

The surface morphologies of the ion milled materials were examined on patterned samples by SEM. Figure 3 show results for InP samples, milled at 45° incidence angle at Ar^+ ion energies of 200 eV (top left and bottom left), 500 eV (top right) and 800 eV (bottom right). The surfaces are reasonably smooth except in the latter case where preferential sputtering of P during the ion milling treatment leads to In droplets remaining on the surface. The GaAs was not as sensitive to variations in the Ar^+ ion energy due to the much smaller mass difference between Ga and As relative to In and P.

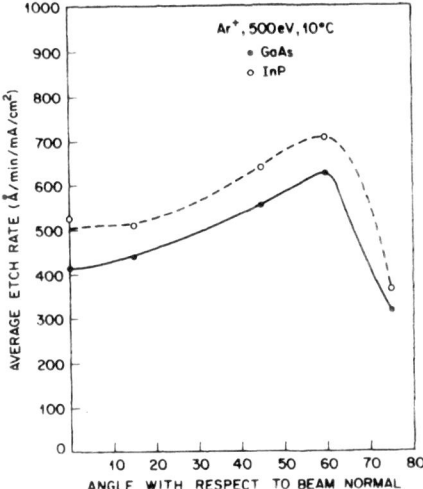

Fig. 2.   Average etch rate normalized to ion current for Ar^+ ion milling of InP and GaAs, as a function of beam incidence angle measured from the normal.

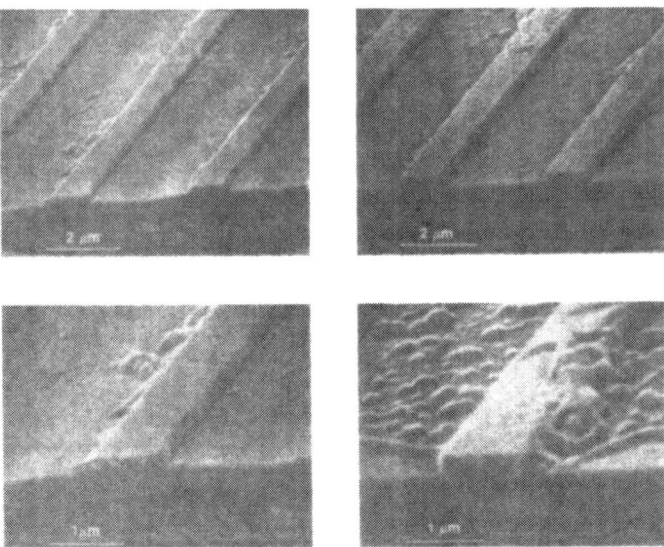

Fig. 3.   SEM micrographs of features ion milled into InP at 45°C incidence angle of the beam with Ar^+ ion energies of 200 eV (top left and bottom left), 500 eV (top right) and 800 eV (bottom right).

Figure 4 shows the 300K PL spectra from Fe-doped InP samples both before and after ion milling at different Ar ion energies. The beam was at 45° incidence angle for each treatment. The PL intensity is reduced by approximately an order of magnitude after even the lowest ion energy (200 eV) milling treatment, and is further reduced for the higher ion energies (500 and 800 eV). These decreases are significantly greater than we have observed for reactive ion etching (RIE) of InP where the maximum ion energies are comparable to those used for the ion milling [20]. This is presumably because the rate is slower for ion milling relative to reactive ion etching in which there are both physical and chemical components to the etching. For ion milling therefore there is more accumulation of lattice disorder because this disordered material is not being removed as quickly as in RIE. The associated reduction in PL intensity is greater in ion milling as a result of the slower etch rates compared to plasma etching techniques.

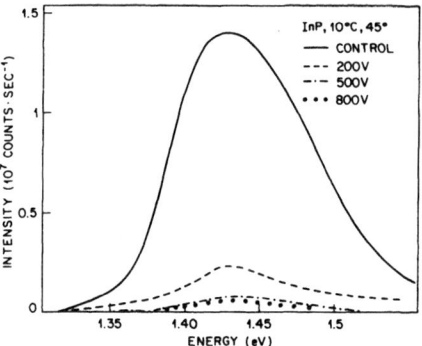

Fig. 4.   PL spectra recorded at 300K of InP (Fe) samples ion milled using 200, 500 or 800 eV Ar⁺ ions.

Ion milling of InP led to surfaces to which we could not fabricate rectifying contacts. Both Hg and evaporated Au metallizations on milled InP samples showed ohmic behavior, as evidenced by the I-V characteristics of Figure 5. Similar data was observed for all ion energies (200-800 eV) and angles of incidence (0-75° from the vertical). The damaged, near-surface region was progressively removed by wet chemical etching in $1HCl:5H_2O$ and the etch depth directly measured by Dektak stylus profilometry after masking a section of the sample with apiezon wax. Evaporated Au Schottky contacts were then deposited onto the etched surface. Figure 5 also shows I-V characteristics from samples from which 390 Å or 945 Å were removed prior to deposition of the Schottky contact. Removal of 390 Å of material was not sufficient to restore the I-V characteristics, while after taking off 945 Å, the I-V data was identical to that of a control sample. It was found that for 500 eV Ar⁺ ion milling at least 600 Å of InP had to be removed by wet chemical etching to restore the initial electrical characteristics of the material as evidenced by the I-V data.

Annealing the ion-milled (500 eV, 45°) GaAs even at 700°C for 30 sec. brought little improvement in the I-V characteristics. The diode was still very leaky with a high reverse current and non-ideal forward characteristic. This is also consistent with previous results indicating that annealing up to 400°C actually leads to a deterioration in the I-V characteristics of diodes fabricated on ion-milled GaAs [7]. Our results show that even 700°C annealing is insufficient to remove the ion-milling damage. We also performed a series of experiments in which various thickness of material were removed by wet chemical etching in a $1H_2SO_4:1H_2:10H_2O$ solution prior to evaporation of the Schottky contacts. Removal of ~235 Å of material from the surface of a sample ion milled with 500 eV Ar^+ ion for four minutes at 45° incidence angle produced a substantial improvement in diode characteristics. We found that under these conditions a minimum of ~485 Å had to be etched off before the I-V characteristics were similar to those of a control sample.

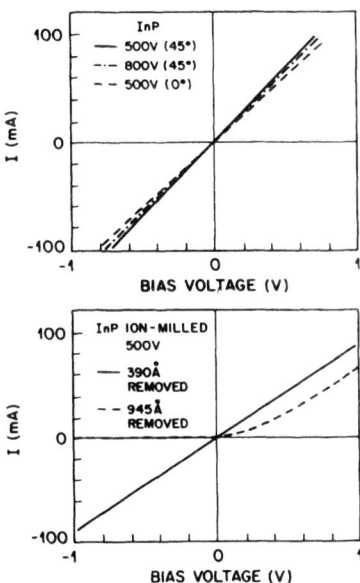

Fig. 5.   I-V data from Au-InP Schottky diodes fabricated on ion milled, n-type InP (top) and after removal of 390 or 945 Å of material prior to evaporation of the Au contact (bottom).

It has been commonly assumed that ion milling of III-V materials produces a thin, amorphous layer at the surface. Ion channelling spectra before and after annealing at 500°C or 800°C, 30 sec of InP ion milled at either 200 eV on 800 eV (45° incidence angle) are shown in Figure 6. The sample annealed at 800°C for 10 sec. has less crystal damage in the upper 100 Å than does the sample ion milled at 200 eV, but the disorder peak after annealing extends to ~800 Å. The damaged region after 500°C/10 sec. annealing of the 500 eV milled sample also extends to greater depths than before annealing, indicating a migration of the disorder.

The sub-surface defects and surface topography of both InP and GaAs samples ion milled at 500 eV (45° incidence angle) were examined by cross-sectional TEM. The annealing for both materials was performed at 800°C for 10 sec. The GaAs showed a layer of extended defects that formed upon annealing. These consisted of a relatively high concentration ($\gtrsim 10^{10}$ cm^{-2}) of dislocation loops (diameter 60-70 Å) at a depth between 130 Å and 460 Å below the surface. This is much greater than the mean range of 500 eV Ar^+ ions in GaAs (24 Å, with a straggle of 12 Å) as determined by a Monte Carlo simulation. The fact that the visible disorder is located deeper than the projected range of the impinging ions is a typical feature of dry etch processes.

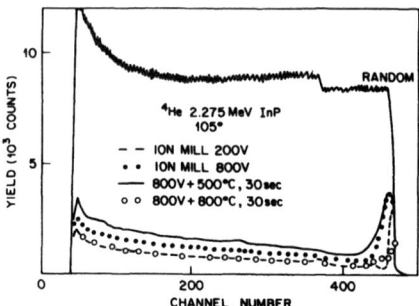

Fig. 6.   Ion channeling spectra recorded at 105° detection angle from ion milled InP before and after annealing.

SUMMARY

Taken together, our data on ion-milling disorder in InP and GaAs illustrate the following key points:

[i]   For low energies (~200 eV) there is introduction of a very near-surface conductive layer in GaAs, with deep acceptor traps at greater depths leading to carrier compensation. Both types of defects can be removed by annealing at ~500°C.

[ii]   For higher ion energies (≥500 eV) the near-surface conductive layer is more prominent, and even annealing at 700°C does not restore the initial characteristics of the material. Annealing at lower temperatures may actually lead to a deterioration of the crystalline quality.

[iii]   For 500 eV Ar$^+$ ion milling at 45° incidence angle, at least 600 Å has to be removed by wet chemical etching of InP and ~485 Å from GaAs before the initial diode characteristics are restored.

[iv]   There is no amorphous layer formation under our ion milling conditions. Dislocations form in GaAs after annealing, but they are not observed in InP. Ion channeling detects more lattice disorder in InP milled under the same conditions in comparison to GaAs.

REFERENCES

1. See for example, C. I. H. Ashbey, EMIS Data Review, RN 15422, December 1985.

2. M. Kawabe, N. Kanzaki, K. Masuda and S. Namba, Appl. Opt. *17*, 2556 (1978).

3. S. W. Pang and W. J. Piancentini, J. Vac. Sci. Technol. *B1*, 1334 (1983).

4. Y. Yuba, T. Ishida, K. Gamo and S. Namba, J. Vac. Sci. Technol. *B6*, 253 (1988).

5. E. D. Cole, S. Sen and L. C. Burton, J. Electron Mater. *18*, 527 (1989).

6. P. J. Smith and D. A. Allan, Vacuum *34*, 209 (1984).

7. P. Kwan, K. N. Bhat, J. M. Borrego and S. K. Ghandi, Solid State Electron. *26*, 125 (1983).

8. Y. G. Wang and S. Ashok, Nucl. Inst. Meth. in Physics Research, B*39*, 461 (1989).

9. P. C. Zalm, Vacuum *36*, 787 (1986).

DOUBLE IMPLANTATION IN GaAs

I.P. KOZLOV , V.B. ODJAEV, V.S. PROSOLOVICH, Yu.N. YANKOVSKY
Byelorussian State University, Lenin Avenue,4, Minsk, USSR 220080

INTRODUCTION

High mobility of electrons, the possibility of formation of semi-insulating layers excites an interest to GaAs. The implantation of donor impurities into semi-insulating GaAs:Cr has a great practical application, allowing to create the conducting submicron layers. To obtain the functional submicron n^+-layers one uses mainly the implantation of ions of Si, that accounts for its small mass and the lack of the necessity for irradiation at higher temperatures. Silicon is the amphoteric impurity in GaAs. At low doses $\leq 10^{13}$ cm^{-2} it behaves as donor impurity with the high level of activation, though at the increase of the implantation dose some of the silicon atoms begin to substitute arsenic and in this position they display the properties of acceptor impurity, leading to self-compensation. Its amphoteric properties were shown with the help of As+Si and Ga+Si double implantation [1]. The increase of activation of Si, as donor impurity, can be obtained by double implantation of $^{31}P + ^{28}Si$ [2]. Phosphorus has atomic mass close to Si, it is isoelectronic to As and does not cause compensation effect in GaAs up to the doses $\approx 10^{15}$ cm^{-2}. Moreover double implantation allows to vary the magnitude of the diffusion coefficients of impurities.

EXPERIMENTAL PROCEDURE

The samples of semi-insulating GaAs:Cr, implanted by silicon ions (E=70 KeV, D=1.8 * 10^{15} cm^{-2}) have been investigated. To decrease complex formation in GaAs, connected with Si implantation, as well as to increase the electrical activation of implanted Si, the preliminary implantation by phosphorus ions (E=75 KeV, D=6 * 10^{15} cm^{-2}) has been made. Then GaAs samples were exposed to thermal annealing during 20 minutes at the temperature of 850°C (in the abundance of As or in H_2 flow) or using annealing by single high-energy (0.4 - 1.2 J*cm^2) pulses (20 - 30 ns) from Q-switched ruby laser.

The investigations were made by EPR-spectroscopy, contactless UHF-photoresistivity, capacitor photo-e.m.f. and Hall effect. The method of UHF-photoconductivity is based on the change of the absorption coefficient of

Mat. Res. Soc. Symp. Proc. Vol. 216. ©1991 Materials Research Society

UHF -power by free charge carriers under the influence of monochromatic radiation. The sample was placed into the loop of the electrical field component in the optical resonator of EPR-spectrometer, operating in X-range.

The method of capacitor photo-e.m.f. constitutes the change of the capacitance between the plates (metal substrate and semitransparent electrode) where sample is placed. During its uniform illumination the electric field appears, caused by the space charge division. The electrical parameters measurements were made using by Van der Pauw method.

RESULTS AND DISCUSSIONS

Spectral curves of UHF-photoconductivity of the implanted GaAs have two strongly marked maximums corresponding to the impurity and interband photoresponses (Fig. 1). The maximum of the signal of interband response in-

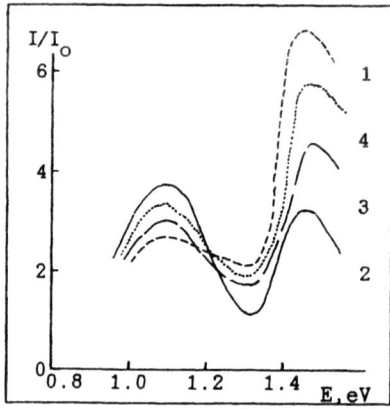

Fig.1. UHF-photoresistivity spectra for GaAs, implanted by Si and P ions:

1 - $D_{Si} = 1.8 * 10^{15} cm^{-2}$,
   $D_P = 1.8 * 10^{15} cm^{-2}$;
2 - $1.8 * 10^{15}$, 0;
3 - $1.8 * 10^{15}$, $3.0 * 10^{15}$;
4 - $1.8*10^{15}$, $6.0 * 10^{14}$.

creases with the increase of phosphorus implantation dose (the dose of Si implantation for all cases is constant) and after reaching the highest value(at $D = 1.8 * 10^{15} P^{+} cm^{-2}$) it starts to decrease; the signal of the impurity photoresponse behaves similarly. The extremal values for the both cases were reached at the equality of the doses of Si and P implantation. It should be noted that the magnitude of the photoresponse of the interband signal was less for the implanted GaAs, then that for the initial samples. Analogous behaviour has the maximum of the impurity photoresponse, which seems to belong to EL 2 centre [3]. Unmonotone dependence of the concentration of EL 2 centre versus phosphorus implantation dose (Fig.1) allows to assume that at the increase of D_P the correlation of the efficiencies of vacancy generation in the sublattices of As and Ga has been changed. These results are in agreement with the data of measurements by the method of contactless photo-e.m.f.. Hall effect measurements, made on the implanted crystals, exposed to the thermal and laser annealing showed that the samples are of n-type conductivity.The maximum electrical activation of Si in GaAs is reached at the equality of implantation doses of Si and P for all types of annealing, that is in good agreement with the data of photoconductivity; the preliminary phosphorus implantation reduces the concentration of As vacancies and results in the increase of Ga vacancies concentration, which ensures the increase of the electrical activation of the implanted silicon. The study of the photoconductivity of the thermal annealing samples, shows the dependence of the signal versus the ratio of the implantation doses of Si and phosphorus and the presence of the extremum at their equality. Besides this the shift of the maximum interband signal occurs versus the dose of phosphorus implantation that is evidently connected with the formation of the triple solutions in the implanted GaAs.

In Si-GaAs:Cr samples, implanted by Si and P ions after the annealing in the abundance of As the new paramagnetic centre($g = 2.0116 \pm 0.0005$; $H = 2.0 \pm 0.5$ Gs) was detected, which was named as AG-M1. This centre does not occur on the analogous samples, annealed in the hydrogen atmosphere. It seems to be connected with the passivation of the broken bonds by hydrogen in GaAs during the annealing and by the evaporating of As. This

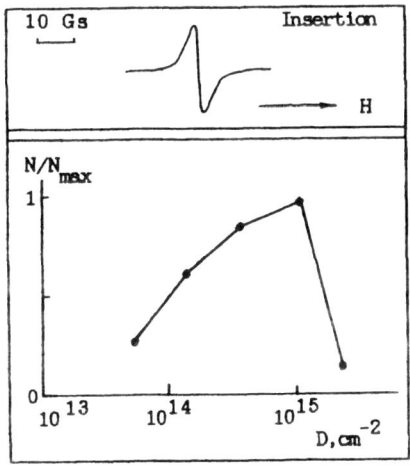

Fig. 2. Dependence of the concentration of AG-M1 centre versus the dose of phosphorus implantation at fixed D_{Si}. There is the spectrum of AG-M1 centre in the insertion.

centre is detected only at the temperature $\leqslant 77$ K [4]. EPR signal is the symmetrical isotopic line of Lorentz form (Fig.2). The shape and the comparatively small linewidth for GaAs allows to suppose the presence of the exchange mechanism of the line narrowing. It should be noted that such narrow spectral lines in GaAs were not observed earlier. The concentration of the given paramagnetic centre depends on the correlation of silicon and phosphorus implantation doses and it has the maximum at their equality (Fig.2). The analysis of the obtained data allows to assume that the paramagnetic AG-M1 centre represents the acceptor type defect including silicon in the sublattice of gallium. It is confirmed by the compensating action of AG-M1 centre in the implanted GaAs samples. This corresponds to the decrease (up to 10%) of the free charge carriers concentration in the implanted samples of GaAs annealing in As atmosphere comparing to the analogous samples treated in the hydrogen flow.

CONCLUSION

It has been estimated that the best results of implanted silicon atoms activation are attained at the dose equality of Si and P for all types of annealing (thermal and laser). The new paramagnetic centre has been detected (g= 2.0116 ±0.0005; H= 2.0 ±0.5 Gs). being supposedly the acceptor type defect, involving silicon in the As sublattice. The formation of the triple compounds is possible during thermal annealing of the implanted GaAs samples.

LITERATURE

[1] E.B.Stoneham,G.A.Patterson,J.M.Gladstone, J.Electr.Mater. 9 , 371(1980)

[2] V. Rybka, V. Odjaev, J. Cervena, V. Hnatowicz, J. Kvitek, J. Jelinkova, Czech.J.Phys. D37, 919 (1987)

[3] J. C. Buorgoin, H. J. von Bardeleben, J.Appl.Phys. 64, R65 (1988)

[4] I.P.Kozlov, V.B.Odjaev, V.S.Prosolovich, V.P.Tolstyh, Proc. Int.Conf. "Microelectronics '90", Minsk, USSR, 60 (1990)

REFLECTION HIGH-ENERGY-ELECTRON DIFFRACTION STUDY OF InP AND InAs (100) IN GAS-SOURCE MOLECULAR BEAM EPITAXY

T. P. Chin, B. W. Liang, H. Q. Hou, and C. W. Tu
Department of Electrical and Computer Engineering 0407, University of California at San Diego, La Jolla, California 92093-0407, U. S. A.

ABSTRACT

InP and InAs (100) were grown by gas-source molecular-beam epitaxy (GSMBE) with arsine, phosphine, and elemental indium. Reflection high-energy-electron diffraction (RHEED) was used to monitor surface reconstructions and growth rates. (2x4) to (2x1) transition was observed on InP (100) as phosphine flow rate increased. (4x2) and (2x4) patterns were observed for In-stabilized and As-stabilized InAs surfaces, respectively. Both group-V and group-III-induced RHEED oscillations were observed. The group-V surface desorption activation energy were measured to be 0.61 eV for InP and 0.19 eV for InAs. By this growth rate study, we are able to establish a precise control of V/III atomic ratios in GSMBE of InP and InAs.

INTRODUCTION

Gas-source molecular-beam epitaxy (GSMBE) is a versatile tool of growing complicate heterostructures not only for its capability of tailoring the material to monolayer preciseness but also the handling of phosphorus, which is relatively more difficult in conventional MBE. Many of the mixed-group-V compounds are of great interests in optoelectronic device applications. For example, $In_xGa_{1-x}As_yP_{1-y}$ lattice-matched to InP has been used for photonic devices operating at 1.3 μm and 1.55 μm for long-wavelength optical communication.[1] Material for even longer wavelength can be achieved by increasing the InAs molar fraction in the alloy.

Reflection high-energy-electron diffraction has been shown to be a powerful in situ diagnostic technique for growth rate and composition control in MBE GaAs and related compounds, but relatively few RHEED studies have been reported for GSMBE of phosphides. Here we report a reflection RHEED study of InP and InAs in GSMBE. Different surface reconstructions were observed under various substrate temperatures and hydride flow rates. By measuring the group-V-limited growth rates under different growth temperatures and hydride flow rates, we were able to determine the V/III atomic ratio on the substrate surface in GSMBE. This is an absolute V/III ratio measurement compare to beam-flux ratio or flow-rate ratio usually used in conventional MBE or metalorganic chemical vapor deposition (MOCVD). Beam-flux ratio or flow-rate ratio measurement is rather system-dependent and easily affected by many unpredictable factors.[2,3] Substrate temperature, which plays a dominant role in the molecule incorporation process, is not considered in the beam-flux or flow-rate measurement, either. However, the ratio of group-V-limited growth rate and group-III-limited growth rate corresponds to the V/III atomic ratio on the growing surface. With the precise control of arsine and phosphine flow rates in GSMBE, we can achieve better control of growth conditions and perform more systematic investigation on the relation between the V/III ratio and layer quality. One purpose of this study is to optimize the growth condition for InP and other III-V compounds in GSMBE. In situ group-V composition control may also be obtained for mixed group-V material, e.g., $GaAs_xP_{1-x}$ or $InAs_xP_{1-x}$.

Mat. Res. Soc. Symp. Proc. Vol. 216. ©1991 Materials Research Society

518

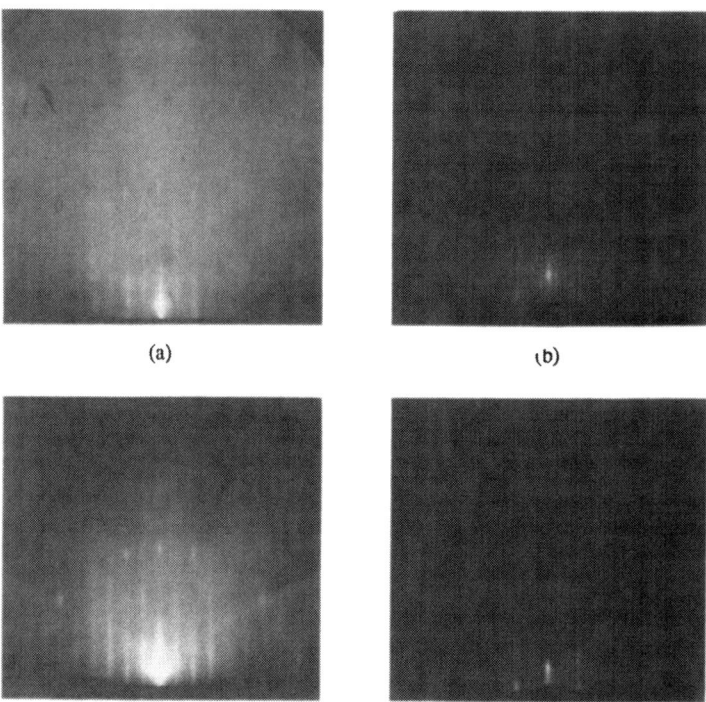

(a)　　　　　　　　　　　　(b)

(c)　　　　　　　　　　　　(d)

Fig.1: RHEED patterns on InP(100) in GSMBE. (a) and (b): (2x4) is observed at In-rich surface. (c) and (d): (2x1) is observed at higher phosphine flow or lower substrate temperature.

EXPERIMENTS AND RESULTS

The GSMBE system consists of a modified Varian Modular GEN-II MBE reactor equipped with a 2200 l/s (H_2) cryopump and an ion pump (which is turned off during growth). Two separate gas cabinets house two gas-source supply systems (100% arsine and 100% phosphine) as well as scrubbers. Both gases are introduced into the growth chamber through the same Varian hydride injector. The normal cracking temperature is 1000°C. The phosphine and arsine flow rate in this study is typically 1-4 sccm. The background pressure ranges between 0.8×10^{-5} and 2×10^{-5} Torr during growth. The growth rates are monitored by the specular-beam intensity oscillation of RHEED. A dual-channel differential amplifier produces clean RHEED signals after subtracting the background noises. The oscillation is then recorded in an IBM PC AT compatible computer to measure the growth rate.[4]

InP (100) substrates were etched in a solution of H_2O_2:NH_4OH:H_2O (2:5:10), rinsed, and blown dry. Each sample was bonded onto a 3-inch Si wafer with In. The Si wafer was then mounted onto a Mo transfer ring with Ta wires. Growth temperatures were calibrated with a pyrometer and the melting point of InSb.

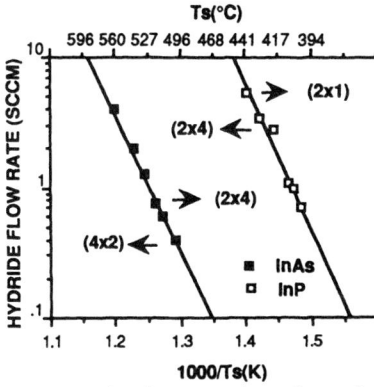

Fig. 2: Hydride flow vs. reciprocal growth temperature at which RHEED pattern occurred on InAs and InP (100).

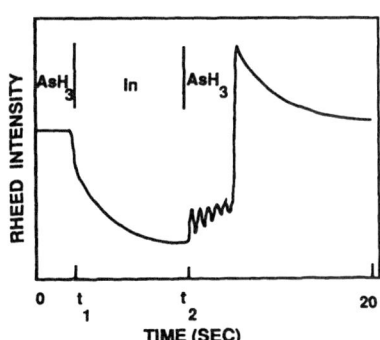

Fig. 3: As-induced RHEED oscillation of on InAs (100).

InAs was grown on InP at 460°C. InAs(100) shows (2x4) and (4x2) reconstructions for As and In stabilized surfaces, respectively. For InP(100), (2x4) was observed right after thermal cleaning. This (2x4) pattern will change to (2x1) under higher phosphine flow rate or lower substrate temperature. Fig. 1 shows RHEED patterns on InP(100). This observation is consistent with that reported by Morishita et al. in metalorganic MBE of InP with phosphine and trimethylindium,[5] however, is different from the conventional MBE results. (2x4) and (4x2) reconstructions were observed on InP(100) under As_4 or P_2 overpressure.[6] The corresponding pattern-transition substrate temperature and arsine flow rate are shown in Fig. 2. According to early studies in GSMBE by Panish et al.,[7,8] this pattern transition indicates the formation of indium droplets on the surface. Thus, one can compare the curves in Fig. 2 with the equilibrium P_2 and As_2 vapor pressure curves in a solid-liquidus system (InP+In and InAs+In) and then obtain the accommodation coefficients of P_2 and As_2 on InP and InAs, respectively. This comparison is not shown here because we need more calibration in our system to convert the hydride flow rate to molecular-beam flux.

Fig. 3 shows a typical arsenic-induced RHEED oscillation on InAs. First, the surface was As-stabilized, showing (2x4) reconstruction. Arsine was then turned off and the In shutter was opened at t_1. The RHEED pattern would switch to (4x2) and the specular-beam intensity decreased as the surface became rougher due to In accumulation. The In shutter was then closed and arsine was turned on at t_2. The RHEED oscillation resumed as the incoming arsenic molecules incorporated with the accumulated In on the surface. After the extra In was consumed, the RHEED pattern changed back to (2x4) and the specular-beam intensity increased abruptly. After this overshoot, the intensity recovered to the original level. Similar experiments were also done on InP (100). The growth rates measured here are limited by the amount of group-V flux and also dependent on substrate temperature. Fig. 4 is a plot of the group-V-limited growth rate as a function of hydride flow rate with substrate temperature as parameters. It is clear that the higher the T_s, the

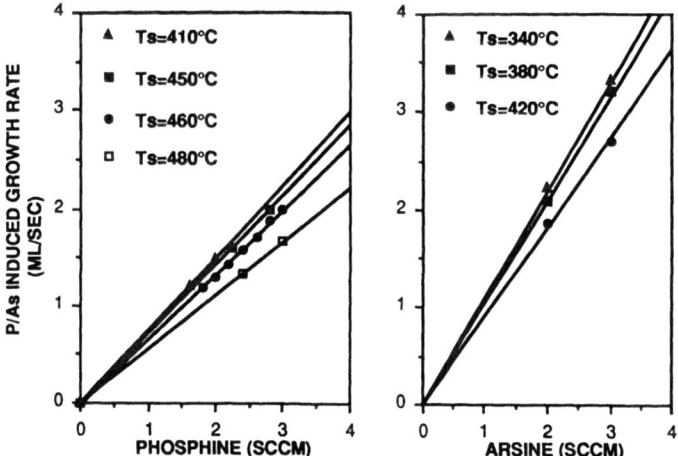

Fig. 4: Group-V-induced growth rate as a function of hydride flow rate at different growth temperature for InP (left) and InAs (right).

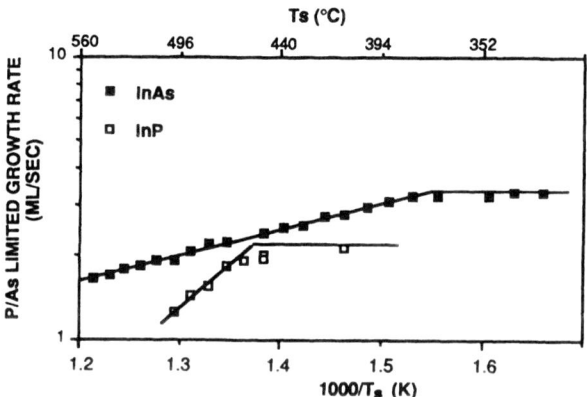

Fig. 5: Group-V-induced growth rate as a function of growth temperature. Arsine and phosphineflow rates are 3 and 2.8 sccm, respectively.

lower the growth rate due to the desorption of group-V molecules. Fig. 5 is the Arrhenius plot of the As- and P-limited growth rates of InAs and InP, respectively, as functions of substrate temperatures. RHEED oscillations were difficult to observe at higher or lower temperatures.

DISCUSSION

Fig. 4 demonstrates one of the advantages of using gas sources, namely, The arsine or phosphine flow can be controlled by a mass flow controller rapidly and accurately. Similar experiments were also performed on GaP, AlP, and GaAs.[9] One can find a curve in Fig. 4 for a corresponding T_s and then adjust the As- or P-limited growth rate by setting the mass flow controller appropriately. The ratio of As- or P-limited growth rate and In-limited growth rate represents the V/III ratio of the atoms incorporated onto the surface. This provides a precise measurement of V/III atomic ratio other than ion-gauge measurements, which is difficult to be used in GSMBE due to the high hydrogen background pressure (typically 1×10^{-5} Torr). Preliminary results shows that higher V/III ratio results in higher impurity concentration and lower mobility. Table 1 shows the Hall measurement results of three undoped 3-μm InP samples grown under different V/III ratios at 450°C. The carrier concentration at 77K increases and mobility decreases as the growth temperature increases. The mobility at room temperature is relatively low probably due to lattice scattering by structural defects. The best result so far is a 3-μm-thick InP layer with a 77K mobility of 62,600 cm^2/V-sec and a background carrier concentration of 5.1×10^{14} cm^{-3}, achieved under V/III$\approx$1 condition.

Group-V molecules will desorb from the surface as the substrate temperature increases. This results in the the group-V-limited growth rate decreasing at higher substrate temperature as shown in Fig. 5. The As-limited growth rate of InAs is constant for $T_s < 400$°C and decreases linearly on an Arrhenius plot as T_s increases. The arsenic and phosphorus surface desorption activation energy (E_a) can be obtained from the slope of the curves. We find $E_a \approx 0.19$ eV for InAs and $E_a \approx 0.61$ eV for InP. Similar experiments in conventional MBE has been done by Chow and Fernandez,[10] they measured E_a to be 0.2 eV for GaAs. These thermodynamic data are currently being used in modeling the $InAs_xP_{1-x}$ growth.

sample #	V/III ratio	300K		77K	
		n	μ	n	μ
C083	1.1	2.3E15	3870	1.7E15	38500
C086	2	5.3E15	3700	3.6E15	31500
C088	3	5.6E15	3780	4.0E15	24000

Table 1: 3μm undoped InP grown under different V/III ratios. The growth temperature was 450°C.

SUMMARY

A RHEED study was performed on InP and InAs (100) grown by GSMBE. (2x4) and (2x1) patterns were observed on InP, while (4x2) and (2x4) patterns were observed on InAs for In-stabilized and for As-stabilized surfaces, respectively. By measuring both group-V and group-III limited growth rates, an accurate V/III ratio control was then established for different III-V compounds in GSMBE. The electrical property of InP was affected by the V/III ratio. As and P surface desorption activation energies on InAs/InP were also measured.

ACKNOWLEDGEMENT

This work is partially supported by the Office of Naval Research and Powell Foundation. The authors would also like to thank C. E. Chang and M. C. Ho for building the RHEED amplifier and valuable discussions.

REFERENCES

1. M.B. Panish and H. Temkin, Annu. Rev. Mater. Sci.19, 209 (1989).

2. R. Fernandez, J. Vac. Sci. Technol. B 6, 745 (1988).

3. R. Chow and R. Fernandez, Mat. Res. Soc. Symp. Proc. 145, 13 (1989).

4. C.E. Chang, T.P. Chin, and C.W. Tu, unpublished.

5. Y. Morishita, S. Maruno, M. Gotoda, Y. Nomura, and H. Ogata, Appl. Phys. Lett. 53 (1), 42, (1988).

6. E. H. C. Parker, *The Technology and Physics of Molecular Beam Epitaxy*, (Plenum, New York, 1985).

7. M.B. Panish and H. Temkin, Annu. Rev. Mater. Sci. 19, 209 (1989).

8. M.B. Panish and S. Sumski, J. Appl. Phys. 55 (10), 3571 (1984).

9. T.P. Chin, B.W. Liang, H.Q. Hou, M.C. Ho, C. E. Chang, and C.W. Tu, Appl. Phys. Lett, Jan. 21, 1991.

10. R. Chow and R. Fernandez, Mat. Res. Soc. Symp. Proc. 145, 13 (1989).

SiGe/Si HETEROJUNCTION INTERNAL PHOTOEMISSION LONG-WAVELENGTH INFRARED DETECTORS

T. L. Lin, A. Ksendzov, S. M. Dejewski, E. W. Jones, R. W. Fathauer, T. N. Krabach, J. Maserjian, and R. W. Terhune
Center for Space Microelectronics Technology, Jet Propulsion Laboratory, California Institute of Technology, 4800 Oak Grove Dr., Pasadena, CA 91109

INTRODUCTION

There is a great need of long-wavelength (8-17μm) infrared (LWIR) focal plane arrays (FPAs) for a variety of space and defense applications. Si-based infrared detectors offers several important advantages, including good uniformity, low cost, and easy integration with Si readout circuitry either monolithically or by indium bump bonding to form large arrays. We report here a novel SiGe/Si heterojunction internal photoemission (HIP) long-wavelength infrared (LWIR) detector fabricated by molecular beam epitaxy [1].

The HIP detector consists of a degenerately doped p^+-SiGe layer as the emitter, and a p-type Si substrate as the collector. The device band diagram and test structure are shown in Fig. 1(a) and (b), respectively. The detection mechanism is infrared absorption in the p^+-SiGe emitter followed by internal photoemission of photoexcited holes over the SiGe/Si heterojunction barrier into the Si substrate, as shown in Fig. 1(a). The cutoff wavelength λ_c of the HIP detector is determined by the energy barrier $q\phi_B$, and is given by

$$\lambda_c \ (\mu m) = 1.24/q\phi_B \ (eV). \qquad (1)$$

The energy barrier $q\phi_B$ is determined by the valence band offset ΔE_v between the SiGe alloy layer and the Si substrate, and the Fermi level in the degenerately doped SiGe layer (Fig. 1(a)), and is given by

$$q\phi_B = \Delta E_v - (E_V - E_F). \qquad (2)$$

The energy band alignment of the SiGe/Si heterostructure has been studied extensively, and the bandgap offset splits approximately 90%/10% between the valence and conduction bands [2]. The bandgap of the commensurately strained SiGe alloy, and correspondingly the SiGe/Si valence band offset ΔE_v, can be tailored by varying the Ge concentration. Figure 2 shows the SiGe/Si valence band offset ΔE_v and corresponding minimum cutoff wavelengths $\lambda_{c,min}$ (assuming $E_v - E_F = 0$)

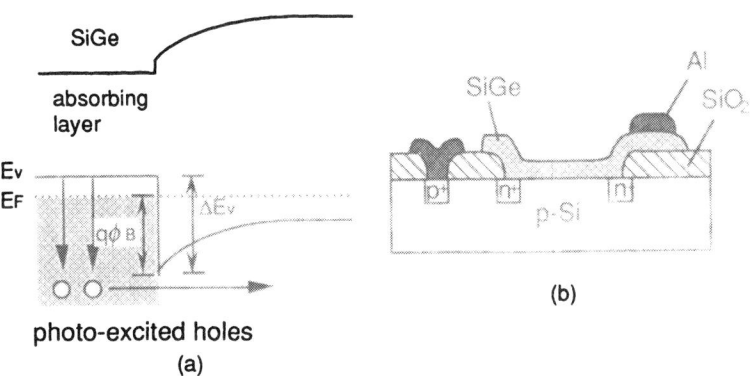

Figure 1. (a) Energy band diagram of the p^+-SiGe/p-Si HIP detector. (b) Structure of the p^+-SiGe/p-Si HIP test device.

Mat. Res. Soc. Symp. Proc. Vol. 216. ©1991 Materials Research Society

thickness (nm)	Ge concentration	hole concentration	mobility at 300K
30	30%	4.4×10^{20} cm^{-3}	19.0 cm^{2}V^{-1}s^{-1}
30	30%	2.9×10^{20} cm^{-3}	23.7 cm^{2}V^{-1}s^{-1}
100	0	1.1×10^{20} cm^{-3}	30.0 cm^{2}V^{-1}s^{-1}

Table 1.   Hole concentrations and hole mobilities of Si and SiGe layers grown by MBE at 540°C using HBO$_2$ as a boron source measured by Hall measurements.

for Ge concentration of 0.1 to 0.4 calculated from calculated and measured SiGe bandgaps reported by People *et al.* [2]. The minimum cutoff wavelength of the SiGe/Si HIP detector $\lambda_{c,min}$ can be tailored over a wide IR range; for example, 5 - 22 μm with x ranging from 0.4 to 0.1. By degenerately doping the SiGe layer ($E_v > E_F$), the cutoff wavelength can be extended to longer wavelength.

EXPERIMENTAL PROCEDURE

SiGe layers were grown by molecular beam epitaxy (MBE) using a Riber EVA 32 Si MBE system with a base pressure of 3×10^{-11} Torr. Prior to MBE growth, the Si wafers were cleaned using the "spin-clean" method, which involves the removal of a chemically grown surface oxide using a HF/ethanol solution in a nitrogen glove box [3]. The substrates were then heated to 500 - 600°C and Ge and Si atoms were co-evaporated from two electron gun sources. Growth rates ranging from 0.5 to 2 Å/s and boron doping concentrations ranging from 10^{19} to 4×10^{20} cm^{-3} were achieved using a HBO$_2$ source from a conventional Knudsen cell [4,5]. SiGe layers for optical and electrical characterization were grown onto double-side polished n-type Si substrates. The p-type SiGe layers are electrically insulated from the conductive n-type substrates by the p-n junction. The SiGe/Si HIP detectors were fabricated by growing p$^+$-SiGe layers on patterned p-type Si (100) wafers with a resistivity of 1-10 Ω cm. N-type guard rings were incorporated at the periphery of the active detector area to minimize edge leakage current. The SiGe layer thicknesses ranged from 10 to 400 nm, and the Ge concentration ranged from 0.2 to 0.4.

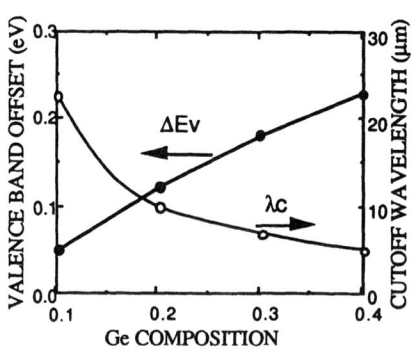
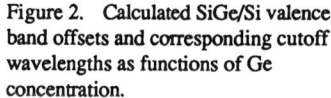
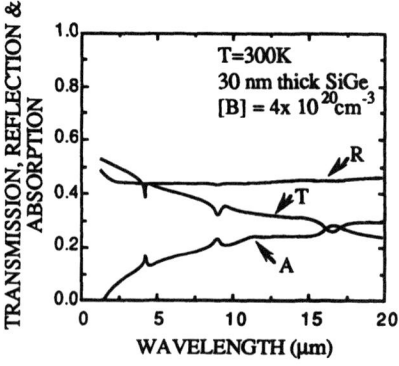

Figure 2.   Calculated SiGe/Si valence band offsets and corresponding cutoff wavelengths as functions of Ge concentration.

Figure 3.   Transmission (T), reflection (R) and absorption (A) of a 30-nm-thick SiGe layer with 30% Ge concentration and [B] = 4×10^{20} cm^{-3} measured by FTIR at 300K.

RESULTS AND DISCUSSION

The electrical and optical properties of degenerately doped p^+-SiGe layers were characterized by Hall measurements and Fourier transform infrared spectrometer (FTIR), respectively. Table 1 shows the boron concentrations and the hole mobilities of Si and SiGe layers (30% Ge concentration) grown at 540°C. Boron concentrations as high as 4×10^{20} cm^{-3} have been achieved using HBO$_2$ as the boron source during MBE growth. The degenerate boron concentrations allow Strong infrared absorption in the SiGe layers. Possible absorption mechanisms for the p^+-SiGe layers are free carrier absorption and intra-valence band transitions. The spectral dependence of transmission, reflection, and absorption of a 30-nm-thick SiGe (30% Ge concentration) layer with a boron concentration of 4×10^{20} cm^{-3} is shown in Fig. 3. The infrared absorption of the p^+-SiGe layer increases monotonically with increasing wavelength, with 5 to 25% absorption in 3-20 μm wavelength range.

There is a trade-off between the infrared absorption and the internal QE (the ratio between collected and photoexcited carriers). A higher absorption can be obtained for HIP detectors utilizing a thicker SiGe layer, but the internal QE will be reduced as photoexcited carriers will suffer more scattering on the average before reaching the SiGe/Si interface. The QEs of two HIP structures with a similar Ge concentration (30%) but different SiGe layer structures are shown in Fig. 4. Both detectors have the same total quantity of boron, and consequently similar infrared absorption. The SiGe layer of detector A (open circles) is thinner (40 nm) and more heavily doped (1×10^{20} cm^{-3}), while that of detector B (filled circles) is thicker (400 nm) and less heavily doped (1×10^{19} cm^{-3}). Because of the reduced layer thickness, detector A allows more photoexcited holes to reach the interface before suffering inelastic scattering, resulting in a higher QE. The photoresponse of detector A at 8 μm is two orders of magnitude higher that that of detector B. Furthermore, the higher doping concentration of detector A reduces the effective potential barrier because its Fermi level moves further below the valence band (Eqn. 2) , and thus extends photoresponse to 10 μm.

Figure 5 shows the photoresponse of two HIP detectors with a higher boron concentration (4×10^{20} cm^{-3}) and two SiGe thicknesses; 10 nm for detector C (open circles) and 30 nm for detector D (filled circles). The active area is 6.25×10^{-4} cm^2 and the Ge concentration is 0.3. The measurement was done at 20K with a -1.5 V bias. QEs of 3-5 % have been achieved in the 8-12 μm region for detector D. Although the SiGe layer thickness of detector D is three times that of detector C, and thus has about three times absorption, the QEs of detector D are only twice those of detector C. This is due to the fact that the internal QE of detector D is reduced because the photoexcited holes suffer more scattering on the average before reaching the SiGe/Si interface due to the thicker SiGe layer. By optimizing the doping profile and thickness of the SiGe layer, and consequently optimizing the trade-off between absorption and the internal QE, the QE of the HIP detector can be further improved.

SUMMARY

In conclusion, a novel p^+-SiGe/p-Si HIP LWIR detector has been demonstrated. Degenerately doped p^+-SiGe layers are utilized for strong IR absorption to generate photoexcited holes. SiGe layers with electrically activated boron concentrations of up to 4×10^{20} cm^{-3} have been grown by MBE at 540°C using an HBO$_2$ source. Photoresponse at wavelengths ranging from 2 to 12 μm has been obtained with QEs of 3-5% in the 8-12 μm range. It should be possible

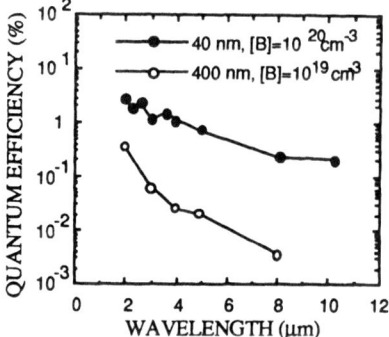
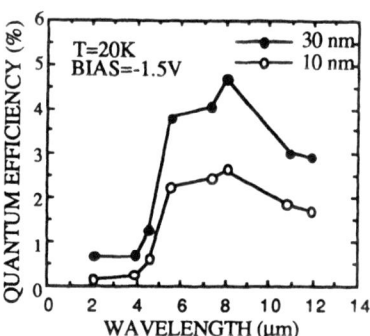

Figure 4.   Quantum efficiencies for two HIP detectors with similar infrared absorption. Detector A (open circles) has a 400-nm-thick SiGe layer doped with $[B]=10^{19}$ cm^{-3} and detector B (filled circles) has a 40-nm-thick SiGe layer doped with $[B] = 10^{20}$ cm^{-3}.

Figure 5.   Photoresponse of two HIP detectors with 30-nm-thick (filled circles) and 10-nm-thick (open circles) SiGe layers doped with the same boron concentration (4×10^{20} cm-3).

to further improve the QE of HIP detectors by optimizing the trade-off between the infrared absorption and the internal QE through optimization of the device structure.

ACKNOWLEDGEMENTS

We would like to thank Dr. Barbara A. Wilson for helpful discussions. The work described in this paper was performed by the Center for Space Microelectronics Technology, Jet Propulsion Laboratory, California Institute of Technology, and was jointly sponsored by the National Aeronautics and Space Administration, Office of Aeronautics, Exploration and Technology and the Strategic Defense Initiative Organization, Innovative Science and Technology Office.

REFERENCE

1. T. L. Lin, J. Maserjian, Appl. Phys. Lett. *57*, 1422 (1990)
2. R. People, J. C. Bean, D. V. Lang, A. M. Sergent, H. L. Stormer, K. W. Wecht, R. T. Lynch, and K. Baldwin, Appl. Phys. Lett. *45*, 1231 (1984).
3. P. J. Grunthaner, F. J. Grunthaner, R. W. Fathauer, T. L. Lin, F. D. Schowengerdt, M. H. Hecht, D. Bell, W. Kaiser, and J. H. Mazur, Thin Solid Films *183*, 197 (1989).
4. T. L. Lin, R. W. Fathauer, and P. J. Grunthaner, Appl. Phys. Lett. *55*, 795 (1989).
5. T. Tatsumi, H. Hirayama, and N. Aizaki, Appl. Phys. Lett. *50*, 1234 (1987).

THE ROLE OF RARE EARTHS IN NARROW ENERGY GAP SEMICONDUCTORS

D.L. PARTIN, J. HEREMANS, D.T. MORELLI AND C.M. THRUSH
Physics Dept., General Motors Research Laboratories, Warren, MI 48090-9055

ABSTRACT

Narrow energy band gap semiconductors are potentially useful for various devices, including infrared detectors and diode lasers. Rare earth elements have been introduced into lead chalcogenide semiconductors using the molecular beam epitaxy growth process. Europium and ytterbium increase the energy band gap, and nearly lattice-matched heterojunctions have been grown. In some cases, valence changes in the rare earth element cause doping of the alloy. Some initial investigations of the addition of europium to indium antimonide will also be reported, including the variation of lattice parameter and optical transmission with composition and a negative magnetoresistance effect.

INTRODUCTION

Narrow energy band gap semiconductors are useful for mid-infrared detectors and sources, by virtue of their small band gaps. Some of them are also useful for magnetic field sensitive transport devices such as magnetoresistance or Hall effect sensors by virtue of their large electron mobilities. Addition of rare earth elements to these materials is potentially useful for changing the energy band gap, carrier scattering mechanism, and other properties of these materials. The greatest progress in heterojunction formation and devices to date has been in the lead-rare earth-chalcogenide systems, where heterojunction and quantum well diode lasers with state-of-the-art performance have been reported [1,2]. While In$_x$Mn$_{1-x}$As is the first magnetic III-V compound studied [3], alloys of rare earth elements with narrow energy gap III-V compounds (InSb, InAs) or II-VI compounds (e.g., Hg$_x$Cd$_{1-x}$Te) have not been previously reported to our knowledge, and we report here on our first preliminary investigations of In$_{1-x}$Eu$_x$Sb. Finally, Bi$_{1-x}$Sb$_x$ is a semi-metal or a narrow energy gap semiconductor, depending upon the value of x. Alloys with rare earth elements have not been previously reported to our knowledge. We have not yet explored such alloys, but we do report here a preliminary investigation of Bi:Eu, with Eu so far only at dopant levels.

LEAD-RARE EARTH-CHALCOGENIDES

The lead chalcogenide compounds (PbTe, PbSe, and PbS) are extensively used for long wavelength diode lasers. Their alloys, principally with Sn and Cd, allow devices to be made with emission wavelengths covering the range from ~ 2.5 to 30 μm [4,5]. Homojunction devices made from these "classic" alloys generally operate only below liquid nitrogen temperatures. However, use of heterojunctions for carrier and optical confinement can significantly lower the threshold current, allowing higher operating temperatures. This led to the study of alloys of these semiconductors with rare earth elements.

The lead chalcogenide compounds have the face-centered cubic (sodium chloride) crystal structure, and hence are often known as "lead salts." The rare earth-monochalcogenide compounds (RE-X) also have the face-centered-cubic crystal structure. They have lattice constants which are comparable to those of the lead salt compounds, and hence alloys of these two material

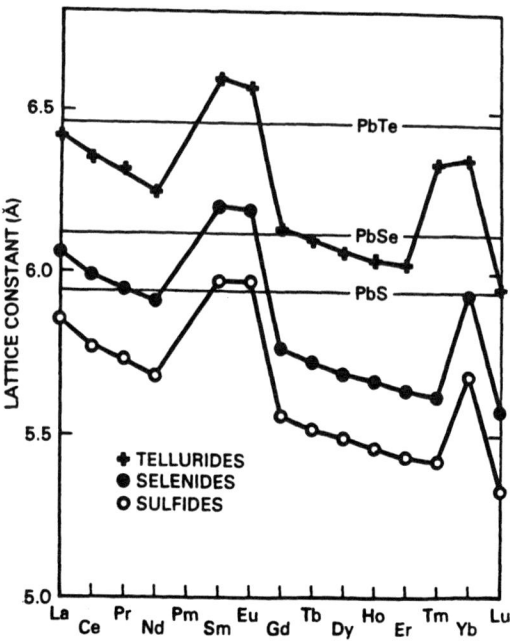

Fig. 1.   Lattice constants of the rare earth monochalcogenide compounds in
the face-centered-cubic crystal structure.

systems (Pb-RE-X) are of interest. The lattice constants of these compounds
are shown in Fig. 1. There is an obvious periodicity in the lattice
constants of the RE-X compounds which is related to the valence of the rare
earth elements. Sm, Eu, Yb, and to some extent Tm are relatively stable in
the divalent state in these compounds. Hence, their chalcogenides tend to
be semiconductors. The other rare earth elements are trivalent. Hence,
their chalcogenides tend to be metallic, with two electrons bonding with
chalcogenide neighbors and one contributed to the conduction band for each
rare earth ion. This has led to the exploration of trivalent rare earth
elements (e.g., La, Gd, Dy, Ho, and Er) as n-type dopants in PbTe and in
PbSe [6-8]. However, the greatest need is for alloys that will appreciably
change the energy band gap without doping the material. Therefore, the
major interest in Pb-RE-X alloys has involved the divalent rare earths. The
energy gaps vs. lattice constants of the tellurides of these and several
other elements are shown in Fig. 2 [1,9,10]. Of the rare earths, Eu and Yb
have so far been most studied for increasing the energy gap of PbTe. This
is important for making heterojunctions between PbTe and higher band gap
material so that, for example, electrons and holes injected into the PbTe
active region of a diode laser will be unable to escape into the higher
energy gap adjacent layers, but will efficiently recombine, producing
lasing. $Pb_{1-x}Eu_xTe$ films have been grown by molecular beam epitaxy which
are single phase up to x ~ 0.3 [11]. The energy band gap vs composition is
shown in Fig. 3 along with data for the competing alloys $Pb_{1-x}Sr_xTe$ and
$Pb_{1-x}Ca_xTe$. Thus, a very useful energy range can be covered, and the small
diffusion coefficient of Eu in PbTe permits very abrupt heterojunctions to
be formed [11]. Since EuTe increases the lattice constant of PbTe whereas
PbSe additions decrease it, $Pb_{1-x}Eu_xSe_yTe_{1-y}$/PbTe heterojunctions may be
lattice-matched for low defect density interfaces. This has led to the

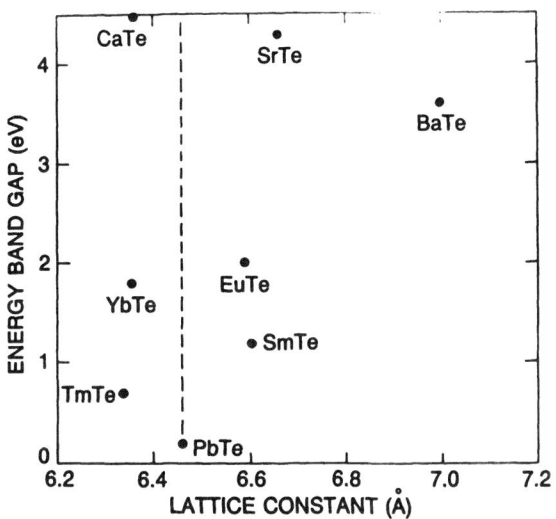

Fig. 2.   Energy band gap vs. lattice constant for several face-centered-cubic
telluride compounds.

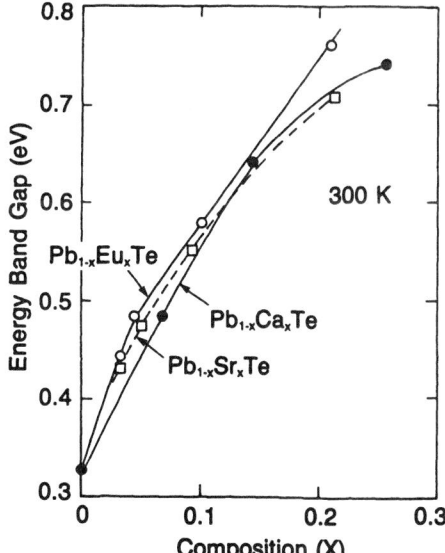

Fig. 3.   Energy band gap vs. composition for PbTe-based ternary compounds at
300K.

fabrication of diode lasers operating at up to 175K under CW conditions, and more recently at 195K CW [1,12]. Slightly lattice-mismatched $Pb_{1-x}Eu_xSe/PbSe$ diode lasers have also been reported [2]. Since the lattice constants of PbS and EuS are almost identical (Fig. 1), $Pb_{1-x}Eu_xS/PbS$ heterojunctions have been grown which are almost lattice-matched [13]. However, europium introduced a deep donor level into PbS which made it difficult to dope the material p-type. This may be caused by some instability in the valence of Eu, leading to the donor-like transition Eu^{+2} + Eu^{+3} + e$^-$. This effectively rules out $Pb_{1-x}Eu_xS$ for diode laser applications, and makes $Pb_{1-x}Sr_xS$ a preferred substitute [13]. A similar donor-like behavior of the "divalent" rare earths has also been observed in $Pb_{1-x}Sm_xTe$ and in $Pb_{1-x}Yb_xTe$ [8, 14]. The fact that Eu^{+2} is unstable in PbS but stable in PbTe is somewhat surprising, since Eu^{+2} is under greater compression in the PbTe lattice, as shown in Fig. 1. This suggests that further study of the stability of Eu^{+2} in lead salt compounds is needed.

INDIUM EUROPIUM ANTIMONIDE

Dopant level concentrations of some rare earths in high energy gap III-V compounds, such as InP:Yb, are currently being studied for their luminescence properties [15]. However, alloys of the rare earth elements and narrow energy gap III-V compounds have not been previously studied to our knowledge. Our interest in $In_{1-x}Eu_xSb$ was due to the need to find new alloys which could provide nearly lattice-matched heterojunctions with InSb. InSb has the zinc blende crystal structure, and no other III-V compound lattice-matches to it. While several phases in the Eu-Sb system have been found, such as Eu_2Sb_3 and $EuSb_2$, the existence of EuSb is controversial [16-18]. The study of the Eu-Sb phase diagram has been difficult because of the extreme reactivity of europium and of europium-rich phases such as EuSb [16]. In the present study, it was hoped that the ultraclean growth environment of a molecular beam epitaxy (MBE) machine and the growth of dilute alloys of $In_{1-x}Eu_xSb$ with x < 0.1 would enable a highly crystalline zinc blende phase to form.

The growth technique of InSb on InP substrates has recently been reported [19, 20]. In the present study, 4-9's pure Eu was used in an additional MBE source oven. Films were grown at 0.8 μm/hr at 360°C substrate temperature or at 0.15 μm/hr at 280°C. In both cases, slightly Sb-rich growth conditions were used, and a total film thickness of 0.5μm was grown. The composition "x" in $In_{1-x}Eu_xSb$ was determined just before or after growth by measuring the In and Eu atomic fluxes onto a quarts crystal deposition monitor, which could be moved into the sample position. The lattice constants of the films, measured by X-ray diffraction, are shown in Fig. 4. While there is relatively little variation with the EuSb mole fraction, the X-ray line widths, shown in Table I, are broad, indicating considerable lattice disorder. The films grown at 360°C had large carrier densities, of order 10^{18} cm^{-3}, whereas those grown at 280°C tended to have carrier densities in the mid-10^{16} cm^{-3} range, which was due to an impurity (Se) in the Sb source material which was in use at the time these films were grown [19]. The cause of these high carrier densities in the films grown at 360°C is not yet known, but may be related in some way to lattice disorder. We note that film properties in $In_{1-x}Mn_xAs$ are also extremely dependent upon growth temperature. The infrared transmittance of $In_{1-x}Eu_xSb$ films is shown in Fig. 5. There is very little difference between films with x=0 and x=0.05, implying that the energy band gap is not a strong function of x, unlike the case for $Pb_{1-x}Eu_xTe$ (see Fig. 3). All of the films listed in Table I had an infrared transmittance similar to those in Fig. 5. This result is disappointing from the point of view of heterojunction formation. However, an unexpected and rather novel finding is that of a negative magnetoresistance (Fig. 6). This effect is perhaps caused by the magnetic moments of europium ions in the InSb lattice. These magnetic moments

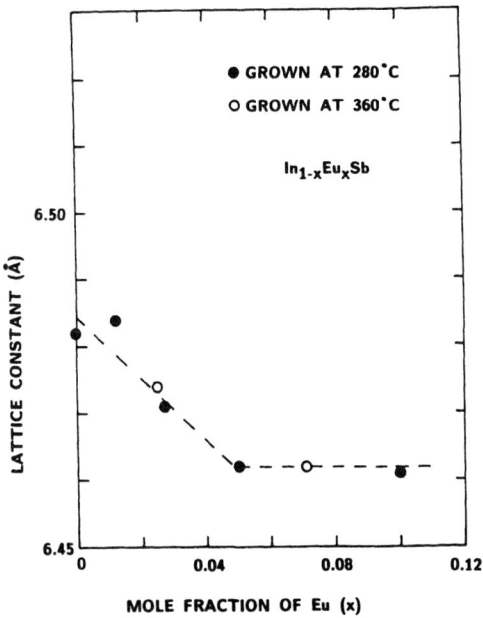

Fig. 4.   Lattice constant of In$_{1-x}$Eu$_x$Sb vs. europium mole fraction. The
dashed curve is an empirical fit to the data.

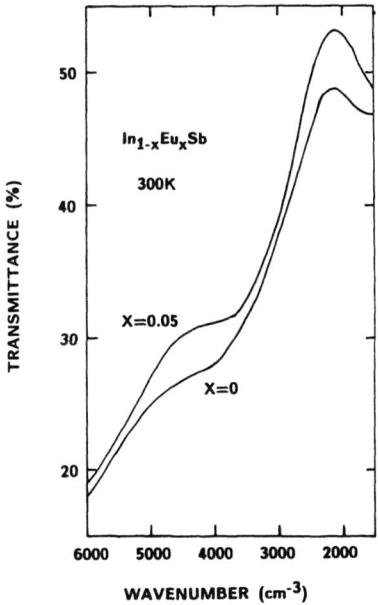

Fig. 5.   Infrared transmittance of InSb and of In$_{0.95}$Eu$_{0.05}$Sb films at 300 K.

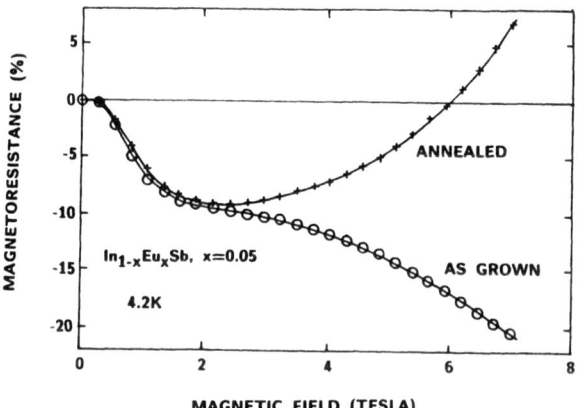

Fig. 6. Magnetoresistance of an $In_{0.95}Eu_{0.05}Sb$ film at 4.2 K with or without annealing.

strongly scatter conduction band electrons, but become aligned in a sufficiently strong magnetic field, thus scattering conduction electrons less. Another sample was made from the same growth after annealing at 360°C for 16 hours. Its low field electron mobility, μ, increased after annealing from 220 to 380 $cm^2V^{-1}s^{-1}$. Since the normal, positive magnetoresistance effect varies as μ^2B^2, this improved mobility is responsible for a positive magnetoresistance at sufficiently large magnetic field (see Fig. 6). We note that these effects were measured with a conventional six probe Hall pattern. The negative magnetoresistance which we observe in $In_{1-x}Eu_xSb$ is not seen in $Pb_{1-x}Eu_xTe$ films which we have examined, possibly because the low temperature electron mobilities in these $Pb_{1-x}Eu_xTe$ films are larger, making the normal positive magnetoresistance effect dominate.

Table I. Properties of $In_{1-x}Eu_xSb$ Films at 300K.

x	Growth Temp. (°C)	Electron Density (cm^{-3})	Electron Mobility ($cm^2V^{-1}s^{-1}$)	XRD Line Width (Deg.)
0	280	4.2×10^{16}	20,300	.13
.012	280	5.0×10^{17}	1,000	.39
.027	280	8.6×10^{16}	200	.88
.050	280	5.0×10^{16}	220	.16
.10	280	9.7×10^{16}	120	.72
.025	360	2.5×10^{18}	1,200	.43
.071	360	2.0×10^{18}	500	.60

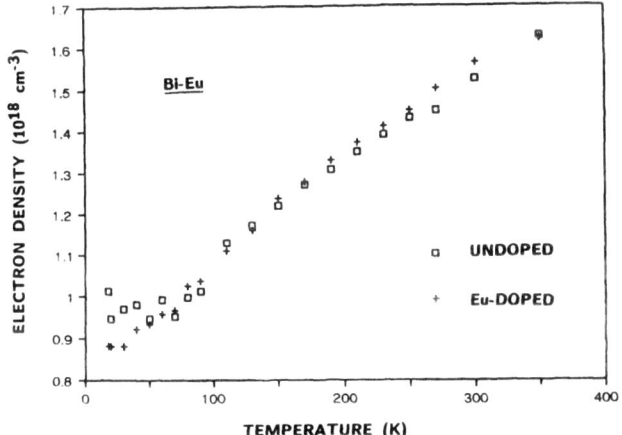

Fig. 7. Electron density of bismuth vs. temperature for a film which has $N_{Eu}=5.1\times10^{18}$ cm^{-3} or which is undoped.

BISMUTH - EUROPIUM

We recently reported on the growth of epitaxial Bi and Bi$_{1-x}$Sb$_x$ films on BaF$_2$ substrates [21,22]. These are the first such epitaxial Bi films which have been grown to our knowledge. Alloying with Sb changes Bi from a semimetal to a semiconductor for a certain composition range [22]. While we have so far only investigated additions of Eu to Bi in the dopant range and not in the "alloy" range (e.g., ~ 1% Eu addition or more), this initial data on transport properties is presented for the sake of completeness. These films were grown at 250°C to a thickness of 0.50 μm using the techniques previously described [21]. The Eu concentration was determined by measurement of the Eu and Bi fluxes with a quartz crystal deposition monitor in the sample position before growth. Other aspects of the growth are the same as those reported previously [21,22]. The transport properties were measured with a conventional 6 probe Hall pattern vs. temperature. The electron densities are shown in Fig. 7 vs. temperature for an undoped and for a heavily doped ($N_{Eu}=5.1\times10^{18}$ cm^{-3}) film. Within experimental error, no difference is seen which can be attributed to europium. The same is true of the hole densities, which are not shown for clarity. The electron and hole mobilities along the <100> direction of the same undoped and Eu-doped films were measured with a conventional 6 probe Hall pattern vs. temperature as shown in Fig. 8 [23]. The mobility is reduced very slightly by addition of Eu, presumably by scattering of electrons by the magnetic moments of europium. We note that the average in-plane carrier mobilities are a factor of 2 less than those along the <100> direction. These preliminary results need to be verified by an independent measurement of the europium concentration in the films, to ensure that all of the measured europium atomic flux incident upon the substrate during growth was incorporated into the film. Higher concentrations of europium also need to be studied for effects on energy band structure as well as on transport properties. The fact that Eu at the concentrations so far studied neither dopes Bi nor appreciably degrades its mobility suggests that Bi$_{1-x}$Eu$_x$ alloys or Bi$_{1-x}$Eu$_x$/Bi heterojunctions may be useful for new device structures.

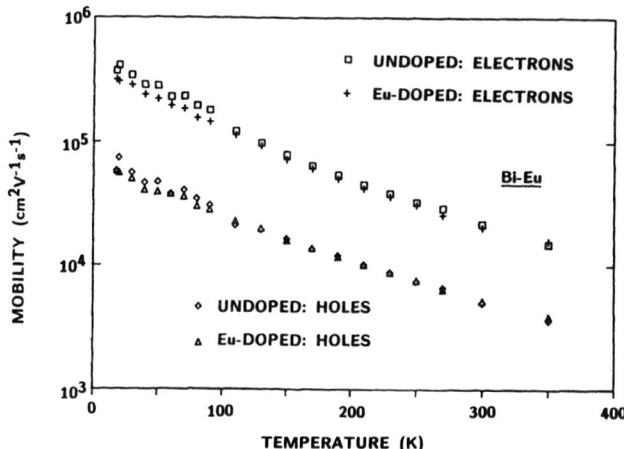

Fig. 8. Electron and hole mobilities of Bi:Eu films vs. temperature which has N_{Eu}=5.1x10^{18} cm^{-3} or which is undoped.

ACKNOWLEDGMENTS

The authors gratefully acknowledge the capable work of D.J. McEwen in performing the infrared transmission measurements and of J.L. Johnson for his X-ray diffraction measurements.

REFERENCES

1. D.L. Partin, IEEE J. Quantum Elec. 24, 1716 (1988).
2. M. Tacke, B. Spanger, A. Lambrecht, P.R. Norton, and H. Bottner, Appl. Phys. Lett. ....
3. H. Munekata, H. Ohno, S. von Molnar, A. Segmuller, L.L. Chang, and L. Esaki, Phys. Rev. Lett. 63, 1849 (1989).
4. H. Holloway and J.N. Walpole, Prog. Cryst. Growth, Charac. 2, 49 (1979).
5. H. Preier, Appl. Phys. 20, 189 (1979).
6. D.L. Partin J. Appl. Phys. 57, 1997 (1985).
7. G.B. Bacskay, P.J. Fensham and I.M. Ritchie, J. Phys. Chem. Solids 29, 1213 (1968).
8. G.T. Alekseeva, M.N. Vinogradova, K.G. Gartsman, A. Yu. Zyuzin, Kh. R. Mailina, L.V. Prokofeva, and L.S. Stilbans, Sov. Phys. Solid State 27, 1953 (1985).
9. R. Suryanarayanan, G. Guntherodt, J.L. Freeouf, and F. Holtzberg, Phys. Rev. B 12, 4215 (1975).
10. R. Suryanarayanan, C. Paparoditis, J. Ferre and B. Briat, J. Appl. Phys.43, 4105 (1972).
11. D.L. Partin, J. Elec. Mater. 13, 493 (1984).
12. Z. Feit, D. Kostyk, R.J. Woods, P. Mak, to be publ., Appl. Phys. Lett.
13. A. Ishida, K. Maramatsu, H. Takashiba, and H. Fujiyasu, Appl. Phys. Lett. 55, 430 (1989).
14. D.L. Partin, J. Electron. Mater. 12, 917 (1983).
15. D. Williams and B. Wessels, Appl. Phys. Lett. 56, 566 (1990).
16. J.B. Taylor, L.D. Calvert, T. Utsunomiya, Yu Wang and J.G. Despault, J. Less Common Metals 57, 39 (1978).

17. F. Hulliger and R. Schmelczer, J. Solid State Chem. 26, 399 (1978).
18. G.S. Viksman, S.P. Gordienko and B.V. Fenochka, Russ. J. Phys. Chem. 53, 290 (1979).
19. D.L. Partin, J. Heremans, D.T. Morelli and C.M. Thrush, to be publ.
20. J.E. Oh, P.K. Bhattacharya, Y.C. Chen, and S. Tsukamoto, J. Appl. Phys. 66, 3618 (1989).
21. D.L. Partin, C.M. Thrush, J. Heremans, D.T. Morelli, and C.H. Olk, J. Vac. Sci. Technol. B 7, 348 (1989).
22. D.T. Morelli, D.L. Partin and J. Heremans, Semicond. Sci. Technol. 5, S257 (1990).
23. D.L. Partin, J. Heremans, D.T. Morelli, C.M. Thrush, C.H. Olk, and T.A. Perry, Phys. Rev. B, 38, 3818 (1988).

Author Index

Subject Index

ISSN 0272 - 9172

Volume 1—Laser and Electron-Beam Solid Interactions and Materials Processing, J. F. Gibbons, L. D. Hess, T. W. Sigmon, 1981, ISBN 0-444-00595-1

Volume 2—Defects in Semiconductors, J. Narayan, T. Y. Tan, 1981, ISBN 0-444-00596-X

Volume 3—Nuclear and Electron Resonance Spectroscopies Applied to Materials Science, E. N. Kaufmann, G. K. Shenoy, 1981, ISBN 0-444-00597-8

Volume 4—Laser and Electron-Beam Interactions with Solids, B. R. Appleton, G. K. Celler, 1982, ISBN 0-444-00693-1

Volume 5—Grain Boundaries in Semiconductors, H. J. Leamy, G. E. Pike, C. H. Seager, 1982, ISBN 0-444-00697-4

Volume 6—Scientific Basis for Nuclear Waste Management IV, S. V. Topp, 1982, ISBN 0-444-00699-0

Volume 7—Metastable Materials Formation by Ion Implantation, S. T. Picraux, W. J. Choyke, 1982, ISBN 0-444-00692-3

Volume 8—Rapidly Solidified Amorphous and Crystalline Alloys, B. H. Kear, B. C. Giessen, M. Cohen, 1982, ISBN 0-444-00698-2

Volume 9—Materials Processing in the Reduced Gravity Environment of Space, G. E. Rindone, 1982, ISBN 0-444-00691-5

Volume 10—Thin Films and Interfaces, P. S. Ho, K.-N. Tu, 1982, ISBN 0-444-00774-1

Volume 11—Scientific Basis for Nuclear Waste Management V, W. Lutze, 1982, ISBN 0-444-00725-3

Volume 12—In Situ Composites IV, F. D. Lemkey, H. E. Cline, M. McLean, 1982, ISBN 0-444-00726-1

Volume 13—Laser-Solid Interactions and Transient Thermal Processing of Materials, J. Narayan, W. L. Brown, R. A. Lemons, 1983, ISBN 0-444-00788-1

Volume 14—Defects in Semiconductors II, S. Mahajan, J. W. Corbett, 1983, ISBN 0-444-00812-8

Volume 15—Scientific Basis for Nuclear Waste Management VI, D. G. Brookins, 1983, ISBN 0-444-00780-6

Volume 16—Nuclear Radiation Detector Materials, E. E. Haller, H. W. Kraner, W. A. Higinbotham, 1983, ISBN 0-444-00787-3

Volume 17—Laser Diagnostics and Photochemical Processing for Semiconductor Devices, R. M. Osgood, S. R. J. Brueck, H. R. Schlossberg, 1983, ISBN 0-444-00782-2

Volume 18—Interfaces and Contacts, R. Ludeke, K. Rose, 1983, ISBN 0-444-00820-9

Volume 19—Alloy Phase Diagrams, L. H. Bennett, T. B. Massalski, B. C. Giessen, 1983, ISBN 0-444-00809-8

Volume 20—Intercalated Graphite, M. S. Dresselhaus, G. Dresselhaus, J. E. Fischer, M. J. Moran, 1983, ISBN 0-444-00781-4

Volume 21—Phase Transformations in Solids, T. Tsakalakos, 1984, ISBN 0-444-00901-9

Volume 22—High Pressure in Science and Technology, C. Homan, R. K. MacCrone, E. Whalley, 1984, ISBN 0-444-00932-9 (3 part set)

Volume 23—Energy Beam-Solid Interactions and Transient Thermal Processing, J. C. C. Fan, N. M. Johnson, 1984, ISBN 0-444-00903-5

Volume 24—Defect Properties and Processing of High-Technology Nonmetallic Materials, J. H. Crawford, Jr., Y. Chen, W. A. Sibley, 1984, ISBN 0-444-00904-3

Volume 25—Thin Films and Interfaces II, J. E. E. Baglin, D. R. Campbell, W. K. Chu, 1984, ISBN 0-444-00905-1

Volume 26—Scientific Basis for Nuclear Waste Management VII, G. L. McVay, 1984, ISBN 0-444-00906-X

Volume 27—Ion Implantation and Ion Beam Processing of Materials, G. K. Hubler, O. W. Holland, C. R. Clayton, C. W. White, 1984, ISBN 0-444-00869-1

Volume 28—Rapidly Solidified Metastable Materials, B. H. Kear, B. C. Giessen, 1984, ISBN 0-444-00935-3

Volume 29—Laser-Controlled Chemical Processing of Surfaces, A. W. Johnson, D. J. Ehrlich, H. R. Schlossberg, 1984, ISBN 0-444-00894-2

Volume 30—Plasma Processing and Synthesis of Materials, J. Szekely, D. Apelian, 1984, ISBN 0-444-00895-0

Volume 31—Electron Microscopy of Materials, W. Krakow, D. A. Smith, L. W. Hobbs, 1984, ISBN 0-444-00898-7

Volume 32—Better Ceramics Through Chemistry, C. J. Brinker, D. E. Clark, D. R. Ulrich, 1984, ISBN 0-444-00898-5

Volume 33—Comparison of Thin Film Transistor and SOI Technologies, H. W. Lam, M. J. Thompson, 1984, ISBN 0-444-00899-3

Volume 34—Physical Metallurgy of Cast Iron, H. Fredriksson, M. Hillerts, 1985, ISBN 0-444-00938-8

Volume 35—Energy Beam-Solid Interactions and Transient Thermal Processing/1984, D. K. Biegelsen, G. A. Rozgonyi, C. V. Shank, 1985, ISBN 0-931837-00-6

Volume 36—Impurity Diffusion and Gettering in Silicon, R. B. Fair, C. W. Pearce, J. Washburn, 1985, ISBN 0-931837-01-4

Volume 37—Layered Structures, Epitaxy, and Interfaces, J. M. Gibson, L. R. Dawson, 1985, ISBN 0-931837-02-2

Volume 38—Plasma Synthesis and Etching of Electronic Materials, R. P. H. Chang, B. Abeles, 1985, ISBN 0-931837-03-0

Volume 39—High-Temperature Ordered Intermetallic Alloys, C. C. Koch, C. T. Liu, N. S. Stoloff, 1985, ISBN 0-931837-04-9

Volume 40—Electronic Packaging Materials Science, E. A. Giess, K.-N. Tu, D. R. Uhlmann, 1985, ISBN 0-931837-05-7

Volume 41—Advanced Photon and Particle Techniques for the Characterization of Defects in Solids, J. B. Roberto, R. W. Carpenter, M. C. Wittels, 1985, ISBN 0-931837-06-5

Volume 42—Very High Strength Cement-Based Materials, J. F. Young, 1985, ISBN 0-931837-07-3

Volume 43—Fly Ash and Coal Conversion By-Products: Characterization, Utilization, and Disposal I, G. J. McCarthy, R. J. Lauf, 1985, ISBN 0-931837-08-1

Volume 44—Scientific Basis for Nuclear Waste Management VIII, C. M. Jantzen, J. A. Stone, R. C. Ewing, 1985, ISBN 0-931837-09-X

Volume 45—Ion Beam Processes in Advanced Electronic Materials and Device Technology, B. R. Appleton, F. H. Eisen, T. W. Sigmon, 1985, ISBN 0-931837-10-3

Volume 46—Microscopic Identification of Electronic Defects in Semiconductors, N. M. Johnson, S. G. Bishop, G. D. Watkins, 1985, ISBN 0-931837-11-1

Volume 47—Thin Films: The Relationship of Structure to Properties, C. R. Aita, K. S. SreeHarsha, 1985, ISBN 0-931837-12-X

Volume 48—Applied Materials Characterization, W. Katz, P. Williams, 1985, ISBN 0-931837-13-8

Volume 49—Materials Issues in Applications of Amorphous Silicon Technology, D. Adler, A. Madan, M. J. Thompson, 1985, ISBN 0-931837-14-6

Volume 50—Scientific Basis for Nuclear Waste Management IX, L. O. Werme, 1986, ISBN 0-931837-15-4

Volume 51—Beam-Solid Interactions and Phase Transformations, H. Kurz, G. L. Olson, J. M. Poate, 1986, ISBN 0-931837-16-2

Volume 52—Rapid Thermal Processing, T. O. Sedgwick, T. E. Seidel, B.-Y. Tsaur, 1986, ISBN 0-931837-17-0

Volume 53—Semiconductor-on-Insulator and Thin Film Transistor Technology, A. Chiang. M. W. Geis, L. Pfeiffer, 1986, ISBN 0-931837-18-9

Volume 54—Thin Films—Interfaces and Phenomena, R. J. Nemanich, P. S. Ho, S. S. Lau, 1986, ISBN 0-931837-19-7

Volume 55—Biomedical Materials, J. M. Williams, M. F. Nichols, W. Zingg, 1986, ISBN 0-931837-20-0

Volume 56—Layered Structures and Epitaxy, J. M. Gibson, G. C. Osbourn, R. M. Tromp, 1986, ISBN 0-931837-21-9

Volume 57—Phase Transitions in Condensed Systems—Experiments and Theory, G. S. Cargill III, F. Spaepen, K.-N. Tu, 1987, ISBN 0-931837-22-7

Volume 58—Rapidly Solidified Alloys and Their Mechanical and Magnetic Properties, B. C. Giessen, D. E. Polk, A. I. Taub, 1986, ISBN 0-931837-23-5

Volume 59—Oxygen, Carbon, Hydrogen, and Nitrogen in Crystalline Silicon, J. C. Mikkelsen, Jr., S. J. Pearton, J. W. Corbett, S. J. Pennycook, 1986, ISBN 0-931837-24-3

Volume 60—Defect Properties and Processing of High-Technology Nonmetallic Materials, Y. Chen, W. D. Kingery, R. J. Stokes, 1986, ISBN 0-931837-25-1

Volume 61—Defects in Glasses, F. L. Galeener, D. L. Griscom, M. J. Weber, 1986, ISBN 0-931837-26-X

Volume 62—Materials Problem Solving with the Transmission Electron Microscope, L. W. Hobbs, K. H. Westmacott, D. B. Williams, 1986, ISBN 0-931837-27-8

Volume 63—Computer-Based Microscopic Description of the Structure and Properties of Materials, J. Broughton, W. Krakow, S. T. Pantelides, 1986, ISBN 0-931837-28-6

Volume 64—Cement-Based Composites: Strain Rate Effects on Fracture, S. Mindess, S. P. Shah, 1986, ISBN 0-931837-29-4

Volume 65—Fly Ash and Coal Conversion By-Products: Characterization, Utilization and Disposal II, G. J. McCarthy, F. P. Glasser, D. M. Roy, 1986, ISBN 0-931837-30-8

Volume 66—Frontiers in Materials Education, L. W. Hobbs, G. L. Liedl, 1986, ISBN 0-931837-31-6

Volume 67—Heteroepitaxy on Silicon, J. C. C. Fan, J. M. Poate, 1986, ISBN 0-931837-33-2

Volume 68—Plasma Processing, J. W. Coburn, R. A. Gottscho, D. W. Hess, 1986, ISBN 0-931837-34-0

Volume 69—Materials Characterization, N. W. Cheung, M.-A. Nicolet, 1986, ISBN 0-931837-35-9

Volume 70—Materials Issues in Amorphous-Semiconductor Technology, D. Adler, Y. Hamakawa, A. Madan, 1986, ISBN 0-931837-36-7

Volume 71—Materials Issues in Silicon Integrated Circuit Processing, M. Wittmer, J. Stimmell, M. Strathman, 1986, ISBN 0-931837-37-5

Volume 72—Electronic Packaging Materials Science II, K. A. Jackson, R. C. Pohanka, D. R. Uhlmann, D. R. Ulrich, 1986, ISBN 0-931837-38-3

Volume 73—Better Ceramics Through Chemistry II, C. J. Brinker, D. E. Clark, D. R. Ulrich, 1986, ISBN 0-931837-39-1

Volume 74—Beam-Solid Interactions and Transient Processes, M. O. Thompson, S. T. Picraux, J. S. Williams, 1987, ISBN 0-931837-40-5

MATERIALS RESEARCH SOCIETY SYMPOSIUM PROCEEDINGS

Recent Materials Research Society Proceedings listed in the front.

CPSIA information can be obtained at www.ICGtesting.com
Printed in the USA
LVOW12s0846230514

386805LV00012BA/579/P